Cerebral Energy Metabolism and Metabolic Encephalopathy

Cerebral Energy Metabolism and Metabolic Encephalopathy

Edited by
DAVID W. McCANDLESS
University of Texas Medical School at Houston
Houston, Texas

PLENUM PRESS • NEW YORK AND LONDON

Library of Congress Cataloging in Publication Data

Main entry under title:

Cerebral energy metabolism and metabolic encephalopathy.

Includes bibliographies and index.
1. Brain—Diseases. 2. Energy metabolism. 3. Metabolism—Disorders. 4. Brain
chemistry. I. McCandless, David W., 1941-   . II. Title: Metabolic encephalopathy.
[DNLM: 1. Brain—metabolism. 2. Brain Diseases—etiology. WM 300 C414]
RC386.2.C36   1985   616.8   84-26443
ISBN-13: 978-1-4684-1211-6   e-ISBN-13: 978-1-4684-1209-3
DOI: 10.1007/978-1-4684-1209-3

To
Sue, Jeff, and Steven

Contributors

Marc S. Abel, Department of CNS Research, Medical Research Division, American Cyanamid Company, Lederle Laboratories, Pearl River, New York 10965

Hajime Arai, Laboratory of Neurochemistry, NINCDS, NIH, Bethesda, Maryland 20205

Allen I. Arieff, Nephrology Section, Department of Medicine, University of California at San Francisco, San Francisco, California 94143

Gorig Brunner, Medizinische Hochschule Hannover, Hannover, West Germany

Roger F. Butterworth, Laboratory of Neurochemistry, Clinical Research Centre, Hôspital St-Luc (University of Montreal), Montreal, Quebec, Canada H2X 3J4

Robert C. Collins, Department of Neurology, Washington University Medical School, St. Louis, Missouri 63110

John A. Dirgo, Edward Mallinckrodt Department of Pediatrics, Division of Neurology, St. Louis Children's Hospital, St. Louis, Missouri 63178

Bogden Djuricic, Laboratory of Neurochemistry, NINCDS, NIH, Bethesda, Maryland 20205

A. A. Farooqui, Department of Physiological Chemistry, Ohio State University, Columbus, Ohio 43210

Gary E. Gibson, Cornell University Medical College, Burke Rehabilitation Center, White Plains, New York 10605

Richard E. Hauhart, Edward Mallinckrodt Department of Pediatrics, Division of Neurology, St. Louis Children's Hospital, St. Louis, Missouri 63178

Richard Hawkins, Departments of Anesthesia and Physiology, The Milton S. Hershey Medical Center, The Pennsylvania State University College of Medicine, Hershey, Pennsylvania 17033

George I. Henderson, Departments of Medicine and Pharmacology, The University of Texas Health Science Center at San Antonio; and Research Service, Audie L. Murphy Memorial Veterans' Hospital, San Antonio, Texas 78284

David Holtzman, Department of Psychiatry and Neurology and Department of Pediatrics, Tulane University School of Medicine, New Orleans, Louisiana 70112

L. A. Horrocks, Department of Physiological Chemistry, Ohio State University, Columbus, Ohio 43210

Anastacio M. Hoyumpa, Jr., Departments of Medicine and Pharmacology, The University of Texas Health Science Center at San Antonio; and Research Service, Audie L. Murphy Memorial Veterans' Hospital, San Antonio, Texas 78284

Frederick C. Kauffman, Department of Pharmacology and Experimental Therapeutics, University of Maryland School of Medicine, Baltimore, Maryland 21201

Alan H. Lockwood, Department of Neurology, University of Texas Health Science Center at Houston, Houston, Texas 77025

W. David Lust, Laboratory of Neurochemistry, NINCDS, NIH, Bethesda, Maryland 20205. *Present address*: Laboratory of Experimental Neurosurgery, Case Western Reserve University, Cleveland, Ohio 44106

David W. McCandless, Department of Neurobiology and Anatomy, University of Texas Medical School at Houston, Houston, Texas 77025

David B. McDougal, Jr., Department of Pharmacology, The Beaumont-May Institute of Neurology, Washington University School of Medicine, St. Louis, Missouri 63110

Alexander L. Miller, Department of Psychiatry, The University of Texas Health Science Center at San Antonio, San Antonio, Texas 78284

Bogomir B. Mrsulja, Laboratory of Neurochemistry, NINCDS, NIH, Bethesda, Maryland 20205

Gerard B. Odell, Department of Pediatrics, University of Wisconsin School of Medicine, Clinical Sciences Center, Madison, Wisconsin 53792

John J. O'Neill, Department of Pharmacology, Temple University School of Medicine, Philadelphia, Pennsylvania 19140

Janet V. Passonneau, Laboratory of Neurochemistry, NINCDS, NIH, Bethesda, Maryland 20205

K. W. Rammohan, Department of Neurology, Ohio State University, Columbus, Ohio 43210

Roderick K. Roberts, Departments of Medicine and Pharmacology, The University of Texas Health Science Center at San Antonio; and Research Service, Audie L. Murphy Memorial Veterans' Hospital, San Antonio, Texas 78284

Steven Schenker, Departments of Medicine and Pharmacology, The University of Texas Health Science Center at San Antonio; and Research Service, Audie L. Murphy Memorial Veterans' Hospital, San Antonio, Texas 78284

Henry S. Schutta, Department of Neurology, University of Wisconsin School of Medicine, Clinical Sciences Center, Madison, Wisconsin 53792

Paul E. Teschan, Division of Nephrology, Department of Medicine, Vanderbilt University School of Medicine, Nashville, Tennessee 37232

Jean Holowach Thurston, Edward Mallinckrodt Department of Pediatrics, Division of Neurology, St. Louis Children's Hospital, St. Louis, Missouri 63178

Tim S. Whittingham, Laboratory of Neurochemistry, NINCDS, NIH, Bethesda, Maryland 20205. Present address: Laboratory of Experimental Neurosurgery, Case Western Reserve University, Cleveland, Ohio 44106

Richard C. Wiggins, Department of Neurobiology and Anatomy, University of Texas Medical School at Houston, Houston, Texas 77025

Yukimasa Yasumoto, Laboratory of Neurochemistry, NINCDS, NIH, Bethesda, Maryland 20205

Leslie Zieve, Hennepin County Medical Center, University of Minnesota, Minneapolis, Minnesota 55415

Preface

In recent years, there has been rapid growth in knowledge pertaining to the nervous system. This has, in some measure, been due to the development and application of a number of techniques such as the 2-deoxyglucose method and microchemical methods for measuring metabolites and regional cerebral blood flow. Data from the application of these techniques are just beginning to be collected, and the next few years promise to bring many new and exciting findings. The study of energy metabolism in brain is particularly interesting due to the fact that although the brain has scant energy reserves (as compared with the liver), it has one of the highest metabolic rates in the body. Recent studies from several laboratories have shown a surprising divergence of responses to metabolic insult in different areas of brain. In this regard, the cerebellum, for example, may have metabolic features which are unique from those of any other region. The high-energy phosphate compounds ATP and phosphocreatine, supplied by the oxidative metabolism of glucose, are necessary for normal cerebral functions such as the maintenance of membrane potentials, transmission of impulses, and synthetic processes. Interruption of substrate or "poisoning" of the system by a variety of means lead to a rapid change in cellular energetics, and ultimately cell death. From the clinical standpoint, an interesting feature of metabolic encephalopathy is that in many cases, early diagnosis and treatment may result in a rapid reversal of symptoms. These observations, coupled with similar ones in experimental models of metabolic encephalopathy, support the concept of the biochemical lesion—a metabolic lesion which precedes structural damage, and from which full recovery may be possible.

This book was written with the idea that the results of studies in animals and in patients with metabolic encephalopathy are similar. The author of each chapter has been asked to consider clinically relevant material and to try to bring together results from both animal and human studies. This dual approach—both basic neurochemistry and the clinical application of that knowledge—should make this book valuable to the clinician as well as the bench scientist.

This book has been subdivided into four sections: an introductory section, a section dealing with direct interruption of substrate, and one section each on intrinsic and extrinsic factors that produce metabolic encephalopathy.

This book would not have been possible without the participation and contributions of the contributors, and I am grateful for their efforts. In addition, I wish to thank Kirk Jensen, Editor, Plenum Publishing Corporation, for his patience and expertise in helping in the preparation of this volume.

David W. McCandless

Houston, Texas

Contents

I. INTRODUCTION

Chapter 1

Cerebral Energy Metabolism

Richard Hawkins

II. METABOLIC ENCEPHALOPATHY ASSOCIATED WITH SEVERE INTERRUPTION OF SUBSTRATE

Chapter 2

Hypoglycemia and Cerebral Energy Metabolism

David W. McCandless and Marc S. Abel

Chapter 3

Hypoxia

Gary E. Gibson

Chapter 4

Ischemic Encephalopathy

W. David Lust, Hajime Arai, Yukimasa Yasumoto, Tim S. Whittingham,
Bogden Djuricic, Bogomir B. Mrsulja, and Janet V. Passonneau

III. METABOLIC ENCEPHALOPATHY RESULTING PRIMARILY FROM INTRINSIC FACTORS

Chapter 5

Pyruvate Dehydrogenase Deficiency Disorders

Roger F. Butterworth

Chapter 6

Carbon Dioxide Narcosis

Alexander L. Miller

Chapter 7

Encephalopathy Due to Short- and Medium-Chain Fatty Acids

Leslie Zieve

Chapter 8

Encephalopathy Due to Mercaptans and Phenols

Leslie Zieve and Gorig Brunner

Chapter 9

Ammonia-Induced Encephalopathy

Alan H. Lockwood

Chapter 10

Bilirubin Encephalopathy

Gerard B. Odell and Henry S. Schutta

Chapter 11

Uremic and Dialysis Encephalopathies

Paul E. Teschan and Allen I. Arieff

Chapter 12

Epilepsy: Pathophysiology of Cerebral Dysfunction

Robert C. Collins

IV. METABOLIC ENCEPHALOPATHY THAT MAY RESULT FROM EXTRINSIC FACTORS

Chapter 13

Niacin–Nicotinamide Deficiency

Frederick C. Kauffman

Chapter 14

Thiamine Deficiency and Cerebral Energy Metabolism

David W. McCandless

Chapter 15

Thiamine Deficiency: Cerebral Amino Acid Levels and Neurologic Dysfunction

*Jean Holowach Thurston, Richard E. Hauhart, John A. Dirgo,
and David B. McDougal, Jr.*

Chapter 16

Alcohol-Induced Encephalopathy

*Roderick K. Roberts, Anastacio M. Hoyumpa, Jr.,   George I. Henderson,
and Steven Schenker*

Chapter 17

Heavy Metal Toxicity and Energy Metabolism in the Developing Brain: Lead as the Model

John J. O'Neill and David Holtzman

LEAD ENCEPHALOPATHY

OTHER METALS WITH AGE-DEPENDENT SUSCEPTIBILITY AND CNS TOXIC MECHANISMS POTENTIALLY SIMILAR TO THOSE OF LEAD

Chapter 18

General Anesthesia

Richard C. Wiggins

Chapter 19

Neurochemical Effects of Viral Infections in the Central Nervous System

K. W. Rammohan, A. A. Farooqui, and L. A. Horrocks

I

Introduction

1

Cerebral Energy Metabolism

RICHARD HAWKINS

1. Introduction

This book discusses a variety of metabolic encephalopathies that may be reflected by changes in cerebral energy metabolism. Whether the primary alteration in energy metabolism results from a direct action on the metabolic pathways of energy production or is mediated indirectly by, for example, abnormal neurotransmitter action, is unknown in most cases. Nevertheless, an appreciation of modern concepts of cerebral energy metabolism is fundamental to a thorough understanding of these disorders. Certainly it is clear that cerebral function and energy metabolism are closely linked; knowledge of the rate of energy utilization can yield valuable insight into cerebral function both at the level of the whole brain and individual structures. Many new concepts regarding energy metabolism as well as new methods of study have emerged in the last decade. The purpose of this chapter is to review some of these with emphasis on those aspects that may facilitate better understanding of the etiology of altered states of consciousness or function.

2. Cerebral Energy Requirements

Although nervous tissue does not participate in processes that require large amounts of energy, such as mechanical work, osmotic work, or extensive biosynthesis, it has almost as high a rate of oxidative metabolism as some tissues that do.* Whereas most vertebrates use 2–8% of their total resting metabolism

---

* The rate of oxygen consumption of rat brain (about 4.4 μmole/minute per gram) (Norberg and Siesjo, 1974), is comparable with that of the heart (8.7 μmole/minute per gram working at 100 torr aortic pressure) (Neely *et al.*, 1969), kidney (7–13 μmole/minute per gram) (Ross, 1978), or liver (5.5 μmole/minute per gram) (Krebs, 1970).

RICHARD HAWKINS • Departments of Anesthesia and Physiology, The Milton S. Hershey Medical Center, The Pennsylvania State University College of Medicine, Hershey, Pennsylvania 17033.

for the central nervous system, higher apes and man use considerably more (Mink *et al.*, 1981). Because of the large brain-to-body weight ratio, the human brain uses a much larger proportion of the whole body energy requirement than does the brain from other species, and the brain of children may demand a very high proportion. In most newborn mammals the rate of oxygen consumption is less than half the adult rate (Cremer and Heath, 1974; Hernandez *et al.*, 1978). However, this does not appear to be the case in humans. In an extensive study of infants and young children, cerebral oxygen consumption ($CMRO_2$) was about the same as in adults, although whole brain cerebral blood flow (CBF) was somewhat higher (Settergren *et al.*, 1980). This finding contrasts with a less extensive study where a somewhat higher $CMRO_2$ (about 24%) was found in children (Kennedy and Sokoloff, 1957).

The central nervous system (CNS) has very little endogenous fuel. There is a steep blood-to-brain gradient of glucose, the brain having only 1–2 μmole/g wet weight compared with blood concentrations of 5–6 μmole/ml (Veech and Hawkins, 1974). The glycogen content of brain is likewise low (2–3 μmole/g wet weight) (Veech and Hawkins, 1974). It has been suggested that, since 3–4 ml of water are required for every gram of glycogen, larger quantities of glycogen would result in volume fluctuations unacceptable in the brain due to the rigidity of the cranium (Cahill and Aoki, 1980). Glycogen granules are much more obvious in astrocytes than other CNS cells, which suggests that glycogen may not be readily available as a reserve fuel for neurons. In any event, the total amount of glucose and glycogen is only sufficient to sustain the normal rate of oxygen consumption for less than 5 min. Although this is sufficient for short periods of excessive energy use (i.e., when demand is greater than the ability of blood to supply glucose, such as during epileptic seizures), it is obviously not adequate during prolonged hypoglycemia. Continuous delivery of oxygen is even more crucial because there is virtually no reserve oxygen supply, and the CNS extracts approximately one third of the total blood oxygen. Therefore, cessation or diminution of blood flow will interfere with cerebral energy metabolism due to an oxygen deficiency before the lack of glucose becomes apparent.

The energy-utilizing reactions of the brain can be divided into two broad categories: biosynthesis and transport. Biosynthesis appears to occur mostly in cell bodies (proteins, polypeptides, lipids, etc.) or in nerve terminals (neurotransmitters). Transport processes, which are especially active in axons and dendrites, include: axoplasmic transport (e.g. enzymes, ionic pumps and other material), the acquisition of essential nutrients, reuptake of neurotransmitters and the transport of ions which are responsible for the maintenance of transmembrane electrical potentials. Some evidence suggests that the major portion of energy is used for ion transport (Crane *et al.*, 1978; Mata *et al.*, 1980; Whittam and Blond, 1964). During barbiturate anesthesia sufficient to cause near-isoelectric encephalographic activity, metabolism may be depressed by about 50–60% (Hawkins *et al.*, 1979b; Crane *et al.*, 1978). If it is assumed that biosynthetic and other energy-consuming processes are not effected, this suggests that more than half of the normal cerebral activity is associated with electrical events (i.e., the maintenance of Na^+ and K^+ gradients). This must of course be regarded as an approximation, because neither the residual ion transport activity nor the effect of biosynthetic processes is taken into account.

3. Blood–Brain Barrier

The blood–brain barrier is of key importance in regulating the entry of essential nutrients and sets limits on the ability of the CNS to meet its energy demands. The CNS is composed of three major interlocking cellular compartments: the vascular, the neuronal, and the glial (Glees, 1972). The capillaries are almost completely ensheathed by glial cells which also line the external surfaces of the brain adjacent to the pia arachnoid. The neurons are therefore isolated, suspended as it were, in a framework of glia. Each capillary endothelial cell is joined to contiguous cells by tight junctions, which are in effect a fusion of the membranes (Reese and Karnovsky, 1967; Brightman and Reese, 1969). Thus, molecules cannot pass between the cells, but must penetrate the luminal and antiluminal membranes. The endothelial cells therefore form a barrier that constitutes the initial point of resistance to the penetration of substrates to neurons. Although lipid soluble molecules such as the gases oxygen and carbon dioxide may pass readily through these membranes, most nutrient molecules are hydrophilic and their passage must be mediated by specific transport systems. Transport systems have now been described for glucose, amino acids, monocarboxylic acids, and other essential nutrients. These transport carriers mediate facilitated diffusion, a process that is equilibratory (not concentrative), sodium- and energy-independent, and stereospecific (Pardridge and Oldendorf, 1977). Contrary to early thoughts, the blood–brain barrier can no longer be thought of as a passive protective mechanism. It is important in the regulation of ionic balance and cerebral volume, and its permeability can be altered by the central adrenergic system (Grubb et al., 1978), tricyclic antidepressants (Preskorn et al., 1981), vasopressin (Raichle and Grubb, 1978), and angiotensin II (Grubb and Raichle, 1981).

Under normal circumstances the brain requires a continuous and balanced supply of essential nutrients in order to maintain function. The rates of substrate transport are equal to, or slightly exceed, brain requirements but, under some circumstances, the balance between need and supply may be disturbed. For instance, during hepatic encephalopathy, the neutral amino acid transport carrier changes in such a way that entry of the amino acids is higher than normal. Moreover, different neutral amino acids are affected to different degrees, i.e., tryptophan > phenylalanine = tyrosine > leucine (Mans et al., 1982). Consequently, the brain is flooded with these neutral amino acids, some of which are neurotransmitter precursors. On the other hand, permeability to essential basic amino acids is decreased by about 80%, as is the permeability to monocarboxylic acids, including the ketone bodies. In the case of the latter two transport systems, the brain may actually be starved for essential nutrients (Mans et al., 1982; Hawkins et al., 1981). Changes in the permeability of the blood–brain barrier have also been found in diabetes (Mans et al., 1981), starvation (Mans et al., 1981; Christensen et al., 1981), hyperglycemia (Gjedde and Crone, 1981), and hypothyroidism (Daniel et al., 1975). Changes in other conditions will undoubtedly be discovered in the future. It is quite conceivable that a serious imbalance in the transport properties of the blood–brain barrier will lead to discernible changes in neural function.

After passing through the endothelial cell layer, metabolites encounter a second barrier: a zone of astrocyte end-foot processes. Astrocytes contain relatively few mitochondria and many glycogen granules, although studies of astrocytes in tissue cultures suggest a more active rate of energy metabolism than this would indicate (Hertz, 1981). There is good evidence that astrocytes serve as a second line of defense in protecting neurons; for instance, astrocytes rapidly convert ammonium ion into glutamine. Studies of labeled ammonium have shown that it is nearly exclusively metabolized into the amide portion of glutamine (Cooper *et al.*, 1979) by a reaction that is catalyzed only within the astrocyte cells (Martinez-Hernandez *et al.*, 1977). There is good evidence that the astrocytes oxidize substrates other than glucose, including amino acids and short- to medium-chain fatty acids (Cremer *et al.*, 1975, 1977; Balazs *et al.*, 1972), perhaps before these compounds can penetrate to the nerve cells. It is an interesting possibility that in some diseases such as Reye's syndrome, where plasma concentrations of fatty acids of various chain lengths are elevated (Trauner *et al.*, 1975, 1977), the astrocytes might become overwhelmed, allowing toxic molecules to reach neurons.

One of the characteristics of cerebral endothelial cells is that pinocytotic vesicles are rarely seen. However, vesicles transporting horseradish peroxidase have been observed within cerebral capillary endothelial cells in rats with portacaval shunts (Laursen and Westergaard, 1977), and small channels have been described during hyperglycemia in lizards (Shivers, 1979). Although it is unlikely that these vesicles could be a physiologically significant source of small metabolite transport, they could be important in moving proteins and other large molecules from the blood to the brain. If the rate of this process were to exceed the rate of phagocytosis by cells such as the pericytes, disturbances in cerebral homeostasis may occur. Perhaps this mechanism contributes to kernicterus when plasma bilirubin concentrations are high.

In addition to changing under pathological circumstances, the blood–brain barrier changes as a function of development and this is undoubtedly important for cerebral energy metabolism. The blood–brain barrier is present at birth, but its permeability characteristics are quite different from that of the adult (Braun *et al.*, 1980; Cornford *et al.*, 1982). The transport capacity for glucose is considerably lower, whereas that for monocarboxylic acids including ketone bodies and lactate is much higher. This presumably reflects a physiological need and may enable the newborn to make use of a wider variety of cerebral fuels (Cremer, 1981).

4. Fuels of Cerebral Energy Metabolism

4.1. Glucose

Glucose is the major but not the only source of cerebral energy. The belief that glucose is the sole source of energy is based on the observations that: (1) the net uptake of glucose, as determined by arteriovenous differences, is sufficient to supply the energy requirement of brain in normal, well-fed individuals (Gibbs *et al.*, 1942), (2) the respiratory quotient was measured to be close to 1 (0.92–0.99), which is compatible with the oxidation of carbohydrate, and

(3) when circulating fuels, including glucose, free fatty acids, ketone bodies, and triglycerides, are decreased by insulin administration, the coma that results can be completely reversed only by glucose or mannose* (for a review, see Sokoloff, 1960).

Although it is clear that normal cerebral function requires glucose, the widely held concept that it is the only fuel requires closer examination. Respiratory quotients are notoriously difficult to measure and are relatively insensitive indicators of the substrates being oxidized. For example, if brain were to oxidize a mixture of 25% free fatty acids and 75% glucose, the respiratory quotient would be 0.925. A respiratory quotient of 0.92 has been observed in patients during diabetic ketoacidosis (when plasma free fatty acids as well as ketone bodies are expected to be elevated). In another group of patients in diabetic coma the respiratory quotient was only 0.87 (Kety et al., 1948), which raises the question whether free fatty acids were being oxidized with possibly important consequences for cerebral function and energy metabolism.

The measurement of arteriovenous differences is likewise technically difficult. Arteriovenous differences are usually very small (the net uptake of glucose is only about 10%), so that highly sensitive techniques must be used. No arteriovenous differences could be found for the ketone bodies during diabetic ketoacidosis in early studies (Sokoloff, 1960), but it is now known that they can supply more than half of the metabolic requirement in starving man (Owen et al., 1967). Thus, data based only on respiratory quotients or arteriovenous differences using older techniques must be interpreted with caution.

4.2. Ketone Bodies

Nervous tissue has the capability of oxidizing ketone bodies (3-hydroxybutyrate and acetoacetate). Although these substrates are normally present in low concentrations in circulation, there are conditions (e.g. starvation, diabetes) in which their concentrations rise and they have the potential of becoming an important respiratory fuel (Robinson and Williamson, 1980). Brain utilization of ketone bodies diminishes the demand for glucose, and thus reduces the need for gluconeogenesis and concomitant degradation of protein.

The metabolism of ketone bodies by human and rat brain is linearly related to their arterial concentrations and the uptake of acetoacetate generally exceeds that of 3-hydroxybutyrate at comparable arterial concentrations (Hawkins et al., 1971). Ketone bodies are consumed as fast as they enter brain, therefore influx is the rate-limiting step (Hawkins and Biebuyck, 1979). Ketone bodies cross the blood–brain barrier by a transport system that also carries other monocarboxylic acids, including short chain fatty acids, lactate and pyruvate (Oldendorf, 1973; Cremer et al., 1976; Cremer et al., 1979). The transport activity varies from region to region with a distribution which is different from that of glucose (Hawkins and Biebuyck, 1979). Monocarboxylic acid transport system

---

* Mannose is presumably phosphorylated by hexokinase and then reacts with phosphomannose isomerase to form fructose-6-phosphate, and then enters the glycolytic pathway (Sloviter and Kamimoto, 1970). However, mannose does not normally occur in the circulation in significant concentrations.

activity is increased moderately by starvation and diabetes and this increases the ability to use ketone bodies in these circumstances (Gjedde and Crone, 1975; Mans et al., 1981).

Ketone body transport activity is greater in young than adult rats (Moore et al., 1976; Cremer et al., 1976). Human infants, likewise, have a significantly greater capacity to extract ketone bodies from the blood (Persson et al., 1972; Kraus et al., 1974). It has been suggested that the capacity of children to utilize ketone bodies may underlie the clinical observation that a ketogenic diet is particularly effective in controlling seizures (DeVivo, 1980).

The ability to use ketone bodies varies in different species. Rats can derive 5–20% of the total brain energy requirement from ketone bodies, whereas adult dogs do not develop significant ketosis. Humans, on the other hand, may depend on ketone bodies for about 60% of the energy requirement after prolonged fasting (Owen et al., 1967). Perhaps ketone bodies are of greatest importance in children in whom brain oxygen consumption may comprise 50% or more of the basal metabolic rate. Young children may develop hypoglycemia and ketoacidosis within several hours and the ability to use fat-derived substrates may become necessary for survival. Although there is circumstantial evidence that ketone bodies can be almost the sole substrate for cerebral metabolism in humans (Cahill and Aoki, 1980), control experiments in animals indicate that some glucose is also necessary (Ide et al., 1969; Patel and Owen, 1977; Rolleston and Newsholme, 1967; Ziven and Snarr, 1972). The utilization of ketone bodies may decrease cerebral glycolysis (perhaps by directly inhibiting phosphofructokinase), increase brain glycogen content, and inhibit the oxidation of pyruvate at the pyruvate dehydrogenase step (Ruderman et al., 1974; DeVivo et al., 1978; Miller et al., 1982).

During hepatic encephalopathy, the transport of ketone bodies is diminished by 70–80% and certain brain regions, notably the lower layers of cortex, are more seriously affected (Hawkins et al., 1981). Thus, it seems that in this particular disease, the brain may be almost incapable of using these substrates. This would be a serious disadvantage in situations where blood glucose is low, such as during starvation.

4.3. Lactate

Due to the relative impermeability of the blood–brain barrier to circulating lactate, it cannot be considered a significant fuel for brain in adult mammals. However, in young mammals the blood–brain barrier is much more permeable to monocarboxylic acids; the V_{max} of lactate entry into brain of suckling rats is many times greater than that of adults (Cremer et al., 1979). When the arterial concentration of lactate exceeds 2 mM in newborn dogs, net uptake occurs, whereas below 2 mM the brain produces lactate (Vannucci et al., 1980; Hernandez et al., 1980; Hellmann et al. 1982). At 10 mM lactate, a concentration that may result from intense exercise, a short period of hypoxia, or excessive insulin administration, it was found that lactate could supply more than 50% of the total metabolic fuel required (Hernandez et al., 1980). Therefore, under abnormal circumstances, lactate can be oxidatively metabolized in lieu of glucose by brains of young mammals, when the circulating concentration rises.

Recently, a child with glucose-6-phosphatase deficiency was described who had hypoglycemia and relatively high concentrations of blood lactate (Fernandez et al., 1982). In this case, the cerebral metabolism of lactate was sufficient to account for 30–40% of the cerebral oxygen consumption, indicating that under special circumstances young humans can also rely on lactate as an important fuel of energy metabolism.

4.4. Amino Acids

The combined unidirectional flux into brain of all the amino acids normally present in the circulation is about 40–45 nmole/min per g in adult rats (Banos et al., 1973; Pardridge, 1977). Considering that the normal rate of glucose utilization is between 700 and 800 nmole/min per g, it is obvious that even the complete oxidation of amino acids would not contribute more than a few percent to the total energy requirement. Several investigators could not demonstrate significant arteriovenous differences of amino acids in well-nourished animals and humans (Pardridge, 1977; Sacks et al., 1982; Settergren et al., 1980) and only a small net uptake during prolonged starvation (Owen et al., 1967). It is likely, however, that some oxidation does occur since it has been shown that ^{14}C or ^{3}H from leucine, proline, and phenylalanine can enter aspartate, glutamate, and glutamine (Cremer et al., 1975, 1977; Van den Berg and Van den Velden, 1970; Berl and Frigyesi, 1968; Roberts and Morelos, 1965). Thus, label from these amino acids was converted to acetyl CoA and further oxidized in the Krebs cycle. The major site of oxidation of the amino acids is believed to be the "small" glutamate compartment because administration of labeled amino acid results in a higher specific activity of glutamine than glutamate. These findings are compatible with the suggestion that glial cells are responsible for the small amount of oxidative metabolism of amino acids that occurs.

4.5. Free Fatty Acids

Considering that free fatty acids are the preferred fuel of oxidative metabolism in many tissues, it is surprising that the CNS does not appear to be more dependent on them. Brain mitochondrial preparations oxidize long-, medium-, and short-chain fatty acids at rates comparable to mitochondria from heart and liver (Vignais et al., 1958; Beattie and Baseford, 1966). The belief that fatty acids are not used by the CNS comes from the observation that the respiratory quotient is close to 1 under a variety of circumstances (although as stated above, in states of altered consciousness this may not be entirely true). Also, attempts to measure arteriovenous differences of free fatty acids, which are difficult to measure, have been unsuccessful (Sokoloff et al., 1977; Owen et al., 1967). On the other hand, there are several reports that indicate that contributions by free fatty acids to cerebral energy metabolism may be significant.

Long-chain fatty acids are acylated to fatty acyl-CoA in the cytoplasm, converted to acyl-carnitine by carnitine palmitoyltransferase, and may then enter the mitochondria where oxidation occurs. The carnitine palmitoyltransferase activity of brain is comparable to that of other tissues (Bremmer, 1963), and the brain content of carnitine is high enough for activity (Babcock and

Hawkins, 1985). Several observations suggest that the blood–brain barrier may limit the oxidation of long-chain fatty acids. When a fatty acid–albumin complex was introduced into dog brain by ventriculocisternal perfusion, it was oxidized (Spitzer, 1973). Also, it has been observed that the incorporation of ^{14}C from palmitate was relatively rapid in areas of the brain that lack of blood–brain barrier, such as the pineal and pituitary glands (Vannucci and Hawkins, 1983). These circumventricular structures utilize palmitate as an energy source and seem to oxidize only small amounts of glucose (Vannucci and Hawkins, 1982; Viña et al., 1984). Direct measurement of albumin-bound palmitate uptake showed that about 5% was extracted on a single pass through the brain in adult rats (Pardridge and Mietus, 1980). This amount of palmitate could account for about 10–15% of the normal fuel requirement, at a blood flow of 1 ml/min per g and circulating free fatty acid concentrations of 0.5 meq/liter. This agrees well with studies of the isolated perfused cat brain that was shown to derive about 10% of its energy requirements from albumin-bound palmitate (Allweis et al., 1966). Fasting puppies, which do not readily become ketotic, can derive about 24% of their brain oxidative fuel from circulating fatty acids (Spitzer, 1973), an amount that is certainly significant.

Short- and medium-chain fatty acid concentrations in the plasma are normally so low [with the possible exception of acetate (Sarna et al., 1979; Buckley and Williamson, 1977)] that they cannot be considered significant fuels, except perhaps in ruminants. The short-chain fatty acids, i.e., acetate and butyrate, enter the brain by the monocarboxylic acid transport system, which is relatively inactive in adult mammals (Sarna et al., 1979; Oldendorf, 1972), but has a higher capacity in the young (Cremer et al., 1976, 1979). It seems possible, therefore, that short-chain fatty acid oxidation by brain may be of some significance in young mammals. Medium-chain fatty acids such as octanoate are much more lipid soluble and their entry into brain is relatively unrestricted (Oldendorf, 1972). In contrast to long-chain fatty acids, short- and medium-chain fatty acids are activated to acyl-CoA esters directly within the mitochondria where they are further metabolized.

Oxidation of long-, medium-, and short-chain fatty acids results in a higher specific activity of label in glutamine than in glutamate. Therefore, their oxidation, similar to that of the amino acids, most likely occurs in non-neuronal cells. This is in contrast to the other substrates (glucose, ketone bodies, lactate, and glycerol) that may be considered to be more general fuels.

In summary, although the major fuel of respiration in the CNS is glucose, the role of other fuels cannot be ignored. In some circumstances, their combined role may be greater than that of glucose. The pineal and pituitary glands, which lack a blood–brain barrier, may rely more heavily on noncarbohydrate fuels of respiration.

5. Study of Cerebral Energy Metabolism at the Tissue Level

The early studies of cerebral energy metabolism were based on measurements obtained from the entire brain usually employing a combination of arteriovenous differences and values of cerebral blood flow. Although a great deal of information was obtained with these techniques, it is likely that significant further advances will be made using methods that can be applied to individual cerebral structures. There are, for instance, indications that states of altered consciousness may involve abnormal energy metabolism in small portions of the CNS such as the reticular formation (McCandless, 1981; McCandless and Schenker, 1981; Hindfelt *et al.*, 1977) or even individual nuclei such as the pontine ventral tegmental nuclei (Miller, 1982).

There are two general approaches to the study of energy metabolism at the tissue level. One is to determine the quantity of specific intermediary metabolites or the activities of enzymes in tissue samples. The other is to measure physiological variables from which inferences about cerebral function may be drawn. Recently, a variety of techniques on quantitative autoradiography have become popular. These newer techniques, which include the regional rates of blood flow, glucose utilization, and blood–brain barrier transport, should be viewed as complementary. They are used to indicate those areas of brain where closer investigation by analytical techniques may be profitable.

5.1. Analytic Techniques

Enzymatic analysis, coupled with either spectrophotometric or fluorometric evaluation, has made it possible to measure the tissue content of almost any intermediary metabolite or enzyme activity. Particularly noteworthy is the work of Lowry and Passonneau (Lowry, 1962; Lowry and Passonneau, 1972), who pioneered elegant techniques of fluorometric analysis that enable quantitative measurements to be made in small portions of tissue, single cells, or even, in some instances, subcellular components. These techniques, developed more than two decades ago, are theoretically powerful enough to measure single molecules. The addition of gas chromatography and high-pressure liquid chromatography should allow the complete evaluation of tissue samples. Critical, however, to the successful application of sophisticated analytical methods is the acquisition of tissue samples by means which prevent postmortem changes. This is a fundamental problem to which a thoroughly satisfactory solution has not yet been found.

A variety of methods have been developed by different investigators for CNS inactivation. The following comments are restricted to three of the more successful and widely used methods: whole brain freezing, "brain blowing," and microwave irradiation.

Perhaps the oldest and most commonly used technique is that of freezing the whole brain *in situ*. This was extensively used by Lowry and his associates who applied this method to the study of mice under relatively well-described conditions. They found that best results were obtained by immersing the whole mouse in liquid N_2, rather than decapitating before immersion. Even so, there is a finite period between the moment of immersion and the freezing of cerebral tissue. This can range from a few seconds at the cortical surface to several minutes for deeper structures (Lust *et al.*, 1980). During this period changes can occur in the pattern of blood flow, local temperature, and intermediary metabolism. The results obtained using animals larger than mice are especially suspect since the brain takes much longer to freeze. Therefore, although whole brain freezing retains structural integrity, the data obtained cannot be viewed as definitive because of the delay in tissue inactivation.

Because of the limits of the whole brain freezing technique, Veech *et al.* (1973) developed the brain blower. This device removes the brain and freezes it within 0.5 sec by propelling it under pressure through a tube that is driven into the skull by an electrically powered solenoid. The brain tissue is blown flat against a disk chilled to the temperature of liquid N_2 whereupon it freezes. Although this technique is the fastest way to inactivate nervous tissue and serves as the benchmark by which intermediary metabolites may be determined, it completely destroys all structural integrity. At present there is no conclusive answer to the question of whether changes in the content of intermediary metabolites occur during the momentary removal and freezing process.

Rapid inactivation by heating from microwave irradiation has been tried by several investigators. However, the goal of rapidly inactivating tissue as well as retaining cerebral anatomy has been elusive (Medina *et al.*, 1980). This is because it is difficult to produce a homogeneously heated brain within such a complex structure. The more rapidly the brain is heated, the more likely that uneven heating will occur with concomitant displacement of softer cerebral structures. Slower heating, on the other hand, more satisfactory from the anatomical viewpoint, is inadequate for the determination of high-energy phosphates such as phosphocreatine and adenine nucleotides. Furthermore, microwave irradiation seems to cause a certain amount of membrane damage so that at least at the cellular level materials are free to diffuse from one area to another (Hampson *et al.*, 1982).

To sum up, sophisticated analytical techniques exist for analysis of the brain. Additional methods are available to indicate the areas of brain most likely to yield valuable information. A major stumbling block is the fundamental problem of obtaining tissue in such a way as to prevent postmortem changes while maintaining tissue structure.

5.2. Complementary Autoradiographic Techniques

It is currently accepted that useful information can be gained about regional cerebral function by measuring the rate of fuel consumption or blood flow in various brain structures. This follows from the logical assumption that the rate of energy utilization by a given portion of cerebral tissue is commensurate with the integrated activity of its components, which together may constitute a functional unit (or multiple of units). The rate of oxidative metabolism is the most satisfactory measure of steady-state aerobic energy metabolism, but it is difficult to measure oxygen utilization in small animals and impossible, at this time, to determine it in individual brain regions or nerve cells *in vivo*. An alternative is to measure cerebral glucose utilization (CMR_{Glc}), which, under most circumstances, is stoichiometrically related to oxygen consumption (6 mole O_2 per mole glucose). Perhaps equally satisfactory is knowledge of regional cerebral blood flow (CBF), which is also believed to be directly related to glucose requirements (Hawkins *et al.*, 1979a,b; Lassen *et al.*, 1978; Reivich, 1974).

In the following discussion several of the more useful techniques are described in some detail. Additional methods, which are based on similar principles but are less widely used, are also available (see Passonneau *et al.*, 1980).

5.2.1. Regional Measurements of Glucose Utilization

The first experiments in which CMR_{Glc} was estimated without determining blood flow or arteriovenous differences were based on the principle of near-complete trapping of ^{14}C from [2-^{14}C]glucose by acid soluble metabolites in the brain (Gaitonde, 1965; Haslam and Krebs, 1963; Krebs, 1965). It was subsequently demonstrated that the integral of blood glucose specific activity closely approximated that of brain after a few minutes and could therefore be used to define the activity of the brain precursor pool with time (Hawkins *et al.*, 1974). With further modifications, this basic method has been used to measure cerebral metabolic rates *in vivo* in several different conditions (Hawkins *et al.*, 1979b; Borgstrom *et al.*, 1976; Raichle *et al.*, 1975, 1980; Lu, *et al.*, 1983; Hawkins *et al.*, 1984).

The usefulness of studying regional glucose utilization by autoradiography was demonstrated by Kennedy *et al.* (1975) using 2-deoxy-D-[1-^{14}C]glucose ([^{14}C]-deoxyglucose), a synthetic analog that competes with glucose for transport and phosphorylation. Subsequently, Sokoloff and his associates (Sokoloff, 1979) derived equations relating [^{14}C]deoxyglucose uptake to CMR_{Glc}. This method has been applied to studies in several species including man using [^{11}C]deoxyglucose and [2-^{18}F]fluorodeoxyglucose (Sokoloff, 1979; Reivich *et al.*, 1982). The results are sometimes striking and seem to agree with our currently held concepts of glucose metabolism. However, the question has arisen of whether the deoxyglucose technique is an accurate measure of CMR_{Glc} or reflects a more complicated metabolic process (Fox, 1984). According to the model, CMR_{Glc} can be calculated according to the formula (Sokoloff, 1979)

$$CMR_{Glc} = \frac{1}{\text{lumped constant}} \times \frac{\text{labeled product formed}}{\text{integrated specific activity of precursors}}$$

where "lumped constant" is a correction factor including several individual kinetic constants.

Among other assumptions, the most important are that:

1. During the period of measurement, CMR_{Glc} is in a steady state
2. The rate of glucose utilization and [^{14}C]deoxyglucose phosphate accumulation can be related by a constant factor that takes into account the transport and phosphorylation kinetics of glucose and deoxyglucose
3. [^{14}C]deoxyglucose phosphate is not dephosphorylated at a significant rate
4. The rate of the hexokinase reaction is essentially equal to the rate of glucose utilization

In the past few years, each of these assumptions has been challenged.

The assumption that CMR_{Glc} is constant during the entire experimental period may not always be valid. This is especially true for longer experiments and during certain states, such as the induction and recovery from anesthesia, cerebral seizures, and so forth. Currently, there is no way of following the moment-to-moment variation in CMR_{Glc} and therefore the results obtained during the 45-min period necessary for measurement reflect the sum of all activities occurring during the experiment. Recently, an extreme situation where steady state could not be assumed and consequently accurate measurements of rates were unobtainable was described by Engel et al. (1982).

The results obtained with [^{14}C]deoxyglucose must be adjusted using a correction factor that includes the different kinetic constants of transport and enzymatic transformation. Pardridge et al. (1982) pointed out that the correction factor or "lumped constant" is not really a constant, but deviates over a fivefold range as a function of plasma and brain glucose concentrations. The deviation appears to be somewhat less in the range of normal plasma and brain glucose concentrations, but nonetheless, errors of 10–30% could easily be made by an unsuspecting investigator.

It has been shown that during the 45-min period required for a single determination, significant quantities of [^{14}C]deoxyglucose phosphate are hydrolyzed and [^{14}C]deoxyglucose is returned to the plasma (Hawkins and Miller, 1978; Huang et al., 1980). Furthermore, there is some indication that the rate at which this occurs may depend on the physiological circumstance. For instance, Karnovsky et al. (1980) found that glucose-6-phosphatase was substantially stimulated during sleep.

The fundamental assumption that the rate of glucose phosphorylation by hexokinase is equal to CMR_{Glc} has been rejected by Huang and Veech (1982). They found that substantial dephosphorylation of glucose-6-phosphate occurs in vivo (about 25% of the rate of phosphorylation). This finding, in addition to the established fact that deoxyglucose phosphate is dephosphorylated in vivo (Hawkins and Miller, 1978), raises the question whether deoxyglucose phosphorylation is a valid measure of CMR_{Glc}. It seems likely that the accumulation of deoxyglucose phosphate is a complex function of two opposing reactions (phosphorylation of deoxyglucose and dephosphorylation of deoxyglucose phosphate) instead of a measure of glucose utilization per se.

Specifically labeled [^{14}C]glucose, which has also been used with autoradiography (Hawkins et al., 1979b, 1984; Lu et al., 1984; Bryan et al., 1983), continues to offer a satisfactory alternative to [^{14}C]deoxyglucose for determination of CMR_{Glc} in small laboratory animals. Recently, this method has been expanded to take into account more rigorously the exchange of glucose between plasma and brain, thus enabling the experiment to be performed in times as short as 5 min (Lu, et al., 1983; Hawkins, et al., 1985).

Despite the uncertainties relating to the mechanism of the [^{14}C]deoxyglucose method, its qualitative use has led to the production of autoradiographs that have excited the imagination of many investigators who have found patterns suggestive of functional relationships. Furthermore, it has renewed interest in the use of autoradiography as a tool to supplement biochemistry and physiology. This interest has not been limited to basic scientists. The rapid evolution of devices that are capable of accurately detecting radiation from structures deep within brain (i.e., positrons and gamma radiation) and the realization that both synthetic analogs and fully metabolizable substrates can be used should allow an immediate transition from basic biochemical studies to clinical investigations of normal and abnormal brain function.

A word of caution is in order relating to studies of CMR_{Glc}, whether performed with deoxyglucose or specifically labeled [^{14}C]glucose, because there are limitations regardless of the isotope used. The basic premise is that CMR_{Glc} is stoichiometrically related to ATP production. Since almost all energy-requiring reactions rely on ATP, a direct link between cellular activity and CMR_{Glc} can be established. An exact relationship, however, may not always exist. Measurement of glucose utilization is primarily useful where animals are considered to be normal and do not have severely disturbed metabolism, injured tissue, or interruption of the blood–brain barrier. There are several circumstances where misinterpretation of results could occur. These include:

1. *Injury*. Warburg (1972) noted that aerobic glycolysis, i.e., lactate production in the presence of oxygen, was a characteristic feature of injured tissues (malignant tumor tissue is an exception in that it is uninjured tissue that manifests aerobic glycolysis). In fact, he considered aerobic glycolysis to be indicative of dying tissue. The occurrence of increased aerobic glycolysis alters the stoichiometric relationship between CMR_{Glc}, oxygen consumption, and ATP production. Without knowledge of the relative proportions, direct determination of a functional relationship is impossible.

2. *Anaerobic glycolysis*. Under most circumstances, brain uses glucose as the predominant respiratory fuel and oxidation of glucose can be considered to be complete. However, there are many circumstances where increased uptake of glucose may not represent oxidative metabolism, but anaerobic glycolysis. These conditions include: injury of ischemia (Levy and Duffy, 1977), abnormally large increases in the rate of metabolism (Chapman et al., 1977; Duffy et al., 1975; Howse et al., 1974), and lowered blood pressure or decreased oxygen tension (Siesjo et al., 1974). Under these circumstances, the rate of glycolysis is generally increased to a greater extent than oxygen consumption. Therefore, the rate of energy metabolism cannot be accurately described by measurement of CMR_{Glc} alone.

3. *Breakdown of the blood–brain barrier*. Loss of blood–brain barrier integrity may increase the permeability to glucose, thereby raising the concentration of background [^{14}C]glucose (or [^{14}C]deoxyglucose). If the breakdown occurs only in specific areas, the assumption that glucose is distributed equally throughout the whole brain will not be valid. A higher brain-to-blood glucose ratio would suggest an increased metabolic rate, where this was, in fact, not so.

4. *Compartmentation*. To date, virtually all the methods used to measure CMR_{Glc} are based on a two- or three-compartment analysis. It would be difficult to convince an anatomist that the brain consisted only of two compartments, yet on the other hand, with respect to glucose, the brain does behave as if it consisted primarily of blood separated from a homogeneous group of cells by the blood–brain barrier (Lund-Anderson,

1979). With current experimental limitations, a more detailed analysis is not feasible, because accurate data available for analyzing more compartments are obtained only with great difficulty and the complexity of systems with more compartments rises almost exponentially.

5. *Steady state*. A fundamental assumption of the methods is that metabolism in all areas is in a steady state during the experimental period. Interpretation becomes difficult if non-steady-state conditions exist. Presently, there is no clear understanding of the moment-to-moment variation in CMR_{Glc}, but it is not difficult to imagine situations where regional activity will change within seconds. Because of this, it is advantageous to shorten the experimental time as much as possible, thus reducing potential errors. In this regard, the use of specifically labeled [^{14}C]glucose has a decided advantage (Hawkins *et al.*, 1984).

6. *Autoradiographs*. It is often claimed that autoradiographs provide a pictorial description of the metabolic and presumably functional relationships. A variety of photographs have been shown and conclusions drawn regarding an increased or decreased rate of glucose metabolism without objective quantitation of glucose consumption. Many authors have presented results where the rates are not given in proper chemical units (e.g., micromoles/minute per gram). Results not supported by calculated rates of metabolism must be interpreted with considerable caution. The appearance of contrast between different photographs may be due to different development times or different blood isotope concentrations.

5.2.2. Regional Cerebral Blood Flow

In trying to determine the changes in the rate of regional cerebral metabolism in very short times (i.e., during a period of seconds), it may be useful to use autoradiographic methods of determining blood flow. It has been established that blood flow and glucose utilization are closely coupled (Hawkins *et al.*, 1979a; Lassen *et al.*, 1978; Reivich *et al.*, 1969). In fact, by plotting currently accepted rates of blood flow against glucose utilization, it can be determined that blood flow is related to glucose consumption by the arteriovenous difference of glucose. Whether the arteriovenous difference is constant in all brain regions remains open to investigation, but first indications are that it may not vary greatly. If this is true, then determination of regional blood flow by accurate methods should enable the measurement of rapid functional changes in various anatomical areas.

The original methodology for determining CBF was established almost 40 years ago by Kety and Schmidt (1948). Their whole brain technique was adapted to regional determinations by Landau *et al.*, (1955) who used quantitative autoradiography and the diffusible gas trifluoromethane. Because of the strong possibility of evaporation, this compound was not further used. A nonvolatile, marginally permeable compound, antipyrine, was introduced by Reivich *et al.*, (1969). Antipyrine, however, was not permeable enough and at the relatively brisk CBF normally found in experimental animals it underestimated the rate of blood flow considerably. A more satisfactory compound, iodoan-

tipyrine, was described by Sakurada *et al.* in 1978. Although the value of io-
doantipyrine has been disputed by Goldman (1980), who suggests that asym-
metrical thioureas would be more satisfactory, it has generally found
acceptance and has been used to measure flows as high as 5 ml/min per g
(Dahlgren *et al.*, 1981).

5.2.3. Regional Ketone Body Utilization

Autoradiographic techniques have also been applied to the determination
of regional rates of ketone body utilization. The basic premises were originally
established by Cremer and Heath (1974) and involve the principle of ^{14}C trapping. The application of quantitative autoradiography has demonstrated that
brain use of ketone bodies is restricted by the blood–brain barrier and that
remarkable changes occur during diabetes, starvation (increased rates), and
during hepatic encephalopathy (decreased rates) (Mans *et al.*, 1981; Hawkins
et al., 1981).

5.2.4. Regional Studies of the Blood–Brain Barrier

Because of the key role of the blood–brain barrier in cerebral metabolism,
it is important to know what the actual rate of penetration of nutrients is under
physiological conditions; that is, in the presence of competitors, hormones,
proteins, and other circulating constituents. Of even greater interest is the determination of the rate of flux into specific anatomical structures of the brain.
Relatively straightforward procedures have been developed for use in rats *in
vivo* and applied to the transport of neutral and basic amino acids during
hepatic encephalopathy (Hawkins *et al.*, 1982; Mans *et al.*, 1982). Glucose transport across the blood–brain barrier has been similarly studied (Hawkins *et al.*,
1983). The results showed that there was a significant correlation, by region,
between glucose influx and utilization, demonstrating that the glucose supply
to individual cerebral structures is closely matched to their metabolic needs.

In conclusion, the concepts of cerebral energy metabolism (and nutrition)
have changed rapidly in the last decade. It is now apparent that normal cerebral
energy metabolism requires a variety of substrates that must be delivered in a
balanced fashion. Disturbances in the supply (either a deficit or an excess) may
have more serious consequences than previously appreciated. With the development of newer and more sophisticated investigative techniques, we can
expect to better define the needs of the CNS even at the structural level, thereby
being in a much better position to intelligently treat disorders of altered consciousness.

References

Allweis, C., Landau, T., Abeles, M., and Magnes, J., 1966, The oxidation of uniformly labelled
    albumin-bound palmitic acid to CO_2 by the perfused cat brain, *J. Neurochem.* **13**:795–804.
Babcock, M., and Hawkins, R. A., 1985, Carnitine content of cerebral tissue (in preparation).
Balazs, R., Patel, A. J., and Richter, D., 1972, Metabolic compartments in the brain: their properties

and relation to morphological structures, in: *Metabolic Compartmentation in the Brain.* (R. Balazs and J. E. Cremer, eds.), John Wiley and Sons, New York, pp. 167–186.

Baños, G. P., Daniel, M., Moorhouse, S. R., and Pratt, O. E., 1973, The influx of amino acids into the brain of the rat *in vivo*: The essential compared with some non-essential amino acids, *Proc. R. Soc. (London) Ser. B.* **183**:59–70.

Beattie, D. S. and Basford, R. E., 1966, Brain mitochondria IV. The activation of fatty acids in bovine brain mitochondria, *J. Biol. Chem.* **241**:1412–1418.

Berl, S., and Frigyesi, T. L., 1968, Metabolism of [¹⁴C]leucine and [¹⁴C]acetate in sensorimotor cortex, thalamus, caudate nucleus and cerebellum of the cat, *J. Neurochem.* **15**:965–970.

Borgstrom, L., Norbert, K., and Siesjo, B. K., 1976, Glucose consumption in rat cerebral cortex in normoxia, hypoxia and hypercapnia, *Acta Physiol. Scand.* **96**:569–574.

Braun, L. D., Cornford, E. M., and Oldendorf, W. H., 1980, Newborn rabbit blood-brain barrier is selectively permeable and differs substantially from the adult, *J. Neurochem.* **34**:147–152.

Bremmer, J., 1963, Carnitine in intermediary metabolism. The biosynthesis of palmitylcarnitine by cell subfractions, *J. Biol. Chem.* **238**:2774–2779.

Brightman, M. W., and Reese, T. S., 1969, Junctions between intimately apposed cell membranes in the vertebrate brain, *J. Cell Biol.* **40**:648–677.

Bryan, R. M., Hawkins, R. A., Mans, A. M., Davis, D. W., and Page, R. B., 1983, Cerebral glucose utilization in awake, unstressed rats, *Am. J. Physiol.* **244**:C270–C275.

Buckley, B. M., and Williamson, D. H., 1977, Origins of blood acetate in the rat, *Biochem. J.* **166**:539–545.

Cahill, G. F., Jr., and Aoki, T. T., 1980, Alternate fuel utilization by brain, in: *Cerebral Metabolsim and Neural Function* (J. V. Passonneau, R. A. Hawkins, D. W. Lust, and F. A. Welsh, eds.), Williams and Wilkins, Baltimore, pp. 234–242.

Chapman, A. G., Meldrum, B. S., and Siesjo, B. K., 1977, Cerebral metabolic changes during prolonged epileptic seizures in rats, *J. Neurochem.* **28**:1025–1035.

Christensen, T. G., Diemer, N. H., Laursen, H., and Gjedde, A., 1981, Starvation accelerates blood–brain glucose transfer, *Acta Physiol. Scand.* **112**:221–223.

Cooper, A. J., McDonald, J. M., Gelbard, A. S., Gledhill, R. F., and Duffy, T. E., 1979, The metabolic fate of [¹³N] labeled ammonia in rat brain, *J. Biol. Chem.* **254**:4982–4992.

Cornford, E. M., Braun, L. D., and Oldendorf, W. H., 1982, Developmental modulations of blood-brain barrier permeability as an indicator of changing nutritional requirements in the brain, *Pediatr. Res.* **16**:324–328.

Crane, P. D., Braun, L. D., Cornford, E. M., Cremer, J. E., Glass, J. M., and Oldendorf, W. H., 1978, Dose dependent reduction of glucose utilization by pentobarbital in rat brain, *Stroke* **9**:12–18.

Cremer, J. E., 1981, Nutrients for the brain: Problems in supplies, *Early Hum. Dev.* **5**:117–132.

Cremer, J. E., and Heath, D. F., 1974, The estimation of rates of utilization of glucose and ketone bodies in the brain of the suckling rat using compartmental analysis of isotopic data, *Biochem. J.* **142**:527–544.

Cremer, J. E., Heath, D. F., Patel, A. J., Balazs, R., and Cavanagh, J. B., 1975, An experimental model of CNS changes associated with chronic liver disease: Portocaval anastomoses in the rat, in: *Metabolic Compartmentation and Neurotransmission. Relation to Brain Structure and Function* (S. Berl, D. D. Clarke, and D. Schnieder, eds.), Plenum Press, New York, pp. 461–478.

Cremer, J. E., Brawn, L. D., and Oldendorf, W. H., 1976, Changes during development in transport processes of the blood–brain barrier, *Biochem. Biophys. Acta* **448**:633–637.

Cremer, J. E., Teal, H. M., Heath, D. F., and Cavanagh, J. B., 1977, The influence of portocaval anastomosis on the metabolism of labeled octanoate, butyrate and leucine in rat brain, *J. Neurochem.* **28**:215–222.

Cremer, J. E., Cunningham, V. J., Pardridge, W. M., Braun, L. D., and Oldendorf, W. H., 1979, Kinetics of blood–brain barrier transport of pyruvate, lactate, and glucose in suckling, weaning and adult rats, *J. Neurochem.* **33**:439–445.

Dahlgren, N., Ingvar, M., and Siesjo, B. K., 1981, Effect of propranolol on local cerebral blood flow under normocapnic and hypercapnic conditions, *J. Cereb. Blood Flow Metab.* **1**:429–436.

Daniel, P. M., Love, E. R., and Pratt, O. E., 1975, Hypothyroidism and amino acid entry into brain and muscle, *Lancet* **2**:872.

DeVivo, D. C., 1980, The effects of ketone bodies on glucose utilization, in: *Cerebral Metabolism* and Neural Function (J. V. Passonneau, R. A. Hawkins, D. W. Lust, and F. A. Welsh, eds.), Williams and Wilkins, Baltimore, pp. 243–254.

DeVivo, D. C., Lechie, M. P., Ferrendelli, J. A., and McDougal, D. B., Jr., 1978, Chronic ketosis and cerebral metabolism, Ann. Neurol. 3:331–337.

Duffy, T. E., Howse, D. C., and Plum, F., 1975, Cerebral energy metabolism during experimental status epilepticus, J. Neurochem. 24:925–934.

Engel, J., Kuhl, D., and Phelps, M. E., 1982, Patterns of human local cerebral glucose metabolism during epileptic seizures, Science 218:64–66.

Fernandez, J., Berger, R., and Smit, G. P. A., 1982, Lactate as energy source for brain in a glucose-6-phosphatase deficient child, Lancet 1:113.

Fox, J. S., 1984, PET scan controversy aired, Science 224:145–146.

Gaitonde, M. K., 1965, Rate of utilization of glucose and compartmentation of α-oxoglutarate and glutamate in rat brain, Biochem. J. 95:803–810.

Gibbs, E. L., Lennox, W. G., Nims, L. F., and Gibbs, F. A., 1942, Arterial and cerebral venous blood. Arterial-venous differences in man. J. Biol. Chem. 144:325–332.

Gjedde, A., and Crone, C., 1975, Induction processes in blood–brain transfer of ketone bodies during starvation, Am. J. Physiol. 229:1165–1169.

Gjedde, A., and Crone, C., 1981, Blood–brain glucose transfer: Repression in chronic hyperglycemia, Science 214:456–457.

Glees, P., 1972, The neuroglial compartments at light microscopic and electron microscopic levels, in: Metabolic Compartmentation in the Brain (R. Balazs and J. E. Cremer, eds.), John Wiley and Sons, New York, pp. 209–234.

Goldman, S. S., 1980, A new family of unsymmetrical thiourea derivatives to measure cerebral blood flow, in: Cerebral Metabolism and Neural Function (J. V. Passonneau, R. A. Hawkins, W. D. Lust, and F. A. Welsh, eds.), Williams and Wilkins, Baltimore, pp. 170–185.

Grubb, R. L., Jr., and Raichle, M. E., 1981, Intraventricular angiotensin II increases brain vascular permeability, Brain Res. 210:426–430.

Grubb, R. L., Jr., Raichle, M. E., and Eichling, J. O., 1978, Peripheral sympathetic regulation of brain water permeability, Brain Res. 144:204–207.

Hampson, R. K., Medina, M. A., and Olson, M. S., 1982, The use of high-energy microwave irradiation to inactivate mitochondrial enzymes, Ann. Biochem. 123:49–54.

Haslam, R. J., and Krebs, H. A., 1963, The metabolism of glutamate in homogenates and slices of brain cortex, Biochem. J. 88:566–578.

Hawkins, R. A., and Biebuyck, J. F., 1979, Ketone bodies are selectively used by individual brain regions, Science 205:325–327.

Hawkins, R. A., and Miller, A. L., 1978, Loss of radioactive 2-deoxy-D-glucose-6-phosphate from brains of conscious rats: Implications for quantitative autoradiographic determination of regional glucose utilization, Neuroscience 3:251–258.

Hawkins, R. A., Williamson, D. H., and Krebs, H. A., 1971, Ketone-body utilization by adult and suckling rat brain in vivo, Biochem. J. 122:13–18.

Hawkins, R. A., Miller, A. L., Cremer, J. E., and Veech, R. L., 1974, Measurement of the rate of glucose utilization by rat brain in vivo, J. Neurochem. 23:917–923.

Hawkins, R. A., Hass, W. K., and Ransohoff, J., 1979a, Cerebral blood flow, glucose utilization, oxidative metabolism, and plasticity after mesencephalic reticular formation lesions, in: Neural Trauma (A. J. Popp, R. S. Bourke, L. R. Nelson, and H. K. Kimelbert, eds.), Raven Press, New York, pp. 9–19.

Hawkins, R. A., Hass, W. K., and Ransohoff, J., 1979b, Measurement of regional brain glucose utilization in vivo using 2-^{14}C glucose, Stroke 10:690–703.

Hawkins, R. A., Mans, A. M., and Biebuyck, J. F., 1981, Regional blood–brain barrier permeability in hepatic encephalopathy, J. Cereb. Blood Flow Metabol. 1(Suppl. 1):385–386.

Hawkins, R. A., Mans, A. M., and Biebuyck, J. F., 1982, Amino acid supply to individual cerebral structures in awake and anesthetized rats, Am. J. Physiol. 242:E1-E11.

Hawkins, R. A., Mans, A. M., Davis, D. W., Hibbard, L. S., and Lu, D. M., 1983, Glucose availability to individual cerebral structures is correlated to glucose metabolism, J. Neurochem. 40:1013–1018.

Hawkins, R. A., Mans, A. M., Davis, D. W., Viña, J. R., and Hibbard, L. S., 1985, Cerebral glucose use measured with [^{14}C]glucose labelled in the 1, 2, or 6 position. Am. J. Physiol. 248 (in press).

Hellmann, J., Vannucci, R. C., and Nardis, E. E., 1982, Blood–brain barrier permeability to lactic acid in the newborn dog: Lactate as a cerebral metabolic fuel, *Pediatr. Res.* **16**:40–44.

Hernandez, M. J., Brennan, R. W., Vannucci, R. C., and Bowman, G. S., 1978, Cerebral blood flow and oxygen consumption in the newborn dog, *Am. J. Physiol.* **234**:R209–R215.

Hernandez, M. J., Vannucci, R. C., Salcedo, A., and Brennan, R. W., 1980, Cerebral blood flow and metabolism during hypoglycemia in newborn dogs, *J. Neurochem.* **35**:622–628.

Hertz, L., 1981, Features of astrocyte function apparently involved in the response of central nervous tissue to ischemia-hypoxia, *J. Cereb. Blood Flow Metabol.* **1**:143–154.

Hindfelt, B., Plum, F., and Duffy, T. E., 1977, Effect of acute ammonia intoxication on cerebral metabolism in rats with portacaval shunts, *J. Clin. Invest.* **59**:386–396.

Howse, D. C., Caronna, J. J., Duffy, T. E., and Plum, T., 1974, Cerebral energy metabolism pH and blood flow during seizures in the cat, *Am. J. Physiol.* **227**:1444–1451.

Huang, M., and Veech, R. L., 1982, The quantitative determination of the *in vivo* dephosphorylation of glucose-6-phosphate in rat brain, *J. Biol. Chem.* **257**:11358–11363.

Huang, S.-C., Phelps, M. E., Hoffman, E. J., Sideris, K., Selin, C. J., and Kuhl, D. E., 1980, Non-invasive determination of local cerebral metabolic rate of glucose in man, *Am. J. Physiol.* **238**:E69–E82.

Ide, T., Stinke, J., and Cahill, G. F., Jr., 1969, Metabolic interaction of glucose, lactate, and beta-hydroxybutyrate in rat brain slices, *Am. J. Physiol.* **217**:784–792.

Karnovsky, M. L., Burrows, B. L., and Zoccoli, M. A., 1980, Cerebral glucose-6-phosphatase and the movement of 2-deoxyglucose during slow wave sleep, in: *Cerebral Metabolism and Neural Function* (J. V. Passonneau, R. A. Hawkins, W. D. Lust, and F. A. Welsh, eds.), Williams and Wilkins, Baltimore, pp. 359–366.

Kennedy, C., and Sokoloff, L., 1957, An adaptation of the nitrous oxide method to the study of cerebral circulation in children: Normal values for cerebral blood flow and cerebral metabolic rate in childhood, *J. Clin. Invest.* **36**:1130–1137.

Kennedy, C., DesRosiers, M. H., Jehle, J. W., Reivich, M., Sharpe, F., and Sokoloff, L., 1975, Mapping of functional neural pathways by autoradiographic survey of local metabolic rate with [2-^{14}C]deoxyglucose, *Science* **187**:850–853.

Kety, S. S., Polis, B. D., Nadler, C. S., and Schmidt, C. F., 1948, The blood flow and oxygen consumption of the human brain in diabetic acidosis and coma, *J. Clin. Invest.* **20**:500–510.

Kety, S. S., and Schmidt, C. F., 1948, The nitrous oxide method for the quantitative determination of cerebral blood flow in man; theory, procedure and normal values, *J. Clin. Invest.* **27**:476–483.

Kraus, H., Schlenker, S., and Schwedesky, D., 1974, Developmental changes of cerebral ketone body utilization in human infants, *Hoppe-Seyler's Z. Physiol. Chem.* **355**:164–170.

Krebs, H. A., 1965, Metabolic interrelations in animal tissues, in: *Selected Topics in Modern Biochemistry* (W. O. Milligan, ed.), Proceedings of the Robert A. Welsh Foundation Conferences on Chemical Research VIII, Houston, Texas, pp. 101–129.

Krebs, H. A., 1970, Rate control of the tricarboxylic acid cycle, in: *Advances in Enzyme Regulation* Volume 8 (G. Weber, ed.), Pergamon Press, Oxford, pp. 335–353.

Landau, W. M., Freygang, W. J., Jr., Rowland, L. P., Sokoloff, L., and Kety S. S., 1955, The local circulation of the living brain: Values for the unanesthetized and anesthetized cat, *Trans. Am. Neurol. Assoc.* **80**:125–129.

Lassen, N. A., Ingvar, D. H., and Skinhoj, E., 1978, Brain function and blood flow, *Sci. Am.* **239**:62–71.

Laursen, H., and Westergaard, E., 1977, Enhanced permeability to horseradish peroxidase across cerebral vessels in the rat after portacaval anastomosis, *Neuropathol. Appl. Neurobiol.* **3**:29–43.

Levy, D. E., and Duffy, T. E., 1977, Cerebral energy metabolism during transient ischemia and recovery in the gerbil, *J. Neurochem.* **28**:63–70.

Lowry, O. H., 1962, The chemical study of single neurons, in: *The Harvey Lectures*, Series 58, Academic Press, New York, pp. 1–19.

Lowry, O. H., and Passonneau, J. V., 1972, *A Flexible System of Enzymatic Analysis*, Academic Press, New York.

Lu, D. M., Davis, D. W., Mans, A. M., and Hawkins, R. A., 1983, Regional cerebral glucose utilization measured with [^{14}C]glucose in brief experiments. *Am. J. Physiol.* **245**:C428–C438.

Lund-Anderson, H., 1979, Transport of glucose from blood to brain, *Physiol. Rev.* **59**:305–352.

Lust, W. D., Murakami, N., DeAzeredo, F., and Passonneau, J. V., 1980, A comparison of methods for brain fixation, in: *Cerebral Metabolism and Neural Function* (J. V. Passonneau, R. A. Hawkins, W. D. Lust, and F. A. Welsh, eds.), Williams and Wilkins, Baltimore, pp. 10–19.

Mans, A. M., Biebuyck, J. F., and Hawkins, R. A., 1981, Regional brain utilization of ketone bodies in starvation and diabetes, *J. Cereb. Blood Flow Metab.* **1**(Suppl. 1):90–91.

Mans, A. M., Biebuyck, J. F., Shelly, K., and Hawkins, R. A., 1982, Regional blood–brain barrier permeability to amino acids after portacaval anastomosis, *J. Neurochem.* **38**:705–717.

Martinez-Hernandez, A., Bell, K. P., and Norenberg, M. D., 1977, Glutamine synthetase: Glial localization in brain, *Science* **195**:1356–1358.

Mata, M., Fink, D. J., Gainer, H., Smith, C. B., Davidsen, L., Savaki, H., Schwartz, W. J., and Sokoloff, L., 1980, Activity-dependent energy metabolism in rat posterior pituitary primarily reflects sodium pump activity, *J. Neurochem.* **34**:213–215.

McCandless, D. W., 1981, Insulin-induced hypoglycemic coma and regional cerebral energy metabolism, *Brain Res.* **215**:225–233.

McCandless, D. W., and Schenker, S., 1981, Effect of acute ammonia intoxication on energy stores in the cerebral reticular activating system, *Exp. Brain Res.* **44**:325–330.

Medina, M. A., Deam, A. P., and Stavinoha, W. B., 1980, Inactivation of brain tissue by microwave irradiation, in: *Cerebral metabolism and Neural Function* (J. V. Passonneau, R. A. Hawkins, W. D. Lust, and F. A. Welsh, eds.), Williams and Wilkins, Baltimore, pp. 56–71.

Miller, A. L., Kiney, C. A., Corddry, D. H., and Staton, D. M., 1982, Interactions between glucose and ketone body use by developing brain, *Dev. Brain Res.* **4**:443–450.

Miller, J. A., 1982, Coma as a legitimate brain activity, *Sci. News* **122**:310.

Mink, J. W., Blumenschine, R. J., and Adams, D. B., 1981, Ratio of central nervous system to body metabolism in vertebrates: Its constancy and functional basis, *Am. J. Physiol.* **241**:R203–R212.

Moore, T. J., Lione, A. P., Sugden, M. C., and Regen, D. M., 1976, β-hydroxy-butyrate transport in rat brain: Developmental and dietary modulations, *Am. J. Physiol.* **230**:619–630.

Neely, J. R., Bowman, R. H., and Morgan, H. E., 1969, Effects of ventricular pressure development and palmitate on glucose transport. *Am. J. Physiol.* **216**:804–811.

Norberg, K., and Siesjö, 1974, Quantitative measurement of blood flow and oxygen consumption in the rat brain. *Acta Physiol. Scand.* **91**:154–164.

Oldendorf, W. H., 1972, Blood–brain barrier permeability to lactate, *Eur. Neurol.* **6**:49–55.

Oldendorf, W. H., 1973, Carrier-mediated blood-brain transport of short-chain monocarboxylic organic acids, *Am. J. Physiol.* **224**:1450–1453.

Owen, O. E., Morgan, A. P., Kemp, H. G., Sullivan, J. M., Herrera, M. J., and Cahill, G. F., Jr., 1967, Brain metabolism during fasting, *J. Clin. Invest.* **46**:1589–1595.

Pardridge, W. M., 1977, Regulation of amino acid availability to the brain, in: *Nutrition and the Brain*, Volume 1 (R. J. Wurtman and J. J. Wurtman, eds.), Raven Press, New York, pp. 141–204.

Pardridge, W. M., and Mietus, L. J., 1980, Palmitate and cholesterol transport through the blood–brain barrier, *J. Neurochem.* **34**:463–466.

Pardridge, W. M., and Oldendorf, W. H., 1977, Transport of metabolic substrates through the blood–brain barrier, *J. Neurochem.* **28**:5–12.

Pardridge, W. M., Crane, P. D., Mietus, L. J., and Oldendorf, W. H. (eds.), 1982, Nomogram for 2-deoxyglucose lumped constant for rat brain cortex, *J. Cereb. Blood Flow Metab.* **2**:197–202.

Passonneau, J. V., Hawkins, R. A., Lust, W. D., and Welsh, F. A., 1980, *Cerebral Metabolism and Neural Function*, Williams and Wilkins, Baltimore.

Patel, M. S., and Owen, O. E., 1977, Development and regulation of lipid synthesis from ketone bodies of rat brain, *J. Neurochem.* **28**:109–114.

Persson, B., Settergress, G., and Dahlquist, G., 1972, Cerebral arteriovenous difference of aceto-acetate and D-β-hydroxybutyrate in children, *Acta Paediatr. Scand.* **61**:273–278.

Preskorn, S. H., Irwing, G. H., Simpson, S., Friesen, D., Rinne, J., and Jerkovich, G., 1981, Medical therapies for mood disorders alter the blood–brain barrier, *Science* **213**:469–471.

Raichle, M. E., and Grubb, R. L., Jr., 1978, Regulation of brain water permeability by centrally-released vasopressin, *Brain Res.* **143**:191–194.

Raichle, M. E., Larson, K. B., Phelps, M. E., Grubb, R. L., Jr., Welch, M. J., and Ter-Pogossian, M. M., 1975, In vivo measurement of brain glucose transport and metabolism employing glucose-^{11}C, *Am. J. Physiol.* **228**:1936–1948.

Raichle, M. E., Laux, B. E., Grubb, R. L., Jr., and Larson, K. B., 1980, Strategies for measurement of cerebral metabolism using positron emission tomography, in: *Cerebral Metabolism and Neural Function* (J. V. Passonneau, R. A. Hawkins, W. D. Lust, and F. A. Welsh, eds.), Williams and Wilkins, Baltimore, pp. 338–397.

Reese, T. S., and Karnovsky, M. J., 1967, Fine structural localization of a blood–brain barrier to exogenous peroxidase, *J. Cell Biol.* **34:**207–217.

Reivich, M., 1974, Blood flow metabolism coupled in brain, *Res. Publ. Assoc. Nerv. Ment. Dis.*, Volume 53, Raven Press, New York, pp. 125–140.

Reivich, M., Jehle, J., Sokoloff, L., and Kety, S. S., 1969, Measurement of regional cerebral blood flow with antipyrine ^{14}C in awake cats, *J. Appl. Physiol.* **27:**296–300.

Reivich, M., Alvi, A., Wolf, A., Greenberg, J. H., Fowler, J., Christman, D., MacGreror, R., Jones, S. C., London, J., Shiue, C., and Yonekura, Y., 1982, Use of 2-deoxy-D [1-^{11}C]glucose for the determination of local cerebral glucose metabolism in humans: Variation within and between subjects, *J. Cereb. Blood Flow Metab.* **2:**307–319.

Roberts, S., and Morelos, B. S., 1965, Regulation of cerebral metabolism of amino acids–IV. Influence of amino acid levels on leucine uptake, utilization and incorporation into protein *in vivo*, *J. Neurochem.* **12:**373–387.

Robinson, A. M., and Williamson, D. H., 1980, Physiological roles of ketone bodies as substrates and signals in mammalian tissues, *Physiol. Rev.* **60:**142–187.

Rolleston, F. S., and Newsholme, E. A., 1967, Effects of fatty acids, ketone bodies, lactate and pyruvate on glucose utilization by guinea pig cerebral cortex slices, *Biochem. J.* **104:**524–533.

Ross, B. D., 1978, The isolated perfused rat kidney. *Clin. Sci. Mol. Med.* **55:**513–521.

Ruderman, N. B., Ross, P. S., Berger, M., and Goodman, M. W., 1974, Regulation of glucose and ketone-body metabolism in brain of unanesthetized rats, *Biochem. J.* **135:**1–10.

Sacks, W., Sacks, S., Brebbra, D. R., and Fleischer, A., 1982, Cerebral uptake of amino acids in human subjects and Rhesus monkeys *in vivo*, *J. Neurosci. Res.* **7:**431–436.

Sakurada, O., Kennedy, C., Jehl, J., Brown, J. D., Carbin, G. L., and Sokoloff, L., 1978, Measurement of local cerebral blood flow with iodo[^{14}C]antipyrine. *Am. J. Physiol.* **234:**H59–H66.

Sarna, G. S., Bradbury, M. W. B., Cremer, J. E., Lai, J. C. K., and Teal, H. M., 1979, Brain metabolism and specific transport at the blood–brain barrier after portocaval anastomosis in the rat, *Brain Res.* **160:**69–83.

Settergren, G., Lindblad, B. S., and Persson, B., 1980, Cerebral blood flow and exchange of oxygen. Glucose ketone bodies, lactate, pyruvate and amino acids in anesthetized children, *Acta Paediatr. Scand.* **69:**457–465.

Shivers, R. R., 1979, The effect of hyperglycemia on brain capillary permeability in the lizard, *Anolis Carolinensis*, A freeze-fracture analysis of blood–brain barrier pathology, *Brain Res.* **170:**509–522.

Siesjo, B. K., Johnnsson, H., Ljunggren, B., and Norbert, K., 1974, Brain dysfunction in cerebral hypoxia and ischemia, *Res. Publ. Assoc. Res. Nerv. Ment. Dis.* **53:**75–112.

Sloviter, H. A., and Kamimoto, T., 1970, The isolated, perfused rat brain preparation metabolizes mannose but not maltose. *J. Neurochem.* **17:**1109–1111.

Sokoloff, L., 1960, Metabolism of the central nervous system *in vivo*, in: *Handbook of Physiology-Neurophysiology*, Volume 3 (J. Field, H. W. Magun, and V. E. Hall, eds.), American Physiological Society, Washington, D.C., pp. 1843–1864.

Sokoloff, L., 1979, The radioactive deoxyglucose method: Theory, procedure and applications for the measurement of local glucose utilization in the central nervous system, in: *Advances in Neurochemistry* (B. W. Agranoff and M. H. Aprison, eds.), Plenum Press, New York, pp. 1–82.

Sokoloff, L., Fitzgerald, G. G., and Kaufman, E. E., 1977, Cerebral nutrition and energy metabolism, in: *Nutrition and the Brain*, Volume 1 (R. J. Wurtman and J. J. Wurtman, eds.), Raven Press, New York, pp. 87–139.

Spitzer, J. J., 1973, CNS and fatty acid metabolism, *Physiologist* **16:**55–68.

Trauner, D. A., Nyhan, W. L., and Sweetman, L., 1975, Short-chain organic acidemia and Reye's syndrome, *Neurology* **25:**296–298.

Trauner, D. A., Sweetman, L., Holm, J., Kulovich, S., and Nyhan, W. L., 1977, Biochemical correlates of illness and recovery in Reye's syndrome, *Ann. Neurol.* **2:**238–241.

Van den Berg, C. S., and Van den Velden, J., 1970, The effect of methionine sulphoximine on the incorporation of labelled glucose, acetate, phenylalanine and proline into glutamate and related amino acids in the brains of mice, *J. Neurochem.* **17:**985–991.

Vannucci, R. C., Hellmann, J., Hernandez, M. J., and Vannucci, S. J., 1980, Lactic acid as an energy source in perinatal brain, in: *Cerebral Metabolism and Neural Function* (J. V. Passonneau, R. A. Hawkins, W. D. Lust, and F. A. Welsh, eds.), Williams and Wilkins, Baltimore, pp. 264–270.

Vannucci, S. J., and Hawkins, R. A., 1983, Substrates of energy metabolism of the pituitary and pineal glands. *J. Neurochem.* **41**:1718–1725.

Veech, R. L., and Hawkins, R. A., 1974, Brain blowing: A technique for *in vivo* study of brain metabolism, in: *Research Methods in Neurochemistry*, Volume 2 (R. Marks and R. Rodnight, eds.), Plenum Press, New York and London, pp. 171–182.

Veech, R. L., Harris, R. L., Veloso, D., and Veech, E. H., 1973, Freeze-blowing: A new technique for the study of brain *in vivo*, *J. Neurochem.* **20**:183–188.

Vignais, P. M., Gallagher, C. H., and Zabini, I., 1958, Activation and oxidation of long chain fatty acids by rat brain, *J. Neurochem.* **2**:283–287.

Viña, J. R., Page, R. B., Davis, D. W., and Hawkins, R. A., 1984, Aerobic glycolysis by the pituitary gland *in vivo*. *J. Neurochem.* **42**:1479–1482.

Warburg, O. (discussed by H. A. Krebs), 1972, The Pasteur effect and relations between respiration and fermentation, in: *Essays in Biochemistry* (P. A. Campbell and F. Dickens, eds.), The Biochemical Society, Academic Press, London and New York, pp. 1–34.

Whittam, R., and Blond, D. M., 1964, Respiratory control by an adenosine triphosphatase involved in active transport in brain cortex, *Biochem. J.* **92**:147–158.

Ziven, J. A., and Snarr, J. F., 1972, Glucose and D(−)-3-hydroxybutyrate uptake by isolated perfused rat brain, *J. Appl. Physiol.* **32**:664–667.

II

Metabolic Encephalopathy Associated with Severe Interruption of Substrate

II

2

Hypoglycemia and Cerebral Energy Metabolism

DAVID W. McCANDLESS and MARC S. ABEL

1. Introduction

The normal adult brain subsists largely on glucose as substrate for energy-requiring processes (Gibbs et al., 1942). Interruption of this supply of glucose to nervous tissue by whatever means is potentially threatening since the energy reserves in brain are scanty. Interruption of the supply of glucose to the brain occurs in insulin overdose. Additionally, certain tumors generate insulin, and hypoglycemia is a frequent feature of liver disease. As is the case with many of the metabolic encephalopathies, an early correction of the defect may lead to complete reversal of altered neurological status. Study of cerebral energy metabolism in hypoglycemia is informative not only in terms of understanding mechanisms of altered brain function, but in terms of defining normal requirements for function. There have been many studies of energy metabolism in hypoglycemia using good models and experimental techniques, and the results from these studies have been helpful in defining the sequence of events in hypoglycemic-induced coma. The purpose of this review is to examine normal glucose metabolism, models of experimental hypoglycemia, and some of the neurochemical aspects of hypoglycemia that have recently been described.

2. Normal Glucose Metabolism

2.1. Glucose Transport

As stated earlier in this chapter, the normal brain of adult humans subsists largely on glucose as its major source of energy (Gibbs et al., 1942). During

DAVID W. McCANDLESS • Department of Neurobiology and Anatomy, University of Texas Medical School at Houston, Houston, Texas 77025.   MARC S. ABEL • Department of CNS Research, Medical Research Division, American Cyanamid Company, Lederle Laboratories, Pearl River, New York 10965.

certain situations, cerebral tissue can sustain activity on other fuels such as ketone bodies, and newborn brain may have even more flexibility for utilization of nonglucose substrates (Hawkins *et al.*, 1971). Under normal resting conditions the brain extracts approximately 10% of the glucose from blood; this amount is not dependent on blood flow since if flow is decreased, glucose extraction increases (Lajtha *et al.*, 1981). Compared with other tissues, the rate of cerebral oxidative metabolism is high.

Although the molecular weight of glucose (180) is relatively small, transport into brain proceeds via facilitated diffusion rather than diffusion (Fishman, 1964; Crone, 1965). The Km for glucose uptake is approximately 8 mM, which is nearly the blood concentration of glucose; however, brain glucose levels are nearly always lower than blood glucose levels. Within certain limits, transport of glucose and therefore brain glucose levels are related to the glucose concentration of blood. Since the glucose concentration is in fact usually about one fourth that of blood, the net flux or transport depends on the gradient of glucose across the blood–brain barrier. Because glucose is constantly being metabolized by the cell, estimates of transport are technically difficult, and the measurement of simple tissue and blood glucose concentrations may be misleading.

There have been somewhat conflicting data regarding the effect of insulin on glucose transport into brain. For example, it has been shown that in mice, insulin administration increases the ratio of brain tissue/blood glucose, as well as the glycogen concentration (Nelson *et al.*, 1968). *In vitro* experiments tend to support this in that rat spinal cord, cerebellum, and cerebral cortex take up glucose at a greater rate in the presence of insulin (Rafaelsen, 1961). Conversely, it has been suggested that the effect on glucose is indirect in that the insulin may affect phosphorylation rates thereby changing the tissue/glucose ratio and altering the transport rate (Daniel *et al.*, 1975).

It is worth noting that other substrates may replace glucose under stress situations. It is beyond the scope of this chapter to discuss this at length, but, for example, it has been shown that mannose can substitute for glucose in perfused rat brain (Sloviter and Kamimoto, 1970). Mannose, however, is not present in significant quantities, and so is not of any importance in normal metabolism. It has been shown that the ketone bodies beta-hydroxybutyrate and acetoacetate are able to substitute for glucose in severe starvation after the body stores of glucose and glycogen are depleted (Owen *et al.*, 1967). It has further been shown that even in rats starved from 24–96 hr there is a significant increase in ketone body uptake and metabolism (Hawkins *et al.*, 1971; Ruderman *et al.*, 1974). Amino acids may also make a contribution to cerebral metabolism during stress, but that contribution is probably not particularly important (Cremer *et al.*, 1977).

2.2. Glucose Pathways

The oxidation of glucose (glucose + $6O_2 \rightarrow 6CO_2 + 6H_2O$) has been shown by many workers to be the major route for metabolism in both man and animals. Estimates of glucose flux through aerobic pathways for man range from about

82–92%, with the balance representing lactate and pyruvate (anaerobic) production (Kety, 1957).

Studies in rats have in general correlated well with those in man. Thus, it has been shown that as much as 93% of glucose is oxidized, the remainder presumably largely going to lactate and pyruvate (Hawkins *et al.*, 1971). Studies on dogs have yielded similar results, with values for glucose oxidation of about 85% and the balance being attributed to lactate and pyruvate production (Drewes and Gilboe, 1973). It should be noted that technically the measurement of arteriovenous differences for lactate/pyruvate is somewhat difficult, and it is not possible to say with certainty that the nonoxidative pathways represent exclusively lactate and pyruvate production.

It is important to remember that there are other pathways that operate from glucose. These include at least two of some importance, namely the formation of glycogen and the operation of the hexosemonophosphate shunt. The net level of glycogen in brain is about 2.5 μmole/g, which is only 1/100 that of liver. This represents a rather small energy reserve, which nevertheless is important in a variety of metabolic perturbations. It has been shown, for example, that in kernicteric Gunn rats there is an increase amounting to about 50% in glycogen in the cerebellum in severely ataxic animals as compared with controls (McCandless and Abel, 1980). Such an increase has been noted in other metabolic perturbations (McCandless and Schwartzenberg, 1981), and may represent a compensatory mechanism by which cerebral tissue attempts to protect itself in stressed metabolic circumstances. Although net glycogen levels are relatively low, they are metabolically active and may serve to afford some degree of protection for the brain.

The hexosemonophosphate shunt is an alternate pathway for glucose metabolism that operates in most tissues including brain. The amount of "flux" through the hexosemonophosphate shunt has been the subject of some controversy over the past 20 years. It now seems certain that flux through the shunt is on the order of 5–8% in adult brain (Hostetler *et al.*, 1970) and somewhat greater than that in newborn brain (Hothersall *et al.*, 1979). The function of the shunt is thought to be twofold: (1) the generation of five carbon moieties for synthetic purposes and (2) the production of reducing power in the form of NADPH. It is because both of these functions are increased in developing and myelinating brain that flux is similarly increased during this time period.

3. Models of Experimental Hypoglycemia

3.1. Methods of Induction

The majority of workers have used insulin to induce experimental hypoglycemia. This, of course, is the obvious method of induction, and in man represents the most commonly encountered cause of hypoglycemic encephalopathy. The dose of insulin administered by various investigators depends somewhat on the relative speed of induction desired. Most workers using mice and rats have used a dose from 50 to 100 units/kg, which induces coma in from 1–2 hr following injections. In larger animals such as dogs, the dose commonly employed is 200–300 units/kg. The mode of administration of insulin is most commonly IP.

3.2. Selection of Animals

The selection of experimental animals depends in many instances on the actual experiment to be performed. Studies that, for example, are designed to examine energy metabolites are usually performed on small mice that freeze rapidly or on rats that can be sacrificed by a method in which metabolites are quickly "fixed." Ordinarily, animals the size of dogs are not used for metabolite studies. Similarly, in studies designated to examine cerebral blood flow, larger animals such as dogs are ideal.

The issue of sacrifice method and the assay of cerebral energy metabolites is critical to an accurate understanding of metabolic events during encephalopathy. The turnover of high-energy phosphates is very rapid and in order to obtain as close to *in vivo* values as possible, sacrifice generally is via microwave irradiation (Medina *et al.*, 1975), freeze blowing (Veech *et al.*, 1973), or submersion of small animals into liquid nitrogen (Ferrendelli *et al.*, 1972). In addition, use of the Ponten method of freezing the brain *in situ* has also been used with success (Ponten *et al.*, 1973). Measurement of labile metabolites necessitates the use of sensitive fluorometric techniques if subtle changes are to be detected. Some early studies by necessity employed less accurate methodologies, and therefore results from these studies need to be carefully interpreted. In some measure, the selection of the animal determines sacrifice method.

3.3. Histologic, EEG, and Behavioral Changes

The cerebral histologic events that accompany hypoglycemia have been studied in primates (Myers and Kahn, 1971) and in rats (Agardh *et al.*, 1980). The studies in rats are particularly important since changes were correlated with biochemical and physiological alterations and adequate fixation and analytical techniques were used. In this study, animals were rendered hypoglycemic by insulin injection. Animals were sacrificed by the Ponten method (Ponten *et al.*, 1973) 30 and 60 min following the onset of isoelectric EEG. Other groups were "reversed" by the IV injection of glucose and sacrificed 30 min or 2 hr later. Samples from the cerebral cortex were examined by both light and electron microscopy.

There were two major types of neuronal injury. In the first instance, cells showed shrinkage and condensation of nuclei and cytoplasm, and the cytoplasmic membrane showed scalloping. This type of damage was seen more than twice as often in rats exposed to 60 min of hypoglycemia as those exposed to 30 min of hypoglycemia. The other type of neuronal injury observed was the appearance of swollen neurons. In addition, empty-looking vacuoles were seen just below the cytoplasmic membrane. As in the case of the first type of damaged cells, swollen neurons were seen more frequently following the longer period of hypoglycemia. During the recovery phase, these changes tended to revert to normal, and following 4 hr of recovery only 2% of the neurons appeared abnormal.

The exact meaning of these changes is hard to determine. It is possible that the two types of injury are related, but no transition cells were seen, and the authors believe that this represented two distinct neuropathological changes. Swelling of neurons could certainly result from a change in the permeability of the cytoplasmic membrane due to an energy deficiency. Similarly, such changes could be secondary to an alteration in the blood–brain barrier, which is known to be rapidly broken down in other metabolic encephalopathies such as ammonia intoxication (Mans *et al.*, 1982). The issue, then, of whether the histological changes noted in hypoglycemia are primary or secondary remains unclear. It is significant that these changes correlate with both EEG and metabolite changes, and following recovery the structural changes diminish.

Electroencephalographic changes accompanying hypoglycemia have been described by several workers (Tews *et al.*, 1965; Hinzen and Muller, 1971; Ferrendelli and Chang, 1973; Feise *et al.*, 1976). In general, there is a direct correlation between the blood glucose levels and the EEG characteristics. In one comprehensive study, it is shown that when blood glucose levels fell from about 8 mM to 2 mM, brain activity was described as slow-wave activity with interspersed normal activity. When blood glucose levels reached 1 mM or less, EEG activity was isoelectric. These studies were carried out on Wistar rats weighing 290–430 g and receiving from 3–35 units/kg insulin (Lewis *et al.*, 1974).

Behavioral changes associated with insulin-induced hypoglycemia range from normal to comatose (Gorell *et al.*, 1976). The behavior of mice and rats correlates closely with blood glucose and the EEG. Some investigators report the onset of coma as falling blood glucose levels approach the 1 mM level (Ghajar *et al.*, 1982). Stuporous behavior, characterized by decreased spontaneous and grooming activities, becomes obvious when blood glucose levels are about 1.5–2.0 mM. In any study of insulin-induced hypoglycemia, it is important to exclude animals that seize or display seizurelike EEG activity since seizures alone alter cellular energy metabolism (McCandless *et al.*, 1979). It is also of interest to study animals in the stuporous or precoma state since this provides a chance to see if changes present in comatose animals precede the onset of the coma. This may provide an indication of the cause/effect significance of biochemical changes. A disadvantage of studies on large animals is that they are usually anesthetized and ventilated, and therefore the behavioral state is not possible to observe. Ultimately, it is important to know what the mental status of the experimental animal is so that it can be correlated with observed neurochemical changes.

3.4. Blood Flow and Oxygen Consumption

Cerebral blood flow as well as oxygen consumption have been studied in insulin-induced hypoglycemia in both man and experimental animals. Even as early as 1948, it was known that insulin-induced hypoglycemic coma was associated with reduced cerebral oxygen consumption (Kety *et al.*, 1948). In a later study, schizophrenic patients undergoing insulin therapy were studied for both cerebral blood flow and cerebral metabolic rate for oxygen (Della Porta *et al.*, 1964). Results from this study showed that when patients were comatose,

11 out of 14 had an average 38% increase in cerebral blood flow as measured by the nitrous oxide method. With glucose administration, values returned to normal. Although there was some variation, oxygen consumption fell during coma.

In an interesting study on hypoglycemia in calves, it was shown that during hypoglycemia (plasma glucose 1.31 mM), blood flow and oxygen consumption were unchanged (Gardiner, 1980). However, when the animals were stuporous (plasma glucose below 1.0 mM), oxygen consumption fell about 20%, but cerebral blood flow remained normal. Only during convulsions was there an increase in cerebral blood flow. At this time, oxygen consumption had doubled as compared with the values obtained in stuporous calves. These dramatic changes during seizure activity emphasize the importance of carefully monitoring behavioral and/or EEG changes and excluding stuporous or comatose animals that seize.

Newborn dogs have also been studied using insulin to induce hypoglycemia (Hernandez *et al.*, 1980; Vannucci *et al.*, 1981). In one of these studies, newborn pups were anesthetized and artifically ventilated to maintain normal oxygen levels (Hernandez *et al.*, 1980). Two groups were studied, those with blood glucose levels above 0.5 mM and those with blood glucose levels below 0.5 mM. Cerebral blood flow was unaltered by hypoglycemia in either group. Cerebral metabolic rate for oxygen was slightly higher in both hypoglycemic groups, but the changes were not statistically significant. The cerebral metabolic rate for glucose was decreased about 50% in the more severely hypoglycemic group. Since EEGs were not recorded to determine the physiological status of the brain and since the animals were anesthetized, it is difficult to relate the biochemical and blood flow studies to the functional status of the pups.

In a recent study utilizing unanesthetized rats, cerebral energy metabolism has been correlated with blood flow in insulin-induced hypoglycemia (Ghajar *et al.*, 1982). Results from this study pertaining to energy metabolism will be discussed in the next section. Rats in this study were categorized as normal controls, lethargic, stuporous, and comatose. These categories were formed using behavioral criteria and EEG. There was no change in cerebral blood flow in any group, using the indicator-fractionation technique. Cerebral metabolic rate for oxygen was decreased in lethargic, stuporous, and comatose rats by about 50% as compared with normal controls. Cerebral metabolic rate for glucose was decreased in all three experimental groups by about 75%. Interestingly, the cerebral metabolic rate for glucose did not decrease until the plasma glucose decreased below 2.5 mM. In fact, glucose transport into brain was such that brain glucose was not in sufficient concentration to saturate hexokinase. Under normal conditions, transport of glucose into brain is two to three times greater than the rate of glucose phosphorylation (Betz *et al.*, 1973).

3.5. Hypothermia

One common physiological consequence of hypoglycemia is the induction of hypothermia. The relationship between hypoglycemia and hypothermia, which occurs in clinical settings and experimental models of hypoglycemia, has been recognized for many years (Gellhorn, 1938; Mayer-Gross and Berliner, 1942). Although there is at least one report of fever associated with severe hypoglycemia (Ramos *et al.*, 1968), the overwhelming majority of literature reports document the production of hypothermia when blood glucose levels are decreased. Under these conditions, body temperature in humans may reach a nadir at 93.5°F. The phenomenon appears to be independent of the etiology of hypoglycemia, suggesting that it is reduced glucose availability per se that is responsible for the lowered body temperature. For example, both alcohol-induced (Freinkel *et al.*, 1963) and insulin-induced (Gale *et al.*, 1983) hypoglycemia produce hypothermia in human subjects. Several clinical reports cite hypothermia as a major component of hypoglycemia (Strouch *et al.*, 1969; Reuler, 1978), and the suggestion has been made that hypothermia be used as an aid in the diagnosis of hypoglycemia (Jaffe, 1966).

One method of experimentally producing hypoglycemia (and the resultant hypothermia) is the administration of 2-deoxy-D-glucose (2-DG), which is an analogue of glucose that is virtually unmetabolized beyond the first step of glycolysis. Thus, when 2-DG is injected, cells become glucose deprived and normal metabolic processes are disrupted. After 2-DG was administered to human subjects intravenously, blood glucose levels were elevated and there was a concomitant fall in body temperature (Landau *et al.*, 1958). Another study reported similar findings as well as an increase in accumulating levels of insulin (Freinkel *et al.*, 1972). Although plasma levels of glucose are high after 2-DG administration, it is important to appreciate the "functional" hypoglycemia to which the intracellular milieu is exposed.

In support of the concept that intracellular glucose deprivation is involved in signaling thermoregulatory mechanisms are the findings that 2-DG-induced hypothermia was completely prevented when fivefold equimolar amounts of fructose were injected simultaneously with 2-DG into the cerebral ventricles of rats, whereas no prevention occurred with fructose and 2-DG injected at equimolar doses (Fiorentini and Muller, 1972). Also, effective inhibition of the 2-DG effect was obtained by stimulation of the Krebs cycle by fumarate and glutamate administration (Fiorentini and Muller, 1972). It is important that central chemoreceptors are not specifically sensitive to glucose, but rather to the rate of oxidative metabolism.

The importance of central chemoreceptor in regulating body temperature has been further investigated in a series of studies where 2-DG was injected either into the lateral cerebral ventricle or directly into selected areas of the conscious rat hypothalamus. In one study, central injections of 2-DG were fivefold more potent at producing hypothermia than injections into the tail vein (Freinkel *et al.*, 1972). The involvement of a specific hypothalamic nucleus in the generation of hypoglycemia-induced hypothermia was investigated by Shiraishi and Mager (1980a). They found that stereotaxic microinjections of 2-DG into selected hypothalamic areas were capable of producing hypothermia. One area of the hypothalamus, the ventral premammillary nucleus (PMV), was particularly sensitive to 2-DG and was the only area in which a dose response effect was elicited. Since administration of 2-DG to all of the sites immediately adjacent to the PMV was ineffective in producing hypothermia, it is tempting to speculate that the PMV is selectively sensitive to decreases in glucose levels (or oxidative metabolism) and may function as a central locus for thermoregulation.

Whereas hypoglycemia results in severe alterations in oxidative metabolism, hypothermia is itself capable of affecting cellular function. There is a plethora of physiological changes that occur as the result of hypothermia (Wong, 1983). With regard to cerebral energy metabolism, hypothermia produces a series of effects that suggest a generalized decrease in cellular function. Plasma glucose levels in hypothermic rabbits were elevated (Bickford and Mottram, 1960), suggesting an impairment of glucose transport or metabolism. A similar increase in glucose levels as well as concomitant decrease in glucose-6-phosphate was seen in mouse brain during hypothermia (Brunner et al., 1971). Glucose utilization in hypothermic rats was demonstrated to be decreased throughout the central nervous system and a positive correlation was shown to exist between body temperature and utilization rates (McCulloch et al., 1982). In addition, hypothermia produced an elevation in phosphocreatine levels in brain and decreased O_2 consumption (Bickford and Mottram, 1960). These data, taken together, imply a decrease in functional activity of the cell as a result of hypothermia. In support of this concept, decreased utilization rates of cerebral high-energy phosphate stores in hypothermic mice have been reported (Brunner et al., 1971; Ferrendelli et al., 1972). This regulation may be important for survival and is consistent with the existence of a self-protective mechanism during metabolic stress. Therefore, the hypothermia associated with hypoglycemia may be an attempt of the organism to reduce energy demands during a period of reduced glucose availability.

4. Neurochemical Aspects of Hypoglycemia

There have been many studies relating cerebral changes to hypoglycemic coma. The present discussion will be limited to those changes involving energy metabolism and some neurotransmitters. No attempt is made to discuss all studies; only a few representative ones in which adequate sacrifice and analytical techniques were used will be discussed.

4.1. Glycolysis

In one study in which cerebral cortical glucose-6-phosphate and glycogen were correlated with blood glucose levels, it was found that both metabolites began to drop when blood glucose reached 2.5 mM (Lewis et al., 1974). Lactate and pyruvate showed a similar decrease in concentration as blood glucose levels reached 2.5 mM. Cortical glucose concentration began to drop at a higher blood glucose level, about 5–6 mM. Control animals showed that these changes were not the result of insulin per se, but resulted from decreased blood glucose levels. These effects on blood and brain glucose levels confirm earlier observations (Mayman et al., 1964; Brunner et al., 1971).

In another study, the effects of insulin hypoglycemia were studied in three cerebral regions: the cortex, cerebellum, and brain stem of mice (Gorell et al., 1976). Results showed a similarity between regions, there being a greater than

95% drop in brain glucose 1–3 hr after insulin injection when the animals were comatose. In this study, recovery was studied by injecting glucose via the tail vein. This treatment resulted in a rapid increase (6 sec) in both blood and brain glucose. Over a 3-min time course, blood glucose gradually fell to about 2 mM. Brain glucose exhibited a biphasic response that consisted of a rapid increase followed by a drop at 20 sec. This was followed by an increase to about 1.0 mM at 2 min, followed by a gradual decline to 0.3 mM by 10 min following glucose administration. Glycogen, on the other hand, fell from about 70–85% in the three regions and rose little when the animals were "reversed" with glucose infusion.

In a subsequent study (Gorell *et al.*, 1977), using insulin to induce hypoglycemia in mice, the glycolytic metabolites glucose-6-PO4, fructose biphosphate, phosphoglycerate, and phosphoryruvate were measured. Except for glucose-6-PO4, all were significantly decreased in control mice. Reversal with glucose infusion acted to return decreased values toward normal within 60 sec. The authors, citing earlier studies showing no change in ATP or phosphocreatine (Ferrendelli and Chang, 1973), suggest that a reduced demand for glucose results in inhibition of phosphofructokinase. Blockage at this step would explain the results in that glucose-6-PO4 was not affected, but metabolites below this step were significantly decreased. Unanswered in these studies is the possible mechanism by which there is a reduced demand for glucose. The initiation of coma clearly places the animal in a lower energy demand state, but the biochemical mechanism for this remained unclear.

4.2. Citric Acid Cycle

Levels of various citric acid cycle metabolites in general fluctuate in a similar manner as those of the glycolytic pathway. In one study, for example (Gorell *et al.*, 1977), cortical citrate, alpha-ketoglutarate, and malate all decreased from 30–60% in comatose mice. When animals were reversed by administering glucose intravenously, levels of these three citric acid cycle metabolites remained low. This was interpreted as meaning that overall cortical citric acid cycle activity remained low even following return of consciousness. It is assumed that levels of alpha-ketoglutarate and malate are valid measures of citric acid cycle metabolism (Goldberg *et al.*, 1966).

In another study, rats that had developed an isoelectric EEG were sacrificed and whole brain was analyzed for the citric acid cycle metabolites citrate, isocitrate, alpha-ketoglutarate, succinate, fumarate, and oxalacetate (Norberg and Siesjo, 1976). Results showed all were significantly decreased except succinate and oxalacetate, which were increased. The increase in succinate levels was attributed to induction of succinate dehydrogenase.

4.3. Energy Metabolites

There have, obviously, been many studies in which high-energy phosphates and metabolites closely associated with them have been measured in hypoglycemic animals. Since ATP is the ultimate energy source for brain and since most energy is ultimately derived from glucose, it seemed logical to assume that ATP would be depleted throughout a hypoglycemic brain. Surprisingly, that concept was not completely correct.

In one comprehensive study (Lewis *et al.*, 1974), ATP and phosphocreatine were measured in rats at various EEG states, and it was found that during hypoglycemic-induced slow-wave activity there was no decrease in high energy phosphates. This confirmed earlier work (Ferrendelli and Chang, 1973). It was also noted that when the EEG was isoelectric, there were dramatic and significant decreases in ATP and phosphocreatine. These studies were carried out in the frontal-parietal portion of the brain. The authors conclude from these data that energy failure per se was not responsible for the initial decrease in the functional activity of the brain.

In a second series of experiments designed to examine recovery following hypoglycemic coma, it was again found that hypoglycemia to the point of isoelectric EEG resulted in severe derangements in energy metabolism (Agardh, *et al.*, 1978). It was further noted that following 5–30 min of isoelectric EEG, recovery induced by glucose administration resulted in restitution of tissue levels of phosphocreatine, ADP, and AMP. ATP values remained only slightly decreased. These data reflect the capacity of the mitochondria for metabolic recovery.

In another study in which the freeze-blowing technique (Veech *et al.*, 1973) was used for rapid sacrifice of rats, energy metabolism was assessed during lethargy, stupor, and coma induced by hypoglycemia (Ghajar *et al.*, 1982). This study showed major decreases in phosphocreatine (40%) and ATP (40%) during coma. There was no change in high-energy phosphates during lethargy; however, ATP was significantly decreased during the stuporous period. These data are important because they suggest changes in energy metabolism that precede coma. These subtle changes may have been detected because of the use of the ultrarapid freeze-blowing technique for sacrifice.

In a recent study (Ratcheson *et al.*, 1981), hypoglycemia was produced using insulin and energy metabolites measured in various regions during early and late slow-wave activity, burst suppression, and early and late isoelectric EEG activity. Results showed glucose and glycogen decreasing throughout the time course about equally in all regions; however, in terms of ATP and PCr there was a selective regional effect. The interesting finding was that the cerebellum did not show a decrease in either ATP or PCr even during the late isoelectric EEG period. The mechanism of this "sparing" of the cerebellum is unclear, but could be related to a decrease in utilization or to an ability of the cerebellum to more effectively extract glucose from the blood. The uniqueness of the cerebellum in metabolic stress has been noted in other experimental models of metabolic encephalopathy (McCandless *et al.*, 1979).

One major problem with the above-cited studies is that they were performed using large pieces of brain for analysis. Since the brain is highly heterogeneous, these samples must include many different cell types. Since consciousness is mediated in the ascending reticular formation (Magoun, 1973), biochemical analysis of this area alone, uncontaminated by adjacent nonaffected tissue is essential. We have undertaken such studies in the reticular formation, and have analyzed high-energy phosphate metabolism in hypoglycemia-induced stupor and coma (McCandless, 1981).

In these studies, we used small 20-g mice and sacrificed them by rapid submersion in liquid nitrogen. Previous studies have shown that sacrifice of small mice in liquid nitrogen is as effective as the use of freeze blowing or microwave irradiation (Lust et al., 1980). Mice were sacrificed and tissue from the brain stem was prepared for microanalysis using the procedures of Lowry and Passonneau (1972). Results showed a preferential effect on cells of the reticular formation. ATP was depleted 30% and phosphocreatine 55% in the precoma stage. During coma, values had returned to normal. High-energy phosphates in an adjacent non-reticular formation area were unaffected. These data are consistent with the concept that a decrease in available energy leads to decreased function in the reticular formation. Decreased reticular formation output in coma has been demonstrated physiologically (Magoun, 1954; Davis et al., 1958). This shutdown in reticular formation activity acts to produce coma wherein the animal is placed in a milieu in which energy demands are lessened. Such a condition is conducive to the possible correction of the deranged metabolic state that has produced the coma. Only during prolonged and severe hypoglycemic coma would a generalized energy failure occur throughout the brain in association with an isoelectric EEG. These changes have been missed in previous studies because large pieces of brain have been analyzed and the relatively small reticular formation has been diluted out by inclusion of contiguous tissue. This concept has been demonstrated in coma produced by ammonia (McCandless and Schenker, 1981) and by octanoic acid (McCandless and Knight, 1983). Further studies of other neurochemical parameters in coma must account for the heterogeneity of brain.

4.4. Acetylcholine

Acetylcholine levels have been examined in hypoglycemic animals and as early as 1955 were found to be decreased in brain (Crossland et al., 1955). Later studies were somewhat more equivocal with some animals showing decreases, whereas others with isoelectric EEGs had normal acetylcholine levels (Tews et al., 1965).

Although the amount of acetylcholine formed from pyruvate represents less than 10% of the oxidized pyruvate (Gibson et al., 1975; Gibson and Blass, 1975), the acetylcholine production seems sensitive to perturbation. For this reason, the synthesis of acetylcholine in vivo has been assessed in mice rendered hypoglycemic (Gibson and Blass, 1976). In these studies, the incorporation of [^{2}H$_4$]choline into acetylcholine was measured after injection of 125 U/kg of insulin. The levels of unlabeled acetylcholine did not change; however, incorporation of labeled choline into acetylcholine declined significantly. This suggests that the decreased incorporation of labeled choline into acetylcholine may occur in some small actively turning over pool. The authors state that these changes may not be due to a nonspecific effect, since other agents with widespread cerebral effects such as cycloheximide and puromycin do not alter acetylcholine synthesis.

4.5. Cyclic Nucleotides

The cyclic nucleotides, cyclic AMP and cyclic GMP, have been measured in the cerebral cortex and cerebellum of mice rendered hypoglycemic by insulin administration (Gorell *et al.*, 1976). There was no effect of hypoglycemia on cyclic AMP; however, cyclic GMP was increased in the cerebral cortex and decreased in the cerebellum. Previous studies of the action of atropine showed a similar elevation of cyclic GMP in the cortex and a decrease in the cerebellum (Ferrendelli *et al.*, 1970). This suggests that hypoglycemia alters cholinergic synapse activity. These authors (Gorell *et al.*, 1976) also measured GABA and did not find the usual inverse relation between cyclic GMP and GABA. In fact, hypoglycemia decreased GABA (and cyclic GMP) in the cerebellum and had no effect on cerebral cortical GABA levels. The results on cyclic nucleotides suggest that their changes may not be the direct result of hypoglycemia, but may be secondary to functional changes.

4.6. Amino Acids

Amino acids have been measured in the brains of hypoglycemic animals because of their close relationship to carbohydrate metabolism and since they can exhibit both excitatory and inhibitory influences on cerebral function. In one study, levels of glutamate, aspartate, glutamine, GABA, and alanine were measured in rats exhibiting either slow-wave EEG, convulsions, or isoelectric EEG induced by insulin (Lewis *et al.*, 1974). There was a significant progressive decrease in amino acids except for aspartate. In a more regional study, aspartate was similarly shown to be increased in both the cerebral cortex and the brain stem (Gorell *et al.*, 1976). In this study, GABA was decreased only in the cerebellum, whereas glutamate was decreased in the cerebral cortex, cerebellum, and brain stem. This pattern was again confirmed (Norberg and Siesjo, 1976) when it was shown that aspartate was increased in hypoglycemia. NH_4 levels were also increased 16-fold. These data all confirm earlier work on amino acids and hypoglycemia (Dawson, 1950; Cravioto *et al.*, 1951). These findings lend credence to the concept that amino acid carbon is made available as substrate for brain by both transamination and oxidative deamination.

5. Summary

In the case of hypoglycemic-induced coma, evidence points to a direct relationship between onset of coma and altered cerebral energy metabolism. The finding of decreased high-energy phosphates throughout the brain at a time when the EEG is isoelectric has been confirmed in several laboratories. The concept of highly selective changes in an area such as the ascending reticular formation emphasizes the need to do studies in as regional a setting as is technically possible. The tremendous heterogeneity of the brain, demonstrated amply by neurophysiologists, should not be overlooked by neurochemists.

ACKNOWLEDGMENTS. The authors wish to express their gratitude to Ms. Diana Parker for her expert secretarial assistance. Supported in part by NIH grant NS 17130 to D.W.M.

References

Agardh, C. D., Folbergergrova, J., and Siesjo, B. K., 1978, Cerebral metabolic changes in profound insulin induced hypoglycemia, and in the recovery period following glucose administration, *J. Neurochem.* **31**:1135–1142.

Agardh, C. D., Kalimo, H., Olsson, Y., and Siesjo, B. K., 1980, Hypoglycemic brain injury, *Acta Neuropathol.* **50**:31–41.

Betz, A. L., Gilboe, D. D., Yudilevich, D. L., and Diewes, L. R., 1973, Kinetics of unidirectional glucose transport into the isolated dog brain, *Am. J. Physiol.* **225**:586–592.

Bickford, A. F., and Mottram, R. F., 1960, Glucose metabolism during induced hypothermia in rabbits, *Clin. Sci.* **19**:345–359.

Brunner, E. A., Passonneau, J. V., and Molstad, C., 1971, The effect of volatile anesthetics on levels of metabolites and on metabolic rate in brain, *J. Neurochem.* **18**:2301–2316.

Cravioto, R. O., Massieu, G., and Izquierdo, J. J., 1951, Free amino acids in rat brain during insulin shock, *Proc. Soc. Exp. Biol.* **78**:856–858.

Cremer, J. E., Teal, H. M., Heath, D. F., and Cavanagh, J. B., 1977, The influence of portocaval anastomosis on the metabolism of labeled octanoate, butyrate, and leucine in rat brain, *J. Neurochem.* **28**:215–222.

Crone, C., 1965, Facilitated transfer of glucose from blood into brain tissue, *J. Physiol.* **181**:103–113.

Crossland, J., Elliott, K. A., and Pappius, H. M., 1955, Acetylcholine content of brain during insulin hypoglycemia, *Am. J. Physiol.* **183**:32–34.

Daniel, P. M., Love, E. R., and Pratt, O. E., 1975, Insulin and the way the brain handles glucose, *J. Neurochem.* **25**:471–476.

Davis, H. S., Dillon, W. H., Collins, W. F., and Randt, C. T., 1958, The effect of anesthetic agents on evoked central nervous system responses, *Anesthesiology* **19**:441–449.

Dawson, R. M., 1950, Studies on the glutamine and glutamic acid content of the rat brain during insulin hypoglycemia, *Biochem. J.* **47**:386–391.

Della Porta, P., Maiolo, A. T., Negri, V. U., and Rossella, E., 1964, Cerebral blood flow and metabolism in therapeutic insulin coma, *Metabolism* **13**:340–343.

Drewes, L. R., and Gilboe, D. D., 1973, Glycolysis and the permeation of glucose and lactate in the isolated, perfused dog brain during anoxia and postanoxic recovery, *J. Biol. Chem.* **218**:2489–2496.

Feise, G., Kogure, K., Busto, R., Scheinberg, P., and Reinmuth, O. M., 1976, Effect of insulin hypoglycemia upon cerebral energy metabolism and EEG activity in the rat, *Brain Res.* **126**:263–280.

Ferrendelli, J. A., and Chang, M. M., 1973, Brain metabolism during hypoglycemia. Effect of insulin on regional central nervous system glucose and energy reserves in mice, *Arch. Neurol.* **28**:173–177.

Ferrendelli, J. A., Steiner, A. L., McDougal, D. B., Jr., and Kipnis, D. M., 1970, The effect of oxotremorine and atropine on cGMP and cAMP levels in mouse cerebral cortex and cerebellum, *Biochem. Biophys. Res. Comm.* **41**:1061–1067.

Ferrendelli, J. A., Gay, M. H., Sedgwick, W. G., and Chang, M. M., 1972, Quick freezing of the murine CNS: Comparison of regional cooling rates and metabolite levels when using liquid nitrogen or Freon 12, *J. Neurochem.* **19**:979–987.

Fiorentini, A., and Muller, E. E., 1972, Sensitivity of central chemoreceptor controlling blood glucose and body temperature during glucose deprivation, *J. Physiol.* **248**:247–271.

Fishman, R. A., 1964, Carrier transport of glucose between blood and cerebrospinal fluid, *Am. J. Physiol.* **206**:836–844.

Freinkel, N., Singer, E. L., Arky, R. A., and Bleicher, C. K., 1963, Alcohol hypoglycemia I. Carbohydrate metabolism of patients with clinical alcohol hypoglycemia and the experimental reproduction of the syndrome with pure ethanol, *J. Clin. Invest.* **42**:1112–1133.

Freinkel, N., Metzger, B. E., Harris, E., Robinson, S., and Mager, M., 1972, The hypothermia of hypoglycemia, *N. Engl. J. Med.* **287**:841–845.

Gale, E. A. M., Bennett, T., McDonald, I. A., Holst, J. J., and Matthews, J. A., 1983, The physiological effects of insulin induced hypoglycemia in man: Response at differing levels of blood glucose, *Clin. Sci.* **65**:263–271.

Gardiner, R. M., 1980, The effects of hypoglycemia on cerebral blood flow and metabolism in the newborn calf, *J. Physiol.* **298:**37–51.

Gellhorn, E., 1938, Effect of hypoglycemia and anoxia on the central nervous system: A basis for rational therapy of schizophrenia, *Arch. Neurol. Psychiatr.* **40:**125–146.

Ghajar, J. B., Plum F., and Duffy, T. E., 1982, Cerebral oxidative metabolism and blood flow during acute hypoglycemia and recovery in unanesthetized rats, *J. Neurochem.* **38:**397–409.

Gibbs, R. L., Lennox W. G., Nims, L. F., and Gibbs, F. A., 1942, Arterial and cerebral venous blood: Arterial-venous differences in man, *J. Biol. Chem.* **144:**325–332.

Gibson, G. E., and Blass, J. P., 1975, Inhibition of acetylcholine synthesis and of carbohydrate utilization by maple syrup urine disease metabolites, *J. Neurochem.* **26:**1073–1078.

Gibson, T. E., and Blass, J. P., 1976, Impaired synthesis of acetylcholine in brain accompanying mild hypoxia and hypoglycemia, *J. Neurochem.* **27:**37–42.

Gibson, G. E., Jope, R., and Blass, J. P., 1975, Decreased synthesis of acetylcholine accompanying impaired oxidation of pyruvic acid in rat brain minces, *Biochem. J.* **148:**17–23.

Goldberg, N. D., Passonneau, J. V., and Lowry, O. H., 1966, Effects of changes in brain metabolism on the levels of citric acid cycle intermediates, *J. Biol. Chem.* **241:**3997–4003.

Gorell, J. M., Dolkart, P. H., and Ferrendelli, J. A., 1976, Regional levels of glucose, amino acids, high energy phosphates, and cyclic nucleotides in the central nervous system during hypoglycemic stupor and behavioral recovery, *J. Neurochem.* **27:**1043–1049.

Gorell, J. M., Law, M. M., Lowry, O. H., and Ferrendelli, J. A., 1977, Levels of cerebral cortical glycolytic and citric acid cycle metabolites during hypoglycemic stupor and its recovery, *J. Neurochem.* **29:**187–191.

Hawkins, R. A., Williamson, D. H., and Krebs, H. A., 1971, Ketone body utilization by adult and suckling rat brain in vivo, *Biochem. J.* **122:**13–18.

Hernandez, M. J., Vannucci, R. C., Salcedo, A., and Brennan, R. W., 1980, Cerebral blood flow and metabolism during hypoglycemia in newborn dogs, *J. Neurochem.* **35:**622–628.

Hinzen, D. H., and Muller, V., 1971, Energies Toffwechsel und Funktion des Kanninchengehirns Wahrend Insulinhypoglykamie, *Pflug. Arch. Eur. J. Physiol.* **265:**328–336.

Hostetler, K. Y., Landau, B. R., White, R. J., Albin, M. S., and Yashon, D., 1970, Contributions of the pentose cycle to the metabolism of glucose in the isolated perfused brain of the monkey, *J. Neurochem.* **17:**33–39.

Hothersall, J., Baquer, N., Breenbaum, A., and McLean, P., 1979, Alternative pathways of glucose utilization in brain, *Arch. Biochem. Biophys.* **198:**478–492.

Jaffe, N., 1966, Hypothermia—A diagnostic aid to hypoglycemia, *S. Afr. Med. J.* **40:**569–572.

Kedes, L. H., and Fields, J. B., 1964, Hypothermia—A clue to hypoglycemia, *N. Engl. J. Med.* **271:**785–787.

Kety, S. S., 1957, The general metabolism of the brain in vivo, in: *The Metabolism of the Nervous System* (D. Richter, ed.), Pergammon Press, London, pp. 221–237.

Kety, S. S., Woodford, R. B., Harmel, M. H., Freylan, F. A., Appel, K. E., and Schmidt, C. F., 1948, Cerebral blood flow and metabolism in schizophrenia, *Am. J. Psychiatry* **104:**765–770.

Lajtha, A. L., Maker, H. S., and Clarke, D. D., 1981, Metabolism and transport of carbohydrates and amino acids, in: *Basic Neurochemistry* (G. J. Siegel, R. W. Albers, B. W. Agranoff, and R. Katzman, eds.), Little Brown and Co., Boston, pp. 329–353.

Landau, B. R., Laszlo, J., Stengel, J., and Burk, D., 1958, Certain metabolic and pharmacologic effects in cancer patients given infusions of 2-deoxy-D-glucose, *J. Natl. Cancer Inst.* **21:**485–494.

Lewis, L. D., Ljunggren, B., Ratcheson, R. A., and Siesjo, B. K., 1974, Cerebral energy state in insulin induced hypoglycemia related to blood glucose and EEG, *J. Neurochem.* **23:**673–679.

Lowry, O. H., and Passonneau, J. V., 1972, *A flexible system of enzymatic analysis*, Academic Press, New York, pp. 129–218.

Lust, W. D., Murakami, N., Azeredo, F., and Passonneau, J. V., 1980, A comparison of methods for brain fixation, in: *Cerebral Metabolism and Neural Function* (J. V. Passonneau, R. A. Hawkins, W. D. Lust, and F. A. Welsh, eds.), Williams and Wilkins, Baltimore, pp. 17–26.

Magoun, H. W., 1954, The ascending reticular system and wakefulness, in: *Brain Mechanisms and Consciousness*, DeLafresnaye, J. F., ed. Charles Thomas, Springfield, ILL., pp. 1–20.

Magoun, H. W., 1973, in *The Waking Brain*, 2nd ed., Charles C Thomas, Springfield, Illinois, pp. 74–97.

Mans, A. M., Biebuyck, J. F., Shelly, K., and Hawkins, R. A., 1982, Regional blood brain barrier permeability to amino acids alter portacaval anastomosis, *J. Neurochem.* **38:**705–717.

Mayer-Gross, W., and Berliner, F., 1942, Observation in hypoglycemia IV: Body temperature and coma, *J. Ment. Sci.* **88**:419–427.

Mayman, C. I., Gatfield, P. D., and Breckenridge, B. N., 1964, The glucose content of brain in anesthesia, *J. Neurochem.* **11**:483–487.

McCandless, D. W., 1981, Insulin induced hypoglycemic coma and regional cerebral energy metabolism, *Brain Res.* **215**:225–233.

McCandless, D. W., and Abel, M. S., 1980, The effect of unconjugated bilirubin on regional cerebellar energy metabolism, *Neurobehav. Toxicol.* **2**:81–84.

McCandless, D. W., and Knight, T., 1983, Octanoic acid induced coma and reticular formation energy metabolism, *J. Cereb. Blood Flow Metab.* **3**(suppl. I):453.

McCandless, D. W., and Schenker, S., 1981, Effect of acute ammonia intoxication on energy stores in the cerebral reticular activating system, *Exp. Brain Res.* **44**:325–334.

McCandless, D. W., and Schwartzenberg, F., 1981, Chronic thiamine deficiency and energy metabolism in Dieter's nucleus, *Res. Commun. Psychol. Psychiatr. Behav.* **6**:183–190.

McCandless, D. W., Feussner, G., Lust, W. D., and Passonneau, J. V., 1979, The sparing of metabolic stress in purkinje cells following maximal electroshock, *Proc. Natl. Acad. Sci. U.S.A.* **76**:1482–1484.

McCulloch, J., Savak, H. E., Jelle, J., and Sokoloff, L., 1982, Local cerebral glucose utilization in hypothermic and hyperthermic rats, *J. Neurochem.* **39**:255–258.

Medina, M. A., Jones, D. J., Stavinoha, W. B., and Ross, D. H., 1975, The levels of labile intermediary metabolites in mouse brain following rapid tissue fixation with microwave irradiation, *J. Neurochem.* **24**:223–227.

Myers, R. E., and Kahn, J. J., 1971, Insulin induced hypoglycemia in non-human primates. II. Long term neuropathological consequences, *Clin. Dev. Med.* **40**:195–206.

Nelson, S. R., Schulz, D. W., Passonneau, J. V., and Lowry, O. H., 1968, Control of glycogen levels in brain, *J. Neurochem.* **15**:1271–1279.

Norberg, K., and Siesjo, B. K. (1976), Oxidative metabolism of the cerebral cortex of the rat in severe insulin induced hypoglycemia, *J. Neurochem.* **26**:345–352.

Owen, O. E., Morgan, A. P., Kemp, H. G., Sullivan, J. M., Herrera, M. J., and Cahill, G. F., 1967, Brain metabolism during fasting, *J. Clin. Invest.* **46**:1589–1595.

Ponten, V., Ratcheson, R. A., and Siesjo, B. K., 1973, Metabolic changes in the brains of mice frozen in liquid nitrogen, *J. Neurochem.* **21**:1121–1126.

Rafaelsen, O. J., 1961, Action of insulin on glucose uptake of rat brain slices and isolated rat cerebellum, *J. Neurochem.* **7**:45–51.

Ramos, E., Zorilla, E., and Hadley, W. B., 1968, Fever as a manifestation of hypoglycemia, *J. Am. Med. Assoc.* **205**:590–592.

Ratcheson, R. A., Blank, A. C., and Ferrendelli, J. A., 1981, Regionally selective metabolic effects of hypoglycemia in brain, *J. Neurochem.* **36**:1952–1958.

Reuler, J. B., 1978, Hypothermia: Pathophysiology, clinical settings, and management, *Ann. Intern. Med.* **89**:519–527.

Ruderman, N. B., Ross, P. S., Berger, M., and Goodman, M. N., 1974, Regulation of glucose and ketone body metabolism in brain of anesthetized rats, *Biochem. J.* **138**:1–10.

Shiraishi, T., and Mager, M., 1980a, 2-deoxy-D-glucose-induced hypothermia: Thermoregulatory pathways in rat, *Am. J. Physiol.* **239**:R270–276.

Shiraishi, T., and Mager, M., 1980b, Hypothermia following injection of 2-deoxy-D-glucose into selected hypothalamic sites, *Am. J. Physiol.* **239**:R265–R269.

Sloviter, H. A., and Kamimoto, T., 1970, The isolated perfused rat brain preparation metabolizes mannose but not maltose, *J. Neurochem.* **17**:1109–1111.

Strouch, B. S., Felig, P., Baxter, J. D., and Schimpff, S. C., 1969, Hypothermia in hypoglycemia, *J. Am. Med. Assoc.* **210**:345–346.

Tews, J. K., Carter, S. H., and Stone, W. E., 1965, Chemical changes in the brain during insulin hypoglycemia and recovery, *J. Neurochem.* **12**:679–693.

Vannucci, R. C., Nardis, E. E., Vannucci, S. J., and Campbell, P. A., 1981, Cerebral carbohydrate and energy metabolism during hypoglycemia in new born dogs, *Am. J. Physiol.* **240**:192–199.

Veech, R. L., Harris, R. L., Veloso, D., and Veech, E. H., 1973, Freeze-blowing: A new technique for the study of brain in vivo, *J. Neurochem.* **20**:183–188.

Wong, K. C., 1983, Physiology and pharmacology of hypothermia, *West. J. Med.* **138**:227–232.

3

Hypoxia

GARY E. GIBSON

1. Introduction

Hypoxia, decreased oxygen availability, may serve as a model to help elucidate the pathophysiological basis of the metabolic encephalopathies. These disorders, in which brain metabolism is altered secondarily to systemic changes, include hyperammonemia, hypoglycemia, nutritional deficiencies such as thiamine or niacin, some inborn errors of metabolism, and heavy metal intoxication (Plum, 1975). The metabolic encephalopathies share similar clinical symptoms: decreased mentation and a loss of attention, alertness, orientation, cognition, memory, and perception, which eventually progress to stupor, coma, and finally death (Plum and Posner, 1980). The similar clinical presentation, despite diverse etiologies, suggests that a common molecular mechanism may underlie the altered brain function in all of them. Furthermore, an understanding of how low oxygen alters brain metabolism may help unravel the complex changes that accompany ischemia. During ischemia, tissue perfusion is also comprised, which reduces the substrate supply and allows accumulation of possibly toxic metabolic products, as well as causing hypoxia or anoxia (complete lack of oxygen). Whether ischemic-induced tissue damage is just an exaggeration of hypoxia's effects or if some unknown variable converts hypoxic changes to ischemic deficits is unknown.

Hypoxia may also have a role in other clinical and nonclinical situations. Anesthetized patients in the postoperative period often experience arterial hypoxia (Diament and Palmer, 1966; Nunn and Payne, 1962). The symptoms of mountain sickness may be induced by hypoxia (Fitch, 1964). Furthermore, airlines pressurize passenger compartments to 8,000 (2438 m), which corresponds to an oxygen tension of about 108 mm Hg, compared with a normal of about 150 mm Hg.

Finally, hypoxia is useful in the study of the biochemical basis of normal

GARY E. GIBSON • Cornell University Medical College, Burke Rehabilitation Center, White Plains, New York 10605.

brain function, since the complex interactions of energy metabolism with biosynthetic pathways, neurotransmitter synthesis, ion homeostasis, and physiological events can be conveniently examined by manipulation of oxygen tensions.

2. Delivery of Oxygen to Tissue *in Vivo* and *in Vitro*

A precise estimate of the oxygen that is available to or required by the brain *in vivo* is difficult to obtain. Whether the brain normally has excess or just sufficient oxygen available is unknown (Thews, 1960, 1963). Both the tissue oxygen tension and redox state must be known to determine this; for example, rapidly metabolizing cells may have low oxygen tensions, which creates a concentration gradient that provides more oxygen. The tissue oxygen tension may be determined with microelectrodes (Crawford *et al.*, 1980) or the fluorescence of an intracellular oxygen detector (Homer *et al.*, 1983). The redox state of cytochrome a,a_3 or the pyridine nucleotides can be used to estimate the oxygenation state of the tissue (Jobsis, 1979; Jobsis and LaManna, 1978).

The numerous factors that regulate the brain's oxygenation can be altered experimentally or by disease: the oxygen tension in the lungs, the oxygen transfer from the lung to the blood, the oxygen transport by blood, heart function, the oxygen diffusion across brain capillaries, and the ability of brain mitochondria to utilize oxygen. The composition of the inspired air can be manipulated to achieve hypoxia or hyperoxia. Breathing rates influence oxygen availability; hypoventilation decreases the oxygen inspired into the lungs. Oxygen diffusion from the lungs into the blood diminishes with pulmonary disease or bronchial infections. The efficiency of oxygen exchange from the lungs to the blood shifts with changes in the oxygen–hemoglobin dissociation curve due to altered temperature, partial pressure of carbon dioxide, pH (Astrup *et al.*, 1965), or hemoglobin concentrations. For example, reduced hemoglobin content (e.g., anemia) diminishes oxygen availability. Different hemoglobin dissociation curves may predict the relative vulnerability of various species to hypoxia (Hall, 1966), since human hemoglobin is 90% saturated at 40 mm Hg, but saturation of rat hemoglobin requires nearly 80 mm Hg. The brain's oxygenation may be compromised by impaired heart function such as that during massive cardiac failure or more subtle insults such as the age-related erratic heart function that leads to cardiogenic dementia. The relative vascular enrichment of a tissue or brain region may alter its response to metabolic insults. Thus, the sensitivity to hypoxia of certain brain regions or of a particular neurotransmitter may be due to poor vascularity rather than to some unique property of the cells. Whether a cell is near the beginning or the end of the capillary may determine its oxygenation, because a significant arteriovenous gradient exists for oxygen (e.g., 95 to 34 mm Hg) (Thews, 1963). Disease or aging may limit oxygen transport across the capillary endothelial cells (Crawford *et al.*, 1980). The diffusion distance from the capillary to the mitochondria (i.e., the primary site of oxygen utilization) may vary between cell types and this distance is greater in brain than in other tissues (Lubbers, 1968). Finally, metabolic toxins, such as potassium cyanide, or disease (e.g., inborn errors of metabolism) may interfere with the mitochondria's ability to effectively utilize oxygen.

In vitro oxygen availability is equivalent to the gas tension of the incubation flask, but the tissue's ability to utilize oxygen may be compromised by numerous factors. Since the thickness of a tissue preparation determines the oxygen diffusion gradient, a preparation with a diffusion distance of 150μ would appear more sensitive to hypoxia than one with a diffusion distance of only 1 μ (Ksiezak and Gibson, 1981a). Both theoretical calculations and experimental measurements show that oxygen cannot penetrate more than 300–400 μ into a slice (Nicholson and Hounsgaard, 1983). Diffusion artifacts may explain why the kinetics of tissue slice oxygen consumption differ from that of mitochondrial suspensions (Buerk and Saidel, 1978) and why the mitochondrial redox state is reduced more *in vivo* than in *vitro* (Rosenthal *et al.*, 1976).

3. Effects of Hypoxia in Man

The well-documented biochemical, physiological, and behavioral changes that occur in man during hypoxia permit rather direct comparison with animal models of hypoxia, where the primary neurochemical deficits can be studied. The respiratory and circulatory systems appear to be the most sensitive to hypoxia, which may be due to the ease with which the changes are quantified compared with an altered EEG or behavioral impairment. The threshold for these alterations may be a reduction in the percent oxygen from 20% (room air) to 18% (Lutz and Schneider, 1919; Ellis, 1919) or 16% (Boothcby, 1945) under hypobaric conditions. At normal atmospheric pressure, 18% oxygen increases the pulse rate of 79% of the patients by 2–8 per min and 16% to 14.5% oxygen elevates the rate in 90% of the patients without alterations in arterial carbon dioxide tensions (Dripps and Comroe, 1947; Kontos *et al.*, 1967). Respiratory minute volume also increases significantly at 16% to 14.5% oxygen.

3.1. Changes in Mental Function

Although severe acute hypoxia is well known to impair mental abilities in man (Haldane *et al.*, 1919), the degree of hypoxia that is required and which processes are most sensitive to decreased oxygen remains controversial. Complex information processing and learning a new skill appear to be the most vulnerable to hypoxia. Performance on spatial transformation tasks (Denison and Ledwith, 1964) and the ability to reproduce a sequence of eight digital operations in response to a light signal decline at 8000 ft (inspired air oxygen tensions of 108 mm Hg; 15.2% oxygen) (Ernsting, 1966; Ernsting *et al.*, 1962; Gedye, 1964). Deficits on complex vigilance tasks occur at simulated altitudes of 5000 (inspired air oxygen tensions of 122 mm Hg) or 8000 ft (inspired air oxygen tensions of 108 mm Hg) and the effects are most pronounced during learning rather than after it (Denison *et al.*, 1966; Gedye, 1964; Crow and Kelman, 1969). Complex card–sorting task (Kelman *et al.*, 1969) and digit span test (Crow and Kelman, 1971) performance, measures of selective attention, are normal at 8000 ft. However, performance on more complicated tasks that include bicycle pedaling and a complex orientation test decline at 8000 ft (Denison et al., 1966). At a simulated altitude of 10,000 ft (inspired air oxygen tensions of 100 mm Hg), repeat testing does not improve performance on numerous psychological measures. At 12,000 ft, short-term memory diminishes with an increased error rate at all testing intervals (Crow and Kelman, 1971). Contrast discrimination and hand steadiness decline in parallel with alveolar oxygen tensions below 50 mm Hg (Otis et al., 1946) and correlate to differences in cerebral-venous oxygen tensions (Ernsting, 1966). Finally, consciousness is lost at cerebral venous oxygen tensions of 17–19 mm Hg (Gibbs et al., 1935), which is equivalent to a calculated gray matter oxygen tension of 4–5 mm Hg (Thews, 1960). Other forms of hypoxia also alter mental function. For example, pernicious anemia diminishes neurologic performance and cerebral oxygen consumption in parallel (Scheinberg, 1951).

3.2. Sensory Function in Man

Hypoxia alters visual and auditory function. Cone and rod, white- and colored-light visual thresholds increase with hypoxia (Ernst and Krille, 1971; Bert, 1878; Wilmer and Berens, 1918; McDonald and Adler, 1939; McFarland and Forbes, 1940; Hecht, 1945; McFarland and Evans, 1939; Sheard, 1945). Twice as much light is required to see a stimulus at a simulated altitude of 4500 m (14,764 ft) than at sea level (McFarland, 1953; McFarland et al., 1944). Changes in foveal differential brightness occur at 1500 to 1800 m (4921–5906 ft). Carbon monoxide intoxication alters brightness thresholds, visual acuity, dark adaptation, vigilance (Otto and Reiter, 1978), and flicker fusion frequency (Lilienthal and Fugitt, 1946; Vollmer, 1946). A 4% increase in carboxyhemoglobin impairs performance on visual brightness discrimination tasks (McFarland et al., 1944), whereas 2% (i.e., 90-min exposure to 50 ppm) alters the ability to perceive differences in the duration of auditory stimuli (Beard and Wertheim, 1967). The latter findings are controversial, since others find no behavioral effects at 11–13% carboxyhemoglobin (Stewart et al., 1970).

3.3. Human EEG and Evoked Potentials

Hypoxic-induced slowing of the EEG was first described in patients that breathed from a closed bag (Berger, 1934; Gibbs et al., 1935). Low oxygen diminishes the 10 per sec waves, then causes slow delta waves and finally leads to a loss of all activity (Davis et al., 1938). Ten percent oxygen slows the EEG from 8–12 Hz to 5–7 Hz in 30% of younger, but only 10% of older men (Rosen et al., 1961). However, the EEG changes do not correlate to the biochemical abnormalities in the healthy male volunteers breathing gas mixtures of 6.9–7.5% oxygen (Cohen, 1967). The EEG alterations are preceded by other behavioral symptoms (dimming of vision, lightheadedness, nausea, and restlessness) and elevated glucose utilization. Gastaut and Meyer (1961) found EEG slowing with arterial oxygen tensions of less than 19 mm Hg and even 33% carboxyhemoglobin does not alter gross spontaneous EEG activity (Groll-Knapp et al., 1978).

Evoked potentials may be a more sensitive indicator of hypoxic-induced changes than the EEG. The response depends on the mode of stimulation and the type of hypoxia. The auditory-evoked brainstem response does not change in humans with 9–13% oxygen (Sohmer *et al.*, 1982) and only anoxia depresses cerebral visually evoked potentials in humans (Fernadez-Guardiola *et al.*, 1959). Although neither the EEG nor the auditory or visually evoked responses are affected by 200 ppm of carbon monoxide (carboxyhemoglobin of 12%), carboxyhemoglobin levels as low as 5.5% decrease the amplitude, but not the latencies of the somatosensory-evoked potentials (Groll-Knapp *et al.*, 1978). Late slow potentials, which are correlates of attention, anticipation, and motor readiness, diminish with moderate carbon monoxide exposure (50, 100, and 150 ppm; Groll-Knapp *et al.*, 1978). Between 20 and 22% carboxyhemoglobin, the amplitude of the 70-msec component increases, but at later stages a negative shift occurs (Hosko, 1970). Carbon monoxide (200 ppm) slightly increases the amplitude of one component of the paired-click-stimuli–evoked potential (Otto *et al.*, 1978). During anemic hypoxia, carboxyhemogloblin has to exceed 20% to depress visual-evoked potentials (Stewart *et al.*, 1970; Hosko, 1970). In severe anemic hypoxia, the auditory-evoked response negative components become larger and the positive components diminish (Sohmer *et al.*, 1982).

3.4. Changes in Cerebral Blood Flow and Cerebral Metabolic Rate for Glucose and Oxygen

Hypoxia alters cerebral blood flow and oxygen uptake in conscious humans (Cohen *et al.*, 1967; Kety and Schmidt, 1948; Burner *et al.*, 1957). Glucose utilization and lactate production increase at levels of normocarbic hypoxic hypoxia that do not change cerebral oxygen uptake or the EEG, although the subjects show mild hypoxic symptoms (Cohen *et al.*, 1967). Ten percent oxygen for 15–30 min increases cerebral blood flow (+ 35%), but does not change the cerebral metabolic rate for oxygen (Kety and Schmidt, 1948). Similar degrees of hypoxia produce dizziness, lightheadedness, confusion, and disorientation. Breathing 6.9–7.5% oxygen for 15 min decreases the arterial oxygen tension from 98.4 to 34.6, almost triples the volume of expired air, nearly doubles the cardiac rate, elevates blood glucose by 9% with normal arterial lactate and pyruvate, and increases cerebral blood flow by 75%, but reduces cerebral vascular resistance by 44%. The cerebral metabolic rate for oxygen remains normal, but glucose utilization increases by 25% and the oxygen/glucose index declines from 91 to 76%. The lactate/glucose index rises from 4.5 to 18.9%, which suggests that hypoxia decreases the proportion of glucose that becomes completely oxidized (Cohen *et al.*, 1967). Carbon-monoxide–induced hypoxia (20% but not 8% carboxyhemoglobin) or hemodilution (28% hematocrit) also increases cerebral blood flow (Paulson *et al.*, 1973).

Cerebrospinal fluid metabolite concentrations reflect hypoxic-induced alterations of brain metabolism. Anoxic insults decrease cisternal pH from 7.3 to 6.95 and double the potassium concentration (Kalin *et al.*, 1975).

3.5. Chronic Hypoxia

Chronic hypoxia is more difficult to study than acute hypoxia because many compensatory physiological mechanisms are activated. Thus, healthy people that live in rarefied atmospheres have an elevated number of red blood cells and normal brain function (Clench *et al.*, 1982). However, chronic brain hypoxia does occur in 92% of the patients with chronic obstructive pulmonary disease, since they often have arterial oxygen tensions of about 51 mm Hg (Dulfano and Sadamu, 1965). Such patients have an impaired ability to acquire and retain new information, to form new concepts and think flexibly, to perform complex perceptual motor maneuvers, and to engage in simple perceptual discriminations (Grant *et al.*, 1982). Furthermore, oxygen therapy improves their performance (Krop *et al.*, 1972). The patients with greater neuropsychological difficulties are more hypoxic, less polycythemic, and transport less oxygen, whereas the arterial carbon dioxide tension and pH are not related to the deficits. In chronic heart disease, which leads to a similar decrease in arterial oxygen tension, auditory reaction time increases (Aisenberg, 1977).

4. Animal Models of Hypoxia

Biochemical analysis of the molecular mechanisms of hypoxic-induced deficits requires animal models. In many ways, animal models of hypoxia mimic the condition in man better than models for any other metabolic encephalopathy, but many differences still exist. For example, the hemoglobin dissociation curves for humans and rodents differ so that 10% oxygen may produce a different degree of brain hypoxia in the two species. The variety of techniques to decrease oxygen availability makes a direct comparison of the results from various laboratories difficult, since the primary biochemical lesion may differ between forms of hypoxia.

Each model of hypoxia has advantages and limitations. The onset rate and duration of the hypoxic episode may be as critical as whether chemical, anemic, or hypoxic hypoxia are examined. In hypoxic or hypobaric hypoxia, blood gases equilibrate rapidly with the oxygen content of the inspired air. However, in sodium nitrite or carbon-monoxide–induced anemic hypoxia, methemoglobin formation requires as much as 40 min to peak before it slowly declines. Slow-onset hypoxia may be inadequate to study deficits such as memory loss. Induction of anemic hypoxia with other chemicals (e.g., 4-dimethylaminophenol) is more rapid, but has only been used minimally (Kiese and Weger, 1969). An animal's age is another important variable, since developing animals are more resistant to hypoxic episodes than mature ones and aged animals may be more sensitive (Gibson *et al.*, 1981b). Hypoxia can permanently alter brain function, if the insult occurs at a particular time during development. Arterial carbon dioxide tension is another unknown variable in many studies. The hyperventilation that accompanies reduced oxygen tensions, diminishes arterial carbon dioxide, which is a potent stimulus of numerous physiological processes. However, to control carbon dioxide requires anesthetized, paralyzed ventilated animals which further complicates interpretation.

4.1. Hypoxic Hypoxia

Hypoxic hypoxia is characterized by decreased oxygen content of the inspired air and is the form of hypoxia to which all others are generally compared.

Hypoxic hypoxia can be conveniently studied in freely moving animals, in paralyzed ventilated animals, or in animals that are restrained in a microwave holder or a brain-blowing device. Each experimental approach has advantages and disadvantages. All methods that utilize restraint obviate any behavioral testing and the accompanying stress response may mask the effect of hypoxia. With the brain blower, nearly instantaneous freezing of the unanesthetized brain is possible and blood gases can be readily manipulated and monitored. However, regional biochemical analysis is not possible and the required restraint stress depresses the arterial carbon dioxide tensions in control rats to 25 mm Hg (Gibson and Duffy, 1981). Although a freely moving animal in a hypoxic environment seems ideal for the simultaneous study of behavior and brain function, this approach has several disadvantages. The animal chamber must be relatively large to accommodate any behavioral-testing apparatus, which makes it difficult to adequately mix the gases, to monitor blood variables, and to rapidly sacrifice the animal. Rodents in a focused microwave apparatus can be conveniently exposed to various gas mixtures and blood gases can be readily monitored. The animal appears relatively unstressed, although blood corticosterone and norepinephrine concentrations are elevated (Freeman and Gibson, 1984). A ventilated animal that is paralyzed with a peripheral neuromuscular blocker, such as curare, allows the best control of blood gases, since the respiration rate and inspired gas composition are easily regulated. However, the accompanying immobilization stress alters cerebral blood flow and other physiological variables (Carlsson *et al.*, 1977). To circumvent this problem and for ethical reasons, such animals are generally treated with anesthetics, such as nitrous oxide. However, both the immobilization stress (Carlsson *et al.*, 1977) and the anesthesia may alter the neurochemical response to hypoxia (Gibson and Duffy, 1981).

4.2. Hypobaric Hypoxia

Hypobaric hypoxia occurs when decreased atmospheric pressure reduces the oxygen content of the inspired air. This approach mimics the hypoxia that accompanies the rapid altitude changes during airplane flight. The complications due to the decline in barometric pressure alone and those from the diminished oxygen content are difficult to distinguish. An additional disadvantage is that behavioral tests and sacrifice must be done in a special chamber.

4.3. Anemic Hypoxia

Anemic hypoxia occurs when the oxygen-carrying capacity of the blood is reduced. This may be induced by a reduction in the hematocrit through hemodilution, by a reduction of hemoglobin's ability to transport oxygen with drugs such as sodium nitrite, carbon monoxide, or 4-dimethylaminophenol, or by interference with the hemoglobin–oxygen dissociation curve by pH or other factors. Arterial oxygen tensions may be normal during anemic hypoxia, but the brain is still hypoxic because hemoglobin's ability to transport oxygen is impaired. The degree of anemic hypoxia can be monitored by measurement of the hematocrit during hemorrhagic anemia or the percent methemoglobin during chemical hypoxia. Chemically induced anemic hypoxia has the advantage that the animal is readily accessible for behavioral testing and rapid sacrifice. Carbon monoxide intoxication limits the accessability of the animal as with hypoxic hypoxia models and, in addition, it is a toxic gas. A disadvantage of chemical hypoxia is a slow rise in methemoglobin formation, although this may be circumvented by the use of 4-dimethylaminophenol, which produces a rapid rise in methemoglobin. Another difficulty is the secondary effects of the compounds that are used to induce methemoglobinemia. For example, sodium nitrite is a weak activator of guanyl cyclase (Kimura *et al.*, 1975). This complication can be partially assessed by *in vitro* incubations. If high *in vitro* concentrations of the same chemicals do not mimic the *in vivo* action, the effects observed *in vivo* are likely due to deficits in oxygen delivery (Gibson and Blass, 1976a). Hemorrhagic hypoxia is often done in paralyzed ventilated animals (McCutcheon *et al.*, 1971), which complicates behavioral testing and interpretation as discussed previously.

4.4. Histotoxic Hypoxia

Histotoxic hypoxia occurs when the tissues' ability to utilize oxygen is impaired, even though adequate oxygen is available. For example, cyanide blocks cytochrome oxidase and the subsequent oxidation of the pyridine or flavin nucleotides. An advantage of histotoxic hypoxia, and the other forms of chemical hypoxia, is the ready access to the animal for behavioral tests and rapid sacrifice. A disadvantage is that the compounds that are used to induce the hypoxia may be nonspecific. For example, cyanide is also a weak activator of guanyl cyclase (Kimura *et al.*, 1975). Conversely, the inhibitors may be too specific for the elucidation of the molecular mechanisms of hypoxia. Thus, if impaired tyrosine hydroxylase activity is a primary pathophysiological mechanism during hypoxia, it would not be detected with compounds that specifically block cytochrome oxidase. A final disadvantage is that cyanide has a narrow safety margin between a dosage with no apparent effect and one that leads to seizures or death.

4.5. Acute versus Chronic Hypoxia

Interpretation of studies on long-term or chronic hypoxia are complicated by secondary and tertiary compensatory mechanisms. For example, increased red blood cell production or shifts in the oxygen dissociation curves allow animals to adjust to chronic hypoxia. Since the focus of this chapter is to describe primary pathophysiological mechanisms of hypoxia, biochemical or physiological results with chronic models will not be discussed.

4.6. Complications Due to Carbon Dioxide

The accelerated respiration that accompanies reduced oxygen lowers arterial carbon dioxide tensions so that an apparent neurochemical effect of low oxygen may instead be due to hypocarbia. At a given alveolar oxygen tension, diminished carbon dioxide further impairs behavioral performance (Rahn and

Otis, 1946). Similarly, during hemorrhagic anemia, many "hypoxic-induced" changes in cerebral metabolism also require reduced arterial carbon dioxide concentrations (Michenfelder and Theye, 1969). Conversely, carbon dioxide may alleviate some hypoxic-induced deficits. Carbon dioxide dilates cerebral vessels, increases pulmonary ventilation, shifts the hemoglobin dissociation curve to deliver more oxygen to the tissue at a given oxygen saturation, prevents the decline in whole brain dopamine levels (Brown *et al.*, 1974), and protects the human brain against anoxia (Gibbs *et al.*, 1943), but does not alter the effects of hypoxia on amino acid or acetylcholine formation from [U-^{14}C]glucose (Gibson *et al.*, 1981a).

4.7. *In Vitro* Hypoxia

In vitro, many of the secondary effects that accompany *in vivo* hypoxia can be carefully controlled. A disadvantage of *in vitro* studies is the additional metabolic insult to the brain due to decapitation from which the brain never fully recovers. Thus, studies where the tissue is not preincubated may examine the effects of hypoxia on recovery, rather than on normal metabolism. Saturating oxygen concentrations are relatively easy to obtain by short-term (i.e., 30 sec) flushing of small incubation flasks with 95 or 100% oxygen. However, reducing the oxygen tensions requires more vigorous flushing (i.e., 10 min). Studies of hypoxia with perfused brain slices are uniquely difficult, since buffers that are flushed continuously with nitrogen still contain 20 mm Hg of oxygen by the time they get to the chamber (Pull *et al.*, 1972; Hirsch and Gibson, unpublished observations). Even direct monitoring of the buffer oxygen tensions does not necessarily indicate the tissue oxygenation because of diffusion artifacts. For example, 2.5% oxygen increases brain slice, but not synaptosomal lactate production (Ksiezak and Gibson, 1981a). Whether the degree of hypoxia may be monitored indirectly by changes in a particular metabolite is uncertain.

4.8. *Ex Vivo* Hypoxia

The effects of *in vivo* hypoxia may be evaluated with subsequent *in vitro* preparations. This approach combines the benefit of an *in vivo* physiological approach to hypoxia with the *in vitro* ability to control variables. The major disadvantage is that the decapitation between the *in vivo* hypoxia and the *in vitro* incubation is a much greater metabolic insult than the *in vivo* hypoxia. A second disadvantage is the difficulty of distinguishing between hypoxic-induced changes and those due to recovery from the tissue preparation for the *in vitro* incubation. The experimental demonstration of mitochondrial and nerve–ending sensitivity to hypoxic insults exemplifies the usefulness of this method (Rafalowska *et al.*, 1980; Booth *et al.*, 1983).

5. Physiological Changes in Animal Models of Hypoxia

5.1. Animal EEG

Hypoxia alters synaptic neurochemistry and membrane polarization (Bradford, 1969; Grossman and Williams, 1971; Silver 1973), which interferes with the electrical activity of the brain. All synapses are not equally sensitive to hypoxia. In cat cortex, moderate anoxia selectively and reversibly blocks spontaneous firing and discharges evoked by acetylcholine, but facilitates those by glutamate (Godfraind et al., 1971). In hippocampal slices, hypoxia causes hyperpolarization that is followed by depolarization and concomitant reduction of the nerve cell input resistance. First spontaneous and then evoked activity disappears (Hansen et al., 1982). Such effects combine to alter the more complex EEG pattern.

The sensitivity of the EEG to hypoxia is well documented (Hockaday et al., 1965; Meyer and Gastaut, 1961; Watanabe et al., 1980; Sotaniemi et al., 1980; Diekman, 1976; Gurvitch and Ginsburg, 1977; Obrist et al., 1973; Ingvar et al., 1976). The initial studies, which followed the EEG after stopping respiration, demonstrate that 11- to 16-Hz waves first activate and then disappear (Prawdicz-Neminski, 1923). The changes occur regionally at different times after respiration is stopped: cerebellar gray (10–12 sec), Ammon's horn (10–12 sec), cerebral gray (14–15 sec), subcortical white (20–22 sec), corona radiata (20–25 sec), caudate nucleus (25–27 sec), lateral geniculate nucleus (28–33 sec), reticular formation (30–40 sec), and other medullary nuclei (over 2 min) (Sugar and Gerard, 1938). Systematically decreasing arterial oxygen tensions from 80 to 20 mm Hg in ventilated paralyzed cats first alters the EEG in the somatosensory and then in the visual cortex. The EEG amplitude initially increases slightly and then slow waves and sharp spikes appear. The slow waves decrease in amplitude at 1–3 min after the onset of hypoxia and then disappear. The small spikes become sporadic and within 5–6 min the EEG flattens (Sohmer et al., 1982). The initial activation, which is followed by depression, may be due to hypoxia's effect on the reticular activating system, since electrophysiologically the reticular substance of the mesencephalon appears to be the most sensitive to controlled hypoxia (Hugelin et al., 1959). In respirated dogs or cats, 7% oxygen causes cortical activation and mesencephalic discharge by 25 sec, generalized arousal at 50–70 sec, deactivation and a depression by 90 sec that is followed by electrical silence (Dell et al., 1961; Baumgartner et al., 1961). The same sequence occurs, but at a faster rate, with pure nitrogen. The initial activation may be due to hypoxia's greater effect on inhibitory neurons, stimulation of carotid body chemoceptors with subsequent reticular and cortical activation (Dell and Bonvallet, 1954), or from electrolyte shifts that partially depolarize neurons (Michael, 1973).

5.2. Evoked Potentials

Hypoxic-induced changes in evoked potentials depend on the severity of the hypoxia and the stimulus type. Various forms of evoked potentials and their individual components may be differentially sensitive to hypoxic insults. Hypoxic hypoxia, ischemia, and asphyxia alter evoked potentials similarly (Naquet and Fernandez-Guardiola, 1960). In general, evoked potentials persist even when the EEG is flat (McCutcheon et al., 1971). Although 10% oxygen does not interfere with the brain stem auditory response, 5.5% oxygen abolishes it (Sohmer et al., 1982). The somatosensory-evoked response is more sensitive than the visually evoked response to hypoxic hypoxia, carbon monoxide intoxication, or hemorrhagic hypoxia (Manil et al., 1977; Colin et al., 1978; Groll-

Knapp *et al.*, 1978). In rabbits, the most sensitive component of the somato-sensory-evoked response is the associative cortex peak, but 6.4% oxygen alters the latency and amplitude of all peaks (Manil *et al.*, 1977). Hemorrhagic hypoxia (hematocrits of 10–15%) increases the latencies of the somatosensory-evoked potentials, whereas hematocrits below 10% depress the amplitude without an effect on neural conduction (Nagao *et al.*, 1978).

More severe hypoxia systematically alters visually evoked potentials. During anoxia, the amplitude increases and then declines (Baumgartner *et al.*, 1961; Hirsch *et al.*, 1960; Noell and Chin, 1950). Furthermore, the initial negativity is more readily altered than the first positive wave (Naquet and Fernandez-Guardiola, 1960, 1961; Noell and Chin, 1950). In ventilated paralyzed cats, hypoxia transiently enhances all peaks, but then reduces the activities in the visual cortex, the lateral geniculate body, and much later in the fiber optic tracts, the optic radiation, and the thalamic neurons (Kayama *et al.*, 1974). In monkeys, latency and amplitude of visually evoked potentials remain normal with arterial oxygen tensions between 30–38 mm Hg, but between 15–25 mm Hg, latency increases and amplitude decreases (Shelbourne *et al.*, 1976).

Chemical and anemic hypoxia also alter visually evoked potentials. Carbon monoxide intoxication [50 ppm (Xintaras *et al.*, 1966); 250 ppm, 25% carboxyhemoglobin; 500 ppm, 40% carboxyhemoglobin; or 1000 ppm, 55% hemoglobin (Dyer and Annau, 1978; Dyer, 1980)] interfers with visually evoked potentials in the cortex and superior colliculus of rats. The magnitude of the superior colliculus response increases at 22–38% carboxyhemoglobin, but returns to normal at 55%, when latencies are also altered (Dyer and Annau, 1978; Dyer, 1980).

5.3. Hypoxia and Peripheral Responses

Severe hypoxia also alters the peripheral nervous system. Oxygen deprivation for up to 4 min does not change the motor neuron membrane potential (Nelson and Frank, 1963). Arterial oxygen tensions between 40–400 mm Hg do not alter the firing of the sympathetic preganglionic neuron, but below 40 mm Hg the firing rate increases as the oxygen tension decreases. During total anoxia, synaptic transmission in the superior cervical ganglion disappears within a minute, but presynaptic and postsynaptic axonal conduction and the postsynaptic response to acetylcholine remain normal for hours (Dolivo, 1974).

Hypoxia, between 10–40% carboxyhemoglobin, progressively reduces the action potential of the rat ventral caudal nerve and at 78% carboxyhemoglobin it disappears (Rebert *et al.*, 1982; Petajan *et al.*, 1976). However, even during moderately severe hypoxia, the intracellular response to single neurons remains normal (Nelson and Frank, 1963). Isovolumic hemodilution in baboons increases the amplitude at 16–20% hematocrits, between 10–15% peripheral nerve and spinal cord hypoxia occur, and below 10% hematocrit the amplitude decreases (Nagao *et al.*, 1978).

5.4. Cerebral Blood Flow and Oxygen Consumption

All forms of brain hypoxia increase cerebral blood flow, which helps maintain a normal or near-normal cerebral energy state even during severe hypoxia

(Johannsson and Siesjo, 1975a). Equivalent degrees of hypoxic hypoxia and carbon monoxide hypoxia, as determined by the arterial oxygen content, increase cerebral blood flow, similarly (Johannsson and Siesjo, 1974; Borgstrom *et al.*, 1975a,b). For example, a decrease in arterial oxygen content from 14% to 4% (Traystman, 1978; Traystman and Fitzgerald, 1981) elevates cerebral blood flow. During hypoxic hypoxia in rats, small continuous increases in cerebral blood flow occur when the arterial oxygen tension declines from 130 to 50 mm Hg. The elevation is more striking below 50 mm Hg so that 20–25 mm Hg increases blood flow to about 500% of normal (Borgstrom *et al.*, 1975a,b). In dogs, an arterial oxygen tension of 50 mm Hg and venous oxygen tension of 30–40 mm Hg mark an acute inflection point for increased blood flow during hypoxic hypoxia (McDowall, 1966). Hemodilution elevates cerebral blood flow in proportion to the reduced hemoglobin concentrations (Borgstrom *et al.*, 1975b). After hemodilution-induced hypoxia in dogs, the increased blood flow to the brain is disproportionately large compared with the increased cardiac output (Fan *et al.*, 1980). Chemical hypoxia also increases blood flow. As low as 2.5% carboxyhemoglobin significantly elevates cerebral blood flow and it increases 200, 300, 400, and 500% of controls with 0.5, 1.0, 1.5, and 2% carbon monoxide gas mixtures, respectively. Subcortical, cerebellar, brain stem, and cortical blood flows respond quantitatively similarly to hypoxia (MacMillan, 1975b).

The mechanisms for the hypoxic-induced increase in cerebral blood flow are not clear. Since arterial oxygen tensions do not decline in carbon-monoxide–induced hypoxia, a stimulation of aortic or carotid body receptors by decreased oxygen cannot produce the increase in cerebral blood flow. Furthermore, denervation does not alter this response in dogs (Traystman and Fitzgerald, 1981) or rabbits (Linton *et al.*, 1975), but can in baboons (Ponte and Purves, 1974). Cervical sympathetic nerves do not have any appreciable effect on cerebral circulation during moderate hypoxia (Linder, 1982). Changes in blood flow precede alterations in lactate, intracellular pH, extracellular pH, or extracellular potassium (Nilsson *et al.*, 1975). Since decreased venous oxygen tensions and increased lactate during hypoxic hypoxia (Bachelard *et al.*, 1974) are greater than during anemic hypoxia, neither these nor blood viscosity trigger the increased cerebral blood flow.

5.5. Behavior and Memory

Hypoxia produces numerous changes in animal behavior, but reported alterations vary considerably, particularly with studies during mild hypoxia. Depressed behavior is detectable on a wide variety of tasks and the decreases approximate the degree of hypoxia (Adler *et al.*, 1950; Birren *et al.*, 1946; Gerathewohl, 1951; Scow *et al.*, 1950). Delayed posthypoxic transient amnesia may be produced in rats by brief anoxic exposure after avoidance training (Tauber and Allweis, 1975). Pure nitrogen-induced hypoxia causes convulsions, but hypoxic gas mixtures minimize the seizures and still interrupt medium (30–180 min) but not long-term memory (> 180 min) (Frieder and Allweis, 1982). When rats are exposed to 8% oxygen for 17-min episodes on each day of a five-day training period, the condition avoidance response declines from 92%

avoidance in controls to 51% in hypoxic animals. Performance by hypoxic rats is also lower on days 1–4 (Boismare *et al.*, 1975; Saligaut *et al.*, 1981) or after a single exposure (Brown *et al.*, 1973). Six percent oxygen increases the condition avoidance latency from 3.4 to 10.2 seconds and the percent avoidance decreases 72% (Brown *et al.*, 1975). Five hours at a simulated altitude of 23,000 ft reduces the number of avoidances from 107 to 22, whether sound or light is the conditioned stimulus (Hurwitz *et al.*, 1971). Similarly, 9% oxygen impairs shuttle-box avoidance acquisition, although repeated testing diminishes the differences (Ledwith, 1967). Furthermore, hypoxia alters maze performance (Hurder, 1950, 1951; Vacher and Miller, 1968), visual discrimination (Shock, 1942), and nondiscriminated avoidance behavior (Buckley *et al.*, 1969).

Exposure during the training–treatment interval may be critical to the production of hypoxic-induced changes in behavior. Thus, exposure to 8% oxygen immediately after a one-trial avoidance list does not produce significant retrograde amnesia (Flohr, 1979), but if the hypoxic insult is between 5 and 420 min after training, amnesia occurs. Pre-exposure to low oxygen does not produce the effect, nor does hypoxia produce amnesia in a multilearning process (Flohr, 1979).

Manipulation of hypoxic-induced changes in memory may be useful in the search for drugs that might stimulate cognition (Gibson and Peterson, 1984).

5.6. Histological Changes

Although numerous reports examine the effects of anoxia, ischemia, or hypoxia/ischemia on histology, few have studied pure hypoxia. Intermittent inhalation of a mixture of 94% nitrogen and 6% oxygen for 7 hr in primates (Van Bogaert *et al.*, 1938) does not produce brain damage. However, periodic exposure of rats to pure nitrogen causes unequivocal brain damage, although this may be due to reduce blood flow (Brown and Brierley, 1968). Severe hypobaric hypoxia (i.e., 60,000 ft) causes damage that is indistinguishable from that due to hypotension and maximal changes occur in cortex and basal ganglia (Brierley *et al.*, 1969; Brierly, 1973). Baboons and monkeys that are exposed to 3.2% oxygen until a cardiorespiratory crisis intervenes develop few pathological alterations (Brierley *et al.*, 1978). In a physiologically controlled model of ischemia (i.e., hypoxia with relative oligemia), hypoxia produces histological damage without apnea, hypotension, convulsions, or reduced blood flow. Microvacuolation of neuronal perikaryon occurs first and the H1 zone (Sommer sector) of the hippocampus is the most vulnerable region (Salford *et al.*, 1973). In other studies, the microvacuoles appear to be swollen mitochondria (McGee *et al.*, 1970).

6. Biochemical Changes during Hypoxia

6.1. Glucose Utilization and Cerebral Metabolic Rate for Oxygen

The cerebral metabolic rate for oxygen generally increases and then decreases with the severity of the hypoxia. Oxygen consumption, which is elevated slightly at 30% carboxyhemoglobin, declines when the cerebral blood flow is insufficient to compensate for the diminished oxygen availability (Traystman, 1978). Arterial oxygen tensions of 25–30 mm Hg stimulate oxygen consumption by 1.5- to 2-fold in paralyzed ventilated rats (Siesjo, 1978). However, in free-breathing rats, neither 15 nor 10% oxygen alter oxygen utilization (Gibson and Duffy, 1981). Hemodilution elevates cerebral blood flow, so that oxygen consumption is maintained at hemoglobin concentrations of 15 g down to 3 g % (Borgstom et al., 1975b).

Even mild hypoxia accelerates glucose utilization. In unanesthetized rats, 15% oxygen stimulates whole brain glucose utilization by 45% and elevates brain lactate concentrations. With 10% oxygen, brain lactate rises further, but glucose utilization reverses to control values (Gibson and Duffy, 1981). In paralyzed respirated animals that are severely hypoxic (arterial oxygen tensions of 25 mm Hg), glucose consumption increases nearly 100% after 1 min, but returns to normal by 15 min (Borgstrom et al., 1976). In mice, 9 or 12% oxygen increases uptake of 2-deoxyglucose in all brain regions. However, the decline in the ratio of deuterated glucose-6-phosphate to deuterated glucose suggests a decline in glucose utilization. The hippocampus, white matter, superior colliculus, and lateral geniculates appear particularly sensitive (Shimada, 1981). In rats, 10% oxygen decreases glucose utilization in the neocortex, thalamus, subthalamus, inferior colliculus, and cochlear nucleus, but increases it in the hippocampus, hypothalamus, amygdala, brain stem, nucleus tractus solitarius, and nucleus ambiguus. However, an average of all brain regions shows no differences (Miyaoka et al., 1979).

Although brain lactate levels increase in early stages of hypoxia, excess arterial lactate concentrations are a poor indicator of hypoxia. In dogs that are lightly anesthetized with chloralose, 7–8% oxygen elevates lactate levels (Huckabee, 1958), but in pentobarbital-anesthetized dogs arterial, oxygen tensions of less than 25 mm Hg are necessary for an increase (Cain, 1956).

6.2. Levels of Energy Metabolites

Although brain glucose utilization and lactate formation accelerate at arterial oxygen tensions of less than 50 mm Hg, the tissue concentrations of ATP, ADP, and AMP remain close to normal even during severe hypoxia (Gurdjian et al., 1944; Schmahl et al., 1966; Siesjo and Nilsson, 1971; Duffy et al., 1972; MacMillan and Siesjo, 1972; Bachelard et al., 1974). Energy metabolite concentrations do not decline before cerebral oxygen consumption decreases (Johannsson and Siesjo, 1975a,b). Brain ATP concentrations are normal even though there is a 40% reduction in glucose and glycogen contents and cerebral oxygen consumption (Duffy et al., 1972). The elevated creatine and reduced phosphocreatine that appear at arterial oxygen tensions of 25 and 35 mm Hg probably reflect a pH effect on the creatine phosphokinase equilibrium (Siesjo et al., 1972). More severe hypoxia in vivo depresses ATP. Below an arterial oxygen tension of 25 mm Hg, ATP declines, while ADP and AMP concentrations rise (Salford et al., 1973; Siesjo and Ljunggren, 1973) and the EEG slows markedly or is nearly isoelectric (Kramer et al., 1968; Nilsson et al., 1975). A 30-min exposure to 2% carbon monoxide (blood carbon monoxide content of 70%) decreases ATP and increases ADP and AMP (MacMillan, 1975a). High dosages of sodium nitrite (> 225 mg/kg) and of potassium cyanide (> 6 mg/kg) are needed to reduce ATP concentrations by chemical hypoxia (Gibson and Blass, 1976a). Although whole brain or gross regional studies (Siesjo et al., 1975b) show normal ATP concentrations during mild to moderate hypoxia, a decline may occur in the synapse or in subcellular fractions. For example, hypoxia diminishes synaptic transmission in hippocampal slices without reducing whole tissue ATP levels. However, ATP in the molecular layer, the synaptic region, decreases by about 15% when the evoked response begins to slow (Lipton and Whittingham, 1982). Isolated mitochondria from anoxic brain show no morphological or biochemical damage (Schultz et al., 1973a,b), so that glycolytic stimulation may maintain ATP levels. Hypoxia may activate glycolysis by stimulation of the rate-limiting enzyme phosphofructokinase (Norberg et al., 1975; Norberg and Siesjo, 1975a).

6.3. Redox States

The oxidation–reduction (redox) state of the tissue is more sensitive to hypoxia than ATP levels. Since cytosolic lactate formation from pyruvate converts NADH to NAD^+, the pyruvate to lactate ratio estimates the cytosolic NAD^+/NADH ratio. The cytosolic NAD^+/NADH ratio may also be approximated from the malate dehydrogenase reaction (i.e., the oxaloacetate/malate ratio). The mitochondrial redox state may be represented by the equilibrium potential of the glutamate ([2-oxoglutarate][NH_4^+]/[glutamate]) or the 3-hydroxybutyrate ([acetoacetate]/[3-hydroxybutyrate]) dehydrogenase reactions. Both the cytosolic and mitochondrial NAD^+/NADH ratios can be converted to potentials with the Nernst equation. The difference between these two potentials represents the transmitochondrial membrane redox potential (Gibson and Blass, 1976a,c). Several pitfalls accompany this approach: (1) Mitochondrial- and cytosolic-reducing equivalents may not be at equilibrium. (2) The reactions whose metabolites are measured are not necessarily at equilibrium either in situ or after various treatments. (3) If pyruvate is formed more rapidly in the cytosol than its transport into the mitochondria, the lactate/pyruvate ratio rises even if the tissue is not hypoxic (Jobsis and LaManna, 1978). (4) Alternatively, elevated pyruvate concentrations may shift the aspartate aminotransferase reaction to replenish glutamate and oxaloacetate with a corresponding decline in aspartate concentrations. The net result is a decrease in aspartate and an increase of malate and alanine, independently of an altered redox state. (5) Since hydrogen ions are part of the dehydrogenase equilibrium, any shift in pH changes the meaning of the redox value. The limitations of the transmitochondrial membrane potential calculations are discussed elsewhere (Blass and Gibson, 1979). Cytosolic redox changes occur within 1 min after a decrease in arterial oxygen tensions to 45 mm Hg (Siesjo et al., 1975a,b), but do not change further. The cytosolic $NADH$-NAD^+ ratio, as calculated from intracellular pH and the malate/oxaloacetate ratio, increases by 50% at 1 min and remains constant during hypoxia (Norberg and Siesjo, 1975a). The transmitochondrial redox potential is sensitive to anemic, histotoxic (Gibson and Blass, 1976a), and hypoxic hypoxia (Blass and Gibson, 1979). It changes before ATP and the alterations are more highly correlated to diminished biosynthetic activity than either the mitochondrial or cytosolic redox potentials (Gibson and Blass, 1976c).

Cellular redox potentials may also be estimated by directly monitoring NADH and cytochrome a,a_3. Some of the earliest studies on *in vivo* NADH metabolism of brain examine the effects of anoxia and demonstrate a maximal reduction in pyridine nucleotides before the EEG is isoelectric (Chance *et al.*, 1962a,b). Low arterial oxygen tensions, which can maintain resting metabolism, increase the steady-state reduction of NAD^+ and limit the NADH oxidation response to further metabolic insults (Rosenthal and LaManna, 1977). Although cytochrome a,a_3 is normally saturated with oxygen *in vitro*, the steady-state redox level of the cerebral cortex *in vivo* is sensitive to the ambient tissue oxygen tensions (Rosenthal *et al.*, 1976). During mild hypoxia, the time course of NAD^+ reduction corresponds to an increased reduction of cytochrome a,a_3. This result and the increased cytochrome a,a_3 oxidation that accompanies hyperoxia suggest that cytochrome a,a_3 is 20% (Jobsis *et al.*, 1977) to 85.5% (Hempel *et al.*, 1977) reduced during normoxia (Jobsis *et al.*, 1977). Hypoxia may damage the synaptosomal plasma and/or mitochondrial membranes and so that mitochondrial substrate oxidation decreases the availability of reducing equivalents for the respiratory chain (Rafalowska *et al.*, 1980).

6.4. Ionic Changes

Altered ion homeostasis during hypoxia is well documented. Whether ionic changes are primary or due to altered oxidative or neurotransmitter metabolism remains to be established. Whether they are primary or secondary does not diminish their pathophysiological importance.

6.4.1. Calcium

Many lines of evidence suggest that hypoxia interferes with calcium homeostasis. Calcium uptake by vascular smooth muscle declines during hypoxia (Ebeigbe, 1982). The hypoxic-induced changes in the cellular redox state and transmitochondrial membrane potential (Gibson and Blass, 1976a,c) may stimulate mitochondrial calcium efflux (Fiskum and Lehninger, 1979). Abolition of the mitochondrial membrane potential by various inhibitors decreases both plasma membrane calcium transport and calcium accumulation by the mitochondrial matrix (Scott *et al.*, 1980). A similar hypoxic-induced depression may diminish calcium-dependent acetylcholine release, since the non-calcium-dependent release is unaltered by hypoxia (Gibson and Peterson, 1982). Furthermore, this hypoxic-induced deficit in release can be overcome by either 4-aminopyridine (Gibson and Peterson, 1982) or 3,4-diaminopyridine (Peterson and Gibson, 1982), which enhance calcium uptake by nerve terminals (Peterson and Gibson, 1984). Low oxygen (2.5% or 0%) diminishes synaptosomal calcium uptake (Peterson and Gibson, 1984) in close parallel with the decline in calcium-dependent acetylcholine release. Furthermore, the reversal of the hypoxic-induced changes in calcium uptake by 3,4-diaminopyridine correlates to a reversal of hypoxia's effect on acetylcholine release (Peterson and Gibson, 1984). Hypoxic-induced oxidation of membrane phospholipids may produce the deficits in calcium uptake (Metsa-Ketala *et al.*, 1981). The relation of this hypoxic-induced decline in calcium uptake to the massive uptake during or following ischemia (Dienel, 1983) is unknown. Upon reoxygenation, calcium gain by hypoxic myocardium is mitochondrial and is supported by proton-donating anions (Harding and Poole-Wilson, 1980; Nakanishi *et al.*, 1982).

6.4.2. Potassium

Interruption of the brain's oxygen supply significantly increases extracellular potassium and the elevation is inversely related to oxygen availability (Morris, 1974). Respiration of rats with nitrogen decreases tissue oxygen tensions to 0 mm Hg and increases extracellular potassium from about 2 mM to more than 50 mM (Sick *et al.*, 1982). Similarly, respiratory arrest in anesthetized rats increases extracellular potassium slowly to 6–10 mM and then abruptly elevates it to 30–50 mM with a doubling time of 5–14 sec in the neocortex, hippocampus, amygdala, caudate nucleus, and thalamus. In the reticular formation the increase is much slower (Bures and Buresova, 1981). Anoxic- or hypoxic-induced decreases in ATP levels cause potassium ions to leak from intracellular to extracellular compartments in synaptosomes (Bito and Myers, 1972; Prince *et al.*, 1973; Zeuthen *et al.*, 1974) with either glucose or lactate/pyruvate as substrate (Pastuszko *et al.*, 1981) or in slices (Li and McIlwain, 1957). Despite this potassium leakage, impulse conduction is nearly normal (Hillman and McIlwain, 1961). The potassium changes are reversible unless glucose is also omitted (Yamamoto and Kurokawa, 1970), so that hypoxia does not alter the final potassium reaccumulation by synaptosomes (Pastuszko *et al.*, 1982).

6.5. Neurotransmitter Metabolism

Although neurotransmitter metabolism is widely regarded to be sensitive to hypoxia, no consensus has developed about which transmitter or metabolic step (e.g., release, uptake, synthesis, or degradation) is the most sensitive. Nor is it clear whether a general metabolic alteration, such as diminished calcium uptake, may differentially interfere with the metabolism of various neurotransmitters. For example, anaerobic preincubation of synaptosomes irreversibly decreases dopamine and serotonin uptake, but does not alter the release of GABA, serotonin, norepinephrine, or dopamine (Pastuszko *et al.*, 1982). Furthermore, in brain slices (Hirsch and Gibson, 1982b; Peterson and Gibson, 1982) or synaptosomes (Gibson and Peterson, 1982), hypoxia reduces the release of acetylcholine, increases that of glutamate, and does not alter norepinephrine release (Hirsch and Gibson, 1982a; McIlwain, 1973). Direct comparison of the various neurotransmitters is difficult because optimal conditions for examination of each transmitter vary and many differences occur between laboratories: type of synaptosomal preparation, slice thickness, buffers, release systems that examine endogenous transmitter release vs. that of radiolabeled compounds as well as choice of hypoxic model. Finally, the pathophysiological importance of any of these changes is difficult to determine and remains questionable.

6.5.1. Acetylcholine

Several lines of evidence suggest that altered acetylcholine metabolism underlies hypoxic-induced behavioral deficits. Acetylcholine synthesis by brain slices (Gibson and Blass, 1976c; Mann et al., 1938, 1939; Quastel et al., 1936), synaptosomes (Ksiezak and Gibson, 1981a), and ganglia (Kahlson and MacIntosh, 1939) require oxygen. The close relation of acetylcholine formation and oxidative metabolism is supported by their proportional inhibition even though less than 1% of the oxidized glucose (Gibson and Blass, 1976b; Ksiezak and Gibson, 1981a,b), pyruvate (Gibson et al., 1975), or 3-hydroxybutyrate (Gibson and Blass, 1979) is eventually converted to acetylcholine. This interaction occurs in synaptosomes and slices whether oxidative metabolism is impaired by decreased oxygen availability (Ksiezak and Gibson, 1981a) or a wide variety of metabolic inhibitors (Gibson et al., 1975; Gibson and Blass, 1976b; Ksiezak and Gibson, 1981a,b). Synaptosomes that are prepared from severely hypoxic rats do not show the usual veratridine-induced stimulation of oxidative metabolism or acetylcholine production (Booth et al., 1983; Harvey et al., 1982).

Oxygen availability also limits acetylcholine synthesis in vivo. Welsh (1943) suggested that hypoxic-induced decreases in brain acetylcholine concentrations are pathophysiologically important. Although acetylcholine concentrations do not accurately reflect functional changes, more recent studies of in vivo synthesis confirm his hypothesis. Anemic (Gibson et al., 1981a), histotoxic (Gibson and Blass, 1976a), and hypoxic (Gibson and Duffy, 1981; Gibson et al., 1981a) hypoxia decrease the synthesis of acetylcholine in rodents, whether it is measured with [U-^{14}C]glucose or [^{2}H$_4$]choline. Anemic hypoxia depresses acetylcholine formation similarly in hippocampus and cortex, but striatum is more sensitive (Peterson and Gibson, 1982). Furthermore, changes in acetylcholine metabolism occur at degrees of hypoxia (e.g., an arterial oxygen tension of 42 or 57 mm Hg or 12–31% methemoglobin) that alter mental function in man, but not energy metabolite concentrations.

The hypoxic-induced inhibition of acetylcholine metabolism is not selective, since the metabolism of other neurotransmitters also changes. However, the cholinergic deficit does appear to be behaviorally and physiologically important. Anoxia blocks cholinergic transmission in the superior cervical ganglia, although axonal conduction is unaffected (Dolivo, 1974). Scopolamine, an anticholinergic drug, depresses mental function in a manner similar to hypoxia (Drachman, 1978). Pretreatment with the acetylcholinesterase inhibitor, physostigmine, prolongs survival time during severe chemical (Gibson and Blass, 1976a) as well as hypoxic hypoxia (Scremin and Scremin, 1979) and improves behavioral performance of hypoxic animals (Gibson et al., 1983). The hypoxic-induced cholinergic deficit appears to be in the central nervous system, since the peripherally acting acetylcholinesterase inhibitor, neostigmine, is ineffective. Both nicotinic and muscarinic components of the cholinergic system are involved, since muscarinic and nicotinic blockers impair physostigmine's beneficial action in hypoxic mice. In addition, nicotine and the muscarinic agonist, arecoline, improve performance and their combination has additive effects (Gibson et al., 1983).

The precise biochemical mechanism by which hypoxia reduces acetylcholine metabolism is unknown. This effect does not appear to be due to inhibition by other neurotransmitters, since *in vitro* where such secondary effects are minimized, mild to moderate hypoxia still impairs potassium-stimulated acetylcholine synthesis (Ksiezak and Gibson, 1981a). The inhibition of metabolism may be due to impaired calcium-dependent release of acetylcholine, since *in vivo*, hypoxia depresses acetylcholine synthesis without a corresponding reduction in acetylcholine levels (Gibson and Duffy, 1981). Anoxia does not impair acetylcholine synthesis by brain slices from regions, such as septum, that make acetylcholine, but do not release it in a calcium-dependent manner (Gibson and Peterson, 1983). Low oxygen (2.5% or 0%) reduces the calcium-dependent release of acetylcholine by brain slices or synaptosomes, whereas non-calcium-dependent release is unaffected (Gibson and Peterson, 1982). Furthermore, the decline in calcium-dependent release can be overcome by 4-aminopyridine (Gibson and Peterson, 1982) or 3,4-diaminopyridine (Peterson and Gibson, 1982), which stimulate calcium uptake by nerve terminals (Peterson and Gibson, 1982). The decline in the calcium-dependent release and its amelioration with 3,4-diaminopyridine parallels the effects of hypoxia on calcium uptake (Peterson and Gibson, 1984). These same compounds ameliorate hypoxic-induced deficits in tight-rope test performance (Gibson *et al.*, 1983; Peterson and Gibson, 1982) and the *in vivo* synthesis of acetylcholine (Peterson and Gibson, 1982).

6.5.2. Amino Acids

The biosynthesis of the amino acid neurotransmitters that are derived from glucose is sensitive to hypoxia. Severe hypoxia decreases aspartate, but elevates GABA, glutamine, and alanine brain concentrations (Duffy *et al.*, 1972; Lovell and Elliot, 1963; MacMillan and Siesjo, 1972; Tews *et al.*, 1963; Norberg and Siesjo, 1975b; Wood, 1967; Wood *et al.*, 1968). The increased pyruvate during hypoxia may shift the alanine aminotransferase reaction toward alanine and 2-oxoglutarate formation, which would then shift the aspartate aminotransferase reaction to replenish glutamate and oxaloacetate, but consume aspartate (Duffy *et al.*, 1972). This hypothesis is difficult to reconcile with the hypoxic-induced decrease in amino acid synthesis from [U-^{14}C]glucose, although metabolic compartmentation may account for some of the differences. Severe hypoxia (5% oxygen) (Yoshino and Elliot, 1970) or potassium cyanide intoxication (Shimada *et al.*, 1974) reduces glucose incorporation into alanine, aspartate, GABA, glutamate, glutamine, and serine. Mild anemic (75 mg/kg of sodium nitrite) or hypoxic (15% oxygen) hypoxia decreases [U-^{14}C]glucose incorporation into alanine (34%), aspartate (49%), GABA (62%), glutamine (59%), glutamate (51%), and serine (37%), but does not alter their concentrations. Nor does hypercarbia (5% carbon dioxide) alter the effects of low oxygen on amino acid formation (Gibson *et al.*, 1981a). More severe hypoxia increases brain GABA levels in mice, hamsters, rats, guinea pigs, and rabbits (Tews *et al.*, 1963; Wood, 1967; Wood *et al.*, 1968; Duffy *et al.*, 1972).

Hypoxia also alters amino acid metabolism *in vitro*. Anoxia impairs GABA uptake by synaptosomes (Pastuszko *et al.*, 1982), but does not alter its calcium-dependent release (Hirsch and Gibson, 1982b; Pastuszko *et al.*, 1982). However, hypoxia stimulates glutamate release from superfused cortical slices (Hirsch and Gibson, 1982b; Pull *et al.*, 1972).

6.5.3. Catecholamines

Catecholamine neurotransmitter systems have been evaluated during hypoxia by measurement of concentrations and turnovers as well as by treatment with dopaminergic or noradrenergic drugs. The precise aspect of catecholamine metabolism that hypoxia alters is particularly difficult to determine, since both the synthetic and degradative enzymes are oxygen dependent.

Studies that examine only catecholamine concentrations during hypoxia are difficult to interpret, since levels are a balance of synthesis and degradation. Thus, altered content may represent either increased or decreased firing of a particular neurotransmitter system. Stupfel and Roffi (1961) first showed that a 1-hr exposure to 5 or 10% oxygen decreases norepinephrine concentrations (− 23%). Since then, reductions in whole brain norepinephrine and dopamine have been found by some (Boismare and LePoncin, 1979), but not others (Prioux-Guyonneau *et al.*, 1979; Davis and Carlsson, 1973a,b; Brown *et al.*, 1975). Exposure of mice to 6–7% oxygen increases dopamine, particularly in the norepinephrine rich areas (Kuno *et al.*, 1981). A transient decrease in norepinephrine and dopamine concentrations occurs with hypobaric hypoxia as moderate as 5200 m (Cymerman *et al.*, 1972). The response of catecholamine metabolite levels to hypoxia is more apparent regionally. Between 1800–7000 m, hypobaric hypoxia does not decrease hypothalamic or the "remainder of brain" dopamine concentrations (Prioux-Guyonneau *et al.*, 1979), whereas at 9000 m hypothalamic levels of norepinephrine decline 50% (Debijadji *et al.*, 1969). During hypobaric oxygen (760–300 mm Hg for 17 min), norepinephrine concentrations are reduced in the medulla oblongata (− 48%) and the thalamus-hypothalamus (− 27%), but are unchanged in the striatum, hippocampus, and the "remainder of the brain." In similarly treated animals, dopamine levels also decline, although not statistically, in the medulla oblongata (− 60%), hypothalamus-thalamus (− 68%), as well as striatum (− 11%), and only the "remainder of the brain" decreases significantly (− 33%) (Saligaut *et al.*, 1981).

In vivo, catecholamine turnover is a sensitive indicator of altered metabolism during hypoxia. Dopamine and norepinephrine synthesis rates can be estimated by measuring the decline in their concentrations after blocking their formation with alpha-methyl-p-DL-tyrosine (AMT). In whole brain, 8% oxygen for 30 min does not alter the depletion rate. However, hypoxia diminishes the increased dopamine and norepinephrine turnover that accompanies foot shock. Hypoxia does alter turnover in selected brain regions of nonshocked rats. In AMT-treated controls, dopamine declines in striatum (− 71%) and limbic forebrain (− 75%), but in hypoxic (6% oxygen for 2 hr) rats, it only decreases 48 and 47%, respectively (Brown *et al.*, 1975), which infers diminished turnover. With hypobaric hypoxia (1800 m), striatal and "rest of brain" dopamine turnover increase 45 and 89%, respectively. However, higher altitudes (5200 m and 7000 m) reduce synthesis in striatum (31 and 72%) and the "rest of the brain" (10 and 45%) (Prioux-Guyonneau *et al.*, 1979). Simulated altitudes of 5200 m or 7000 m decrease norepinephrine turnover in all other regions, except for the hypothalamus (Prioux-Guyonneau *et al.*, 1979). In hypoxic (6–7% oxygen) mice, dopamine turnover, by the AMT method, decreases in norepinephrine- and dopamine-enriched areas. Further evidence of decreased catecholamine turnover are the reductions in homovanillic acid after probenecid treatment (Davis and Carlsson, 1975a,b) and diminished accumulation of norepinephrine and dopamine after monoamine oxidase inhibition (Brown *et al.*, 1975).

The precise mechanism for the hypoxic-induced depression of catechol- amine metabolism is unclear. The lack of oxygen may impair tyrosine hy- droxylase, which is the rate-limiting step in catecholamine synthesis (Uden- friend, 1966). The rate of tyrosine hydroxylation can be estimated *in vivo* by measuring the accumulation of 3,4-dihydroxyphenylalanine (L-dopa) after in- hibition of aromatic amino acid decarboxylase. Whole brain L-dopa accumu- lation is reduced by breathing 9.4% (-18%), 7.5% (-34%), or 5.6% (-38%) oxygen (Davis and Carlsson, 1973a,b; Hedner and Lundborg, 1979). Hypoxia (8% oxygen) reduces L-dopa accumulation in unstressed (-22%) and stressed (-15%) rats (Brown *et al.*, 1974). Thus, tyrosine hydroxylase appears mod- erately altered by hypoxia.

Other catecholamine enzymes may also be directly altered by hypoxia. Hypoxia's impairment of the L-dopa-induced increase in norepinephrine (Brown *et al.*, 1973) suggests that low oxygen inhibits dopamine-beta-hydrox- ylase. Hypoxia does not appear to alter dopa transport into the brain or aromatic amino acid decarboxylase activity, since low oxygen does not reduce the L- dopa-induced increase in dopamine (Brown *et al.*, 1975). Nor does hypoxia appear to interfere with catechol-O-methyltransferase, since 3-methoxytyra- mine formation increases, decreases or is uneffected during hypoxia (Brown *et al.*, 1975). An hypoxic-induced elevation of dopamine after L-dopa loading suggests a partial inhibition of monoamine oxidase (Brown *et al.*, 1975). The decreased synthesis may be due to reduce substrate concentrations, since L- dopa administration increases dopamine synthesis and overcomes the hypoxic- induced inhibition (Brown *et al.*, 1975).

Altered release or uptake may reduce catecholamine turnover during mod- erate to severe hypoxia. Unchanged concentrations despite depressed synthesis suggests either that release is impaired or that synthesis and degradation are altered similarly. Since catechol-O-methyltransferase occurs extraneurally, its product, 3-methoxytyramine, estimates dopamine release. A hypoxic-induced decline in striatal 3-methoxytyramine may reflect reduced dopamine release (Brown *et al.*, 1975). However, *in vitro* dopamine release from hypoxic (20 mm Hg) perfused cortical slices (Hirsch and Gibson, 1982a,b) or anoxic synapto- somes (Pastuszko *et al.*, 1982) is unaffected.

Hypoxia also alters peripheral catecholamine metabolism (Snider *et al.*, 1974). The responses primarily reflect sympathetic stimulation, which may alter the brain's response to hypoxia. Low oxygen decreases catecholamine concentrations, but accelerates their turnover in heart (Goldman and Harrison, 1970; Prioux-Guyonneau *et al.*, 1976a,b) as well as adrenals (Steinsland *et al.*, Snider *et al.*, 1974) and elevates the urinary concentrations of catecholamine metabolites (Myles and Ducker, 1973; Daniels and Chosey, 1972).

The hypoxic-induced changes in dopamine or norepinephrine metabolism may be pathophysiologically important. The dependence of the conditioned avoidance response on an intact catecholaminergic system (Seiden et al., 1973) and its decline during hypoxia stimulated interest in the catechols and behavior during hypoxia. Reserpine, which disrupts catecholamine storage (Carlsson, 1966), and alpha-methyl-p-tyrosine, which inhibits catecholamine synthesis (Nagatsu et al., 1964), block the conditioned avoidance response and L-dopa prevents this disruption (Corrodi and Hanson, 1966; Seiden and Carlsson, 1964) by a central nervous system mechanism (Seiden and Martin, 1971). The similarity of these pharmacologically and hypoxic-induced alterations in the conditioned avoidance response strengthens the association of the hypoxic-induced deficits with the catecholaminergic systems.

Hurwitz et al. (1971) first suggested a role for catecholamines in hypoxic-induced behavioral disruptions. A lack of correlation with catecholamine levels does not support this hypothesis. Hypobaric hypoxia (7000 m) causes decreases in dopamine and norepinephrine that do not parallel changes in the Sidman avoidance behavior (Hurwitz et al., 1971) and large deficits in the conditioned avoidance response occur without alterations in brain dopamine or norepinephrine concentrations (Davis and Carlsson, 1973a,b; Brown et al., 1975).

Reversal of hypoxic-induced neurochemical and behavioral changes by catecholaminergic drugs supports an important role for catechols during hypoxia. Tranylcypromine, a monoamine oxidase inhibitor, prevents the hypoxic-induced (7000 m) decrease in dopamine and norepinephrine concentrations, but does not improve performance by hypoxic animals (Hurwitz et al., 1971). L-dopa, with or without a peripheral dopa-decarboxylase inhibitor, reduces the effects of hypobaric or hypoxic hypoxia (Brown et al., 1975) on spontaneous mobility and the conditioned avoidance response (Brown et al., 1973; Boismare et al., 1979). Low oxygen (6% oxygen) increases latency from 3 to 10 sec and L-dopa (100 mg/kg) administration during hypoxia reduces the latency to 5.5 sec. Hypoxic animals show a median avoidance of 25%, as compared with their prehypoxic avoidance of 100%, and performance of hypoxic animals reverses to 80 or 90% with 50 or 100 mg/kg of L-dopa, respectively. Apomorphine (mg/kg), a potent dopaminergic agonist, improves learning by hypobaric hypoxic animals without any alterations in cortical oxygen tensions. Avoidance learning, which is 93% in nonhypoxic controls, declines to 52% during hypoxia and morphine raises this to 72% (Saligaut et al., 1981). Apomorphine also increases avoidance response of hypoxic-hypoxic animals from 12.5 to 85%, although mean latencies are not altered (Brown et al., 1975). High morphine dosages (10 mg/kg) depress learning and catecholamine levels similarly to hypoxic animals, and hypoxia does not depress norepinephrine or dopamine levels or behavioral performance further in morphine-treated animals (Saligaut et al., 1981), which suggests they act at similar sites. Finally, the amphetamine-induced stimulation of locomotor activity in mice, which is due to increased release of catecholamines, is depressed by 12% oxygen, and the response is restored by L-dopa (Brown and Engle, 1973).

The hypoxic-induced decrease in conditioned avoidance response may not be solely due to impaired catecholamine metabolism. Inhibitors of aromatic amino acid decarboxylase do not impair the response during normoxia. Furthermore, other catecholaminergic drug treatments and hypoxia inhibit the response more than drug treatment alone (Brown et al., 1975). These findings imply that effects of hypoxia are not solely due to impaired monoamine metabolism.

6.5.4. Serotonin

Serotonin metabolism is also altered by hypoxia, but the behavioral consequences are unknown. Even prolonged hypoxic hypoxia (2 hr; 6% oxygen) does not alter brain serotonin concentrations (Brown *et al.*, 1975; Hedner and Lundborg, 1979). Although moderate hypobaric hypoxia (1800–5200 m) fails to modify serotonin levels, severe hypoxia reduces them about 30% (Prioux-Guyonneau *et al.*, 1982). Synthesis studies with parachlorophenylalanine, an inhibitor of tryptophan hydroxylase, reveal increased degradation at 1800 m and decreased synthesis at 5200 and 7000 m (Prioux-Guyonneau *et al.*, 1982). Decreased (-20%) 5-hydroxyindole acetic acid levels in the cerebral spinal fluid after probenecid treatment support the postulate of impaired synthesis (Davis and Carlsson, 1973a,b).

The biochemical mechanisms that lead to the decline in synthesis are unknown. *In vitro*, serotonin release from cortical slices (Hirsch and Gibson, 1982b) or synaptosomes (Pastuszko *et al.*, 1982) is unaffected by hypoxia. However, anoxic preincubation of synaptosomes irreversibly decreases the rate of serotonin accumulation (Pastuszko *et al.*, 1982). *In vivo* tryptophan hydroxylase activity, the rate-limiting step in serotonin synthesis (Lovenberg *et al.*, 1968), as assessed by 5-hydroxytryptamine accumulation, decreases with hypoxia (Davis and Carlsson, 1973a,b; Brown *et al.*, 1975). Hypoxia reduces activity by 46% in striatum, limbic forebrain, and cerebral hemispheres and by 36% in diencephalon and lower brain stem (Brown *et al.*, 1974).

The behavioral importance of the hypoxic-induced changes in serotonin metabolism is unclear. Several dosages of 5-hydroxytryptophan do not reverse hypoxia's effects on either latency or avoidance response. Nor are the hypoxic-induced deficits in tryptophan hydroxylase activity altered by foot shock (Brown *et al.*, 1974, 1975).

6.5.5. Adenosine

Although adenosine's role as a neuromodulator is not as well established as the transmitters discussed above, adenosine metabolism is sensitive to hypoxia. Decreasing the percent oxygen from 29.7 to 20.0, 10.7 or 5.5% elevates adenosine's levels from 9.4 to 6.4, 30.0 or 63.3 nmole/mg, respectively. The addition of carbon dioxide to the hypoxic gas mixture reduces the increase (Rubio *et al.*, 1975). When animals are sacrificed by freeze blowing, adenosine concentrations are normal between arterial oxygen tensions of 200 and 100 mm Hg, double at 50 mm Hg, and increase sevenfold at 30 mm Hg (Winn *et al.*, 1981). However, with *in situ* freezing methods that also require anesthesia, adenosine elevations do not occur (Rehncrona *et al.*, 1978).

6.6. Cyclic Nucleotide Alterations

Since cyclic GMP and cyclic AMP may be second messengers of neurotransmitters, alterations in their concentrations may reflect functional changes in neuronal systems. Neither severe anemic, histotoxic (Gibson et al., 1978, 1979), or hypoxic hypoxia (Folbergrova et al., 1981; Rehncrona et al., 1978) alter cyclic AMP levels. In marked contrast, ischemia rapidly elevates cyclic AMP (Burkard, 1972). The degree of hypoxia that is required to elevate cyclic GMP levels is controversial. Increased cyclic GMP occurs with mild degrees of sodium-nitrite-induced anemic hypoxia or potassium-cyanide-induced histotoxic hypoxia in unanesthetized animals that are sacrificed by focused microwave irradiation (Gibson et al., 1978). However, sodium nitrite and potassium cyanide, at concentrations much higher than the brain is exposed to during anemic or histotoxic hypoxia, are activators of purified guanyl cyclase in vitro (Kimura et al., 1975). In nitrous-oxide-anesthetized animals that are sacrificed by the Ponten method, cyclic GMP increases only in severely hypoxic animals with decreased adenylate energy charge and massive lactic acidosis (Folbergrova et al., 1981). However, the nitrous oxide may diminish the effect of hypoxia (Gibson and Duffy, 1981) on cyclic GMP.

6.7. Fatty Acids

Arachidonate and other fatty acids may be important in the pathophysiology of hypoxia, brain edema, ischemia, and seizures (Bazan, 1970; Fishman and Chan, 1981; Furlow and Bass, 1975; Gardiner et al., 1981). Fifteen minutes of 7% oxygen elevates brain free fatty acids, especially the polyunsaturated ones (Strosznajder, 1979). The response is not due to increased synthesis, because both hypoxia and ischemia decrease lipid concentrations and the incorporation of linoleate, glycerol, or palmitate into membrane phospholipids (Strosznajder, 1980; Alberghina and Guiffrida, 1981). [^{14}C]Glucose incorporation into phosphatidylinositol is specifically inhibited by hypoxia (Strosznajder and Domanska-Janik, 1980). The hypoxic-induced release of arachidonate is mainly from cellular membrane inosine and choline phosphoglyceride (Marion and Wolfe, 1979) by activation of phospholipase A_2 (Bazan, 1976; Sun et al., 1980) or from plasmalogens (Horrocks et al., 1978).

The elevated arachidonate during ischemia or hypoxia may be toxic to cells. Free fatty acids are potent uncouplers of oxidative phosphorylation and may normally regulate the coupling of oxidation and phosphorylation (Lazarewicz et al., 1972; Lehninger and Remmert, 1959; Ozawa et al., 1967). Their uncoupling action correlates to mitochondrial swelling (Chan and Fishman, 1978). Furthermore, fatty acids increase intracellular sodium (Chan and Fishman, 1978; Chan et al., 1980), inhibit amino acid uptake (Rhoads et al., 1982), and stimulate lipid peroxidation by cortical slices (Chan et al., 1980). Arachidonate at concentrations that occur in severely hypoxic animals (Gardiner et al., 1981) can impair mitochondrial calcium uptake and stimulate mitochondrial calcium efflux (Denzlinger et al., 1982). During reversible complete ischemia, hypoglycemia, or bicuculline-induced seizures, arachidonate accumulation occurs before extracellular calcium decreases.

7. Conclusion

Hypoxia is an interesting problem from both a basic science and a clinical point of view. Although a large amount of data has accumulated over the last few decades, a clear mechanistic understanding of hypoxia's action is still elusive. Similarly, the usefulness of hypoxia in understanding the molecular basis of the other metabolic encephalopathies and the pathophysiological basis of tissue damage after stroke still needs to be established.

ACKNOWLEDGMENTS. Supported in part by grants NS03346 and AG04171, the Winifred Masterson Burke Relief Foundation, and the Brown and Williamson Company. The author is grateful to Christine Peterson for her valuable and constructive criticisms of this review.

References

Adler, H. F., Burhardt, W. L., Ivy, A. L., and Atkinson, A. J., 1950, Effect of various drugs on psychomotor performance at ground level in simulated altitudes at 18,000 ft in a low pressure chamber, *J. Av. Med.* **231**:221–236.

Aisenberg, T., 1977, Hypoxemia and auditory reaction time in congenital heart disease, *Percept. Mot. Skills* **45**:595–600.

Alberghina, M., and Guiffrida, A. M., 1981, Effect of hypoxia on the incorporation of [2-^{3}H]glycerol and [1-^{14}C]palmitate into lipids of various brain regions, *J. Neurosci. Res.* **6**:403–419.

Astrup, P., Engel, K., Sezeringhaus, W., and Munson, E., 1965, Influence of temperature and pH on the dissociation curve of oxyhemoglobin of human blood, *Scand. J. Lab. Clin. Invest.* **17**:515–523.

Bachelard, H. S., Lewis, L. D., Ponten, U., and Siesjo, B. K., 1974, Mechanisms activating glycolysis in the brain in arterial hypoxia, *J. Neurochem.* **22**:395–401.

Baumgarten, G., Creutzfeldt, O., and Jung, R., 1961, Microphysiology of cortical neurons in acute anoxia and retinal ischemia in: *Cerebral Anoxia and the Electroencephalogram* (H. Gastaut and T. S. Meyer, eds.), Charles C Thomas, Springfield, Illinois, pp. 5–34.

Bazan, N. G., 1970, Effect of ischemia and electroconvulsive shock on the free fatty acid pool in brain, *Biochem. Biophys. Acta* **7**:403–413.

Bazan, N. G., 1976, Free arachidonate and other lipids in the nervous system during early ischemia and after electroshock, in: *Function and Metabolism of Phospholipids in the Central and Peripheral Nervous System* (G. Porcellati, L. Amaducci, and C. Galli, eds.), Plenum Press, New York, pp. 317–356.

Beard, R. R., and Wertheim, G. A., 1967, Behavioral impairment associated with small doses of carbon monoxide, *Am. J. Pub. Health* **57**:2012–2022.

Berger, H., 1934, Uber das elekeroenkephalogramm des menschen IX, *Arch. Psychiatr.* **102**:538–557.

Bert, P., 1878, La pression barometrique, Paris, Masson et Cie., p. 347.

Birren, J. E., Fisher, M. B., Vollmer, E., and King, B. G., 1946, Effects of anoxia on performance at several simulated altitudes, *J. Exp. Physiol.* **36**:36–39.

Bito, L. S., and Myers, R. E., 1972, On the physiological response of the cerebral cortex to acute stress (reversible asphyxia), *J. Physiol.* (London) **221**:349–370.

Blass, J. P., and Gibson, G. E., 1979, Consequences of mild, graded hypoxia, in: *Advances in Neurology*, Volume 26 (S. Fahn, J. N. Davis, and L. P. Rowland, eds.), Raven Press, New York, pp. 229–254.

Boismare, F., LePoncin, M., Belliard, J. P., and Hacpille, L., 1975, Reduction of hypoxia induced disturbances by previous treatment with benserazide and L. Dopa in rats, *Experientia* **31**:1190–1192.

Boismare, F., LePoncin-Lafitte, M., and Rapin, J. R., 1979, Blockade of the different enzymatic steps in the synthesis of brain amines and memory (CAR) in hypoxic hypobaric rats treated or not with L-DOPA, in: *Catecholamines: Basic and Clinical Frontiers* (E. Usdin, I. J. Kopin, and J. Barchas, eds.), Pergamon Press, New York, pp. 1726–1728.

Booth, R. F. G., Harvey, S. A. K., and Clark, J. B., 1983, Effects of in vivo hypoxia on acetylcholine synthesis by rat brain synaptosomes, *J. Neurochem.* **40**:106–110.

Bootheby, W. M., 1945, Effects of high altitude on the composition of air: Introductory remarks, *Proc. Staff Meet. Mayo Cl.* **20**:209.

Borgstrom, L., Johannson, H., and Siesjo, B. K., 1975a, The relationship between arterial PO_2 and cerebral blood flow in hypoxic hypoxia, *Acta Physiol. Scand.* **93**:423–432.

Borgstrom, L., Johannson, H., and Siesjo, B. K., 1975b, Influence of acute normovolemic anemia on cerebral blood flow and O_2 consumption in anesthetized rats, *Acta Physiol. Scand.* **93**:505–514.

Borgstrom, L., Norberg, K., and Siesjo, B. K., 1976, Glucose consumption in rat cerebral cortex in normoxia, hypoxia and hypercapnia, *Acta Physiol. Scand.* **96**:569–574.

Bradford, H. F., 1969, Respiration *in vitro* of synaptosomes from mammalian cerebral cortex, *J. Neurochem.* **16**:675–684.

Brierley, J. B., 1973, Pathology of cerebral ischemia, in: *Cerebral Vascular Disease 8th Conference* (H. F. McDowell and R. W. Brennan, eds.), Grunne and Stratton, New York, pp. 59–75.

Brierley, J. B., Brown, A. W., Excell, B. J., and Meldrum, B., 1969, Brain damage in the Rhesus monkey resulting from profound arterial hypotension - I: Its nature, distribution and general physiological correlates, *Br. Res.* **13**:68–83.

Brierley, J. B., Prior, P. F., Calverely, J., and Brown, A. W., 1978, Profound hypoxia in Papio Anuris and Macaca Mulatta—Physiological and neuropathological effects, *J. Neurol. Sci.* **37**:1–29.

Brown, A. W., and Brierley, J. B., 1968, The natural distribution in earliest stages of anoxic ischaemic nerve cell damage in the rat brain as defined by the optical microscope, *Br. J. Exp. Pathol.* **49**:87–106.

Brown, R. M., and Engle, J., 1973, Evidence for catecholamine involvement in the suppression of locomotor activity due to hypoxia, *J. Pharm. Pharmacol.* **25**:815–819.

Brown, R. M., Davis, J. N., and Carlsson, A., 1973, Dopa reversal of hypoxic-induced disruption of the conditioned avoidance response, *J. Pharm. Pharmacol.* **25**:412–414.

Brown, R. M., Snider, S. R., and Carlsson, A., 1974, Changes in biogenic amine synthesis and turnover induced by hypoxia and/or foot shock stress. II. The central nervous system, *J. Neural Transm.* **35**:293–305.

Brown, R. M., Kehr, W., and Carlsson, A., 1975, Functional and biochemical aspects of catecholamine metabolism in brain under hypoxia, *Brain Res.* **85**:491–509.

Buckley, J., Solaro, R., and Barry, H., 1969, Effects of phenformin HCl on rats subjected to stimulated high altitude, *J. Pharm. Sci.* **58**:348–351.

Buerk, D. G., and Saidel, G. M., 1978, Local kinetics of oxygen metabolism in brain and liver tissues, *Microvasc. Res.* **16**:391–405.

Bures, J., and Buresova, O., 1981, Cerebral $[K^+]_e$ increase as an index of the differential susceptibility of brain structures to terminal anoxia and electroconvulsive shock, *J. Neurobiol.* **12**:211–220.

Burkard, W. P., 1972, Catecholamine-induced increase in cyclic adenosine 3′,5′-monophosphate in rat brain *in vivo*, *J. Neurochem.* **19**:2615–2619.

Burner, J., Lambertsen, C. J., Owen, S. G., Wendel, H., and Chiedi, H., 1957, Effect of 0.08 and 0.8 atmospheres of inspired PO_2 on cerebral hemodynamics at a "constant" alveolar PCO_2 of 43, *Fed. Proc.* **16**:130.

Cain, S. M., 1956, Appearance of excess lactate in anesthetized dogs during anemic and hypoxic-hypoxia, *Am. J. Physiol.* **209**:604–610.

Carlsson, A., 1966, Pharmacological depletion of catecholamine stores, *Pharmacol. Rev.* **18**:541–549.

Carlsson, C. A., Hagerdal, M., Kaasik, A. E., and Siesjo, B. K., 1977, A catecholamine mediated increase in cerebral oxygen uptake during immobilization stress in rats, *Brain Res.* **119**:223–231.

Chan, P. H., and Fishman, R. A., 1978, Brain edema: Induction in cortical slices by polyunsaturated fatty acids, *Science* **201**:358–360.

Chan, P. H., Fishman, R. A., Lee, J. L., and Quan, S. C., 1980, Arachidonate acid-induced swelling in incubated rat brain cortical slices: Effects of bovine serum albumin, *Neurochem. Res.* **5**:629–636.

Chance, B., and Schoener, B., 1962a, Correlation of oxidation-reduction changes of intracellular reduced pyridine nucleotide and changes in electroencephalogram of the rat in anoxia. *Nature* **195**:956–958.

Chance, B., Cohen, P., Jobsis, F., and Schoener, B., 1962b, Intracellular oxidation-reduction states *in vivo*, *Science* **137**:499–508.

Clench, J., Ferrell, R. E., and Schull, W. J., 1982, Effect of chronic altitude hypoxia on hematologic and glycolytic parameters, *Am. Physiol. Soc.* **1982**:r447.

Cohen, M. M., 1962, The effect of anoxia on the chemistry and morphology of cerebral cortex slices in *vitro*, *J. Neurochem.* **9**:337–344.

Cohen, P. J., 1967, Effects of hypoxia and normocarbia on cerebral blood flow and metabolism in conscious man, *J. Appl. Physiol.* **23**:183–189.

Cohen, P. J., Alexander, S. C., Smith, T. C., Reivich, M., and Wollman, H., 1967, Effects of hypoxia and normocarbia on cerebral flood flow and metabolism in conscious man, *J. Appl. Physiol.* **23**:183–189.

Colin, F., Bourgain, R., and Manil, J., 1978, Progressive alteration of somatosensory evoked potential waveforms with lowering of cerebral tissue PO_2 in the rabbit, *Arch. Int. Physiol. Biochim.* **86**:677–679.

Corrodi, H., and Hanson, L. C. F., 1966, Central effects of an inhibitor of tyrosine hydroxylation, *Psycho Pharmacologia* (Berl.) **10**:116–125.

Crawford, D. W., Back L. H., and Cole, M. A., 1980, In *vivo* oxygen transport in the normal rabbit femoral arterial wall, *J. Clin. Invest.* **65**:1498–1508.

Crow, T. J., and Kelman, G. R., 1969, Impairment of mental performance at a simulated altitude of 8000 feet, *Aerospace Med.* **40**:981.

Crow, T. J., and Kelman, G. R., 1971, Effect of mild acute hypoxia on human short-term memory, *Br. J. Anaesth.* **43**:548–552.

Cymerman, A., Robinson, S. M., and McCullough, D., 1972, Alteration of rat brain catecholamine metabolism during exposure to hypobaric hypoxia, *Can. J. Physiol.* **50**:321–327.

Daniels, J., and Chosey, J., 1972, Epinephrine and norepinephrine excretion during running training at sea level and altitude, *Med. Sci. Sports* **4**:219–224.

Davis, P. A., Davis, H., and Thompson, J. W., 1938, Progressive changes in the human electroencephalogram under low oxygen tension, *Am. J. Physiol.* **1234**:51–52.

Davis, J. N., and Carlsson, A., 1973a, Effect of hypoxia on tyrosine and tryptophan hydroxylation in unanesthetized rat brain, *J. Neurochem.* **20**:913–915.

Davis, J. N., and Carlsson, A., 1973b, The effect of hypoxia on monoamine synthesis, levels and metabolism in rat brain, *J. Neurochem.* **21**:783–790.

Debijadji, R., Perovic, L., Varagic, V., and Stosic, N., 1969, Effect of hypoxic-hypoxia on the catecholamine content and some cytochemical changes in the hypothalamus of the cat, *Aerospace Med.* **40**:445–449.

Dell, P., and Bonvallet, M., 1954, Controle direct et reflexe de l'activite du systeme reticulaire activateur ascendant du tronc cerebral par l'oxygene et le gaz carbonique du sang, *C. R. Seances Soc. Biol.* **148**:885–858.

Dell, P., Hugelin, A., and Bonvallet, M., 1961, Effects of hypoxia on the reticular and cortical diffuse systems, in: *Cerebral Anoxia and the Electroencephalogram* (H. Gastaut and J. S. Meyer, eds.), Charles C Thomas, Springfield, Illinois, pp. 46–58.

Denison, D., and Ledwith, F., 1964, Complex reaction times at simulated cabin altitude of 8000 feet, R.A.F. Institute of Aviation Medicine Report No. 284 (1964), Ministry of Defense, London.

Denison, D. M., Ledwith, F., and Poulton, E. C., 1966, Complex reaction times at stimulated altitudes of 5000 feet and 8000 feet, *Aerospace Med.* **37**:1010.

Denzlinger, C., Hertting, G., and Jackisch, R., 1982, Synaptosomal calcium uptake systems: Prostaglandins are probably not involved in the regulation of calcium fluxes into and within the nerve endings, *J. Neurochem.* **39**:499–506.

Diament, M. L., and Palmer, K. N. V., 1966, Postoperative changes in gas tensions of arterial blood and ventilatory function, *Lancet* **2**:180.

Diekman, V., 1976, EEG analysis by a autoregressive moving average model application to the study of EEG changes measured under various blood gas levels, in: *Proceedings of the 2nd Symposium of the Study Group for EEG Methodlogy*, Kinstanz, Jogny sur Vevey, AEG-Telefunken, EDP, A-510, pp. 369–378.

Dienel, G., 1983, Accumulation and retention of ^{45}calcium in postischemic rat brain, (Abstr.) *J. Neurochem.* **41**(suppl.):S22.

Dolivo, M., 1974, Metabolism of mammalian sympathetic ganglia, *Fed. Proc. Am. Soc. Exp. Biol.* **3**:1043–1048.

Drachman, D. A., 1978, Central cholinergic system, in: *Psychopharmacology: A Generation of*

Progress (M. A. Lipton, A. Dimascio, and K. F. Killaam, eds.), Raven press, New York, pp. 651–662.

Dripps, R. D., and Comroe, J. H., 1947, The effect of the inhalation of high and low oxygen concentrations on respiratory, pulse rate, ballistocardiogram and arterial oxygen saturation (oximeter) of normal individuals, *J. Am. Physiol.* **149**:277–291.

Duffy, T. E., Nelson, S. R., and Lowry, O. H., 1972, Cerebral carbohydrate metabolism during acute hypoxia and recovery, *J. Neurochem.* **19**:959–977.

Dulfano, M. J., and Sadamu, I., 1965, Hypercapnia: Mental changes and extrapulmonary complications: An expanded concept of the CO_2 intoxication syndrome, *Ann. Intern. Med.* **63**:829–841.

Dyer, R. S., 1980, Effects of prenatal and postnatal exposure to carbon monoxide on visually evoked responses in rats, in: *Neurotoxicity of the Visual System* (W. H. Merigan and B. Weiss, eds), Raven Press, New York, pp. 17–33.

Dyer, R. S., and Annau, A., 1978, Carbon monoxide and superior colliculus evoked potentials, in: *Multidisciplinary Perspectives in Event-Related Brain Potential Research* (D. A. Otto, ed.), U.S. Government Printing Office, Washington, D.C., pp. 417–419.

Ebeigbe, A. B., 1982, Influence of hypoxia on contractility and calcium uptake in rabbit aorta, *Experientia* **38**:935–937.

Ellis, M. M., 1919, Respiratory volumes of men during short exposures to constant low oxygen tensions attained by rebreathing, *Am. J. Physiol.* **50**:267.

Ernest, J. T., and Krill, A. E., 1971, The effect of hypoxia on visual function (psychophysiological studies), *Invest. Ophthalmol.* **10**:323–328.

Ernsting, J., 1966, Effects of hypoxia upon human performance and electroencephalogram, in: *Oxygen Measurements in Blood and Tissues and Their Significance* (J. P. Payne and W. B. Hill, eds.), Churchill Ltd., London, pp. 245–259.

Ernsting, J., Gedye, J. L., and McHardy, G. J. R., 1962, Anoxia subsequent to rapid decompression, in: *Human Problems of Supersonic and Hypersonic Flight* (A. Buckanan-Barbour and H. F. Whittinghan, eds.), Pergamon press, Oxford, p. 359.

Fan, F.-C., Chen, R. Y. Z., Schuessler, G. B., and Chien S., 1980, Effects of hematocrit variations on regional hemodynamics and oxygen transport in the dog, *Am. Physiol. Soc.* **1980**:H545.

Fernadez-Guardiola, A., Bostem, F., Naquet, R., and Gastaut, H., 1959, Effets de l'anoxie sur les potentiels evoques par la lumiere chez l'homme et chez l'animal, *J. Physiol.* (Paris) **51**:463.

Fishman, R. A., and Chan, P. H., 1981, Hypothesis: Membrane phospholipid degradation and polyunsaturated fatty acids play a key role in the pathogenesis of brain edema, *Ann. Neurol.* **10**:75.

Fitch, R. F., 1964, Mountain sickness. A cerebral form, *Ann. Intern. Med.* **60**:871–876.

Fiskum, G., and Lehninger, A. L., 1979, Regulated release of Ca^{2+} from respiring mitochondria by $Ca^{2+}/2H^+$ antiport, *J. Biol. Chem.* **254**:6236–6239.

Flohr, H., 1979, Hypoxia-induced retrograde amnesia, in: *Brain Mechanisms in Memory and Learning: From the Single Neuron to Man* (M. A. B. Brazier, ed.), Raven Press, New York, pp. 277–291.

Folbergrova, J., Nilsson, B., and Sakabe. T., 1981, The influence of hypoxia on the concentrations of cyclic nucleotides in the rat brain, *J. Neurochem.* **36**:1670–1674.

Freeman, G. B., and Gibson, G. E., 1984, Stress indices in blood in animals restrained for focussed microwave irradiation, *Fed. Proc.* **43**:772.

Frieder, B., and Allweis, C., 1982, Delayed post hypoxic transient amnesia is not associated with electrical brain seizures, *Physiol. Behav.* **29**:1059–1064.

Furlow, T. W., and Bass, N. H., 1975, Stroke in rats produced by carotid injection of sodium arachidonate, *Science* **187**:658–660.

Gardiner, M., Nilsson, B., Rehncrona, S., and Siesjö, B. K., 1981, Free fatty acids in rat brain in moderate and severe hypoxia, *J. Neurochem.* **36**:1500–1505.

Gastaut, H., and Meyer, J. S. (eds.), 1961, *Cerebral Anoxia and the Electroencephalogram*, Charles C. Thomas, Springfield, Illinois.

Gedye, J. L., 1964, Transient changes in the ability to reproduce a sequential operation following rapid decompression, R.A.F. Institute of Aviation Medicine Report No. 271 (1964), Ministry of Defense (Air), London.

Gerathewohl, S., 1951, Methods of the analysis of psychomotor performance under hypoxia, *J. Aviat. Med.* **22**:196–206.

Gibbs, F. A., Davis, H., and Lennox, W. G., 1935, Electroencephalogram in epilepsy and in conditions of impaired consciousness, *Arch. Neurol. Psychiatr.* **34**:1133–1148.

Gibbs, F. A., Williams, D., and Gibbs, E. L., 1940, Modification of the cortical frequency spectrum by changes in CO_2, blood sugar and O_2, *J. Neurophysiol.* **3**:49–58.

Gibbs, F. A., Gibbs, E. L., and Lennox, W. G., 1943, The value of carbon dioxide in counteracting the effects of low oxygen, *J. Aviat. Med.* **14**:250.

Gibson, G. E., and Blass, J. P., 1976a, Impaired synthesis of acetylcholine in brain accompanying hypoglycemia and mild hypoxia, *J. Neurochem.* **27**:37–42.

Gibson, G. E., and Blass, J. P., 1976b, Inhibition of acetylcholine synthesis and carbohydrate utilization by maple-syrup-urine disease metabolites, *J. Neurochem.* **26**:1073–1078.

Gibson, G. E., and Blass, J. P., 1976c, A relation between $[NAD^+]/[NADH]$ potential and glucose utilization in rat brain slices, *J. Biol. Chem.* **25**:4127–4130.

Gibson, G. E., and Blass, J. P., 1979, Proportional inhibition of acetylcholine synthesis accompanying impairment of 3-hydroxybutyrate oxidation in rat brain slices, *Biochem. Pharmacol.* **28**:133–139.

Gibson, G. E., and Duffy, T. E., 1981, Impaired synthesis of acetylcholine by mild hypoxic hypoxia or nitrous oxide, *J. Neurochem.* **36**:28–33.

Gibson, G. E., and Peterson, C., 1982, Decreases in the release of acetylcholine *in vitro* with low oxygen, *Biochem. Pharmacol.* **31**:111–115.

Gibson, G. E., and Peterson C., 1983, Acetylcholine metabolism in septum and hippocampus *in vitro*, *J. Biol. Chem.* **258**:1142–1145.

Gibson, G. E., and Peterson, C., 1984, Pharmacological approaches to age-related deficits in oxidative metabolism, in: *Assessment in Geriatric Psychopharmacology* (T. Crook, S. Ferris, and R. Bartus, eds.), Mark Powley Associates, New Canaan, Connecticut, pp. 323–343.

Gibson, G. E., Jope, R., and Blass, J. P., 1975, Decreased synthesis of acetylcholine accompanying impaired oxidation of pyruvic acid in rat brain minces, *Biochem. J.* **148**:17–23.

Gibson, G. E., Shimada, M., and Blass, J. P., 1978, Alterations in acetylcholine synthesis and in cycle-GMP in mild cerebral hypoxia, *J. Neurochem.* **31**:757–760.

Gibson, G. E., Shimada, M., and Blass, J. P., 1979, Protection by tris(hydroxymethyl)aminomethane against behavioral and neurochemical effects of hypoxia, *Biochem. Pharmacol.* **28**:747–750.

Gibson, G. E., Peterson, C., and Sansone, J., 1981a, Decreases in amino acid and acetylcholine metabolism during hypoxia, *J. Neurochem.* **37**:192–201.

Gibson, G. E., Peterson, C., and Sansone, J., 1981b, Neurotransmitter and carbohydrate metabolism during aging and mild hypoxia, *Neurobiol. Aging* **2**:165–172.

Gibson, G. E., Pelmas, C. J., and Peterson, C., 1983, Cholinergic drugs and 4-aminopyridine alter hypoxic-induced behavioral deficits, *Pharmacol. Biochem. Behav.* **18**:909–916.

Godfraind, J. M., Kawamura, H., Krnjevic, K., and Pumain, R., 1971, Actions of dinitrophenol and some other metabolic inhibitors on cortical neurones, *J. Physiol.* **213**:199–222.

Goldman, R. H., and Harrison, D. C., 1970, The effects of hypoxia and hypercapnia on myocardial catecholamines, *J. Pharmacol. Exp. Ther.* **174**:307–314.

Grant, I., Heaton, R. K., McSweeny, A. J., Adams, K. M., and Timms, R. M., 1982, Neuropsychologic findings in hypoxemic chronic obstructive pulmonary disease, *Arch. Intern. Med.* **142**:1470–1476.

Groll-Knapp, E., Haider, M., Hoeller, H., Jenkner, G., and Stidl, H. G., 1978, Neuro- and psychophysiological effects of moderate carbon monoxide exposure, in: *Multidisciplinary Perspectives in Event-Related Brain Potential Research* (D. A. Otto, ed.), U.S. Government Printing Office, Washington, D.C., pp. 424–430.

Grossman, R. G., and Williams V. F., 1971, Electrical activity and ultrastructure of cortical neurones and synapses in ischemia, in: *Brain Hypoxia* (J. B. Brierley and B. S. Meldrum, eds.), Spastics International Medical Publication/Lippincott, Philadelphia, pp. 61–75.

Gurdjian, E. S., Stone, W. E., and Webster, J. E., 1944, Cerebral metabolism in hypoxia, *Arch. Neurol. Psychiatr.* **51**:472–477.

Gurvitch, A. M., and Ginsburg, D. A., 1977, Types of hypoxic and posthypoxic delta activity in animals and man, *Electroencephalogr. Clin. Neurophysiol.* **42**:297–308.

Haldane, J. S., Kellas, A. M., and Kennaway, E. L., 1919, Experiments on acclimatisation to reduced atmospheric pressure, *J. Physiol. (London)* **53**:181.

Hall, F. G., 1966, Minimal utilizable oxygen and the oxygen dissociation curve of blood of rodents, *J. Appl. Physiol.* **21**:375–378.

Hansen, A. J., Hounsgaard, J., and Jahnsen, H., 1982, Anoxia increases potassium conductance in hippocampal nerve cells, *Acta Physiol. Scand.* **115**:301–310.

Harding, D. P., and Poole-Wilson, P. A., 1980, Calcium exchange in rabbit myocardium during and after hypoxia: Effect of temperature and substrate, *Cardiovas. Res.* **14**:435–445.

Harvey, S. A. K., Booth, R. F. G., and Clark, J. B., 1982, The effects *in vitro* of hypoglycaemia and recovery from anoxia on synaptosomal metabolism, *Biochem. J.* **206**:433–439.

Hecht, S., 1945, Anoxia and brightness discrimination, *J. Gen. Physiol.* **29**:335.

Hedner, T., and Lundborg, P., 1979, Regional changes in monoamine synthesis in the developing rat brain during hypoxia, *Acta. Physiol. Scand.* **106**:139–143.

Hempel, F. G., Jobsis, F. F., La Manna, J. C., Rosenthal, M., and Saltzman, H. A., 1977, Oxidation of cerebral cytochromes a,a_3 by oxygen plus carbon dioxide at hyperbaric pressures, *J. Appl. Phyiol.* **43**:872–877.

Hillman, H. H., and McIlwain, H., 1961, Membrane potentials in mammalian cerebral tissues *in vitro*: Dependence on ionic environment, *J. Physiol. (London)* **157**:263–278.

Hirsch, H., Bange, F., Pulver, G., and Steffens, I., 1960, Evoked responses of the cat's visual cortex to optic tract stimulation at temperatures between 39 and 15 °C, *Electroencephalogr. Clin. Neurophysiol.* **12**:679–684.

Hirsch, J. A., and Gibson, G. E., 1982a, Anoxia inhibits release of acetylcholine but not of norepinephrine from rat-brain slices, *Fed. Proc.* **41**:8738.

Hirsch, J. A., and Gibson, G. E., 1982b, The selective alteration of neurotransmitter release by anoxia, *Soc. Neurosci. Abst.* **8**:794.

Hockaday, J. M., Potts, F., Epstein, E., Bonazzi, A., and Schwab, R. S., 1965, Electroencephalographic changes in acute cerebral anoxia from cardiac or respiratory arrest, *Electroencephalogr. Clin. Neurophysiol.* **18**:575–586.

Homer, L. D., Shelton, J. B., and Williams, T. J., 1983, Diffusion of oxygen in slices of rat brain, *Amer. J. Physiol.* **244**:15–22.

Horrocks, L. A., Spanner, S., Mozzi, R., Chun Fu, S., D'Amato, R. A., and Krakowa, S., 1978, Advances in experimental medicine and biology, in: *Myelination and Demyelination*, Volume 100 (J. Palo, ed.), Plenum Press, New York, pp. 423–438.

Hosko, M., 1970, The effect of carbon monoxide on the visual evoked potential and the spontaneous electroencephalogram, *Arch. Environ. Health* **21**:174–180.

Huckabee, W. E., 1958, Relationship of pyruvate and lactate during anaerobic metabolism. Effect of breathing low oxygen gases, *J. Clin. Invest.* **37**:264–271.

Hugelin, A., Bonvallet, M., and Dell P., 1959, Activation reticulaire et corticale d'origine chemoceptive au cours de hypoxic, *Electroencephalogr. Clin. Neurophysiol.* **11**:325–340.

Hurder, W., 1950, Relations between brain and behavior in rats following exposure to anoxia, *Am. Psychol.* **5**:225.

Hurder, W., 1951, Changes in maze performance in rats following exposure to anoxia, *J. Comp. Physiol. Psychol.* **44**:473–478.

Hurwitz, D. A., Robinson, S. M., and Barofsky, I., 1971, Behavioral decrements and brain catecholamine changes in rats exposed to hypobaric hypoxia, *Psychopharmacology* **19**:26–33.

Ingvar, D. H., Sjolund, B., and Ardo, A., 1976, Correlation between dominant EEG frequency, cerebral oxygen uptake, and blood flow, *Electroencephalogr. Clin. Neurophysiol.* **41**:268–276.

Jobsis, F. F., 1979, Oxidative metabolic effects of cerebral hypoxia, in: *Advances in Neurology*, Volume 26 (S. Fahn, J. N. Davis, and L. P. Rowland, eds.), Raven Press, New York, pp. 299–318.

Jobsis, F. F., and LaManna, J. C., 1978, Kinetic aspects of intracellular redox reactions. *In vivo* effects during and after hypoxia and ischemia, in: *Extrapulmonary Manifestations of Respiratory Disease* (E. D. Robin, ed.), Marcel Dekker, Inc., New York, pp. 63–106.

Jobsis, F. F., Keizer, J. H., LaManna, J. C., and Rosenthal, M., 1977, Reflectance spectrophotometry of cytochrome a,a_3 in vivo, *J. Appl. Physiol. Respir. Environ. Exercise Physiol.* **43**(5):1977.

Johannsson, H., and Siesjo, B. K., 1974, Blood flow and oxygen consumption in the rat in hypoxic hypoxia, *Acta Physiol. Scand.* **91**:136–138.

Johannsson, H., and Siesjo, B. K., 1975a, Cerebral blood flow and oxygen consumption in the rat in hypoxic hypoxia, *Acta Physiol. Scand.* **93**:269–276.

Johannsson, H., and Siesjo, B. K., 1975b, Brain energy metabolism in anesthetized rats in acute anemia, *Acta Physiol. Scand.* **93**:515–525.

Kahlson, G., and MacIntosh, F. C., 1939, Acetylcholine synthesis in a sympathetic ganglion, *J. Physiol.* **97**:408–416.

Kalin, E. M., Tweed, W. A., Lee, J., and MacKeen, W. L., 1975, Cerebrospinal-fluid acid-base and electrolyte changes resulting from cerebral anoxia in man, *N. Engl. J. Med.* **293**:1013–1016.

Kayama, Y., 1974, Evoked potentials of the cortical visual system during and after hypoxia in cats, *Electroencephalogr. Clin. Neurophysiol.* **36**:619–628.

Kelman, G. R., Crow, T. J., and Bursill, A. E., 1969, Effect of mild hypoxia on mental performance assessed by a test of selective attention, *Aerospace Med.* **40**:301–303.

Kety, S. S., and Schmidt, C. F., 1948, The effects of altered arterial tensions of carbon dioxide and oxygen on cerebral blood flow and cerebral oxygen consumption of normal young men, *J. Clin. Invest.* **27**:484–492.

Kiese, M., and Weger, N., 1969, Formation of ferri-haemoglobin with aminophenols in the human for the treatment of cyanide poisoning, *Eur. J. Pharmacol.* **7**:97–105.

Kimura, H., Mittal, C., and Murad, F., 1975, Activation of guanylate cyclase from rat liver and other tissues by sodium azide, *Nature* **257**:700–702.

Kontos, H. A., Levasseur, J. E., Richardson, D. W., Mauck, H. P., Jr., and Patterson, J. L., Jr., 1967, Comparative circulatory responses to systemic hypoxia in man and in unanesthetized dog, *J. Appl. Physiol.* **3**:381–386.

Kramer, R. S., Sanders, A. P., Lesage, A. M., Woodhall, B., and Sealy, W. C., 1968, The effect of profound hypothermia on preservation of cerebral ATP content during circulatory arrest, *J. Thorac. Cardiovasc. Surg.* **56**:699–709.

Krop, H. D., Block, A. J., and Cohen, E., 1972, Neuropsychologic effects of continuous oxygen therapy in chronic obstructive pulmonary disease, *Chest* **64**:317–322.

Ksiezak, H., and Gibson, G. E., 1981a, Oxygen dependence of glucose and acetylcholine metabolism in slices and synaptosomes from rat brain, *J. Neurochem.* **30**:305–324.

Ksiezak, H. J., and Gibson, G. E., 1981b, Acetylcholine synthesis and CO_2 production from variously labelled glucose in rat brain slices and synaptosomes, *J. Neurochem.* **37**:88–94.

Kuno, T., Marukawa, A., Fujiwara, H., and Tanaka, C., 1981, Dopamine accumulation in the mouse brain under hypoxia, *Jpn. J. Pharmacol.* **31**:503–509.

Lazarewicz, J. W., Strosznajder, J., and Gromek, A., 1972, Effects of ischemia and exogenous fatty acids on the energy metabolism in brain mitochondria, *Bull. Acad. Pol. Sci. (Biol.)* **20**:599–606.

Ledwith, F., 1967, The effects of hypoxia on shuttle avoidance in the rat, *Psychonomic Sci.* **8**:203–204.

Lehninger, A. L., and Remmert, L. F., 1959, An endogenous uncoupling and swelling agent in liver mitochondria and its enzymatic formation, *J. Biol. Chem.* **234**:2459–2464.

Li, C. L., and McIlwain, H., 1957, Maintenance of resting membrane potentials in slices of mammalian cerebral cortex and other tissues *in vitro*, *J. Physiol. (London)* **139**:178–190.

Lilienthal, J. S., and Fugitt, C. H., 1946, The effect of low CO concentrations on the altitude tolerance of man, *Am. J. Physiol.* **145**:359–364.

Linder, J., 1982, Effects of cervical sympathetic stimulation on cerebral and ocular blood flows during hemorrhagic hypotension and moderate hypoxia, *Acta Physiol. Scand.* **114**:379–386.

Linton, R. A. F., Miller, R., and Camerson, I. R., 1975, The effect of hypercapnia, hypoxia, and carotid sinus nerve section on hypothalamic blood flow in anesthetised rabbits, in: *Blood Flow and Metabolism in the Brain* (M. Harper, B. Jennet, D. Miller, and J. Rowan, eds.) Churchill Livingstone, London, pp. 232–234.

Lipton, P., and Whittingham, T. S., 1979, The effect of hypoxia on evoked potentials in the *in vitro* hippocampus, *J. Physiol. (London)* **287**:427–438.

Lipton, P., and Whittingham, T. S., 1982, Reduced ATP concentration as a basis for synaptic transmission failure during hypoxia in the *in vitro* guinea pig hippocampus, *J. Physiol. (London)* **325**:51–65.

Lovell, R. A., and Elliott, K. A. C., 1963, The gamma-aminobutyric acid and factor I content of brain, *J. Neurochem.* **10**:382–414.

Lovenberg, W., Jequier, E., and Sjoerdsma, A., 1968, Tryptophan hydroxylation in mammalian systems, *Adv. Pharmacol.* **6A**:21–26.

Lubbers, D. W., 1968, The oxygen pressure field of the brain and its significance for the normal and critical oxygen supply of the brain, in: *Oxygen Transport in Blood and Tissue* (D. W.

Lubbers, U. C. Luft, G. Thews, and E. Witzleb, eds.), Georg Thieme Verlag, Stuttgart, pp. 124–139.

Lutz, B. R., and Schneider, E. C., 1919, Alveolar air and respiratory volume at low oxygen tensions, *Am. J. Physiol.* **50**:280.

MacMillan, V., 1975a, The effects of acute carbon monoxide intoxication on the cerebral energy metabolism of the rats, *Can. J. Physiol. Pharmacol.* **53**:354–362.

MacMillan, V., 1975b, Regional cerebral blood flow of the rat in acute carbon monoxide intoxication, *Can. J. Physiol.* **53**:644–650.

MacMillan, V., and Siesjo, B. K., 1972, Brain energy metabolism in hypoxemia, *Scand. J. Clin. Lab. Invest.* **30**:127–136.

Manil, J., Colin, F., and Bourgain, L., 1977, Modifications of somatosensory evoked cortical potentials during hypoxia in the awake rabbit, *Adv. Exp. Med. Biol.* **94**:509–516.

Mann, P. J. G., Tennenbaum, M., and Quastel, J. H., 1938, On the mechanism of acetylcholine formation in brain *in vitro*, *Biochem. J.* **32**:243–246.

Mann, P. J. G., Tennenbaum, M., and Quastel, J. H., 1939, Acetylcholine metabolism in the central nervous system. The effects of potassium and other cations on acetylcholine liberation, *Biochem. J.* **33**:822–835.

Marion, J., and Wolfe, L. S., 1979, Origin of the arachidonic acid released postmortem in rat forebrain, *Biochem. Biophys. Acta.* **574**:25–32.

McCutcheon, E. P., Frazier, D. T., and Boyarsky, L. L., 1971, Changes in the somatosensory cortical evoked potential produced by hypovolemic shock, *Proc. Soc. Exp. Biol. Med.* **136**:1063–1071.

McDonald, R., and Adler, F. H., 1939, Effect of anoxemia on the dark adaptation of the normal and of the vitamin A-deficient subject, *Arch. Ophthalmol.* **22**:980.

McDowall, D. G., 1966, Interrelationships between blood oxygen tension and cerebral blood flow, in: *Oxygen Measurement in Blood and Tissues* (J. F. Nunn, ed.) Churchill-Ltd., London, pp. 205–219.

McFarland, R. A., 1953, Stimuli primarily related to high altitude flight, in: *Human Factors in Our Transportation*, McGraw-Hill, New York, pp. 153–169.

McFarland, R. A., and Evans, J. N., 1939, Alterations in dark adaptation under reduced oxygen tensions, *Am. J. Physiol.* **127**:37.

McFarland, R. A., and Forbes, W. H., 1940, The effects of variation in the concentration of oxygen and of glucose on dark adaptation, *J. Gen. Physiol.* **24**:69.

McFarland, R. A., Roughton, F. J. W., and Halperin, M. H., 1944, The effects of CO and altitude on visual thresholds, *J. Aviat. Med.* **15**:381–348.

McGee, S. M., Brown, A. W., and Brierley, J. B., 1970, A combined light and electron microscope study of early anoxic-ischemic cell change in rat brain, *Brain Res.* **20**:193–200.

McIlwain, H., 1973, Consequences of cerebral hypoxia examined at tissue-metabolic level, in: *Monographs in Neural Sciences*, Volume 1 (M. M. Cohen, ed.), Karger Basel, Chicago, pp. 122–129.

Metsa-Ketala, T., Laustiola, K., Lilius, E. M., and Vapaatalo, H., 1981, On the role of cyclic nucleotides in the regulation of cardiac contractility and glycolysis during hypoxia, *Acta Pharmacol. Toxicol.* **48**:311–319.

Michael, J. A., 1973, Neurophysiological effects of hypoxia, *Monogr. Neurol. Sci.* **1**:65–121.

Michenfelder, J. D., and Theye, R. A., 1969, The effects of profound hypocapnia and dilutional anemia on canine metabolism and blood flow, *Anesthesiology* **31**:449–457.

Miyaoka, M., Shinohara, M., Kennedy, C., and Sokoloff, L., 1979, Alterations in local cerebral glucose utilization (LCGU) in rat brain during hypoxemia, *Trans. Amer. Neurol. Assoc.* **104**:104.

Morris, M. E., 1974, Hypoxia and extracellular potassium activity in the guinea-pig cortex, *Can. J. Physiol.* **52**:872–882.

Myles, A. S., and Ducker, A. J., 1973, The role of the sympathetic nervous system during exposure to altitude in rats, *Int. J. Biometeor* **17**:51–58.

Nagao, S., Roccatorte, P., and Moody, R. A., 1978, The effects of isovolemic hemodilution and reinfusion of packed erythrocytes on somatosensory and visual evoked potentials, *J. Surg. Res.* **25**:530–537.

Nagatsu, T., Levett, M., and Udenfriend, S., 1964, Tyrosine hydroxylase: The initial step in norepinephrine biosynthesis, *J. Biol. Chem.* **239**:2910–2917.

Nakanishi, T., Nishroka, K., and Jarmakani, J. M., 1982, Mechanism of tissue calcium gain during reoxygenation after hypoxia in rabbit myocardium, *Am. J. Physiol.* **242**:H437–H449.

Naquet, R., and Fernandez-Guardiola, A., 1960, Effets de differents types d'anoxie sur l'activite electrophysiologique cerebale spontanee et evoguee chez le chat, *J. Physiol.* (Paris) **62**:885.

Naquet, R., and Fernandez-Guardiola, A., 1961, Effects of various types of anoxia on spontaneous and evoked cerebral activity in the cat, in: *Cerebral Anoxia and Electroencephalogram* (H. Gastaut and J. S. Meyer, eds.) Charles C. Thomas, Springfield, Illinois, pp. 43–51.

Nelson, P. G., and Frank, K., 1963, Intracellularly recorded responses of nerve cells to oxygen deprivation, *Am. J. Physiol.* **205**:208–211.

Nicholson, C., and Hounsgaard, J., 1983, Diffusion in the slice microenvironment and implication for physiological studies, *Fed. Proc.* **12**:2865–2868.

Nilsson, B., Norberg, K., Nordstrom, C.-H., and Siesjo, B. K., 1975, Influence of hypoxia and hypercapnia on CBF in rats, in: *International Symposium on Cerebral Blood Flow and Metabolism* (M. Harper, B. Jennett, D. Miller, and J. Rowan, eds.), Churchill Livingstone, Edinburgh-London-New York, pp. 9.19–9.23.

Noell, W., and Chin, H. I., 1950, Failure of the visual pathway during anoxia, *Am. J. Physiol.* **161**:573.

Norberg, K., and Siesjo, B. K., 1975a, Cerebral metabolism in hypoxic hypoxia. I. Pattern of activation of glycolysis, a re-evaluation, *Brain Res.* **86**:31–44.

Norberg, K., and Siesjo, B. K., 1975b, Cerebral metabolism in hypoxic hypoxia. II. Citric acid cycle intermediates and associated amino acids, *Brain Res.* **86**:45–54.

Norberg, K., Quistorff, B., and Siesjo, B. K., 1975, Effects of hypoxia of 10–45 seconds duration on energy metabolism in the cerebral cortex of unanaesthetized and anaesthetized rats, *Acta Physiol. Scand.* **95**:301–310.

Nunn, J. F., and Payne, J. P., 1962, Hypoxemia after general anaesthesia, *Lancet* **2**:631.

Obrist, W. D., Saltzman, H. A., Sulg, I. A., Thompson, L. W., and Townsend, R. E., 1973, The quantitative EEG in hypoxia and hyperbaric conditions, *Swed. J. Defense Med.* **9**:446–471.

Otis, A. B., Rhan, H., Epstein, M. A., and Fenn, W. O., 1946, Performance as related to composition of alveolar air, *Am. J. Physiol.* **146**:207–221.

Otto, D. A., and Reiter, L., 1978, Neurobehavioral assessment of environmental insult, in: *Multidisciplinary Perspectives in Event-Related Brain Potential Research* (D. A. Otto, ed.), U.S. Government Printing Office, pp. 409–416.

Otto, D., Benignus, V., Prah, J., and Converse, B., 1978, Paradoxical effects of carbon monoxide on vigilance performance and event-related potentials, in: *Multidisciplinary Perspectives in Event-Related Brain Potential Research* (D. A. Otto, ed.), U.S. Printing Office, pp. 440–443.

Ozawa, K., Seta, K., Araki, H., and Handa, H., 1967, Rapid liberation of potassium ions from brain mitochondria, *J. Biochem.* **62**:584–590.

Pastuszko, A., Wilson, D. F., Erecinska, M., and Silver, I. A., 1981, Effects of in vitro hypoxia and lowered pH on potassium fluxes and energy metabolism in rat brain synaptosomes, *J. Neurochem.* **36**:116–123.

Pastuszko, A., Wilson, D. F., Erecinska, M., and Silver, I. A., 1982, Neurotransmitter metabolism in rat brain synaptosomes: Effect of anoxia and pH, *J. Neurochem.* **38**:1657–1667.

Paulson, O. B., Parving, H.-H., Olesen, J., and Shinhoj, E., 1973, Influence of carbon monoxide and of hemodilution on cerebral blood flow and blood gases in man, *J. Appl. Physiol.* **35**:111–116.

Petajan, J. H., Packham, S. C., Frens, D. B., and Dinger, B. G., 1976, Sequelae of carbon monoxide-induced hypoxia in the rat, *Arch. Neurol.* **33**:152–157.

Peterson, C., and Gibson, G. E., 1982, 3,4-Diaminopyridine alters acetylcholine metabolism and behavior during hypoxia, *J. Pharmacol. Exp. Ther.* **222**:576–582.

Peterson, C., and Gibson, G. E., 1984, Synaptosomal calcium metabolism during hypoxia and 3,4-diaminopyridine treatment, *J. Neurochem.* **42**:248–253.

Plum, F., 1975, The metabolic encephalopathies, in: *The Nervous System*, Volume 2, *The Clinical Neurosciences* (D. Tower, ed.), Raven Press, New York, pp. 193–201.

Plum, F., and Posner, J. B., 1980, *The Diagnosis of Stupor and Coma*, 3rd ed., F. A. Davis Co., Philadelphia, Pennsylvania.

Ponte, J., and Purves, M. J., 1974, The role of the carotid body chemoreceptors and carotid sinus baroreceptors in the control of cerebral blood vessels, *J. Physiol.* **237**:315–340.

Prawdicz-Neminski, W. W., 1923, Zur Kentniss der elektrischen und der innervationsvorgange in den funtionellen elementen and geweben des tierischen organismus, elektrocerebrogramm der saugetiere, *Pflug. Arch. Eur. J. Physiol.* **209**:362–382.

Prince, D. A., Lux, H. D., and Neher, E., 1973, Measurement of extracellular potassium activity in cat cortex, *Brain Res.* **50**:489–495.

Prioux-Guyonneau, M., Durand, J., Rapin, J. R., and Cohen, Y., 1976a, High altitude influence on the level and turnover time of cardiac norepinephrine in rats, *J. Physiol. (Paris)* **72**:579–587.

Prioux-Guyonneau, M., Jacquot, C., Cohen, Y., and Rapin, J. R., 1976b, Influence del' hypoxie normobare et hypobare sur le taux de renouvellement de la noradrenaline cardiaque, *Experientia* **32**:1024–1025.

Prioux-Guyonneau, M., Cretet, E., Jacquot, C., Rapin, J. R., and Cohen, Y., 1979, The effect of various simulated altitudes on the turnover of norepinephrine and dopamine in the central nervous system of rats, *Pflug. Arch. Eur. J. Physiol.* **380**:127–132.

Prioux-Guyonneau, M., Mocafr-Cretet, E., Redjimi-Hafsi, F., and Jacquot, C., 1982, Changes in brain 5-hydroxytryptamine metabolism induced by hypobaric hypoxia, *Gen. Pharmacol.* **13**:251–254.

Pull, I., Jones, D. A., and McIlwain, H., 1972, Superfused cerebral tissues in hypoxia: Neurotransmitter and amino acid retention; Labile constituents and response to excitation, *J. Neurobiol.* **3**:311–323.

Quastel, J., Tennenbaum, M., and Wheatley, A. H. M., 1936, Choline ester formation in and choline esterase activities of tissues *in vitro*, *Biochem. J.* **30**:1668–1681.

Rafalowska, U., Erecinska, M., and Wilson, D. F., 1980, The effect of acute hypoxia on synaptosomes from rat brain, *J. Neurochem.* **34**:1160–1165.

Rahn, H., and Otis, A. B., 1946, Alveolar air during simulated flights to high altitudes. *Am. J. Physiol.* **150**:202–221.

Rebert, C. S., Houghton, P. W., Howd, R. A., and Pryor, G. T., 1982, Effects of hexane on the brainstem auditory response and caudal nerve action potential, *Neurobehav. Toxicol. Teratol.* **4**:79–85.

Rehncrona, S., Siesjo, B. K., and Westerberg, E., 1978, Adenosine and cyclic AMP in cerebral cortex of rats in hypoxia, status epilepticus and hypercapnia, *Acta Physiol. Scand.* **164**:453–463.

Rhoads, D. E., Kaplan, M. A., Peterson, N. A., and Raghupathy, E., 1982, Effects of free fatty acids on synaptosomal amino acid uptake systems, *J. Neurochem.* **38**:1255–1260.

Rosen, R., Simonson, E., and Baker, J., 1961, Electroencephalograms during hypoxia in healthy men, *Arch. Neurol.* **5**:648–654.

Rosenthal, M., and LaManna, J. C., 1977, Oxidative metabolism and electrophysiological activity in intact central nervous system, in: *Oxygen and Physiological Function* (F. F. Jobsis, ed.), Professional Information Library, Dallas, Texas, pp. 515–530.

Rosenthal, M., LaManna, J. C., Jobsis, F. F., Levasseur, J. E., Kontos, H. A., and Patterson, J. L., 1976, Effects of respiratory gases on cytochrome *a* in intact cerebral cortex. Is there a critical PO_2?, *Brain Res.* **108**:143–154.

Rubio, R., Berne, R. M., Beckman, E. L., and Curnish, R. R., 1975, Relationship between adenosine concentration and oxygen supply in rat, *Am. J. Physiol.* **228**:1896–1902.

Saligaut, C., Moore, N., Lerclerc, J. L., and Boismare, F., 1981, Hypobaric hypoxia: Central catecholamine levels, cortical PO_2 and avoidance response in rats treated with apomorphine, *Aviat. Space Environ. Med.* **52**:166–170.

Salford, L. G., Plum, F., and Siesjo, B. K., 1973, Graded hypoxic-oligemia in rat brain. II. Neuropathological alterations and their implications, *Arch. Neurol.* **29**(4):234–238.

Scheinberg, P., 1951, Cerebral blood flow and metabolism in pernicious anemia, *Blood* **6**:213.

Schmahl, F. W., Betz, E., Dettinger, E., and Hohorst, H. J., 1966, Energiestoffwechsel der Grosshirnrinde und elektroencephalogramm bei sauerstoffmangel, *Pflug. Arch. Eur. J. Physiol.* **292**:46–59.

Schultz, H., Silverstein, P. R., Vapalahti, M., Bruce, D. A., Mela, L., and Langfit, T., 1973a, Brain mitochondrial function after ischemia and hypoxia. I. Ischemia induced by increased intracranial pressure, *Arch. Neurol.* **29**:408.

Schultz, H., Silverstein, P. R., Vapalahti, M., Bruce, D. A., Mela, L., and Langfit, T., 1973b, Brain mitochondrial function after ischemia and hypoxia. II. Normotensive systemic hypoxia, *Arch. Neurol.* **29**:417.

Scott, I. D., Akerman, K. E. O., and Nicholls, D. G., 1980, Calcium ion transport by intact synaptosomes, *Biochem. J.* **192**:873–880.

Scow, J., Krasna, L., and Ivy, A., 1950, The immediate and accumulative effect in psychomotor performance of exposure to hypoxia, high altitude and hyperventilation, *J. Aviat. Med.* **21**:79–81.

Scremin, A. M. E., and Scremin, O. U., 1979, Physostigmine induced cerebral protection against hypoxia, *Stroke* **10**:142–143.

Seiden, L. S., and Carlsson, A., 1964, Brain and heart catecholamine levels after L-DOPA administration in reserpine treated mice correlations with a conditioned avoidance response, *Psychopharmacologia* **5**:178–181.

Seiden, L. S., and Martin, T. W., 1971, Potentiation of effects of L-DOPA on conditioned avoidance behavior by inhibition of extracerebral DOPA decarboxylase, *Physiol. Behav.* **6**:453–458.

Seiden, L. S., Brown, R. M., and Levy, A. J., 1973, Brain catecholamine and conditioned behavior mutual interactions, in: *Chemical Modulation of Brain Function* (H. C. Salelli, ed.), Raven Press, New York, pp. 261–275.

Sheard, C., 1945, Effect of anoxia, oxygen and increased intrapulmonary pressure on dark adaptation, *Mayo Clin. Proc.* **20**:230.

Shelbourne, S. A., Jr., McLaurin, A. N., and McLaurin, R. L., 1976, Effects of graded hypoxia on visual evoked responses of rhesus monkeys, in: *Head Injuries* (R. L. McLaurin, ed.), Grune and Stratton, New York, pp. 89–93.

Shimada, M., 1981, Glucose uptake in mouse brain regions under hypoxic hypoxia, *Neurochem. Res.* **6**:993–1003.

Shimada, M., Kihara, T., Kurimoto, K., and Watanabe, M., 1974, Incorporation of ^{14}C from [U-^{14}C]glucose into free amino acids under cyanide intoxication, *J. Neurochem.* **23**:379–384.

Shock, N. W., 1942, The effect of learning on repeated exposures to lowered oxygen tension of the inspired air, *J. Comp. Physiol. Psychol.* **34**:55–63.

Sick, T. J., Rosenthal, M., LaManna, J. C., and Lutz, P. L., 1982, Brain potassium ion homeostasis, anoxia, and metabolic inhibition in turtles and rats, *Am. Physiol. Soc.* **243**:R281–R288.

Siesjo, B. K., 1978, Brain energy and catecholaminergic activity in hypoxia, hypercapnia and ischemia, in: *Neurotransmitters in Cerebral Coma and Stroke* (K. Jellinger, I. Klatzo, and P. Riederer, eds.), Springer-Verlag, New York, pp. 17–22.

Siesjo, B. K., and Ljunggren, B., 1973, Cerebral energy reserves after prolonged hypoxia and ischemia, *Arch. Neurol.* **29**:400–407.

Siesjo, B. K., and Nilsson, L., 1971, The influence of arterial hypoxemia upon labile phosphates and upon extracellular and intracellular lactate and pyruvate concentrations in the rat brain, *Scand. J. Clin. Lab. Invest.* **27**:83–96.

Siesjo, B. K., Folbergrova, J., and MacMillan, V., 1972, The effect of hypercapnia upon intracellular pH in the brain evaluated by the bicarbonate carbonic acid method and from the creatine phosphokinase equilibrium, *J. Neurochem.* **19**:2483–2495.

Siesjo, B. K., Johannsson, H., Norberg, K., and Salford, L. G., 1975a, Brain function metabolism and blood flow in moderate and severe arterial hypoxia, in: *Brain Work Alfred Benzon Symposium VIII*, Munksgaard, Kopenhamn, pp. 101–125.

Siesjo, B. K., Norberg, K., Ljunggren, B., and Salford, L. G., 1975b, Hypoxia and cerebral metabolism in a basis and practice of a neuroanesthesia, in: *Monography in Anaesthesiology*, Volume 2 (E. Gordon, ed.), Excerpta Medica, Amsterdam, pp. 47–82.

Silver, I. A., 1973, Local PO_2 in relation to intracellular pH, cell membrane potential and potassium leakage in hypoxia and shock, *Adv. Exp. Biol. Med.* **37A**:223–231.

Snider, S. R., Brown, R. M., and Carlsson, A., 1974, Changes in biogenic amine synthesis and turnover induced by hypoxia and/or foot shock stress. I. The adrenal medulla, *J. Neural Transm.* **35**:283–291.

Sohmer, H., Gafni, M., and Chisin, R., 1982, Auditory nerve-brain stem potentials in man and cat under hypoxic and hypercapnic conditions, *Electroencephalogr. Clin. Neurophysiol.* **53**:506–512.

Sotaniemi, K. A., Sulg, I. A., and Hokkanen, T. E., 1980, Quantitative EEG as a measure of cerebral dysfunction before and after open-heart surgery, *Electroencephalogr. Clin. Neurophysiol.* **50**:81–95.

Steinsland, O. S., Passo, S. S., and Nahas, G. G., 1970, Biphasic effects of hypoxia on adrenal catecholamine content, *Am. J. Physiol.* **218:**995–998.

Stewart, R. D., Peterson, J. E., Varetta, E. D., Bachand, R. T., Hosko, M. J., and Herrmann, A. A., 1970, Experimental human exposure to carbon monoxide, *Arch. Environ. Health* **21:**154–164.

Strosznajder, J., 1979, The effects of hypoxic-hypoxia on phospholipids in brain subcellular membranes, *Int. Soc. Neurochem.* **7:**599.

Strosznajder, J., 1980, Incorporation of linoleic acid into membrane glycerophospholipids from rat brain submitted to ischemia and hypoxia, *Neurochem. Res.* **5:**1265–1277.

Strosznajder, J., and Domanska-Janik, K., 1980, Effect of anoxia and hypoxia on brain lipid metabolism, *Neurochem. Res.* **5:**583–589.

Stupfel, M., and Roffi, J., 1961, Effect of anoxia and different levels of carbon dioxide on the noradrenaline and adrenaline content of rat brain, *C. R. Seances Soc. Biol.* **155:**237–240.

Sugar, O., and Gerard, R. W., 1938, Anoxia and brain potentials, *J. Neurophysiol.* **1:**558–572.

Sun, G. Y., Manning, R., and Strosznajder, J., 1980, Effects of postdecapitative ischemia and hypoxia on the phosphoglyceride acyl groups of rat brain membranes, *Neurochem. Res.* **5:**1211–1219.

Tauber, B., and Allweis, C., 1975, Effects of acute hypoxia on memory, *Isr. J. Med. Sci.* **11:**71.

Tews, J. K., Carter, S. H., Roa, P. D., and Stone, W. E., 1963, Free amino acids and related compounds in dog brain: Post-mortem and anoxic changes, effects of ammonium chloride infusion, and levels during seizures induced by picrotoxin and by pentylenetetrazol, *J. Neurochem.* **10:**641–653.

Thews, G., 1960, Die sauerstoffdiffusion in gehirn, *Pflug. Arch. Eur. J. Physiol.* **271:**197.

Thews, G., 1963, Implications to physiology and pathology of oxygen diffusion at the capillary level, in: *Selective Vulnerability of the Brain in Hypoxaemia* (J. P. Schade and W. H. McMenemey, eds.), Blackwell Scientific Publications, Oxford, England, pp. 27–35.

Traystman, R. J., 1978, Effect of carbon monoxide hypoxia and hypoxic hypoxia on cerebral circulation, in: *Multidisciplinary Perspectives in Event-Related Brain Potential Research* (D. A. Otto, ed.), U.S. Government Printing Office, Washington, D.C., pp. 453–457.

Traystman, R. J., and Fitzgerald, S. R., 1981, Cerebrovascular response to hypoxia on baroreceptor- and chemoreceptor-denervated dogs, *Am. Physiol. Soc.* **10:**H724.

Undenfriend, S., 1966, Tyrosine hydroxylase, *Pharmacol. Rev.* **18:**43–51.

Vacher, J., and Miller, A., 1968, Altitude-acclimatization: Its effect on hypoxia-induced performance decrements, *Psychopharmacologia* **12:**250–257.

Van Bogaert, J., Dallemagne, M. J., and Wigria, R., 1938, Recherches sur le besoin d'oxygene chronique et aigu chez *Macaca rhesus*, *Arch. Intern. Med.* **13:**335–378.

Vollmer, E. P., 1946, The effects of carbon monoxide on three types of performance at simulated altitudes of 10,000 and 15,000 feet, *J. Exp. Psychol.* **36:**244–251.

Watanabe, K., Miyazaki, S., Hara, K., and Hakamade, S., 1980, Behavioral state cycles, background EEGs and prognosis of newborns with perinatal hypoxia, *Electroencephalogr. Clin. Neurophysiol.* **49:**618–625.

Welsh, J. H., 1943, Acetylcholine levels of rat cerebral cortex under conditions of anoxia and hypoglycemia, *J. Neurophysiol.* **6:**329–336.

Wilmer, W. H., and Berens, C., Jr., 1918, Medical studies on aviation. V. The effect of altitude on ocular functions, *J. Am. Med. Assoc.* **71:**1394.

Winn, H. R., Rubio, R., and Berne, R. M., 1981, Brain adenosine concentration during hypoxia in rats, *Am. J. Physiol.* **241:**H235–H242.

Wood, J. D., 1967, A possible role of gamma-aminobutyric acid in the homeostatic control of brain metabolism under conditions of brain hypoxia, *Exp. Brain Res.* **4:**81–84.

Wood, J. D., Watson, V. J., and Ducker, A. J., 1968, The effect of hypoxia on brain gamma-aminobutyric acid levels, *J. Neurochem.* **15:**602–608.

Xintaras, C., Johnson, B. l., Ulrich, C. W., Terrill, R. E., and Sobecki, F., 1966, Application of the evoked response technique in air pollution toxicology, *Toxicol. Appl. Pharmacol.* **8:**77–87.

Yamamota, C., and Kurokawa, M., 1970, Synaptic potentials recorded in brain slices and their modification by changes in the level of tissue ATP, *Exp. Brain Res.* **10:**159–170.

Yoshino, Y., and Elliot, K. A. C., 1970, Incorporation of carbon atoms from glucose into free amino acids in brain under normal and altered conditions, *Can. J. Biochem.* **48:**228–235.

Zeuthen, T., Hiam, R. C., and Silver, I. A., 1974, Microelectrode registration of ion activity in brain, *Adv. Exp. Med. Biol.* **50:**145–146.

4

Ischemic Encephalopathy

W. DAVID LUST, HAJIME ARAI, YUKIMASA YASUMOTO,
TIM S. WHITTINGHAM, BOGDEN DJURICIC,
BOGOMIR B. MRSULJA, and JANET V. PASSONNEAU

1. Introduction

Stroke is one of the leading causes of death in the United States, surpassed only by heart disease and cancer. Perhaps more important is the fact that greater than 50% of stroke victims survive the episode with varying degrees of disability. Although it is very difficult to find reliable figures on the incidence of stroke, almost 300,000 persons are hospitalized each year for initial stroke and there are nearly 2 million survivors of stroke in the United States today. Thus, the prevention of strokes is obviously a major public health concern. Indeed, with the advances in the detection of stroke and the improvement in the diagnostic procedures, the number of stroke-related deaths has been decreasing in the last ten years. However, the treatment of stroke also deserves a great deal of attention, not only to reduce the number of deaths, but also to optimize the quality of life in those patients who survive. It is apparent that a more enlightened treatment of stroke will only evolve when a greater understanding is attained of the pathophysiological events that occur during and after a stroke. A variety of experimentally induced models of ischemia are now being used to investigate the pathophysiological basis of strokes. To date, attempts to describe a specific pathogenic phenomenon that would account for loss of brain function have generally failed. However, new and exciting approaches to the investigation of stroke offer some hope for the future.

The study of mechanisms involved in stroke is complicated by its varying etiology. Stroke is defined clinically as a syndrome characterized by a host of

W. DAVID LUST, HAJIME ARAI, YUKIMASA YASUMOTO, TIM S. WHITTINGHAM, BOGDEN DJURICIC, BOGOMIR B. MRSULJA, and JANET V. PASSONNEAU • Laboratory of Neurochemistry, National Institute of Neurological and Communicative Disorders and Stroke, National Institutes of Health, Bethesda, Maryland 20205. Present address for W.D.L. and T.S.W.: Laboratory of Experimental Neurosurgery, Case Western Reserve University, Cleveland, Ohio 44106.

neurological events that have a rapid onset and that usually progress over a 24-hr period. The cause is generally attributed to an interruption in blood flow to the brain in the form of either (1) an occlusion due to a thrombus or an embolus, or (2) a subarachnoid or intracerebral hemorrhage. Perhaps the best information on the incidence and prevalence of stroke in the United States comes from a recently published National Survey of Stroke (Weinfeld, 1981), which provides statistics for the years 1971, 1973, 1975, and 1976. The study reports that in a population sample of nearly 1900 stroke victims, almost 50% could not be specifically categorized. However, of the remaining group about 87% of the strokes were attributable to an occlusion, whereas only 13% were due to hemorrhage. Other statistics have relevance to patient care and survival. For example, the fatality rate in the hemorrhagic group was almost twofold greater than in stroke due to occlusion. The incidence of stroke over the age of 45 is higher in men than women. Additional factors such as coma, seizures, and headaches are associated with a less optimistic prognosis for the patient. Further, the mortality rate is highest in the first month and survival is longer in younger patients. With the increasing use of CT and PET scans for more reliable diagnosis, future surveys on stroke will yield data with greater precision. In spite of some uncertainties, interested readers will find the National Survey of Strokes to be an invaluable source of statistics on the overall impact of stroke in the United States.

On the basis of both clinical and experimental evidence, it is generally accepted that the brain can survive between 5 and 10 min of ischemia before permanent neurological deficits are manifested. As will be discussed later, some investigators consider this duration to be a conservative estimate. Nevertheless, certain facts about the brain serve to explain why ischemia has such a profound effect on its function. The brain weight is about 2% of the total body weight and yet receives 15% of the basal cardiac output and consumes about 20% of the resting oxygen consumption of the body. Under normal circumstances, the brain relies solely on the metabolism of glucose to maintain the high metabolic rate. The energy reserves in the brain are limited and only capable of sustaining the needed ATP for about 1 min in the absence of blood flow. These factors collectively suggest that there is a delicate balance between a continual supply of nutrients from the blood and the energy demands of the brain. The dependence of the brain on an uninterrupted blood flow is further illustrated by the rapid and marked changes that occur during ischemia. On cessation of blood flow to the brain, spontaneous electrical activity is lost within 15 sec. The energy status of the tissue as reflected by the high-energy phosphates, ATP and P-creatine, is severely depressed within 1 min (Lowry et al., 1964). Thus, the brain becomes quiescent and unable to support active processes such as transport and synthesis in a very short period of time after the onset of the insult. Although these changes clearly demonstrate the need for a patent circulation to the brain to maintain function, brief interruptions of circulation to the brain with the concomitant shutdown of physiological and biochemical processes do not solely account for the development of brain dysfunction. Brain damage, as reflected by histological evidence and/or neurological signs, is usually manifested only after periods of ischemia of 5 min or more.

In this article, the emphasis will be on the biochemical events that take place in the adult brain both during and after an ischemic episode. The neonatal brain is apparently more resistant to ischemia (Himwich, 1951) and will not be considered in this review. One reason for examining brain metabolites during an ischemic insult is to look for a pathogenic event that could account for the onset of experimental brain damage and/or neurological dysfunction with increasing periods of ischemia. It is obvious that sustained ischemia will eventually lead to necrosis of the tissue. Between the onset of ischemia and development of necrosis, there is an obvious demarcation into two groups: (1) ischemic periods up to 5 min, after which functional recovery appears essentially complete when circulation is re-established; and (2) ischemic periods of 10 min or longer, after which neurological deficits arise and the mortality rate increases with the duration of ischemia. Changes in metabolites evident during the shorter ischemic period are thus probably not related to irreversible loss of function, whereas changes arising after longer periods might be. Of course, the early loss of ATP may inhibit an energy dependent process necessary to the survival of the tissue. However, metabolite alterations that occur during both brief and prolonged ischemic periods have not yielded information to explain the differential in survival.

In contrast to ischemic events, the metabolic changes that occur during recirculation have proved to be quite intriguing. Before the brain can begin to function again, certain prerequisites have to be met. Some recovery of ATP is necessary for synthetic and transport processes; without ATP, the disturbed cellular gradients for Na^+, K^+, and Ca^{2+} would not be re-established. Further, the compartmentation and metabolism of neurotransmitters would not be normalized. Although metabolic recovery is apparently a permissive step to the restoration of function, restitution of the energy status, alone, does not guarantee functional recovery. Another interesting aspect of recovery is that ischemia apparently sets the stage for pathological events to occur on the reintroduction of oxygen and glucose. The experimental findings indicate that metabolic recovery is not a simple reversal of the processes that occurred during ischemia; it extends far beyond the time necessary for the gross normalization of energy metabolism. Lastly, neuropathologists have clearly demonstrated the phenomenon of selective vulnerability, that some neurons are more susceptible to ischemia than others. Recently, several investigators have developed models compatible with neurochemical techniques allowing investigations to be done on specific areas and neurons. Given the constraints on this review, some of the older biochemical information relating to stroke will not be presented in detail. For interested readers, the book, *Brain Energy Metabolism*, by Siesjo (1978) is an excellent source of information. It is the purpose of this review to present some of the newer and, in some cases, more controversial aspects of stroke research.

2. Models of Ischemia

There are many experimental models of cerebral ischemia that differ both in the species chosen as well as in the method used to produce the desired cessation of blood flow (Moossy, 1979). The ischemia may be either focal or global, and the ischemic insult may be transient or permanent depending on the model. The selection of a model must take into account many different factors, not the least of which are the goals set forth for the experiment. In the study of the pathology arising from an ischemic episode, the use of nonhuman primates and embolic methods appear to be more relevant to the human condition (Moossy, 1979). However, such an approach for biochemical studies would be technically quite difficult.

Gerbils and rats are currently the most commonly used animals in biochemical studies of brain ischemia, although mice were used in the pioneer work by Lowry et al. (1964). The obvious considerations in developing a model of ischemia are ease of handling, cost, maintenance, and generation of sufficient data for the application of statistics. However, other factors, centering on the techniques used for the occlusion of blood flow, have dictated the popularity of a given model. Decapitation is obviously the easiest way to produce complete ischemia, but studies on recovery are then precluded. Similarly, injection of microspheres into the internal carotid artery permits the investigation of ischemia but not recovery (Kogure and Schwartzman, 1979). Kabat and Dennis (1938) perfected the use of the tourniquet method that prevents blood flow to the brain by inflating a neck cuff to a pressure substantially greater than that of the blood pressure. This approach has also been applied in combination with hypotension in the rat (Diemer and Siemkowicz, 1980). Another way to prevent blood flow to the brain is to enhance intracranial pressure by increasing the volume of the cerebrospinal fluid (Kramer and Tuynman, 1967). Unilateral ischemia has also been produced in the rat by using a combination of common carotid artery occlusion and a brief exposure to nitrogen atmosphere (Levine, 1960). Each of these methods had particular drawbacks that, although it did not diminish their usefulness, prevented their general acceptance.

A major breakthrough came with the demonstration by Levine and Payan (1966) that there was an anomaly of the circle of Willis in the gerbil such that unilateral ligation of the common carotid artery resulted in ischemia in the ipsilateral forebrain in approximately 40% of the animals. Simplicity of the surgical procedures led to its general acceptance, and subsequently it was demonstrated that almost all the gerbils exhibited neurological signs of ischemia if both common carotid arteries were ligated. However, there is some controversy surrounding the value of this model. Since the gerbil is prone to spontaneous seizures (Cox and Lomax, 1976), it has been suggested that many of the biochemical sequelae to an ischemic insult may, in fact, be due to the seizures associated with ischemia. The latest development in models of ischemia is a four-vessel occlusion in the rat (Pulsinelli and Brierley, 1979): electrocauterizing the two vertebral arteries and occlusion of the two common carotid arteries. The last two models have found the greatest acceptability by the neurochemist. There are some dissimilarities between the two models that make direct comparisons on the basis of ischemic duration quite difficult. For example, the rats will survive up to 72 hr after 30 min of four-vessel occlusion (Pulsinelli and Brierly, 1979), whereas only about 12% of the gerbils will survive 24 hr after 20 min of ischemia (Conger et al., 1981). In spite of the inherent differences in the models of ischemia, however, it is truly surprising how similar many of the experimental findings have been.

3. Physiological Response to Ischemia

Brain function is dependent on a number of factors that give rise to the excitability of the tissue. These include unique properties of membranes, electrochemical gradients, neurotransmitter metabolism, and ionic pumps. It is indeed important to know how the various components of excitability respond to an ischemic episode. Rossen *et al.* (1943) described in some detail the response of conscious human subjects to short periods of ischemia induced by a cervical inflatable cuff to arrest blood flow to the brain. Loss of consciousness occurred within about 7 sec after the onset of ischemia at which time a delta wave appeared in the EEG. In the same study, circulatory arrest to the brain was extended to 100 sec in another group with apparently no ill effects once the individual recovered consciousness. EEG recordings were not monitored in this group; however, EEG has been shown experimentally to be suppressed after 20 sec of ischemia in the cat (Hossmann and Olsson, 1970), after 10 sec in the rat (Swaab and Boer, 1972), and after 30 sec in the gerbil (Cohn, 1979). Thus, spontaneous excitability as indicated by EEG recordings does persist, albeit with changes, for some time after the loss of consciousness.

Other parameters of excitability have also been examined during ischemia. Hossmann and Olsson (1970) demonstrated that the I and D waves recorded from the pyramidal tract after stimulation in the sensorimotor cortex persisted for approximately 4 and 6 min, respectively. Therefore, some of the basic building blocks of excitability persist for a substantially longer period of ischemia than does the EEG.

Some recent work by Hansen and Zeuthen (1981) provides additional insight into the alterations of electrolyte gradients during ischemia. After the onset of ischemia, there is a small but gradual increase in the extracellular concentration of potassium. The increase occurs over a 2-min period and appears to be independent of other relevant electrolytes. The enhanced potassium efflux is attributed to an increase in potassium conductance, the cause of which is presently unclear (Hansen *et al.*, 1982). At approximately 2 min of ischemia the levels of calcium and sodium decrease and those of potassium increase further in the extracellular space. Although the initial changes in potassium are associated with a hyperpolarization, the secondary large increase and the changes in sodium and calcium result in about a 15-mV depolarization of the cells from the resting state. Thus, almost from the start of ischemia, the excitability of the cells would be affected; first by the initial hyperpolarization and then the secondary depolarization. The changes in electrolyte concentrations do not readily explain the rapid suppression of the EEG in the brain, although it is possible that the activation of potassium conductance is involved. Since the integrity of the membrane is apparently intact during ischemia as reflected by the existence of a membrane potential, the loss of energy-dependent pumping may be responsible for the new steady-state condition. Because most of these physiological responses to ischemia occur well within the first 5 min of ischemia, it does not appear that the shutting down of brain function per se is responsible for the brain malfunction induced by ischemia. The entire process, in fact, may be a protective mechanism whereby the brain becomes quiescent until the insult abates.

4. Metabolic Changes with Ischemia

4.1. Energy Metabolism

The energy derived from the hydrolysis of ATP supports a variety of transport and biosynthetic processes in the brain. The original work by Lowry *et al.* (1964) on the decapitated mouse brain clearly demonstrated that ATP was rapidly lost during ischemia. Since oxidative phosphorylation is dependent on oxygen, the production of ATP during ischemia is entirely dependent on the less efficient anaerobic glycolysis and on the activities of adenylate kinase and creatine kinase. The creatine kinase equation yields one ATP for each P-creatine and in its predominant ionic form is as follows (Veech, 1980):

$$P^{2-}\text{-creatine} + Mg\ ADP^- + H^+ \leftrightharpoons Mg\ ATP^{2-} + \text{creatine} \qquad (1)$$

Since the metabolic rate of brain has been determined to be approximately 25 mM high-energy phosphates/kg per min (Lowry *et al.*, 1964) and the P-creatine levels are 4 to 5 mM in most brain regions, the production of ATP at the expense of P-creatine would only support total brain function for less than 30 sec. The yield of ATP from the adenylate kinase reaction (Eq. 2) would be even less:

$$2\ Mg\ ADP^- \leftrightharpoons Mg\ ATP^{2-} + Mg^{2+} \qquad (2)$$

Even with the inclusion of anaerobic glycolysis, which will be discussed below, it is readily apparent that most of the ATP in brain would be consumed well within the first 2 min of ischemia, as has been shown experimentally (Lowry *et al.*, 1964).

Another aspect of adenylate metabolism that is generally ignored is the loss of the adenylate pool (ATP + ADP + AMP) during ischemia. As the 5′ AMP accumulates during ischemia, a portion of it is dephosphorylated forming adenosine. There apparently are two pools of adenylates: the first is lost with a half-time of about 4 min and the second about 40 min (unpublished observation). The time-dependent decrease in the adenylate pool during ischemia may be of great importance to the survival of the tissue during recovery and will be discussed later.

4.2. Glycolysis

Glucose is oxidized via glycolysis to yield two pyruvate molecules that in turn are metabolized to carbon dioxide and water in the tricarboxylic acid (TCA) cycle. More than 90% of the ATP normally produced by the metabolism of glucose comes from the TCA cycle, which ceases during ischemia. Some ATP can be produced from anaerobic glycolysis, the metabolism of glucose to lactate. Since the supply of substrate from the blood is also lost during ischemia, the only available glucose for metabolism comes from an intracellular pool or from the storage form, glycogen. The net ATP yield per glucose entity from glucose is 2 and from glycogen is 3 (the phosphorylation of glucose requires ATP, whereas the phosphorolysis of glycogen does not). Even if the glucose and glycogen collectively equaled 5 mM in concentration, the yield of ATP would only be sufficient to sustain activity for less than a minute. In spite of the limited substrate reserves, glycolysis is increased by as much as seven-fold during ischemia which tends to maintain ATP by balancing energy production with the energy demands of the tissue (Lowry et al., 1964). From the analysis of the intermediates, the authors concluded that the enhanced glycolysis was due to the activation of hexokinase and phosphofructokinase (Lowry and Passonneau, 1964). Despite an increase in the rate of glucose use, brain ATP levels are not effectively maintained in the absence of a blood-borne supply of substrate and oxygen which may explain the rapid response of the brain to an ischemic insult.

4.3. Lactate

One of the major theories concerning the development of brain injury during ischemia centers on the production of lactate. Lactate is the end product of anaerobic glycolysis and, as expected, increases greatly during ischemia. Several investigators have shown that preloading experimental animals with glucose results in a greater amount of morphological damage in the brain for a given period of ischemia (Myers and Yamaguchi, 1976; Kalimo et al., 1981). In both of these studies, hyperglycemia resulted in a significantly greater lactate accumulation during ischemia. The decrease in pH concomitant with the increase in lactate alone cannot fully account for the greater amount of injury, since lowering the cellular pH with hypercapnia has only minimal effects on tissue morphology (Folbergrova et al., 1974). One current hypothesis is that a lower pH in conjunction with another ischemia-induced event, such as energy depletion, may account for deleterious effects on the tissue (Rehncrona et al., 1981). Although elevated lactate may be a factor in exaggerating the impact of ischemia, several facts suggest that lactate may only have a modulatory effect. Most of the neuropathological evidence on brain injury induced by ischemia indicates that certain regions of the brain are more susceptible to an ischemic episode than others. Generally, in regional analyses, lactate production appears to be rather uniform. Therefore, it is unlikely that the elevated lactate with or without energy depletion could account for the heterogeneous pattern of injury noted after ischemia unless some brain regions are more sensitive than others to lactate accumulation. Also, in complete ischemia, the increase in lactate is quite rapid and peaks after the glucose and glycogen are consumed. In that situation, where lactate plateaus relatively early in ischemia, the severity of brain injury still increases with longer periods of ischemia. Thus, the role of elevated lactate in the development of brain damage remains unclear.

4.4. Tricarboxylic Acid Cycle and Respiration

The production of energy by the oxidation of the metabolic products of glycolysis depends on the TCA cycle, the electron transport chain, and oxidative phosphorylation. In the absence of molecular oxygen, both the electron transport chain and oxidative phosphorylation cease to function and therefore are not a factor in continued energy production during ischemia. Several laboratories have examined the concentration of the TCA cycle intermediates during ischemia and their findings are summarized in Table I. Goldberg et al. (1966) concluded from their results that the observed changes reflect a change from a flux state to equilibrium conditions. Certain facts about the oxidation of pyruvate will serve to explain the changes that occur in the TCA cycle intermediates. The major substrates for the electron transport system are the reduced pyridine nucleotides that are generated by the TCA cycle. In the absence of molecular oxygen, as occurs in complete ischemia, the pyridine nucleotides in the mitochondria shift toward the reduced form. This effect would inhibit the oxidation of pyruvate via pyruvate dehydrogenase which requires NAD as a substrate. In addition, the decrease in NAD would also tend to reduce the activity of the several other dehydrogenases found in the mitochrondria (at least in the desired direction of the cycle). The accumulation of pyruvate, succinate, and fumarate in the initial stages of ischemia is consistent with a shift from oxidized to reduced pyridine nucleotides.

The delayed changes in the TCA cycle intermediates are a little more difficult to explain. When the anaplerotic reactions that feed into the TCA cycle were included, Siesjo (1978) was able to offer a reasonable interpretation of the data. The secondary loss of pyruvate, for example, could be attributed to the activity of glutamate-pyruvate transaminase:

$$\text{Glutamate} + \text{Pyruvate} \rightleftharpoons \text{Alanine} + \alpha\text{-Ketoglutarate } (\alpha KG) \qquad (3)$$

In fact, Folbergrova et al. (1974) have reported a significant increase in alanine after 5 min of ischemia, although a similar increase in the levels of αKG was not noted.

The depletion of αKG, instead of the predicted increase, is not too surprising, since αKG is a substrate for many enzymatic reactions such as other transaminases as well as reductive amination. In addition, the relatively low concentration of oxaloacetate and αKG compared with other mitochondrial metabolites indicate that these substrates could be enzymatically converted to another mitochondrial metabolite without any detectable change in the latter. Although the TCA cycle intermediates do undergo a significant amount of rearrangement with no apparent loss of concentration in total carbon skeletons (Folbergrova et al., 1974), experiments on mitochondria isolated from ischemic brains indicate that normal respiration can be achieved on the return of oxygen and substrate (Ginsberg et al., 1977). Although there may be secondary events associated with the changes in TCA cycle intermediates, mitochondrial dysfunction due to derangement of the TCA cycle, the respiratory chain, or oxidative phosphorylation does not appear to be the direct cause of cellular damage during ischemia.

TABLE I. Relative Changes in Tricarboxylic Acid Cycle Intermediates at Various Times of Ischemia

Metabolite	Ischemic duration (percent of control)		
	30 sec[a]	5 min[b]	10 min[c]
Pyruvate	151	8.3	12.4
Citrate	75	32.7	27.1
α-Ketoglutarate	40	0	2.9
Succinate	145	270	—[d]
Fumarate	134	85	—
Malate	108	65.7	67.3
Oxaloacetate	50	0	2.5

[a] Goldberg et al. (1966).
[b] Folbergrova et al. (1974).
[c] Ljunggren et al. (1974).
[d] The dashes indicate that data were not available for those metabolites.

4.5. Neurotransmitters and Cyclic Nucleotides

The levels of the monoaminergic neurotransmitters have been the subject of investigation by several laboratories (Faiman et al., 1973; Kogure et al., 1975; Robinson et al., 1975; Lavyne et al., 1975; Lust et al., 1975: Mrsulja et al., 1976c; Welch et al., 1978). The results do not offer a clear-cut picture of the fate of the monamines during ischemia. Five of the papers (Faiman et al., 1973; Lavyne et al., 1975; Robinson et al., 1975; Lust et al., 1975; Mrsulja et al., 1976c) show a significant decrease in norepinephrine with ischemia. Welch et al. (1978) presented evidence, however, that the decrease in norephinephrine is only evident when the animals are overtly seizing. Dopamine has also been shown to decrease during ischemia (Lavyne et al., 1975; Robinson et al., 1975; Lust et al., 1975; Mrsulja et al., 1976c); however, two other reports indicate either there is no change (Welch et al., 1978) or an increase (Kogure et al., 1975). Serotonin also appeared to decrease during ischemia in two studies, but the rate and extent of the decrease is somewhat less than that for the catecholamines (Lust et al., 1975; Maruki et al., 1984). In another report, the serotonin levels were elevated from 15 min to 6 hr of unilateral ischemia (Mrsulja et al., 1976c). However, the preponderance of the evidence indicates that the monaminergic neurotransmitters are diminished during ischemia. This finding is not particularly surprising, since the critical biosynthetic enzymes, as well as certain degradative enzymes for these monoaminergic neurotransmitters, require molecular oxygen. The discrepancies, where they arise, are probably due either to the time of sampling, the brain region examined, or the ischemic model used.

The effects of ischemia on the levels of glutamate and GABA have also been determined. The concentrations of glutamate either do not change during ischemia (Tews et al., 1963; Folbergrova, 1974) or are slightly depressed (Lust et al., 1975). In contrast, the levels of GABA increase gradually with longer periods of ischemia (Tews et al., 1963; Folbergrova et al., 1974; Lust et al., 1975). The changes in these amino acids may simply indicate a shift from a flux to an equilibrium condition; however, if the observed changes occurred similarly in neurotransmitter pools, there could be an effect on function during recirculation.

The first demonstration that ischemia stimulated the accumulation of cyclic AMP was made by Breckenridge (1964). The cyclic AMP concentrations in brain increased approximately two-fold at 20 sec after decapitation. Subsequently, several laboratories have demonstrated that cyclic AMP increases rapidly in the first minute of ischemia and thereafter falls off toward control values with time (Steiner et al., 1972; Lust and Passonneau, 1976, 1979). It is interesting to note that the time to achieve peak cyclic AMP levels is considerably longer during unilateral rather than bilateral ischemia (Lust et al., 1975; Watanabe and Ishii, 1976). In contrast to cyclic AMP, cyclic GMP decreases with ischemia, but the rate of change is substantially slower than that for cyclic AMP (Steiner et al., 1972; and Passonneau et al., 1977).

4.6. Status of the Brain at the Onset of Recirculation

It seems worthwhile at this time to summarize the status of the tissue during ischemia. Within minutes of the onset of ischemia, all energy production has essentially stopped and all those processes dependent on the hydrolysis of ATP have similarly ceased. The rapid loss of excitability attests to the fact that the brain can only function if a precise set of conditions is met. The maintenance of electrochemical gradients is perhaps the most critical and with a decrease in (Na^+,K^+)ATPase activity, the sodium and potassium gradients are rapidly lost (Astrup et al., 1977). In addition, calcium homeostasis is also affected by ischemia (Harris et al., 1981). Secondary to the changes in the electrolyte milieu of the tissue are the relatively slow changes observed in neurotransmitter systems. The magnitude of the alterations in these systems probably would not prevent excitability, but would have a modulatory effect on normal brain function. In addition to these derangements, there is a significant decrease in intracellular pH (Ljunggren et al., 1974), a suspension of protein synthesis (Kleihues and Hossmann, 1973), and a rapid increase in the release of membrane-bound free fatty acids (Aveldano and Bazan, 1975). Thus, the brain is exposed to a variety of abnormal chemical conditions, any one of which could have an effect on function. How the brain recovers from the collective effect of these disturbances is the subject of the next several sections. However, it should be iterated that the specific cause of irreversible brain damage is not readily apparent from the description of the events that occur during ischemia, although it could be as simple as the absence of ATP which results in the loss of an energy-dependent process, as yet undetermined, that is fundamental to the survival of the tissue.

In spite of the large number of metabolic alterations that occur during ischemia, it would appear that the brain reaches a new steady-state condition that apparently is an intermediate stage between normality and infarction. Although the brain has lost excitability, the neurons maintain a membrane potential, albeit somewhat depolarized from normal values (Hansen et al., 1982). Thus, the integrity of the membrane must be preserved to some degree to allow the demonstration of an electrochemical gradient. In addition, Clendenon et *al.* (1971) have demonstrated that the release of lysosomal enzymes, a good indicator of ongoing cell injury or even cell death, only occurs after 3 hr of anoxia-ischemia in the rat. According to the authors, the release does not appear to be a critical factor in the development of cell injury. The integrity of the lysosome indicates that cell destruction is not evident during periods of ischemia compatible with recovery of function. The duration of the ischemic insult will ultimately determine if reversibility is possible.

5. Recovery after an Ischemic Episode

5.1. Cerebral Circulation

Discussion of blood flow to the brain has been intentionally avoided, principally because blood flow should not be a factor during ischemia. It should be stated that some blood flow to the ischemic zone of the brain is evident in many of the stroke models; however, the amount is minimal and usually not sufficient to support brain function. Since the purpose of this paper is to present changes in the brain induced by ischemia, the preponderance of the information comes from experiments on global rather than focal ischemia. In the latter, there are usually complications with the border zone around the ischemic core where, because of collateral circulation, the critical decrease in blood flow has not been achieved. For a detailed discussion of focal versus incomplete ischemia, see Siesjo (1978).

Blood supply to the brain following ischemia is an important consideration to the restoration of brain function. Ames *et al.* (1968) demonstrated that there were localized areas of the brain that were not perfused following ischemia and the extent of tissue involvement increased with longer periods of ischemia. Clearly, once the means of preventing blood flow is removed and the brain or a portion of the brain does not receive a blood supply, then ischemia persists. Under those conditions, it is unlikely that recovery will ensue and the infarcts that occur with prolonged ischemia will arise. The importance of the "no-reflow" phenomenon to the onset of brain damage is the subject of some controversy. A number of other investigators have confirmed the presence of the "no-reflow" phenomenon in other models of ischemia (Osburne and Halsey, 1975; Levy *et al.*, 1975). However, Levy *et al.* (1975) have demonstrated that ischemic brain damage does occur even when reflow occurs and Harrison *et al.* (1975) showed that increasing the blood pressure in the postischemic period minimized no-reflow and yet the mortality rate in the gerbils was unchanged. Thus, although the lack of adequate recirculation may be a factor in the recovery process, the appearance of brain damage despite recirculation suggests that other mechanisms may be more critical to the evolution of injury.

Cerebral blood flow and blood pressure do appear to be critical to the recovery of the animals. Cantu *et al.* (1969) reported that reversal of postischemic hypotension in rabbits permitted the restoration of neural function. Hossmann *et al.* (1973) demonstrated that a period of hyperemia during recirculation occurred primarily in those animals that exhibited signs of functional recovery. In another study, the survival rate of gerbils after ischemia increased in those animals that showed a complete restoration of blood flow to the brain (Osburne and Halsey, 1975). In spite of these reports that normal or enhanced perfusion during recirculation improves the likelihood of recovery, studies in humans with acute stroke indicate that cortical infarcts can be associated with areas of focal hyperemia (Olsen et al., 1981). Recently, Pulsinelli et al. (1982b) demonstrated that there were two periods of hyperemia: the first occurred shortly after the onset of recirculation and was observed in all regions examined, whereas the second occurred two days after ischemia and only in those regions exhibiting histological evidence of brain damage. Thus, it is quite possible that the hyperemia associated with the onset of reperfusion is beneficial to the recovery of brain function and is quite distinct from the hyperemia observed in infarcted regions of the brain.

5.2. Physiological Recovery after Ischemia

The survival rate of the animals is perhaps the easiest approach to evaluating the extent of ischemic brain damage. For example, all the gerbils generally survived for 24 hr after 5 min of bilateral ischemia, whereas, almost 90% of the gerbils died within 24 hr after 20 min of bilateral ischemia in the study by Conger et al. (1981). It would appear that between 5 and 20 min of bilateral ischemia certain events occurred that led to irreversible brain damage. However, there are a number of confounding factors, such as edema, seizures, and so forth, in such an approach. Indeed, Wexler (1972) demonstrated that there were marked changes in blood constituents during ischemia that were attributed to the stress of the insult. These findings clearly indicate that there are deleterious peripheral effects resulting from brain ischemia. Ideally, the restoration of function would be best studied by recording a component of brain excitability (i.e., EEG). A systematic study of the relationship between the restoration of activity and the duration of ischemia induced by compression has been described by Ljunggren et al. (1974). There was a consistent loss of EEG within 15 to 20 sec of ischemia. Following 1 min of ischemia, brain activity was evident at 5 min of recovery and essentially normal by 25 min. After longer periods of compression ischemia (5 and 10 min), the return of activity was first evident after about 20 min of recirculation, but the EEG recordings in these two groups exhibited persistent abnormalities over the 3 hr of investigation. It would appear from these results that the restoration of function, as reflected by the EEG, is determined by the duration of the ischemic insult. Another approach used to assess the extent of recovery is the measurement of evoked potentials. Hossmann and Sato (1970) were able to show a total recovery of the evoked potential following 1 hr of ischemia. This duration of ischemia far exceeds the limits of reversibility that most investigators have found both clinically and experimentally. The importance of these findings cannot be overstated, since it demonstrates that many of the building blocks (i.e., ionic gradients, membrane potentials, etc.) that give rise to excitability of the brain can apparently survive prolonged periods of ischemia. The explanation for the recovery of these animals after such a long period of ischemia is presently unclear, but may be related to the experimental paradigm used (i.e., species, anesthesia, mechanical ventilation, vasopressor agents, etc.). Whatever the reason, these results indicate that under certain circumstances, the brain apparently can "survive" longer periods of ischemia than previously thought.

Extracellular electrolyte concentrations are also being used to assess the severity of ischemic episodes. As this methodology is perfected, it may be possible to determine at what step in the hierarchy of excitability that failure initially occurs. Already, Hansen and Zeuthen (1981) have been able to demonstrate the changes in extracellular K^+, Na^+, Cl^-, and Ca^{2+} that occur during ischemia. What is clearly lacking is the demonstration of the time course of restoration of these electrolytes in tissues that exhibit restoration of function from those that do not.

In a recent report, Djuricic et al. (1984) demonstrated that total potassium content was depressed after periods of bilateral ischemia longer than 5 min and in the early stages of recirculation. The loss of potassium by as much as 40% obviously could have a profound effect on the re-establishment of electrochemical gradients and on osmoregulation. Such an effect should also be considered in the interpretation of the extracellular potassium results.

5.3. Metabolic Recovery

The restoration of energy metabolism is a prerequisite for the recovery of brain function following ischemia. In the last ten years, many laboratories have examined metabolic recovery by measuring key brain metabolities after varying periods of ischemia (Ljunggren et al., 1974; Kleihues et al., 1975; Mrsulja et al., 1976a; Kobayashi et al., 1977; Levy and Duffy, 1977; Conger et al., 1981). In this section, the nature of the metabolic restoration will be discussed in some detail with an emphasis on those events that reflect a persisting derangement in metabolism during the recovery period. Generally, those postischemic events that either arise or become more pronounced with longer periods of ischemia should provide some insight into the biochemical systems that may be involved in the development of brain injury.

5.3.1. Energy Metabolism

The restoration of the high-energy phosphates following ischemia has been one of the major arguments that failure of energy metabolism is not the principle cause of irreversible brain damage. However, in spite of the normalization of both ATP and P-creatine during recirculation, there are other indications that energy metabolism is not completely restored to its preischemic condition. Only in the most extreme cases (i.e., 60 min of ischemia) is the magnitude of the ATP restoration severely compromised (Fig. 1). Even in the 30-min group where survival is unlikely, the ATP levels in the gerbil brain recover to the same extent as those for shorter periods of ischemia. Thus, it would appear that the ability to produce ATP, most likely by oxidative phosphorylation, is spared during periods of ischemia of up to 30 min. These findings are consistent with those of Ginsberg et al. (1977), who demonstrated that the respiration in mitochondria isolated from ischemic brains was seemingly unaffected. However, the rate of ATP regeneration and the slight but consistent depression of ATP during recirculation represent postischemic abnormalities in energy metabolism that could have an impact on the recovery of function.

As shown in Fig. 1, the rate of ATP restoration decreases with increasing periods of ischemia. Clearly, the recovery of ATP in the 1-min ischemic group is essentially complete within 1 min of reperfusion, whereas after longer periods of ischemia the levels of ATP remain significantly lower than those of control even after 5 min of recirculation. Therefore, the rate of ATP restoration, unlike the magnitude of ATP recovery, appears to be a good indicator of the magnitude of the ischemic insult. A slower rate of restoration would also tend to prolong the episode. Since the progression from reversible to irreversible stages of ischemia occurs in minutes, the delay in the restoration of the energy status of the tissue could be an important factor in the survival of the animal. The explanation for the depressed rate of restoration with longer periods of

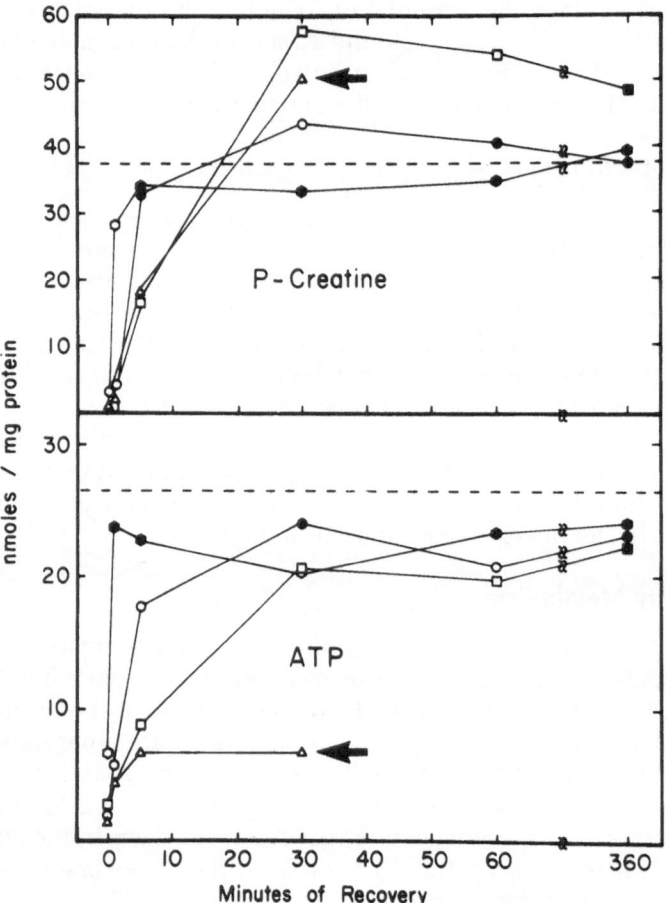

FIGURE 1. Comparison of high-energy phosphate levels in the cerebral cortex during recirculation after varying periods of bilaterial ischemia in the gerbil. The various times of ischemia are designated as follows: (○) 1 min, (⊙) 5 min, (□) 30 min, and (△) 60 min. Each value is the mean of at least five determinations and the open symbols represent values significantly different ($p < 0.05$) from those of control (dashed line). The gerbils were prepared and the assays performed as described by Kobayashi *et al.* (1977). The arrows indicate the level of high-energy phosphates at 30 min of recirculation after 60 min of bilateral ischemia. At this time, P-creatine levels were significantly greater than control, even though the concentrations of ATP were only 25% of control.

ischemia is not immediately obvious; however, since similar effects are seen with P-creatine, glucose, glycogen, and pyruvate (Ljunggren et al., 1974; Mrsulja et al., 1976a; Kobayashi et al., 1977), the delayed recovery of the metabolites may merely reflect the time necessary to prime the various metabolic pathways (i.e., a reversal from an equilibrium state to a condition of metabolic flux).

Tissue ATP levels approached but never quite reached full restoration until almost 1 day of recirculation, as shown in Fig. 1 and by others (Ljunggren et al., 1974; Mrsulja et al., 1976a; Levy and Duffy, 1977). Given the time-dependent loss of the adenylates (ATP + ADP + AMP) during ischemia and the relatively slow de novo and salvage pathway synthesis of ATP, the depression of ATP undoubtedly reflects the diminished adenylate pool. The importance of the lower ATP levels has generally been dismissed since the energy charge of the tissue is restored. However, a major drawback to the measurement of brain metabolites is the inability to distinguish between the various cellular compartments (i.e., glia, neurons, and endothelial cells). For example, the overall depression of ATP in the tissue could be as much as 20%. If the ATP decrease was localized to just one cellular compartment, the reduction of ATP within that component could significally affect survival, despite the apparently normal energy charge. Until such time that the depression in ATP levels is shown to be uniform in all cells, the importance of this effect to the recovery of the brain cannot be answered.

The changes in P-creatine during recirculation also provide some interesting insights into the metabolic status of the tissue. These observations emphasize the need to examine the entire metabolite profile before concluding that energy metabolism has recovered. The rate of P-creatine restoration decreases with increasing periods of ischemia (Fig. 1). At first glance, the levels of this labile high-energy phosphate appear to have recovered and one could conclude that the energy status of the tissue was normalized after all periods of ischemia. However, the levels of ATP in the 60-min ischemia group clearly indicate that this is not the case. The production of P-creatine when the levels of ATP are dangerously low appears to be a pathogenic event (Fig. 1, note arrows). Further evidence for this comes from the greater overshoots of P-creatine with increasing periods of ischemia. Since a decrease in pH that persists during recovery would favor a breakdown of P-creatine (see Eq. 1), it is likely that other factors are involved. Other reactants such as creatine and ADP do not change in a way to account for this phenomenon (Kobayashi et al., 1977), so it would appear that the accumulation of P-creatine may be related to a perturbed internal milieu. Veloso et al. (1973) have demonstrated a magnesium effect on the creatine kinase equilibrium; alterations in the concentration of intracellular Mg^{2+} could perhaps explain the changes in P-creatine.

As alluded to earlier, there is a coupling of energy production to energy consumption. Generally in the discussion of the recovery process, a derangement in a given metabolite has been attributed to problems in energy production. However, the exquisite work by Sokoloff and co-workers (1977) using the 2-deoxyglucose method has provided an alternative explanation to the perturbation in energy production. Quite simply, they found that the rate of glucose metabolism is directly linked to the functional activity of the brain. Thus, a change in the concentration of an energetic metabolite could reflect an altered state of excitability rather than a problem with energy production. Although it may be difficult to distinguish exactly what is affecting energy metabolism, especially at early periods of recirculation, the reader should be cognizant that recent work with long-term recovery has demonstrated that electrical activity in certain brain regions does change during recirculation (Suzuki *et al.*, 1983b). The altered brain activity and not an abnormality in energy production induced by ischemia may account for the metabolite derangements that occur concurrently. Our findings with the high-energy phosphates suggest that the metabolic machinery for the production of ATP is relatively resistant to ischemia, although the availability of substrates including ADP and oxygen may be rate limiting to the regeneration of ATP. Failure of oxidative phosphorylation, therefore, does not appear to be a factor in either the onset of neurological dysfunction or the early manifestation of brain damage.

5.3.2. Glycolysis

The concentration of glucose, as the major metabolic substrate for the brain, is particularly important to the recovery process, and since glucose synthesis in the brain is minimal, its presence undoubtedly indicates the restoration of blood flow. The recovery of glucose after varying periods of bilateral ischemia in the gerbil is presented in Fig. 2 and is representative of results found by a number of other laboratories (Ljunggren *et al.*, 1974; Kleihues *et al.*, 1975; Mrsulja *et al.*, 1976a). As with the high-energy phosphates, the rate of glucose restoration decreases with longer times of ischemia. Glucose levels also exhibit an overshoot after all periods of bilateral ischemia which, as with P-creatine, appears to be greater with extended ischemic periods. The inability to restore ATP after 60 min of ischemia (Fig. 1) does not appear to be due to the absence

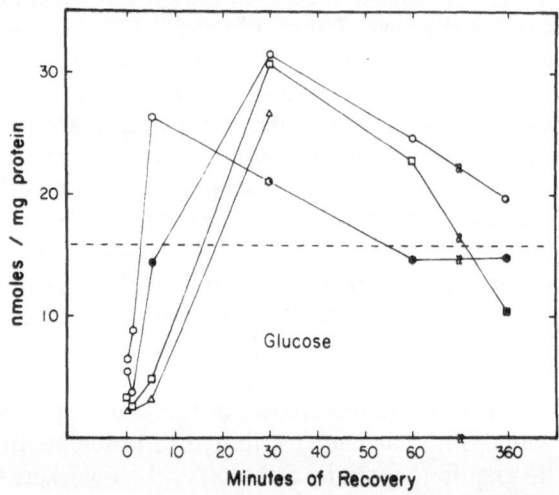

FIGURE 2. Time course of glucose restitution after 1, 5, 30, and 60 min of bilateral ischemia in the gerbil. Symbols represent 1 min (◐), 5 min (○), 30 min (□), and 60 min (△) of bilateral ischemia. For details, see Fig. 1.

of either blood flow to the region or the supply of glucose. It is presently unclear if the elevated glucose is due to hyperglycemia, since Ljunggren *et al.* (1974) reported an enhanced plasma glucose during recirculation, whereas Pulsinelli *et al.* (1982b) did not. Another possibility is that the elevated glucose reflects a diminished utilization during recirculation. In support of this, Pulsinelli *et al.* (1982b) have shown that glucose phosphorylation, as measured by the 2-deoxyglucose method, is decreased during recirculation and in another paper Pulsinelli and Duffy (1983) showed an elevation of brain glucose concurrent with the depressed glucose utilization. Thus, the elevated glucose may reflect a diminished rate of glucose metabolism or hypometabolism at various stages of recirculation depending on the ischemic insult. Although the levels of glucose remained depressed at 5 min of recirculation following both 30 and 60 min of ischemia, the production of both ATP and P-creatine at the corresponding times indicates that glucose was available to the tissue, but probably was being rapidly consumed at the onset of recirculation. It would appear that the ability to transport and metabolize glucose during recirculation is not compromised and that the elevated glucose reflects a lower demand for its metabolism imposed by the metabolic needs of the tissue during recirculation.

The fate of the elevated lactate generated during ischemia provides additional information about the recovery process. Lactate levels do not increase further during recovery, indicating that aerobic glycolysis is once again operational (Ljunggren *et al.*, 1974; Kleihues *et al.*, 1975; Kobayashi *et al.*, 1977). In addition, the rate of lactate disappearance during recirculation does not appear to be dependent on the duration of ischemia, being nearly complete by 30 min of recirculation after either 5 or 60 min of ischemia (Kobayashi *et al.*, 1977). It is presently unknown how the lactate is lost from the tissue (i.e., by washout or metabolism). However, if the accumulated lactate is metabolized, this might account, in part, for the decrease in glucose utilization.

Some other metabolites associated with glycolysis have also been measured with some representative values noted in Table II. After 3 and 7.5 min of ischemia, the levels of pyruvate rapidly rebounded to values near those of control by 0.1 min of recirculation (Ljunggren *et al.*, 1974) and then increased to values greater than control by 15 min of recirculation. If the source of the pyruvate is glucose, then the priming time to resume the glycolytic pathway is rather fast. In contrast to pyruvate, the levels of glycogen are very slow to recover (Kobayashi *et al.*, 1977). As with other types of brain injury, there is a large delayed accumulation of glycogen. For example, the levels of glycogen increased by twofold in the cerebral cortex at 1 week after 1 and 3 hr of unilateral ischemia in the gerbil (Mrsulja *et al.*, 1976a). The significance of the glycogen findings will be discussed in more detail in Section 6 on regional ischemia.

5.3.3. Tricarboxylic Acid Cycle

The normalization of the TCA cycle intermediates has been principally investigated by Siesjo and co-workers after various periods of compression ischemia (Folbergrova *et al.*, 1974; Ljunggren *et al.*, 1974; Nordstrom *et al.*, 1978). The depressed levels of both citrate and α-ketoglutarate at the end of the ischemic episode increase rapidly and are greater than control by 15 min of recirculation. Malate also exhibits a twofold increase at 5 min of recirculation after both 3 and 7.5 min of ischemia. From almost a threefold increase after 5 min of ischemia, succinate levels drop during reperfusion, but are still significantly elevated (130% of control) at 15 min. The changes in the TCA cycle intermediates during recirculation support the concept that the recovery process is not a mere reversal of the ischemia-induced events. As shown in Table II, three out of the four TCA cycle intermediates measured exhibited an overshoot during recirculation, as did both glycogen, P-creatine, and pyruvate. Thus, metabolic recovery appears to be a unique process that arises from the reintroduction of glucose and oxygen to a physiologically and biochemically quiescent brain. Although the derangements in the glycolytic and TCA cycle intermediates do not appear to be pathogenic, given the restoration of the energy status of the tissue, the changes in amino acids and cyclic nucleotides, because of their functional significance, could be.

Although it has been shown that respiration of mitochondria isolated from ischemic brains is essentially normal (Ginsberg et al., 1977), a recent report by Duckrow et al. (1981) indicates that oxidative metabolism does exhibit certain abnormalities during recirculation. Using reflection spectrometry to measure the redox state of cytochrome oxidase, the investigators noticed a metabolic

TABLE II. Additional Examples of Postischemic Changes in Metabolites

Metabolite	Ischemia[a]	Recovery[b]
Glycogen	$\downarrow$ (2, 14%)[c]	$\uparrow$ (5, 6hr, 173%)[e]
Pyruvate	$\downarrow$ (5, 8%)[d]	$\uparrow$ (5, 15, 145%)[d]
Lactate	$\uparrow$ (10, 6X)[c]	—[c]
ADP	$\uparrow$ (3, 3X)[f]	—[f]
5'AMP	$\uparrow$ (5, 24X)[e]	—[f]
Citrate	$\downarrow$ (5, 33%)[d]	$\uparrow$ (5, 15, 137%)[d]
α-Ketoglutarate	$\downarrow$ (5, 0%)[d]	$\uparrow$ (3, 5, 2.4X)[f]
Succinate	$\uparrow$ (5, 3X)[d]	—[d]
Malate	$\downarrow$ (5, 66%)[d]	$\uparrow$ (5, 15, 112%)[d]
Alanine	$\uparrow$ (5, 140%)[d]	$\uparrow$ (5, 15, 292%)[d]
Aspartate	—[d]	$\downarrow$ (5, 15, 83%)[d]
Glutamate	—[g]	$\downarrow$ (60, 5, 60%)[g]
GABA	$\uparrow$ (60, 2.5X)[g]	—[g]
Cyclic AMP	$\uparrow$ (1, 14X)[e]	$\uparrow$ (5, 5, 41X)[e]
Cyclic GMP	$\downarrow$ (10, 42%)[e]	$\uparrow$ (5, 5, 2X)[e]

[a] The first number in parenthesis is the time of ischemia in minutes and the second, the magnitude of the change either in percent of control (%) or magnitude greater than control (X). The dashed lines indicate those metabolites that do not change during ischemia. The upward arrow reflects significant increases in the particular metabolite, whereas the downward arrow represents a significant decrease.

[b] The first number in parenthesis is the length of ischemia in minutes, the second is the duration of recirculation in minutes except for hr, which represents hours, and the third is the magnitude of the change as described under ischemia. The lines denote the metabolites that are directly restored to preischemic values. The arrows are as described for ischemia.

[c] Lowry et al. (1964).

[d] Folbergrova et al. (1974).

[e] Kobayashi et al. (1977).

[f] Ljunggren et al. (1974).

[g] Mrsulja et al. (1976a).

dysfunction when the cerebral cortex was stimulated. This finding is particularly intriguing since it suggests that there may be residual deficits in the metabolic machinery induced by ischemia that are masked until the brain experiences another insult or stimulus of some type. Thus, using the available criteria, the margin of safety in the affected brain may have been reduced, even in those animals exhibiting complete recovery. This would have tremendous implications in a condition such as transient ischemic accidents where there may be multiple short-term insults. Although there has been some experimental work on repeated ischemic episodes (Mrsulja et al., 1977), further studies are needed to determine if tolerance of the brain to a second insult has been altered.

5.3.4. Amino Acids

The levels of alanine remain elevated during recirculation, being about threefold and twofold greater than control at 15 min of recirculation after 5 min of ischemia and at 90 min of recovery after 30 min of ischemia, respectively (Folbergrova et al., 1974; Nordstrom et al., 1978). GABA concentrations decrease gradually during recirculation and are only restored to control levels after about 1 hr of recirculation (Folbergrova et al., 1974; Mrsulja et al., 1976a). Although the levels of both aspartate and glutamate were unchanged during ischemia, both amino acids decreased during recirculation (Ljunggren et al., 1974; Mrsulja et al., 1976a). It is unclear how these changes occur, although they may be related to the anaplerotic reactions associated with the TCA cycle. More importantly, the latter three amino acids, as putative neurotransmitters, may have an effect on the functional recovery of the brain, as suggested by Ljunggren et al. (1974). What is particularly attractive about this possibility is that under normal circumstances the feedback of metabolism on functional activity is probably minimal; however, it is quite possible that the perturbations in neurotransmitter systems induced by ischemia could have an effect on the levels of excitability of the brain. This conceivably could be a basis for the phenomenon of selective vulnerability of certain neurons to ischemia, which will be discussed in a future section.

5.3.5. Monoaminergic Neurotransmitters and Cyclic Nucleotides

The monoamines have also been examined during recirculation (Mrsulja et al., 1976c; Gaudet et al., 1978; Cvejic et al., 1980). In the first two studies, the concentrations were determined in brains from unilaterally ligated gerbils. In spite of using the same model, there are a number of discrepancies between the results. Since the only ischemic duration that was common to both papers was 1 hr, a valid comparison of the recovery results can be made only at this interval. The elevated levels of serotonin after 1 hr of ischemia decreased slowly, but were still greater than control after 1 week of recirculation (Mrsulja et al., 1976c). In contrast, Gaudet et al. (1978) demonstrated a significant decrease in serotonin in both the ipsilateral and contralateral hemisphere after 1 hr of ischemia that persisted for 30 min of recirculation (the longest period of recirculation sampled in this study). The reasons for the differences are presently unclear.

The results for norepinephrine and dopamine are equally difficult to interpret. In the study by Gaudet et al. (1978), the levels of both neurotransmitters in the ipsilateral hemisphere were not significantly different from control at 30 min of reflow, whereas in the study by Mrsulja et al. (1976c), the levels of dopamine decreased to approximately 30% of control at 5 hr of recirculation. In addition, norepinephrine increased almost twofold at 1 hr of recirculation and thereafter decreased to 50% of control at 5 hr. Given the information provided by these papers, it is difficult to account for the observed differences unless the experimental procedures (i.e., depth of anesthesia, assay method, or time of sampling, etc.) are responsible. In that context, Cvejic et al. (1980) demonstrated a diurnal oscillation in the levels of both dopamine and norepinephrine during recirculation and this may be a factor in these somewhat confusing findings.

The recovery of the cyclic nucleotides after ischemia are particularly interesting for a number of reasons. Since a relationship between the steady-state levels of neurotransmitters and functional activity of the system has not been established, the measurement of the steady-state levels of neurotransmitters may be misleading. The cyclic nucleotides, in contrast, are neuroeffectors that mediate the actions of a number of neurotransmitters (for review, see Phillis, 1977). Therefore, the changes in the levels of cyclic nucleotides may reflect altered functional states within the CNS. Additional evidence for this comes from the work of Stone et al. (1975), who demonstrated that the cyclic nucleotides have a physiological effect on excitable cells; i.e., the iontophoretic application of cyclic AMP inhibited the firing rate of pyramidal tract neurons of the cerebral cortex, whereas cyclic GMP was excitatory. The cyclic nucleotides have also been shown to be potent activators of protein kinases which in turn have a pronounced influence on both metabolic and physiological processes within the brain (Greengard, 1981).

The concentrations of cyclic AMP, which were elevated during unilateral ischemia, increased even further during recirculation (Mrsulja et al., 1976a). Subsequent studies using bilateral ischemia have shown that the magnitude of the response was greater after longer periods of ischemia and was maximal at about 5 min of recirculation, after which the levels decreased toward those of control (Fig. 3). Since a number of adenylate cyclase agonists are known to be released during ischemia when the levels of ATP are essentially depleted, a unique situation occurs at the onset of recirculation; there is an increased pool of agonists poised at the receptor but no substrate available to the enzyme. The restoration of oxygen and glucose results in an immediate rise in ATP concentration that triggers a burst of cyclic AMP accumulation. A similar cyclic AMP increase can be mimicked in brain slices by adding oxygen and glucose to a medium previously devoid of oxygen and glucose (Lust et al., 1982). The characteristics of the response are quite similar to those observed in vivo. Using the brain slice model, attempts to identify the agonist(s) that mediates the postischemic rise in cyclic AMP have failed. For example, adenosine, an activator of adenylate cyclase (for review, see Daly, 1976), has been shown to increase dramatically during ischemia (Nordstrom et al., 1978). Therefore, adenosine seemed like a plausible mediator for the cyclc AMP response; however, the postischemic rise in cyclic AMP was not inhibited in the presence of theophylline, a potent adenosine antagonist. At present, other agonists including serotonin and the catecholamines have also been excluded and the mechanism of this phenomenon remains unclear. Other investigators have examined the activity and hormone responsiveness of the adenylate cylcase both during and after ischemia (Schwartz *et al.*, 1976; Taylor *et al.*, 1982) and have not found changes that would explain the large accumulation of cyclic AMP. Given the physiological potency of cylic AMP and also the nature of the postischemic response (i.e., the magnitude as well as the time course), it would appear that the modulation of its accumulation during recirculation would offer an interesting insight into the recovery process.

Cyclic GMP also increases during the recovery period (Kobayashi *et al.*, 1977), but the peak increase is only about sixfold greater than control. There is some indication that the magnitude of the increase is greater after longer periods of ischemia, as is the case with cyclic AMP. The metabolism of cyclic GMP has not been studied as extensively as that for cyclic AMP; however, it is known that the accumulation of cyclic GMP is closely associated with calcium-dependent events (Ferrendelli *et al.*, 1976). The results for cyclic GMP are, therefore, of some interest in light of recent suggestions that calcium is involved in the manifestation of brain injury (Siesjo, 1981).

The changes in the cyclic nucleotides during recirculation could be interpreted as being pathogenic; in the case of cyclic AMP as a result of pertur-

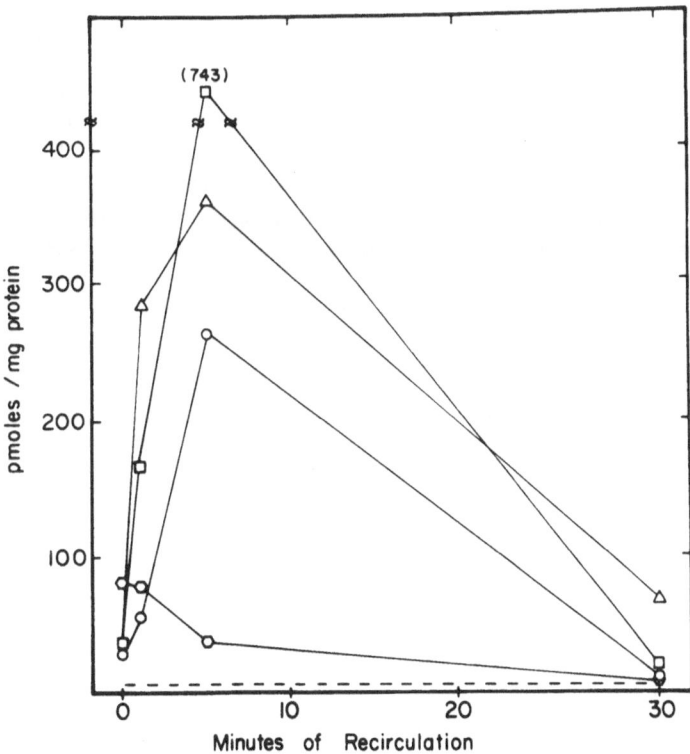

FIGURE 3. Postischemic accumulation of cyclic AMP after 1, 5, 30, and 60 min of bilateral ischemia in the gerbil. The symbols represent the following periods of bilateral ischemia: 1 min (○), 5 min (○), 30 min (□), and 60 min (△). The animals were prepared and the assays performed as described by Kobayashi *et al.* (1977). Each value represents the means of four or more determinations.

bations in the adenylate cyclase agonists and ATP, and in the case of cyclic GMP due to derangements in calcium metabolism and GTP. The normal relationship between the cyclic nucleotides and synaptic transmission would have been short-circuited in such a way that the cyclic nucleotides generated were not due to synaptic activity, but rather either to a spontaneous release of a neurotransmitter during ischemia or to the liberation of a cyclase agonist induced by ischemia. Thus, the elevated cyclic AMP, generated in the absence of synaptic activity, might have an inhibitory effect on neuronal excitability during the early stages of recirculation. It is interesting to note that spontaneous brain activity is not evident during the early stages of recovery when cyclic AMP is elevated. Thus, metabolic perturbations may have a marked effect on brain function by affecting the levels of metabolites common to both systems. For this reason, further studies on the cyclic nucleotides as well as the amino acid neurotransmitters should be pursued.

5.3.6. Summary of the Recovery Process

The list of changes that occur on reflow is long and it is hoped that the nature of these changes convinces the reader that the recovery process is far more complex than a simple reversal of the events induced by ischemia (Table II). It is unlikely that the brain has within its homeostatic mechanisms a plan to deal with extreme energy deprivation. What apparently happens is that the brain reaches a lower state of order that is readily reversible once the metabolic processes are restored. If the ischemic episode is extended, then reversibility is lost. From the results described above, it would appear the activity of the various metabolic pathways during recirculation remains operable, in spite of the abnormal intracellular environment at the onset of recirculation. The way the metabolites are restored is quite varied, from the gradual decline of such metabolites as lactate and GABA to an additional accumulation of cyclic AMP. There are also those metabolites that are unaffected by ischemia, but which change during recirculation. It is proposed that those metabolite changes that do not take a direct path to recovery are unique to the recovery process and are potentially pathogenic. The possibility that a perturbed metabolic status could have an effect on function also exists. The measurement of metabolites during recovery has been useful in assessing the extent of injury induced by ischemia. The rate of restoration of many metabolites decrease with increasing periods of ischemia. In addition, there are many events that occur during recirculation (i.e., overshoots in glucose and P-creatine and the rise in cyclic AMP) that are greater with increasing durations of ischemia. By most criteria, the recovery of metabolism in the gerbil model appears to be complete by 1 hr after up to 30 min of bilateral ischemia. However, the long-term effects of ischemia on metabolic restoration examined in various regions of the brain will clearly demonstrate that the recovery process continues for a much longer time than previously thought.

6. Regional Response to Ischemia

Most of the information on the biochemical response to ischemia presented in the earlier parts of this chapter were either performed on extracts from the cerebral cortex or from the entire hemisphere. It is becoming increasingly evident that such an approach, while being convenient, ignores a very important lesson taught by the neuropathologist. Histological techniques have been used for years to assess the extent of cell damage induced by ischemia, but the literature on the subject is far too extensive to be presented here. A major problem, at least to the neurochemist, has been that histological evidence of injury requires hours if not days before an effect of ischemia is manifested and then the response in a given region may vary even though the duration of ischemia was the same. The latency in the structural changes of the affected cells and the absence of predictability are just two reasons why an integrated approach between the neuropathologist and neurochemist has been lacking. However, in the last five years, the phenomenon of selective vulnerability, as originally demonstrated histologically, has found a place in the experimental paradigm of the neurochemist. There are clearly regional differences in the response of neurons to ischemia; certain populations of neurons in the cerebral cortex, hippocampus, striatum, and cerebellum have been shown to be more susceptible to an ischemic episode than others (Brierley *et al.*, 1973). A quote from that paper written more than ten years ago deserves repeating since it gives the basis for a currently popular trend in research on ischemia:

> No hypothesis purporting to account for the neuropathology of any type of hypoxia merits serious consideration unless it can offer a reasonable explanation for the pattern of selective vulnerability. (Brierley *et al.*, 1973)

Now, instead of sampling the brain at random, only those areas that show the greatest susceptibility to ischemia are being examined. Such an approach should add some insight into the much larger question of what causes brain dysfunction: are the neurological deficits induced by ischemia a result of localized loss of function or are they due to a uniform deficit throughout the brain? If the former is true, then much of the biochemical information on ischemia and recovery may, in fact, be misleading.

6.1. Biochemical Models of Selective Vulnerability

Several laboratories have examined the selective vulnerability of certain neurons to ischemia (Brierley *et al.*, 1973; Diemer and Siemkowicz, 1980), but in this review we will only consider those studies that have evolved into neurochemical investigations. Using the four-vessel occlusion method in rats, Pulsinelli *et al.* (1982a) examined the time course of neuronal damage after varying periods of ischemia. Of the regions examined, the hippocampus and striatum exhibited the greatest amount of damage. The severity of the injury was most pronounced in the hippocampus where most CA 1 neurons of the hippocampus were damaged in 85% of the rats examined after 20 min of ischemia. The threshold for ischemic injury was somewhat longer for the striatum than for the hippocampus, requiring at least 30 min of ischemia before neuronal damage was evident. Although the thalamus, neocortex, and cerebellum also exhibited moderate injury, the response was less severe and more random than that for the striatum and hippocampus. The results indicate that there is a regional progression of injury with increasing periods of ischemia. Considering both the high degree of reproducibility and the severity of the response, certain regions such as the hippocampus and striatum lend themselves to neurochemical investigations of events that lead up to the manifestation of neuronal damage. The differential time course in the development of cellular injury for the various regions should also be useful in determining if a common biochemical lesion is responsible for the cell damage.

Another breakthrough came from the observation of Kirino (1982), who demonstrated the loss of CA 1 neurons of the hippocampus four days after 5 min of bilateral ischemia in the gerbil (Fig. 4). Although there were other changes in the hippocampus, the nearly complete loss of the CA 1 neurons in the anterior portion of the hippocampus provides an ideal model for the neurochemical investigation of both ischemia-induced cell death and the phenomenon of selective vulnerability.

6.2. Physiological Response and Blood Flow

Many of the changes observed in the rat and gerbil models are similar. In the hippocampus, there is an initial hyperemia following the onset of circulation followed by a period of hypoperfusion (Pulsinelli *et al.*, 1982b; Suzuki *et al.*, 1983a). The duration of the hypoperfusion was not evident in either model at 6 hr of recirculation. This period of low blood flow may be critical to the tissue, since at this time the reduced supply of nutrients from the blood could effect a secondary insult to the tissue.

Another advantage to the protocol used for the studies on selective vulnerability is that the animals survive the ischemic insult. Although the rats appear to survive for at least 3 days after 30 min of four-vessel occlusion, specifics on the mortality rate were not given (Pulsinelli and Brierley, 1979). However, the animals exhibited little or no spontaneous movement during the first 6 hr of recirculation following 30 min of ischemia and normal motor ac-

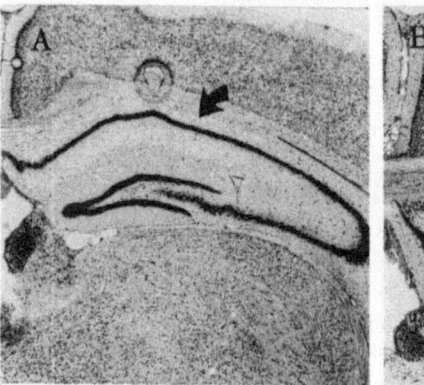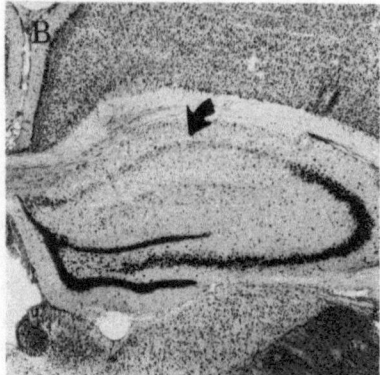

FIGURE 4. Cross section of the hippocampus in control (A) and experimental (B) gerbils at four days after 5 min of bilateral ischemia. As designated by the arrows, the pyramidal neurons in the CA 1 region of hippocampus have essentially disappeared by the fourth day after 5 min of bilateral ischemia. The gerbils were anesthetized with 45 mg/kg (IP) of thiamylal and the common carotid arteries were exposed and looped. On the following day, the arteries were occluded for 5 min with aneurysm clips. On the fourth day, the animals were frozen *in situ* with liquid nitrogen and the brain sectioned at a thickness of 20 μm. The hippocampal sections were then stained with thionin.

tivity was never fully restored. After 20 min of ischemia, a time that resulted in hippocampal damage, the animals had little or no evidence of neurological deficits. In the gerbil model, the mortality rate during recirculation was quite low and behavioral evidence of long-lasting neurological damage was lacking. In our laboratory, gerbils have survived up to six months after 5 min of bilateral ischemia with no detectable alterations in behavior being noted.

Suzuki *et al.* (1983b) have also examined spontaneous neuronal activity in the cerebral cortex and hippocampus at various stages of recirculation following 5 min of bilateral ischemia. As would be expected, detectable electrical activity ceased in the CA 1 region of the hippocampus and the cerebral cortex during ischemia. During recirculation, spontaneous activity of the neurons occurred after about 20 min. The firing rate of the CA 1 neurons was increased during the first day of recirculation, whereas it was essentially normal in the cerebral cortex. By the second day, the activity of the CA 1 neurons was isoelectric. Therefore, ischemia had an effect on the activity of the CA 1 neurons not observed in the cerebral cortex. The phenomenon of selective vulnerability is thus clearly demonstrable by electrophysiology as well as histology. In addition, these findings provide neurophysiological correlates for the biochemical changes that will be described below.

6.3. Regional Metabolic Changes during Ischemia

Based on measurements made during an ischemic insult, the regional changes in metabolites appear to be quite uniform (Abel and McCandless, 1982), although some regional differences have been noted. For example, the cyclic AMP levels increased to a greater extent in the septum and hippocampus than other regions examined (Arai *et al.*, 1981). However, the time courses of the regional metabolite alterations during ischemia have not been rigorously studied and it is quite possible that the rates at which the metabolites change may differ. The decision to emphasize the recovery period stems from the inability to detect a pathogenic event during ischemia that would account for the onset of cellular damage.

6.4. Glucose Utilization

The 2-deoxyglucose method has been applied to both models of selective vulnerability and it provides some interesting informaton on the status of metabolism during recirculation. In the four-vessel occlusion model in the rat, glucose metabolism was reduced at 1 hr of recirculation in all regions except the hippocampus, and then either remained depressed for up to two days as in the cerebral cortex or was gradually restored in that time period (white matter). A decrease in glucose metabolism in the hippocampus was only evident either after 6 hr of recirculation or when histological damage appeared (Pulsinelli *et al.*, 1982b). In the gerbil model, Suzuki *et al.* (1983a) found an initial elevation in glucose metabolism at 10 min of recirculation following 5 min of bilateral ischemia. Glucose metabolism was apparently normal by 6 hr, then was quite variable up to two days and thereafter was depressed in the hippocampus. The precise quantitation of glucose metabolism is somewhat doubtful since ischemic recovery appears to be a dynamic process with many phases. However, the dramatic changes observed in both models clearly indicates that hippocampal metabolism is markedly altered compared with other regions between the onset of recirculation and cell death.

6.5. Regional Metabolites

Several laboratories have examined the levels of various metabolites prior to and at the onset of cellular damage in the hippocampus (Pulsinelli and Duffy, 1983; Arai et al., 1982). In spite of the different models used, the findings are surprisingly similar. Although the paper by Pulsinelli and Duffy (1983) examined the cerebral cortex, hippocampus, and striatum in rats, the discussion will focus on the former two regions since they are also common to the gerbil study. In general, the response in the striatum was essentially similar to that of the other affected region, the hippocampus. In the study on gerbils, the animals were frozen in situ with liquid nitrogen and the brain was then sectioned. After freeze drying the tissue, the CA 1 and CA 3 regions of the hippocampus and the cerebral cortex were free-hand dissected and the samples (< 2 µg) were weighed on a quartz-fiber balance. Metabolites were measured by quantitative histochemistry (Lowry and Passonneau, 1972). In the four-vessel occlusion model, the entire CA 1 zone was sampled, in contrast to the somal layers in the gerbil study.

6.5.1. Energy Metabolites

The concentrations of ATP and P-creatine were maintained at or near control until morphological changes were observed in the CA 1 region at which time there was a significant decrease in both metabolites (Pulsinelli and Duffy, 1983; Arai et al., 1982). There was a depression of ATP by as much as 20% during recirculation in the rat, but this occurred in both regions examined and is typical of previous results demonstrating that complete recovery of ATP is quite slow. There was also a depression in ATP levels in the gerbil; however, the effect was not consistent. In the rat model, there was also an overshoot in the levels of P-creatine during recirculation that was apparent in the hippocampus at 1 and 3 hr, whereas in the cerebral cortex it was evident up to two days of recirculation. A similar response was observed in the gerbil model. The high-energy phosphates decreased by 30 and 38% in the hippocampus of the rat and gerbil, respectively, at a time when cell death was marked. In the gerbil model, the energy charge was examined in the CA 1 subfield at a time when the high-energy phosphates were depressed. The calculated ratio was within 5% of control, indicating that the energy status of the surviving cells was essentially normal. From these data, it would appear that energy metabolism in the affected region is quite like that of the unaffected areas, at least until neuronal death. There does not appear to be a secondary anoxic/ischemic insult during recirculation; the loss of high-energy phosphates in the presence of a normal energy charge probably reflects the loss of the neurons and not an anoxic episode. In spite of the pronounced hypoperfusion at certain stages of recirculation, the brain is able to maintain its energy status. The concept that there is a period where energy production cannot meet energy demand does not appear to be valid in this instance.

6.5.2. Glycolysis

In both studies, the recovery period was generally characterized by a significant increase in glucose in the regions examined. However, in the gerbil model, the glucose levels subsequently decreased to or below control values at 6 hr of recirculation in both the hippocampus and cerebral cortex. The significance of the initial glucose overshoot is presently unclear, but may reflect a period of hypometabolism. The lactate levels were not noticeably different from control during recirculation, except in the cerebral cortex of the gerbil where the levels were significantly depressed from 1.5 to 12 hr of recirculation (Fig. 5). When cell damage becomes evident, the lactate levels are increased in the hippocampus of both the rat and gerbil, which may reflect the predom-

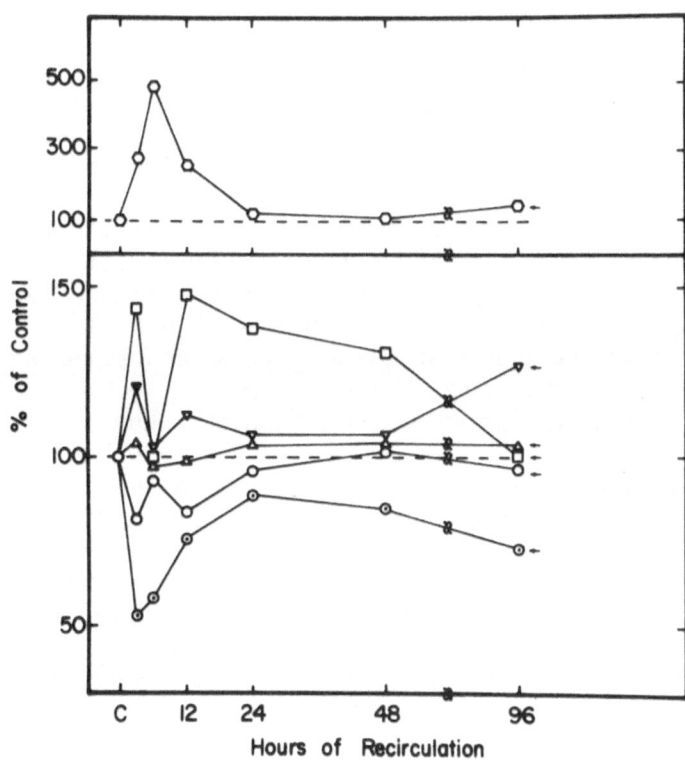

FIGURE 5. Delayed changes of metabolites in the cerebral cortex following 5 min of bilateral ischemia in the gerbil. The tissues were prepared as described in Fig. 4 except the sections were freeze dried at − 40 °C and 1-μg pieces of the dried cerebral cortex were freehand dissected and weighed on a quartz-fiber balance. The high-energy phosphates were measured by the luciferin-luciferase technique (Lust *et al.*, 1981) and the remaining metabolites by enzymatic cycling technique described by Lowry and Passonneau (1972). The experimental values are presented as percent of those from sham-operated control and represent triplicate determinations from at least four animals. Glycogen (○), PCr (▽), ATP (△), glucose (□), glutamate (C), and lactate (◉).

inance of glial metabolism at that time. In the gerbil model, the elevation of glucose and depression of lactate in the cerebral cortex is consistent with diminished metabolism, which was also indicated by the 2-deoxyglucose method in rats. Generally, the metabolite changes described thus far indicate that there are delayed metabolic perturbations following ischemia that are not limited solely to the susceptible areas. It would not appear that secondary energy failure is responsible for neuronal death in the hippocampus and that the observed changes in energy status undoubtedly are an effect of cell death.

6.5.3. Glycogen

Brain glycogen, the storage form of glucose, has been shown to accumulate in large quantities following trauma to the brain (Miquel and Haymaker, 1965; Clendenon et al., 1971; Watanabe and Passonneau, 1974; Mrsulja et al., 1976b). Glycogen concentrations increase during recirculation, as shown in Fig. 5, but in contrast to previous reports the response is apparently biphasic. There is an initial large accumulation of glycogen that is maximal at 6 hr of recirculation in all three regions examined. The graded response (cortex > CA 3 > CA1) in the accumulation of glycogen may reflect a differential need for glucose metabolism in the three regions (i.e., the increased excitability of the CA 1 neurons at 6 hr of recirculation in turn increases the energy demands of the tissue which partially inhibits the accumulation of glycogen). Subsequently, the levels of glycogen decrease to those of control in all areas. There is a second increase in the CA 1 region that is apparently due to glial infiltration concomitant with the loss of the neurons. The activities of glycogen synthase and phosphorylase have been examined in the three regions (unpublished observation). The activity of the "on" form of the enzymes did change, but not in the direction needed to explain the increase of glycogen. Thus, the mechanism leading to the accumulation of glycogen remains a mystery.

The changes in glycogen are a good example of another aspect of ischemia not yet considered in this review. Although glycogen has been demonstrated in neurons (Cammermeyer and Fenton, 1982; Passonneau and Lowry, 1971), most of the glycogen has been found in glia (Klatzo et al., 1970). Attempts to demonstrate the presence of glycogen in neurons of the gerbil brain during recirculation using the PAS technique have generally failed (I. Klatzo, personal communication). Thus, unlike the other metabolite measurements, the localization of glycogen to a particular compartment or cell type broadens the emphasis to include the supporting role of glia. The results suggest a close metabolic interaction between the two cell types. For example, the increased excitability of the CA 1 cells at 6 hr of recirculation results in a diminished glial accumulation of glycogen in the CA 1 region as compared with the cerebral cortex. The role that glia play during and after ischemia has generally been ignored since there are no simple experimental approaches to the problem. The changes in glycogen, ostensibly in the glia, could reflect derangement in glial metabolism, but also could be an indirect effect due to altered neuronal function.

The time course of ischemia-induced generation of glycogen has previously been shown to be consistent with the "maturation" phenomenon (Mrsulja et rate of maturation of a pathophysiological event is inversely related to the duration of ischemia (i.e., the onset of pathophysiology is delayed after shorter periods of ischemia). Of course, this only applies once the threshold for a given event has occurred. The biosynthesis and degradation of glycogen are exquisitely regulated by a set of enzymes which in turn are controlled by effectors such as cyclic AMP and calcium. Since glycogen levels apparently follow this relationship, then glycogen metabolism should provide a good model system whereby the pathomechanisms involved in the "maturation" phenomenon could be examined.

6.5.4. Other Metabolites

The cyclic nucleotides as well as the amino acids, GABA, and glutamate have also been measured in the gerbil model. The changes were generally characterized by a similarity between regions, although there were some differences including a significant increase in cyclic GMP in the CA 1 region at 6 hr of recovery. More importantly, the changes indicate that the recovery process extends for a much longer period than previously thought. This opens up the possibility that novel approaches to the treatment of stroke can be administered hours, if not days, after the ischemic event and which could either improve the likelihood of survival or minimize the neurological deficits arising from the stroke.

The measurements of the amino acids indicate that neurotransmitter systems may be affected for extended periods following ischemia. One possible explanation for selective vulnerability of certain neuronal populations is that the neurotransmitter system specific to that group of neurons is perturbed (i.e., synthesis, degradation, reuptake, binding, etc). For the CA 1 neurons that are thought to receive glutaminergic inputs, the significant changes in glutamate during recirculation may offer a clue to the biochemical lesion responsible for cell death. In addition, the changes in glutamate observed in other regions may be benign since glutamate may not be the neurotransmitter for that population of neurons. If the measured changes in glutamate do not occur in the functional pool, then such a proposal has little validity. The apparent susceptibility of the hippocampal neurons to kainic acid, a glutamate analogue, as well as to glutamate itself, strengthens the possibility that selective vulnerability may, in part, rest with the neurotransmitter system specific to a neuronal type.

6.5.5. Summary of Regional Studies

The predictable loss of CA 1 neurons of the hippocampus has proved to be an ideal model for the biochemical investigations of the events that lead up to cell death induced by ischemia. Although there are some differences between the responses of the affected and unaffected regions, many of the changes are common to all areas examined. The delayed metabolic changes observed after relatively short periods of ischemia are a good indication that recovery proceeds for hours, if not days, after the insult. This finding increases the likelihood that posttraumatic therapy of stroke may be of some value and research to this end is warranted. In a supplement to the *Journal of Cerebral Blood Flow and Metabolism* (1982) entitled "Protection of the Brain from Hypoxia," both experimental and accepted methods of stroke treatment are described and it is clear that therapy in the context of selective vulnerability is minimal.

Another finding common to both the rat and gerbil models is that the loss of high-energy phosphates does not precede the death of the neurons. Thus, energy failure does not appear to be the cause of CA 1 neuronal loss, although it remains a possibility that the metabolism of dendrites and/or presynaptic terminals, areas not specifically examined in these studies, could have been affected. Ekstrom von Lubitz and Diemer (1982) have shown early morphological changes in the stratum radiatum of the hippocampus, a region rich in dendrites and nerve terminals. Perhaps, the site of selective vulnerability is to be found within specific cell processes.

Regional studies have also created as many new questions as they have answered. One distinct possibility is that the vulnerability of certain neurons is responsible for the less severe neurological dysfunction following an ischemic episode. As an ischemic episode progresses, less susceptible populations of neurons are affected and there is an enhanced deficit of neurological function. In the case of the gerbil model, the CA 1 neurons are lost by four days after 5 min of ischemia and yet the behavior of the animals appears normal. Perhaps the threshold necessary for the onset of neurological dysfunction requires the loss of CA 1 neurons plus an as yet unidentified population of neurons that are affected only by longer periods of bilateral ischemia. In the rat model, neurological deficits occurred only after 30 min when the striatum as well as the hippocampus exhibited morphological damage. Since most of the metabolite measurements following ischemia have historically been performed on samples from the cerebral cortex, it is quite possible that the results from this region of intermediate susceptibility could account for the lack of correlation between metabolic perturbations and neurological dysfunction.

7. Miscellaneous

7.1. Edema

The accumulation of water in tissue is a common occurrence in ischemia and recirculation. The extent of edema is obviously a confounding factor in the recovery of function following an ischemic episode. Brain edema has been extensively investigated for many years, primarily because it is a common sequelae to a number of neurological diseases. It would be impossible to adequately survey the literature here, but for interested readers the volume by Cervos-Navarro and Ferszt (1980) will provide information on the pathology, diagnosis, and therapy of brain edema.

The edema process is quite complex and the underlying pathomechanisms involved are a subject of some controversy. However, since much of this review focused on biochemical changes both during and after ischemia, it is rather interesting to find that perturbations in serotonin and cyclic AMP described above have been implicated in the development of edema (Mrsulja *et al.*, 1980).

In another study the rate of brain swelling after 15 min of bilateral ischemia was maximal during the first hour of recirculation when the total brain content of potassium was between 61 and 83% of control values and the activity of (Na^+, K^+)-ATPase was also less than 50% of control (Djuricic et al., 1984). The relationship between metabolite changes and the development of edema is very promising; however, further investigations are needed.

7.2. Treatment Affecting the Outcome of Ischemia

No review on ischemia would be complete without a section on the experimental approaches that have been tried to protect the brain from the effects of ischemia. The literature is replete with studies on the effects of various agents during ischemia; however, the results are often confusing. The problems that arise in establishing the efficacy of an agent in the treatment of stroke are illustrated by the results obtained with barbiturates. Early studies on the effects of barbiturates demonstrated that treated mice can survive anoxia for a considerably longer period of time than can control mice (Arnfred and Secher, 1962; Wilhjelm and Arnfred, 1965). Subsequent investigations have not always supported the efficacy of the barbiturates. Michenfelder and Milde (1975) found no effect of pentobarbital in gerbils, dogs, and cats, but the metabolic alterations of ischemia were significantly reduced in the squirrel monkey. The effect on monkeys was confirmed by Moseley et al. (1975). In contrast, Smith et al. (1974) were able to demonstrate a protective effect in dogs. The reason for the discrepancy in the canine studies may rest with the dosages used and time of administration or may be related to the different surgical procedures employed. In another study on primates, Hoff et al. (1975) showed that relatively high doses of pentobarbital were required for a protective effect. Although the preponderance of evidence suggests a protective effect of barbiturates, further studies are needed to clarify the conflicting reports. How the barbiturates affect the ischemic process is presently unclear, but may be related to the vasoconstrictor activity of pentobarbital (Hanson et al., 1975), to the ability of barbiturates to reduce the metabolic rate of the brain (Folbergrova et al., 1970), or to the partial reduction in the ischemia-induced increase in free fatty acids by pentobarbital (Nemoto et al., 1982).

The difficulty in assessing the value of an agent on ischemia is, in part, due to (1) the different models used, (2) the criteria selected to assess ischemic damage, (3) the use of irreversible insults, and (4) the varying times of drug administration. To date, the models of ischemia as well as the species used are probably as great, if not greater, than the number of agents that have been tested. The list of species includes mice, rats, gerbils, cats, dogs, monkeys, baboons, and humans. The methods for the prevention of blood flow are also as varied, including decapitation, unilateral and bilateral carotid ligation, middle cerebral artery occlusion, compression ischemia, and a combination of internal carotid artery and MCA occlusion. Another problem is that the parameters used to assess the extent of brain injury vary from the traditional morphological examination of the tissue, which is difficult to quantitate, to the measurement of free fatty acids in the decapitated rat brain which has not proven to be a mediator of cell damage during an ischemic episode. It is probably safe to say that no agent, to date, has been tested in the same species with the same method of occlusion and criteria of brain injury by two different laboratories. Another drawback is that most of the studies have been performed on irreversible brain ischemia, which most investigators would concede results in the death of the tissue no matter what the treatment. The survival of the tissue adjacent to the infarct, however, may be influenced by the appropriate treatment. A major problem with this type of model is that the extent of the ischemic insult cannot be readily assessed. Finally, in the search for an agent effective against stroke, it is important to define the goal of the treatment. In certain cases when a stroke is anticipated for whatever reason, treatment prior to the insult is possible and drugs should be tested for this purpose. However, of equal importance is the treatment of the stroke, either in progress or during reflow, to minimize the neurological sequelae. In light of recent evidence suggesting that the recovery period extends for hours if not days after an insult, a greater emphasis should be placed on the development of novel therapeutic approaches once the insult has occurred. The previously described temporal changes in metabolites will provide a time frame for the administration of the experimental agents, and the metabolic systems perturbed during recovery will provide a clue as to the type of agent to be used. A systematic investigation of available drugs in the treatment of stroke is needed, but it will not be a simple task considering all the variables that will be encountered.

Wauquier (1982) describes current methods for protecting the brain from hypoxia. Included in this report are some exciting approaches to the treatment of stroke including the use of a calcium channel blocker, flunarizine, and a nonbarbiturate hypnotic, etomidate. In addition, efforts to improve the hemodynamic properties both during and after an ischemic insult continue (Hossmann, 1982). The investigation of these agents not only will be useful in the treatment of stroke, but also should provide interesting information on the stroke process.

7.3. Brain Injury Induced by Ischemia: Is There a Culprit?

Over the years, there have been a number of theories regarding the biochemical basis for ischemic brain damage. When considering the etiology of brain injury, it is important to distinguish between the damage resulting from a continual deprivation of blood flow and the damage developing during reflow. In this section, we are concerned with the injury developing during recirculation, because in the absence of reflow there is no hope for recovery. Since energy production is a necessary first step in the functional recovery of the tissue, it was postulated that postischemic brain damage resulted from an impaired ability to generate ATP. Subsequent investigations indicated that the energy charge recovered even in tissues which eventually demonstrated histological evidence of injury (Siesjo, 1978). It has also been suggested by Myers and Yamaguchi (1976) that lactate accumulation during ischemia may be harmful to the brain. However, more recent investigations suggest that lactate may contribute to, but is not the sole cause of, brain damage (Rehncrona et al., 1981). It has also been proposed that free radical formation mediates the brain damage observed following ischemia (Siesjo, 1978). Unfortunately, the enhanced accumulation of free radicals has not been demonstrated *in vivo*, and subsequent studies with free radical scavengers would tend to minimize the role of free radicals in the evolution of brain injury. A possible abnormality in calcium homeostasis is the theory that is currently in vogue. This is a most attractive hypothesis given the importance of calcium to a variety of brain functions. Siesjo (1981) has described how a perturbed calcium metabolism could have a deleterious effect on brain tissue. One major problem with this proposal is that hard evidence for derangements in calcium metabolism is lacking. Recently, however, Hossmann *et al.* (1983) demonstrated that calcium levels in cat brain at 3 hr of recirculation after 1 hr of ischemia increased by 35%. Obviously, more studies of this type are needed to give credence to the calcium hypothesis.

One of the major drawbacks to all the theories presented is the existence of selective vulnerability of certain neuronal populations to ischemia. A preponderance of the neurons exhibit minimal damage even after periods of ischemia that result in marked neurological deficits. And yet, most if not all of the cells would have been exposed to similar conditions; i.e., increases in lactate, free radicals, and calcium or decreases in high-energy phosphates. Thus, these changes alone do not appear to be responsible for the onset of brain damage. If, however, the selectively vulnerable neurons have unique characteristics, it may well be that the combination of the metabolite changes along with the inherent properties of a particular cell could account for the greater susceptibility to ischemia. For instance, there is a relatively high calcium conductance in the dendrites of the CA 1 pyramidal neurons of the hippocampus. The synapses in this region appear to be primarily glutaminergic and the glutamate levels in the dendritic layer do change during postischemia. Thus, CA 1 pyramidal neurons may suffer a synergistic insult generated by the metabolic perturbation and the specific biophysical characteristics of the dendritic membrane (i.e., glutamate receptors and high calcium conductance). Future investigators undoubtedly will resolve these issues, but it will not be too surprising to find some validity to all the hypotheses set forth and that, in reality, there is no single culprit.

8. Summary

Stroke is one of the leading causes of death in the United States today and is therefore a major public health concern. In recent years, the number of stroke deaths has been decreasing, while the number of stroke survivors is steadily increasing. Only through a better understanding of the ischemic process will treatments be developed that either will reduce the deleterious effects of an impending stroke or will minimize the neurological deficits arising after stroke. Although there are other confounding factors such as edema or altered blood flow, the metabolic response of the tissue to an ischemic episode would appear to be the primary concern.

The biochemical changes that occur both during and after ischemia have been described. Ischemia is characterized by a rapid loss of high-energy metabolites and a somewhat slower transition of the metabolic pathways to an equilibrium state. To date, no pathogenic event has been described during ischemia that would account for the loss of brain function during reflow. The reintroduction of oxygen and glucose to ischemic tissue triggers a host of responses. From the overall profile of metabolites during recirculation, it would appear that the recovery process is more than a mere reversal of the ischemia-induced events. There are changes in metabolite concentrations specific to the reflow period (i.e., a large secondary rise in cyclic AMP, a decrease in glutamate and aspartate, and an increase in cyclic GMP). In addition, the rates of restoration of ATP, P-creatine, and glucose decrease with increasing periods of ischemia. Recent work indicates that the recovery process may extend for days after an ischemic episode. Collectively, the results indicate that the recovery of ischemic tissue is indeed a unique process. It is also quite possible that the pathogenic events that give rise to irreversible brain damage occur during recirculation.

The selective vulnerability of neurons to ischemia was discussed in some detail. The loss of CA 1 neurons of the hippocampus days after the ischemic episode has provided an excellent model for the investigation of the biochemical events that precede cell death. Although a specific biochemical lesion has not been identified, it is evident that these cells, which are destined to die, initially recover and maintain their energy status despite a period of hypoperfusion during recirculation. Further investigations of the response of the selectively vulnerable areas of the brain should provide an interesting insight into the loss of neural function, as well as being a basis for developing new approaches to the treatment of stroke.

This chapter is intended only as an overview of ischemic encephalopathy and many of the topics have not been covered in detail. it is hoped that the bibliography, although certainly not complete, may serve as a starting point for more extensive literature searches. The authors suggest that the interested reader will find the monographs, *Brain Energy Metabolism* (Siesjo, 1978) and *Cerebral Metabolism and Neural Function* (Passonneau et al., 1980) to be quite useful. In addition, the journals, *Stroke* and *Journal of Cerebral Blood Flow and Metabolism*, are particularly good sources of current information on ischemia and related subjects.

ACKNOWLEDGMENT. The authors gratefully acknowledge the editorial assistance of Mrs. Lois V. Trigg.

References

Abel, M. S., and McCandless, D. W., 1982, Metabolic profile of hippocampal regions after bilateral ischemia and recovery, *Neurochem. Res.* **7**:789–797.

Ames, A., Wright, R. L., Kowada, M., Thurston, J. M., and Majno, G., 1968, Cerebral ischemia. II. The no-reflow phenomenon, *Am. J. Pathol.* **52**:437–453.

Arai, H., Lust, W. D., and Passonneau, J. V., 1981, Adenylate metabolism in regions of the gerbil brain after ischemia, *Trans. Am. Soc. Neurochem.* **12**:229.

Arai, H., Lust, W. D., and Passonneau, J. V., 1982, Delayed metabolic changes induced by 5 min of ischemia in gerbil brain, *Trans. Am. Soc. Neurochem.* **13**:177.

Arnfred, I., and Secher, O., 1962, Anoxia and barbiturates, *Arch. Int. Pharmacodyn.* **139**:67–74.

Astrup, J., Symon, L., Branstron, N. M., and Lassen, N. A., 1977, Cortical evoked potential and extracellular K^+ and H^+ at critical levels of brain ischemia, *Stroke* **8**:51–57.

Aveldano, M. I., and Bazan, N. G., 1975, Differential lipid deacylation during brain ischemia in a homeotherm and poikilotherm. Content and composition of free fatty acids and triacylglycerols, *Brain Res.* **100**:99–110.

Breckenridge, B. McL., 1964, The measurement of cyclic adenylate in tissue. *Proc. Natn. Acad. Sci., U.S.A.,* **52**:1580–1586.

Brierley, J. B., Meldrum, B. S., and Brown, A. W., 1973, The threshold and neuropathology of cerebral "anoxic-ischemia" cell change, *Arch. Neurol.* **29**:367–374.

Cammermeyer, J., and Fenton, I. M., 1982, Factors restricting maximal preservation of neuronal glycogen after perfusion fixation with dimethyl sulfoxide and iodoacetic acid in Bouin's solution, *Histochemistry* **76**:439–456.

Cantu, R. C., Ames, A., Di Giacinto, G., and Dixon, J., 1969, Hypotension: A major factor limiting recovery from cerebral ischemia, *J. Surg. Res.* **9**:525–529.

Cervos-Navarro, J., and Ferszt, R., 1980, Brain edema: Pathology, diagnosis and therapy, in: *Advances in Neurology,* Volume 28, Raven Press, New York.

Clendenon, N. R., Allen, N., Komatsu, T., Liss, L., Gordon, W. A., and Heimberger, K., 1971, Biochemical alterations in the anoxic-ischemic lesion of rat brain, *Arch. Neurol.* **25**:432–448.

Cohn, R., 1979, Convulsive activity in gerbils subjected to cerebral ischemia, *Exp. Neurol.* **65**:391–397.

Conger, K. A., Garcia, J. H., Kauffman, F. C., Lust, W. D., Murakami, N., and Passonneau, J. V., 1981, Alanine: Glutamate ratios as an index of reversibility of cerebral ischemia in gerbils, *Exp. Neurol.* **71**:370–382.

Cox, B., and Lomax, P., 1976, Brain amines and spontaneous epileptic seizures in the mongolian gerbil, *Pharmacol. Biochem. Behav.* **4**:263–267.

Cvejic, V., Djuricic, B. M., and Mrsulja, B. B., 1980, Oscillatory pattern of catecholamine metabolism following transient cerebral ischemia in gerbils, in: *Circulatory and Developmental Aspects of Brain Metabolism* (M. Spatz, B. B. Mrsulja, Lj. M. Rakic, and W. D. Lust, eds.), Plenum Press, New York, pp. 97–102.

Daly, J. W., 1976, The nature of receptors regulating the formation of cyclic AMP, *Life Sci.* **18**:1349–1358.

Diemer, N. H., and Siemkowicz, E., 1980, Regional glucose metabolism and nerve cell damage after cerebral ischemia in normo- and hypoglycemic rats, in: *Circulatory and Developmental Aspects of Brain Metabolism* (M. Spatz, B. B. Mrsulja, Lj. M. Rakic, and W. D. Lust, eds.), Plenum Press, New York, pp. 23–32.

Djuricic, B. M., Micic, D. V., and Mrsulja, B. B., 1984, Phasic recognition of edema caused by ischemia. In: *Recent Progress in the Study and Therapy of Brain Edema,* (K. G. Go, and A. Baethman, eds.), Plenum Press, New York, pp. 491–498.

Duckrow, R. B., LaManna, J. S., and Rosenthal, M., 1981, Disparate recovery of resting and stimulated oxidative metabolism following transient ischemia, *Stroke* **12**:677–686.

Ekstrom von Lubitz, D. K. J., and Diemer, N. H., 1982, Complete cerebral ischaemia in the rat: An ultrastructural and stereological analysis of the distal stratum radiatum in the hippocampal CA-1 region, *Neuropathol. Appl. Neurol.* **8**:197–215.

Faiman, M. D., Myers, M. B., and Schowen, R. L., 1973, Post-mortem degradation kinetics of brain norepinephrine, *Biochem. Pharmacol.* **22**:2171–2181.

Ferrendelli, J. A., Rubin, E. H., and Kinscherf, D. A., 1976, Influence of divalent cations on regulation of cyclic GMP and cyclic AMP levels in brain tissue, *J. Neurochem.* **26**:741–748.

Folbergrova, J., Lowry, O. H., and Passonneau, J. V., 1970, Changes in metabolites of the energy reserves in individual layers of mouse cerebral cortex and subjacent white matter during ischemia and anesthesia, *J. Neurochem.* **17**:115–1162.

Folbergrova, J., Ponten, U., and Siesjo, B. K., 1974, Patterns of changes in brain carbohydrate metabolites, amino acids and organic phosphates at increased carbon dioxide tensions, *J. Neurochem.* **22**:1115–1125.

Gaudet, R., Welch, K. M. A., Chabi, E., and Wang, T.-P., 1978, Effect of transient ischemia on monoamine levels in the cerebral cortex of gerbils, *J. Neurochem.* **30**:751–757.

Ginsberg, M. D., Mela, L., Wrobel-Kuhl, K., and Reivich, M., 1977, Mitochondrial metabolism following bilateral cerebral ischemia in the gerbil, *Ann. Neurol.* **1**:519–527.

Goldberg, N. D., Passonneau, J. V., and Lowry, O. H., 1966, Effects of changes in brain metabolism on the levels of citric acid cycle intermediates, *J. Biol. Chem.* **241:**3997–4003.

Greengard, P., 1979–80, Intracellular signals in the brain, *Harvey Lect.* **75:**277–331.

Hansen, A. J., and Zeuthen, T., 1981, Extracellular ion concentrations during spreading depression and ischemia in the rat brain cortex, *Acta Physiol. Scand.* **113:**437–445.

Hansen, A. J., Hounsgaard, J., and Jahnsen, H., 1982, Anoxia increases potassium conductance in hippocampal nerve cells, *Acta Physiol. Scand.* **115:**301–310.

Hanson, E. J., Anderson, R. E., and Sundt, T. M., 1975, Influence of cerebral vasoconstricting and vasodilating agents on blood flow in regions of focal ischemia, *Stroke* **6:**642–647.

Harris, R. J., Symon, L., Branstron, N. M., and Bayhan, M., 1981, Changes in extracellular calcium activity in cerebral ischemia, *J. Cereb. Blood Flow Metabol.* **1:**203–209.

Harrison, M. J. G., Sedal, L., Arnold, J., and Ross Russell, R. W., 1975, No-reflow phenomenon in the cerebral circulation of the gerbil, *J. Neurol. Neurosurg. Psychiatry* **38:**1190–1193.

Himwich, H. E., 1951, *Brain Metabolism and Cerebral Disorders*, Williams and Wilkins, Co., Baltimore.

Hoff, J. T., Smith, A. L., Hankinson, H. L., and Nielsen, S. L., 1975, Barbiturate protection from cerebral infarction in primates, *Stroke* 6:28–33.

Hossmann, K.-A., 1982, Treatment of experimental cerebral ischemia, *J. Cereb. Blood Flow Metab.* **2:**275–297.

Hossmann, K.-A., and Olsson, Y., 1970, Suppression and recovery of neuronal function in transient cerebral ischemia, *Brain Res.* **22:**313–325.

Hossmann, K.-A., Lechtape-Gruter, H., and Hossmann, V., 1973, The role of cerebral blood flow for the recovery of the brain after prolonged ischemia, *Z. Neurol.* **204:**281–299.

Hossmann, K.-A., Paschen, W., and Csiba, L., 1983, Relationship between calcium accumulation and recovery of cat brain after prolonged cerebral ischemia, *J. Cereb. Blood Flow Metab.* **3:**346–353.

Hossmann, K.-A., and Sato, L., 1970, Recovery of neuronal function after prolonged cerebral ischemia. *Science.* **168:**375–376.

Kabat, H., and Dennis, C., 1938, Decerebration in the dog by complete temporary anemia of the brain, *Proc. Soc. Exp. Med.* **38:**864–865.

Kalimo, H., Rehncrona, S., Soderfeldt, B., Olsson, Y., and Siesjo, B. K., 1981, Brain lactic acidosis and ischemic cell damage: 2. Histopathology, *J. Cereb. Blood Flow Metab.* **1:**313–327.

Kirino, T., 1982, Delayed neuronal death in the gerbil hippocampus following ischemia, *Brain Res.* **239:**57–69.

Klatzo, I., 1975, Pathophysiologic aspects of cerebral ischemia, in: *The Nervous System, The Basic Neurosciences*, Volume 1 (D. W. Tower, ed.), Raven press, New York, pp. 313–322.

Klatzo, I., Farkas-Bargeton, E., Guth, L., Miquel, J., and Olsson, Y., 1970, Some morphological and biochemical aspects of abnormal glycogen accumulation in the glia, in: *Sixth International Congress of Neuropathology*, Masson et Cie, Paris, pp. 351–365.

Kleihues, P., and Hossmann, K.-A., 1973, Regional incorporation of L-[3-H] tyrosine into cat brain proteins after 1 hour of complete ischemia, *Acta Neuropathol.* **25:**313–324.

Kleihues, P., Hossmann, K.-A., Pegg, A. E., Kobayashi, K., and Zimmermann, V., 1975, Resuscitation of the monkey brain after one hour complete ischemia. III. Indications of metabolic recovery, *Brain Res.* **95:**61–73.

Kobayashi, M., Lust, W. D., and Passonneau, J. V., 1977, Concentrations of energy metabolites and cyclic nucleotides during and after bilateral ischemia in the gerbil cerebral cortex, *J. Neurochem.* **29:**53–59.

Kogure, K., and Schwartzman, R. J., 1980, Seizure propagation and ATP depletion in the rat stroke model, *Epilepsia* **12:**62–72.

Kogure, K., Scheinberg, P., Matsumoto, A., Busto, R., and Reinmuth, O. M., 1975, Catecholamines in experimental brain ischemia, *Arch. Neurol.* **32:**21–24.

Kramer, W., and Tuynman, J. A., 1967, Acute intracranial hypertension—An experimental investigation, *Brain. Res.* **6:**686–705.

Lavyne, M. H., Moskowitz, M. A., Larin, F., Zervas, N. T., and Wurtman, R. J., 1975, Brain H[3]-catecholamine metabolism in experimental cerebral ischemia, *Neurology* **25:**483–485.

Levine, S., 1960, Anoxic-ischemic encephalopathy in rats, *Am. J. Pathol.* **36:**1–17.

Levine, S., and Payan, H., 1966, Effects of ischemia and other procedures on the brain and retina of the gerbil (*Meriones unguiculatus*), *Exp. Neurol.* **16:**255–262.

Levy, D. E., and Duffy, T. E., 1977, Cerebral energy metabolism during transient ischemia and recovery in the gerbil, *J. Neurochem.* **28**:63–70.

Levy, D. E., Brierley, J. B., and Plum, F., 1975, Ischemic brain damage in the gerbil in the absence of 'no-reflow', *J. Neurol. Neurosurg. Psychiatry* **38**:1197–1205.

Ljunggren, B., Schutz, H., and Siesjo, B. K., 1974, Changes in energy state and acid-base parameters of the rat brain during complete compression ischemia, *Brain Res.* **73**:277–289.

Lowry, O. H., and Passonneau, J. V., 1964, The relationships between substrates and enzymes of glycolysis in brain, *J. Biol. Chem.* **239**:31–42.

Lowry, O. H., and Passonneau, J. V., 1972, *A Flexible System of Enzymatic Analysis*, Academic Press, New York.

Lowry, O. H., Passonneau, J. V., Hasselberger, F. X., and Schulz, D. W., 1964, Effect of ischemia on known substrates and cofactors of the glycolytic pathway in brain, *J. Biol. Chem.* **239**:18–30.

Lust, W. D., and Passonneau, J. V., 1976, Cyclic nucleotides in murine brain: Effect of hypothermia on adenosine 3', 5' monophosphate, glycogen phosphorylase, glycogen synthase and metabolites following maximal electroshock or decapitation, *J. Neurochem.* **26**:11–16.

Lust, W. D., and Passonneau, J. V., 1979, Cyclic nucleotide levels in brain during ischemia and recirculation, in: *Neuropharmacology of Cyclic Nucleotides* (G. C. Palmer, ed.), Urban and Schwarzenberg, Baltimore-Munich, pp. 228–252.

Lust, W. D., Mrsulja, B. B., Mrsulja, B. J., Passonneau, J. V., and Klatzo, I., 1975, Putative neurotransmitters and cyclic nucleotides in prolonged ischemia of the cerebral cortex, *Brain Res.* **98**:394–399.

Lust, W. D., Feussner, G. K., Barbehenn, E. K., and Passonneau, J. V., 1981, The enzymatic measurement of adenine nucleotides and P-creatine in picomole amounts, *Anal. Biochem.* **110**:258–266.

Lust, W. D., Arai, H., and Passonneau, J. V., 1982, Pharmacology of the postischemic rise of cyclic AMP in brain slices, *Trans. Am. Soc. Neurochem.* **13**:117.

Maruki, C., Merkel, N., Rausch, W. D., and Spatz, M., 1984, Brain monoamines in cerebral ischemic edema, the effect of gamma-hydroxybutyrate, in: *Recent Progress in the Study and Therapy of Brain Edema.* (K. G. Go, and A. Baethmann, eds.) Plenum Press, New York, pp. 673–681.

Michenfelder, J. D., and Milde, J. H., 1975, Influence of anesthetics on metabolic, functional and pathological responses to regional ischemia, *Stroke* **6**:405–410.

Miquel, J., and Haymaker, W., 1965, Astroglial reactions to ionizing radiation: with emphasis on glycogen accumulation, in: *Biology of Neuroglia, Progress in Brain Research*, Volume 15 (E. D. P. De Robertis and R. Carrea, eds.), Elsevier, Amsterdam, pp. 89–114.

Moossy, J., 1979, Morphological validation of ischemic stroke models, in: *Cerebrovascular Diseases* (T. R. Price and E. Nelson, eds.), Raven Press, New York, pp. 3–10.

Moseley, J. I., Laurent, J. P., and Molinari, G. F., 1975, Barbiturate attenuation of the clinical course and pathologic lesions in a primate stroke model, *Neurology* **25**:870–874.

Mrsulja, B. B., Lust, W. D., Mrsulja, B. J., Passonneau, J. V., and Klatzo, I., 1976a, Post-ischemic changes in certain metabolites following prolonged ischemia in the gerbil cerebral cortex, *J. Neurochem.* **26**:1099–1103.

Mrsulja, B. B., Lust, W. D., Mrsulja, B. J., Passonneau, J. V., and Klatzo, I., 1976b, Brain glycogen following experimental cerebral ischemia in gerbils, *Experientia* **32**:732–733.

Mrsulja, B. B., Mrsulja, B. J., Spatz, M., and Klatzo, I., 1976c, Brain serotonin after experimental vascular occlusion, *Neurology* **26**:785–787.

Mrsulja, B. B., Lust, W. D., Mrsulja, B. J., and Passoneau, J. V., 1977, Effect of repeated cerebral ischemia on metabolites and metabolic rate in gerbil cortex, *Brain Res.* **119**:480–486.

Mrsulja, B. B., Djuricic, B. M., Cvejic, V., Mrsulja, B. J., Abe, K., Spatz, M., and Klatzo, I., 1980, Biochemistry of experimental ischemic brain edema, in: *Advances in Neurology*, Volume 28 (J. Cervos-Navarro and R. Ferszt, eds.) Raven Press, New York, pp. 217–230.

Myers, R. E., and Yamaguchi, M., 1976, Effects of serum glucose concentration on brain response to circulatory arrest, *J. Neuropathol. Exp. Neurol.* **35**:301.

Nemoto, E. M., Shiu, G. K., Nemmer, J. P., and Bleyaert, A., 1982, Free fatty acids in the pathogenesis and therapy of ischemic brain injury, *J. Cereb. Blood Flow Metab.* **2**:S59–S61.

Nordstrom, C.-H., Rehncrona, S., and Siesjo, B. K., 1978, Restitution of cerebral energy state, as well as of glycolytic metabolites, citric acid cycle intermediates and associated amino acids after 30 minutes of complete ischemia in rats anaesthetized with nitrous oxide or phenobarbital, *J. Neurochem.* **30**:479–486.

Olsen, T. S., Larsen, B., Bech Skriver, E., Herning, M., Enevoldsen, E., and Lassen, N. A., 1981, Focal cerebral hyperemia in acute stroke. Incidence, pathophysiology and clinical significance, *Stroke* **12**:598–607.

Osburne, R. C., and Halsey, J. H., 1975, Cerebral blood flow. A predictor of recovery from ischemia in the gerbil, *Arch. Neurol.* **32**:457–461.

Passonneau, J. V., and Lowry, O. H., 1971, Metabolite flux in single neurons during ischemia and anesthesia, in: *Recent Advances in Quantitative Histo- and Cytochemistry*, Hans Huber, Bern, pp. 198–212.

Passonneau, J. V., Kobayashi, M., and Lust, W. D., 1977, The effect of bilateral ischemia and recirculation on energy reserves and cyclic nucleotides in the cerebral cortex of gerbils, in: *Alcohol and Aldehyde Metabolizing Systems: Intermediary Metabolism and Neurochemistry*, Volume III (R. G. Thurman, J. R. Williamson, H. R. Drott, and B. Chance, eds.), Academic Press, New York, pp. 485–498.

Passonneau, J. V., Hawkins, R. A., Lust, W. D., and Welsh, F. A. (eds.), 1980, *Cerebral Metabolism and Neural Function*, Williams and Wilkins, Baltimore.

Phillis, J. W., 1977, The role of cyclic nucleotides in the CNS, *Can. J. Neurol. Sci.* **4**:151–193.

Pulsinelli, W. A., and Brierley, J. B., 1979, A new model of bilateral hemispheric ischemia in the unanesthetized rat, *Stroke* **10**:267–271.

Pulsinelli, W. A., and Duffy, T. E., 1983, Regional energy balance in rat brain after transient forebrain ischemia, *J. Neurochem.* **40**:1500–1503.

Pulsinelli, W. A., Brierley, J. B., and Plum, F., 1982a, Temporal profile of neuronal damage in a model of transient forebrain ischemia, *Ann. Neurol.* **11**:491–498.

Pulsinelli, W. A., Levy, D. E., and Duffy, T. E., 1982b, Regional cerebral blood flow and glucose metabolism following transient forebrain ischemia, *Ann. Neurol.* **11**:499–509.

Rehncrona, S., Rosen, I., and Siesjo, B. K., 1981, Brain lactic acidosis and ischemic cell damage: 1. Biochemistry and neurophysiology, *J. Cereb. Blood Flow Metab.* **1**:297–312.

Robinson, R. G., Shoemaker, W. J., Schlumpf, M., Valk, T., and Bloom, F. E., 1975, Effect of experimental cerebral infarction in rat brain on catecholamines and behavior, *Nature* **255**:332–334.

Rossen, R., Kabat, H., and Anderson, J. P., 1943, Acute arrest of cerebral circulation in man, *Arch. Neurol. Psychiatry* **50**:510–528.

Schwartz, J. P., Mrsulja, B. B., Mrsulja, B. J., Passonneau, J. V., and Klatzo, I., 1976, Alterations of cyclic nucleotide-related enzymes and ATPase during unilateral ischemia and recirculation in gerbil cerebral cortex, *J. Neurochem.* **27**:101–107.

Siesjo, B. K., (ed.), 1978, *Brain Energy Metabolism*, John Wiley and Sons, Chichester, New York, Brisbane, Toronto.

Siesjo, B. K., 1981, Cell damage in the brain: A speculative synthesis, *J. Cereb. Blood Flow Metab.* **1**:155–185.

Smith, A. L., Hoff, J. T., Nielsen, S. L., and Larson, C. P., 1974, Barbiturate protection in acute focal cerebral ischemia, *Stroke* **5**:1–7.

Sokoloff, L., 1977, Relation between physiological function and energy metabolism in the central nervous system, *J. Neurochem.* **29**:13–26.

Sokoloff, L., Reivich, M., Kennedy, C., Des Rosiers, M. H., Patlak, C. S., Pettigrew, K. D., Sakurada, O. and Shinohara, M., 1977, The [14-C]deoxyglucose method for the measurement of local cerebral glucose utilization: Theory, procedure, and normal values in the conscious and anesthetized albino rat. *J. Neurochem.* **28**:897–916.

Steiner, A. L., Ferrendelli, J. A., and Kipnis, D. M., 1972, Radioimmunoassay for cyclic nucleotides: Effects of ischemia, changes during development and regional distribution of adenosine 3', 5'-monophosphate and guanosine 3', 5'-monophosphate in mouse brain, *J. Biol. Chem.* **247**:1121–1124.

Stone, T. W., Taylor, D. A., and Bloom, F. E., 1975, Cyclic AMP and cyclic GMP may mediate opposite neuronal responses in the rat cerebral cortex, *Science* **187**:845–846.

Suzuki, R., Yamaguchi, T., Kirino, T., Orzi, F., and Klatzo, I., 1983a, The effects of 5-minute ischemia in mongolian gerbils: I. Blood-brain barrier, cerebral blood flow, and local cerebral glucose utilization changes, *Acta Neuropathol.* **60**:207–216.

Suzuki, R., Yamaguchi, T., Li, C.-L., and Klatzo, I., 1983b, The effects of 5-minute ischemia in mongolian gerbils: II. Changes of spontaneous neuronal activity in cerebral cortex and CAl sector of hippocampus, *Acta Neuropathol.* **60**:217–222.

Swaab, D. F., and Boer, K., 1972, The presence of biologically labile compounds during ischemia and their relationship to the EEG in rat cerebral cortex and hypothalamus *J. Neurochem.* **19**:2843–2853.

Taylor, M. D., Palmer, G. C., and Callahan, A. S., 1982, Alterations of catecholamine-sensitive adenylate cyclase in gerbil cerebral cortex after bilateral ischemia, *Exper. Neurol.* **76**:495–507.

Tews, J. K., Carter, S. H., Roa, P. D., and Stone, W. E., 1963, Free amino acids and related compounds in dog brain: Post-mortem and anoxic changes, effects of ammonium chloride infusion, and levels during seizures induced by picrotoxin and by pentylenetetrazol, *J. Neurochem.* **10**:641–653.

Veech, R. L., 1980, Freeze-blowing of brain and the interpretation of the meaning of certain metabolite levels, in: *Cerebral Metabolism and Neural Function* (J. V. Passonneau, R. A. Hawkins, W. D. Lust, and F. A. Welsh, eds.), Williams & Wilkins. Baltimore, pp. 34–41.

Veloso, D., Guynn, R. W., Oskarsson, M., and Veech, R. L., 1973, The concentrations of free and bound magnesium concentration in rat tissues, *J. Biol. Chem.* **248**:4811–4819.

Watanabe, H., and Ishii, S., 1976, The effect of brain ischemia on the levels of cyclic AMP and glycogen metabolism in gerbil brain *in vivo, Brain Res.* **102**:385–389.

Watanabe, H., and Passonneau, J. V., 1974, The effect of trauma on cerebral glycogen and related metabolites and enzymes, *Brain Res.* **66**:147–159.

Wauquier, A., 1982, Brain protective properties of etomidate and flunarizine, *J. Cereb. Blood Flow Metabl.* **3**:S53–S56.

Weinfeld, F. D., 1981, The national survey of stroke, *Stroke* **12**:113–168.

Welch, K. M. A., Wang, T.-P. F., and Chabi, E., 1978, Ischemia-induced seizures and cortical monoamine levels, *Ann. Neurol.* **3**:152–155.

Wexler, B. C., 1972, Pathophysiological responses to acute cerebral ischemia in the gerbil, *Stroke* **3**:71–78.

Wilhjelm, B. J., and Arnfred, I., 1965, Protective action of some anesthetics against anoxia, *Acta Pharmacol. Toxicol.* **22**:93–98.

III

Metabolic Encephalopathy Resulting
Primarily from Intrinsic Factors

5

Pyruvate Dehydrogenase Deficiency Disorders

ROGER F. BUTTERWORTH

1. Introduction

The pyruvate dehydrogenase complex (PDHC) catalyzes the irreversible oxidation of pyruvate to acetyl CoA according to the reaction:

$$\text{Pyruvate} + \text{NAD}^+ + \text{CoASH} \rightarrow \text{Acetyl CoA} + \text{NADH} + \text{CO}_2$$

The acetyl CoA formed is then either oxidized in the tricarboxylic acid cycle or used for biosynthetic purposes.

PDHC can be resolved into three catalytic components and two regulatory enzymes (Fig. 1). The three catalytic enzymes are pyruvate decarboxylase, E_1 (EC: 4.1.1.1.), dihydrolipoyl transacetylase, E_2 (EC: 2.3.1.12), and dihydrolipoyl dehydrogenase, E_3 (EC: 1.6.4.3). The two regulatory enzymes are pyruvate dehydrogenase kinase (EC: 2.7.1.99), which catalyzes the Mg-ATP-dependent phosphorylation of E_1 with subsequent inactivation, and pyruvate dehydrogenase phosphate phosphatase (EC: 3.1.3.43), which is responsible for the dephosphorylation of the phosphorylated enzyme with concomitant activation (Sheu et al., 1981). PDHC activity is subject to regulation, in most tissues studied including brain, at two levels. First, acetyl CoA and NADH exert product inhibition by competing with the substrates CoA and NAD^+, respectively. Second, PDHC activity is determined by the degree of PDHC phosphorylation. This "phosphorylation state" of the enzyme complex is controlled by the relative activities of the two regulatory enzymes PDH kinase and PDH phosphate phosphatase, both of which are subject to control by a number of agents. PDHC kinase is activated by acetyl CoA and NADH and is inhibited by pyruvate, CoA, NAD^+, TPP, and ADP. The enzyme is also inhibited by dichloroacetate (DCA).

ROGER F. BUTTERWORTH • Laboratory of Neurochemistry, Clinical Research Centre, Hôpital St-Luc (University of Montreal), Montreal, Quebec, Canada H2X 3J4.

PDH phosphate phosphatase is activated by Mg^{2+} and Ca^{2+} and is inhibited by fluoride ion.

Hereditary deficiencies of PDHC were suggested in the late 1960s in cases of familial lactic acidosis (Lonsdale *et al.*, 1969). In the last 15 years, hereditary neurological diseases have been identified in which there is convincing evidence for genetic abnormalities of PDHC or one of its constituent enzymes. These findings are due to a substantial degree to the pioneering efforts of Blass and co-workers in the United States. For a fuller description of the clinical aspects of these disorders, the reader is referred to two recent reviews (Blass, 1980, 1981).

2. Inherited Disorders of PDHC

Accumulation of excessive quantities of pyruvate, lactate, and alanine in blood, urine, or CSF of affected patients is consistent with, although not in itself indicative of PDHC deficiency. The diagnosis of PDHC deficiency disease is best demonstrated in biopsied or cultured tissues from suspected patients using adequate assay techniques to measure PDHC activity, or the activity of one of the constituent enzymes of the complex, in disrupted cells. Demonstration then of defective pyruvate oxidation by intact cell preparations provides supportive evidence of PDHC deficiency disease. Measurement of PDHC activity in brain tissue is likely to yield less definitive results since the cerebral enzyme has been found to be deficient in neurological diseases in which no primary *inherited* defect of PDHC has been postulated (see Section 2.6).

Generally, the more severe the PDHC defect, the more severe is the accompanying clinical illness, the earlier its onset, the more rapid its progression, and the more widespread the signs of CNS damage. Thus patients with less than 15% of normal PDHC generally have presented with severe neurological disease and lactic acidosis beginning in infancy, whereas patients with 20–40% of normal PDHC have a milder illness in which intermittent ataxia is the most prominent neurological sign (Blass *et al.*, 1976c). Specific abnormalities of E_1, E_2, and E_3 components of PDHC have been reported (Table I). In addition,

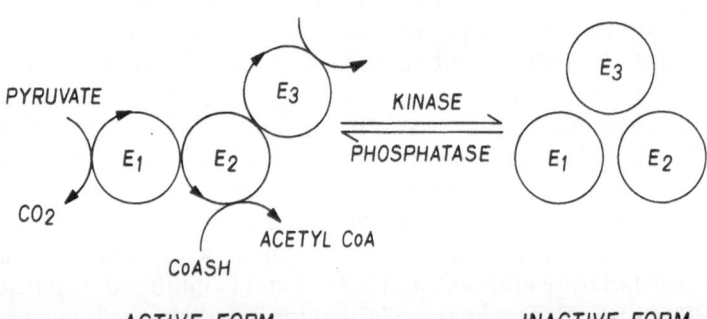

FIGURE 1. Constituent catalytic and regulatory enzymes of the pyruvate dehydrogenase complex (PDHC). E_1 [pyruvate decarboxylase (EC: 4.1.1.1.)]; E_2 [dihydrolipoyl transacetylase (EC: 2.3.1.12)]; and E_3 [dihydrolipoyl dihydrogenase (EC: 1.6.4.3)].

there is recent evidence from three laboratories to suggest that Leigh's subacute necrotizing encephalomyelopathy may be associated with a *regulatory* defect of PDHC (De Vivo *et al.*, 1979; Butterworth, 1982b; Sorbi and Blass, 1982).

2.1. Inherited Disorders of Pyruvate Decarboxylase (E_1; EC: 4.1.1.1.)

The most severely affected patient with an inherited defect of E_1 reported to date is the patient of Farrell *et al.* (1975). In this case, E_1 was virtually undetectable. The patient presented in the first days of life with profound *lactic acidosis* and severe generalized neurological disease. Two other studies have reported similar fatal congenital lactic acidosis associated with deficiencies of E_1 of the order of 10% (Stromme *et al.*, 1976; Willems *et al.*, 1974). In contrast, patients from several other studies (see Table I) have presented with deficiencies of E_1 of the order of 20–40% residual activity and, in most cases, these patients showed less drastic neurological symptoms. These patients have been generally classified as having *intermittent ataxia of childhood*.

TABLE I. Patients with Deficiency of a PDHC-Constituent Enzyme

Defective enzyme	Percent normal activity	Age	Congenital lactic acidosis	Intermittent cerebellar ataxia	References
Pyruvate	0	Neonate	√	.	Farrell *et al.* (1975)
decarboxylase	2	Neonate	√	.	Wick *et al.* (1977)
E_1 (EC: 4.1.1.1)	4	Neonate	√	.	Stromme *et al.* (1976)
	10	9 months	√	.	Willems *et al.* (1974)
	10	11 months	√	.	Toshima (1980)
	12	28 months	.	(√)	Falk *et al.* (1976)
	15	16 days	√	.	Robinson *et al.* (1980)
	17	5 days	.	√	Blass *et al.* (1971)
	17	9 years	.	√	Blass *et al.* (1971)
	20	11 years	.	√	Falk *et al.* (1976)
	22	2 years	√	.	Robinson *et al.* (1980)
	23	4½ years	.	√	Robinson *et al.* (1980)
	39	3 years	.	√	Wick *et al.* (1977)
	40	2½ years	.	.	Robinson *et al.* (1980)
	46	2 years	√	.	Robinson *et al.* (1980)
	—	5 years	.	√	Oka *et al.* (1976)
Dihydrolipoyl	15	3 years[a]	√	.	Blass *et al.* (1972)
transacetylase	15	8½ years	√	.	Cederbaum *et al.* (1976)
E_2 (EC: 2.3.1.12)					
Dihydrolipoyl	3	2 years	√	.	Robinson *et al.* (1980)
dehydrogenase	5	7 months	√	.	Robinson *et al.* (1977b)
E_3 (EC: 1.6.4.3)	13	Neonate	√	.	Haworth *et al.* (1976)
	—	1 month	√	.	Haworth *et al.* (1976)
	—	3 days	√	.	Haworth *et al.* (1976)

[a] May also be a defect of EC: 1.6.4.3.

The reason for the two types of presentation (congenital lactic acidosis of infancy and intermittent ataxia of childhood) is not clear. However, it has recently been suggested that since the E_1 component is an $\alpha_2\beta_2$ subunit tetramer, it is possible that the two syndromes represent a defect in different subunits of the enzyme (Robinson *et al.*, 1980). There is no experimental evidence to date, however, to support such a contention.

2.2. Inherited Disorders of Dihydrolipoyl Transacetylase (E_2; EC: 2.3.1.12)

A presumptive but unproved diagnosis of inherited deficiency of E_2 has been reported (Cederbaum *et al.*, 1976). The case presented as congenital lactic acidosis with severe mental retardation. PDHC activity was reduced to 15% of normal values in the presence of normal activities of both E_1 and α-ketoglutarate dehydrogenase.

2.3. Inherited Disorders of Dihydrolipoyl Dehydrogenase (E_3; EC: 1.6.4.3)

Five cases of inherited defects of E_3 have been reported. In the most severe cases (3% and 5% of control values), the presentation was chronic acidosis, optic atrophy, and hypoglycemia (Robinson *et al.*, 1980). Since E_3 is common to both the pyruvate and α-ketoglutarate dehydrogenase complexes, these patients also showed significantly reduced activities of both enzyme complexes (below 10% in all tissues studied) and elevated serum pyruvate and α-ketoglutarate in their urine. Enzyme assays revealed a low activity of both the pyruvate and α-ketoglutarate dehydrogenase complexes in cultured fibroblasts of one of the siblings (Haworth *et al.*, 1976) suggesting that these patients had an inherited defect of E_3.

2.4. PDHC Abnormalities in Spinocerebellar Disorders

Since the initial finding of diminished activity of E_1 in disrupted skin fibroblasts from a patient with a spinocerebellar disorder (Blass *et al.*, 1971), results of several studies on larger numbers of patients with hereditary spinocerebellar disease have been published. In one such study, biopsied muscle from a number of patients was shown to oxidize pyruvate to CO_2 more slowly than did muscle from controls and diminished pyruvate oxidation was confirmed in intact cultured skin fibroblasts (Kark *et al.*, 1974). Subsequent studies produced evidence of PDHC abnormalities in disrupted fibroblasts from four patients with the clinical diagnosis of Friedreich's ataxia (Blass *et al.*, 1976a). Abnormal pyruvate metabolism was confirmed in some patients with Friedreich's ataxia (Barbeau *et al.*, 1978; Livingstone *et al.*, 1980), but the finding of abnormal activities of PDHC in cultured skin fibroblasts was not confirmed by others (Melancon *et al.*, 1979; Stumpf and Parks, 1979; Bertagnolio *et al.*, 1980). The reasons for these discrepancies undoubtedly include difficulties of both a clinical diagnostic and biochemical methodological nature. The spinocerebellar degenerations are a heterogeneous group of neurodegenerative disorders; at least 20 inborn errors (including E_1 deficiency) have been described to date in association with the clinical diagnosis of spinocerebellar degeneration (Plaitakis et al., 1980; Blass, 1981; Stumpf et al., 1982). In addition, there are documented case reports where one sibling has one of the commonly accepted syndromes, while a second sibling with presumably the same genetic abnormality has another of the so-called "classic" spinocerebellar degenerations (Blass, 1980). To add to these difficulties, the use of cultured skin fibroblasts for diagnosis of inherited defects of PDHC in which activity of the whole complex is measured will require, in the light of recent findings by Sheu et al. (1981), modifications of previously described experimental procedures. PDHC activity in human skin fibroblasts is regulated by a phosphorylation-dephosphorylation mechanism as is the case with other tissues. The enzyme can be activated by preincubation with dichloroacetate, an inhibitor of PDH kinase, and when so treated, PDHC activity in human cultured skin fibroblasts is in the range of 3–6 nmole/min per mg protein for infants. Such values are 5- to 20-fold higher than those previously reported using untreated cells (Sheu et al., 1981), suggesting that PDHC in fibroblasts exists largely in the inactive (phosphorylated) state. Unfortunately, no studies of PDHC activity in fibroblasts using such a modified procedure have been reported (at least at the time of writing) in patients with spinocerebellar degeneration.

In the meantime, another pyruvate-metabolizing enzyme, namely malic enzyme (EC: 1.1.1.40), has been found to be seriously deficient (activities of less than 10% of normal) in fibroblasts from eight Friedreich's ataxia patients (Stumpf et al., 1982).

2.5. PDHC Deficiency in Leigh's Subacute Necrotizing Encephalomyelopathy

Leigh's subacute necrotizing encephalomyelopathy (Leigh's disease) is a familial disorder of infancy and early childhood. Macroscopic examination of the brain reveals characteristic necrotizing lesions of deep midline structures; microscopically, the necrotic lesions show demyelination, capillary proliferation, and cell loss (Leigh, 1951). In view of its histological resemblance to Wernick's encephalopathy and since pyruvic acidemia and lactic acidemia are present in many patients, a disturbance of pyruvate metabolism in Leigh's disease has been suspected for some time. In a brief report (Farmer et al., 1973), a child with suspected Leigh's disease was reported to have a deficiency of E_1. This finding was confirmed in two other patients with Leigh's disease (Blass et al., 1976b); both patients developed ataxia in the first three years of life, both had siblings who died of autopsy-proved Leigh's disease, and studies of disrupted fibroblasts from both patients revealed diminished E_1 activity (less than 25%). In a more recent study, an E_1 defect of less than 20% has been reported in a muscle biopsy sample from a patient with Leigh's disease (Evans, 1981a).

In addition, defective *activation* of PDHC in Leigh's disease has been reported from three laboratories (De Vivo et al., 1979; Butterworth, 1982b; Sorbi and Blass, 1982) (see Table II). In one of these studies, autopsy examination confirmed the diagnosis of Leigh's disease in a 7-month-old infant who showed intermittent lactic acidemia. Enzyme studies of postmortem samples revealed low nonactivated PDHC in liver and brain, the lowest cerebral values being found in pons and midbrain. Spontaneous reactivation of the complex following *in vitro* inactivation with ATP was significantly less in liver homogenates from the patient as compared with controls and an *in vivo* defect of PDHC regulation was postulated (De Vivo *et al.*, 1979). In a subsequent study, a 17-month-old boy with lactic acidemia was investigated. A liver biopsy sample revealed normal basal PDHC activity and normal Km for pyruvate, but defective activation of the enzyme complex following incubation with $MgCl_2$ (10 mM) and $CaCl_2$ (2 mM) in the presence of ATP (Butterworth and Melancon, unpublished study; Butterworth, 1982b), in support of a defect of PDHC activation in this patient in whom the diagnosis of Leigh's disease was later confirmed at autopsy. Abnormal PDHC activation has since been confirmed in two other autopsy-proved cases of Leigh's disease (Sorbi and Blass, 1982). In this latter study, incubation with the PDH kinase inhibitor, dichloroacetate, increased PDHC activity in disrupted fibroblasts from controls, but not from two patients with Leigh's disease. It might be reasonable, in light of these findings, to advance the hypothesis that a deficiency of the regulatory enzyme PDH phosphate phosphatase may be a primary defect in some patients with Leigh's disease. However, results of a recent study in which phosphatase was directly measured on cultured fibroblasts from two PDHC-deficient Leigh's disease patients revealed activities of the activation enzyme within normal limits (Sheu and Blass, 1983).

2.6. Other Neurological Disease Associated with PDHC Deficiency

A recent report describes a child with a rapidly progressive neurological disorder, psychomotor retardation, seizures, and respiratory disturbances. CSF pyruvate and lactate were elevated without notably elevated levels in serum. Studies in liver, muscle, and cultured fibroblasts revealed no abnormality in pyruvate oxidation, whereas biochemical studies of a brain biopsy revealed a deficiency of PDHC with the morphological picture of *progressive poliodystrophy with hypomyelinization* (Prick *et al.*, 1981). In another study, PDHC activity was measured in regions of the human brain from controls without neurological disease and from patients with Huntington's disease (HD). PDHC activity was reduced to 40% and 50% of control values in HD putamen and caudate, respectively, and it was suggested that these decreases may prove to

TABLE II. PDHC Abnormalities in Leigh's Subacute Necrotizing Encephalomyelopathy

Enzyme defect	% Normal value	Onset	Reference
E_1	—	12 months	Farmer *et al.* (1973)
E_1	15	5 months	Blass *et al.* (1976b); Kohlschutter *et al.* (1978)
E_1	25	<36 months	Blass *et al.* (1976b)
E_1	30	9 months	Evans (1981a)
PDHC activation		4 months	De Vivo *et al.* (1979)
PDHC activation		17 months	Butterworth (1982b)
PDHC activation		<24 months	Sorbi and Blass (1982)
PDHC activation		12 months	Sorbi and Blass (1982)

play a role in the pathophysiology of Huntington's disease (Sorbi *et al.*, 1981). Choline acetyl transferase (CAT) activities were also reduced in these brain structures in support of other data indicating a close relationship between cholinergic function and the capacity for pyruvate oxidation in brain. A more complete discussion of this interrelationship can be found in Section 4.4.1. It has been suggested that the degenerative process per se may be related to abnormalities of PDHC. Thus there is evidence to suggest that disturbances in PDHC activity (either in total enzyme activity or in the phosphorylation of the α-subunit) may result in changes in the regulation of cytosolic free calcium resulting in the activation of calcium-sensitive proteases and subsequent neuronal degeneration (Beaudry *et al.*, 1983). In another report, PDHC activity was found to be less than 12% of normal in liver samples obtained at autopsy from six patients with Reye's syndrome (Robinson *et al.*, 1977a). It has been suggested that this finding was likely secondary to mitochondrial damage and a general decrease of mitochondrial enzyme activity in this disorder (Blass, 1980).

3. Acquired Disorders of PDHC

Several reports have appeared, based on studies in experimental animals, of PDHC inhibition by exposure to toxic metals. For example, it has been reported that *ethyl* mercury caused marked inhibition of $^{14}CO_2$ production from [1-^{14}C]pyruvate and [2-^{14}C]pyruvate (Cremer, 1962) in rat brain slices consistent with inhibition of PDHC. Reduced cerebral pyruvate oxidation was later found in the brains of rabbits treated chronically with low doses of *methyl* mercury (Menon and Kark, 1976).

Mitochondrial respiration studies using liver mitochondria from rats chronically exposed to *arsenic* indicated decreased state 3 respiration rates and decreased respiration control ratios for pyruvate/malate-mediated respiration (Fowler *et al.*, 1977). Measurement of PDHC in liver homogenates from these rats treated chronically showed inhibition of the enzyme complex both before and after in vitro activation with Mg^{2+}, and it was suggested that arsenic's toxic effect may be mediated either by the regulatory PDHC kinase or phosphatase (Schiller *et al.*, 1977).

Trialkyltin compounds have efficient biocidal properties and are used as insecticides, fungicides, and wood preservatives. Triethyltin compounds are neurotoxic, producing hind-limb paralysis and CNS demyelination in rats chronically administered 1–10 mg/kg body weight of triethyltin sulphate. Within 1 hr, the brain contains 20 nmole of triethyltin per gram tissue (Rose and Aldridge, 1966), and on subfractionation of brain homogenates, 40% of the triethyltin is recovered from the mitochondrial fraction. Subsequent in vivo studies showed that the main action of triethyltin in the brain is to decrease the rate at which pyruvate formed from glucose is oxidized (Cremer, 1970). No direct effect on PDHC, however, has been reported.

Other studies have shown that lead (Pb^{2+}) binds to rat brain mitochondria, significantly reducing tricarboxylic acid cycle activity and energy production (Gremek *et al.*, 1981). When animals were exposed to 600 ppm Pb^{2+},

[^{14}C]glucose incorporation into citrate and acetylcholine (Ach) was diminished (Sterling *et al.*, 1982). Since glucose and pyruvate are reportedly the major precursors of Ach, this is consistent with a deficiency of cerebral PDHC.

Another important acquired disorder of PDHC is *thiamine deficiency encephalopathy*. There is now a substantial body of evidence to support the hypothesis of impaired cerebral PDHC accompanying chronic thiamine deficiency (Dreyfus and Hauser, 1965; McCandless and Schenker, 1968; Butterworth, 1982a, 1982b). Some aspects of PDHC deficiency accompanying thiamine deficiency are discussed in Section 4. The subject of the metabolic encephalopathy accompanying thiamine deficiency is covered more fully in Chapter 14 of this book.

4. Pathophysiology of PDHC Deficiency Disorders

Substantial progress has been made in recent years toward the understanding of the mechanisms responsible for the particular sensitivity of the CNS to inherited and acquired disorders of PDHC. The first possible explanation to be considered was that of tissue-specific isoenzymes of PDHC. Three distinct lines of evidence, however, suggest that this may not be the case. First, studies of purified PDHC proteins did not provide evidence of differences between brain and other tissues (Berrera *et al.*, 1972). Second, kinetic studies of partially purified PDHC from brain and other tissues provided no evidence of tissue-specific differences (Seiss *et al.*, 1971; Blass and Lewis, 1973). Third, several cases of inherited deficiencies of PDHC have been reported in which the enzyme deficiency was apparent in all tissues examined including brain (Farrell *et al.*, 1975; Stromme *et al.*, 1976; Blass and Gibson, 1978). Thus one probably cannot invoke isoenzymes to explain the selective vulnerability of brain to PDHC deficiency.

4.1. PDHC: The Rate-Limiting Step in Cerebral Glucose Oxidation

Studies in experimental animals have shown that there may be little, if any, excess of PDHC activity in mammalian brain over that required to maintain pyruvate flux. In the normal conscious adult rat, the rate of glucose utilization by whole brain is of the order of 6.5 nmole/min per mg protein (Hawkins *et al.*, 1974). Each mole of glucose is converted (almost completely) into two moles of pyruvate and then oxidized so that pyruvate flux through PDHC is approximately 12 nmole/min per mg protein. PDHC in whole adult rat brain is reportedly in the range of 45 nmole/min per mg protein (Ksiezak-Reding *et al.*, 1982; Butterworth and Giguère, 1984) (Fig. 2), of which approximately 30% (i.e., 13.5 nmole per min per mg protein) appears to be in the active form (Ksiezak-Reding *et al.*, 1982). Thus, the resultant activity of the enzyme complex is little in excess of the pyruvate flux under normal physiological conditions. This finding, which has been replicated in several laboratories, has led to the generally expressed view that PDHC is rate limiting (Cremer and Teal, 1974; Jope and Blass, 1976; Stumpf and Kraus, 1979). This implies that any impairment of PDHC activity will likely have functional consequences

(Blass and Gibson, 1978). Furthermore, there is evidence to suggest a variation of PDHC activity and of the ratio of PDHC activity and of the ratio of PDHC activity to pyruvate flux in different brain structures in the CNS (Reynolds and Blass, 1976; Butterworth and Giguère, 1984) (Fig. 2).

PDHC activity in normal adult rat brain varies from a high of 52.24 nmole/min per mg protein in hippocampus to a low of 34.96 nmole/min per mg protein in cerebellum (Fig. 2). A previous study revealed a similar distribution pattern in cat brain, lowest values of PDHC being found in an area of anterior cerebellar vermis (Reynolds and Blass, 1976). In this area of brain, activity of PDHC was only just in excess of pyruvate flux as calculated from oxygen uptake measurements, and it was suggested that if these measurements reflect in vivo conditions, flux of pyruvate to acetyl CoA might be selectively impaired in anterior cerebellar vermis by PDHC deficiencies. In support of such a hypothesis, other studies have shown that homogenates of rat cerebellum oxidized pyruvate more slowly than did those of brain stem or cerebrum (Dreyfus and Hauser, 1965; McCandless and Schenker, 1968). This apparent selective vulnerability of cerebellum to PDHC deficiency is particularly interesting; the most common clinical abnormality observed in mild PDHC deficiencies (both inherited and acquired) is ataxia, consistent with impaired cerebellar function.

4.2. PDHC in Mammalian Brain during Postnatal Development

Certain species, the so-called "nonprecocial species," such as mouse and rat, are born in a relatively poor state of neurological development. It is into this category that man fits. It has recently been demonstrated that the devel-

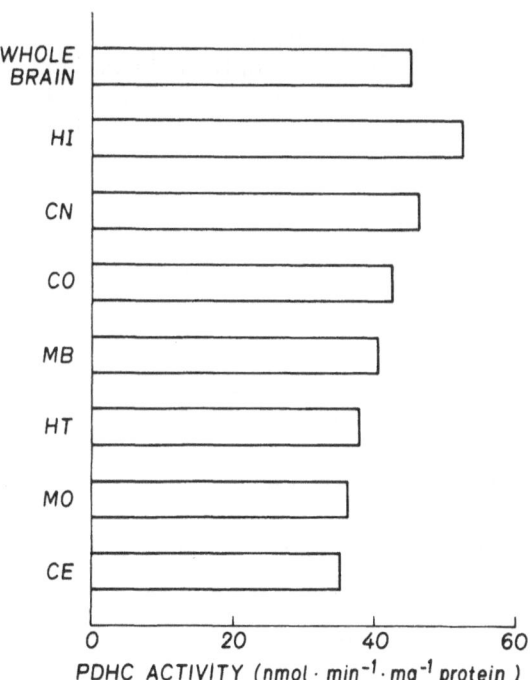

FIGURE 2. PDHC activity in regions of the adult rat brain: Hippocampus (HI), caudate nucleus (CN), midbrain (MB), medulla oblongata/pons (MO), hypothalamus (HT), cerebral cortex (CO), and cerebellum (CE).

opment of enzymes associated with the complete aerobic utilization of glucose shows a definite close correlation with the onset of neurological competence (Booth *et al.*, 1980). PDHC, for example, is found to develop postnatally with threefold to fourfold increases occurring during the suckling period to finally attain adult levels at 20–25 days (Cremer and Teal, 1974; Stumpf and Kraus, 1979; Booth *et al.*, 1980; Butterworth and Giguère, 1984). This developmental profile correlates well with the age at which the rat attains neurological competence with fully coordinated motor control. The rate of pyruvate oxidation by brain mitochondria prepared from immature rats at different ages closely parallels the developmental pattern of PDHC (Land *et al.*, 1977). Furthermore, it has been reported that the ratio PDHC (active)/PDHC (total) does not change during brain maturation (Stumpf and Kraus, 1979), suggesting that the PDHC delay in maturation is due to a delay in synthesis of enzyme protein rather than inactivation (by phosphorylation) of existing enzyme molecules.

A recent study of regional ontogenic changes of PDHC in rat brain revealed delayed maturation of the enzyme in cerebellum, reflecting the delayed maturation of cerebellar function in the rat (Butterworth and Giguère, 1984).

4.3. PDHC and Cerebral Energy Metabolism

In discussing the effects of inherited and acquired deficiencies of PDHC on cerebral energy metabolism, it is important that two general principles be considered, namely the degree of the PDHC deficiency and the changing pattern of cerebral energy substrates during development.

4.3.1. The Degree of the PDHC Deficiency

Normal brain function in the adult requires, except during severe ketonemia, an adequate supply of glucose and oxygen for the synthesis of ATP and other high-energy compounds. Since PDHC is probably a rate-limiting enzyme in cerebral glucose utilization, inherited or acquired deficiencies of PDHC are likely to have an adverse effect on cerebral energy metabolism. However, it is important to differentiate between very severe (or total) PDHC deficiency and deficiencies of a milder nature. Very severe impairment of glucose oxidation, as exemplified by ischemia following occlusion of a major blood vessel, is accompanied by rapid loss of ATP (and other high-energy compounds) in brain tissue and cell death results. On the other hand, following mild impairments of glucose oxidation, the homeostatic mechanisms of the brain appear to be capable of maintaining levels of high-energy phosphates. For example, hypoxia can cause impairment of neurological function without concomitant changes in ATP and related compounds (Gibson, *et al.*, 1981). Similarly, in insulin-hypoglycemic encephalopathy, it has been shown that, at least up to the onset of stupor associated with isoelectric EEG traces, there are no significant changes in the levels of high-energy phosphates in the CNS (Siesjo, 1978). Similar results have been obtained in experimental thiamine deficiency encephalopathy, a metabolic encephalopathy associated with impairments of cerebral PDHC activity of 50–60% (McCandless and Schenker, 1968). Thus, there is reason to believe that mild impairments of PDHC may not produce significant changes in high-energy phosphates and that synthetic pathways in brain may be more sensitive to mild impairments of PDHC than are energy-producing pathways (Blass and Gibson, 1978).

4.3.2. The Changing Pattern of Cerebral Energy Substrates during Brain Development

In fetal brain, glucose is metabolized primarily to pyruvate and lactate through the glycolytic pathway and it is the predominance of this pathway in the brain of newborn rats that permits short-term survival under anoxic conditions (Benjamins and McKhann, 1972). Postnatally, aerobic metabolism predominates with oxidation of cerebral energy substrates to CO_2 and increased energy production. It is now well established that the normal immature mammalian brain utilizes both ketone bodies (acetoacetate and β-hydroxybutyrate) and glucose as energy substrates. Comparison of sucking and adult rats shows greater ketone body utilization in the young; arteriovenous differences in ketone bodies are four times higher in the young than in the adult. This greater utilization is determined by the higher blood concentrations of ketone bodies in young rats. In addition, the developmental pattern of brain mitochondria to oxidize ketone bodies is consistent with the observed changes in levels of ketone-body-metabolizing enzymes (Land and Clark, 1976; Booth et al., 1980). After weaning, levels of ketone-body-metabolizing enzymes fall to attain adult levels.

In the light of these findings, the development of enzymes involved in pyruvate metabolism suggests that glucose utilized by immature brain is not primarily oxidized by the tricarboxylic acid cycle (as it is in adult brain), but is preferentially used in the pentose shunt pathway and in an "anaplerotic" role by conversion to oxaloacetate, to supplement the loss of tricarboxylic acid cycle intermediates used for synthetic purposes. The lag in PDHC development is most likely a key control factor in this regard (Land and Clark, 1976; Booth et al., 1980). The relative importance of alternative energy sources in neonatal vs. adult brain is shown schematically in Fig. 3.

Thus, the effect of PDHC deficiency on cerebral energy metabolism will depend on the "maturational status" of the brain and on the degree of the PDHC deficiency. Mild deficiencies of PDHC might be expected to have more adverse effects on cerebral energy metabolism after weaning. In fact, it has been suggested that this changing energy substrate phenomenon may account for the observation that many patients with mild to moderate inherited defects of PDHC appear to be relatively well early in life, showing neurological symptoms only after the age of 2 years or so (Blass and Gibson, 1978).

An attempt has been made to exploit the observation that ketone bodies are alternative cerebral energy substrates by treating some patients with inherited PDHC deficiencies with ketogenic diets (Falk et al., 1976) (see Section 5).

4.4. PDHC Deficiency and Neurotransmitter Function

4.4.1. Acetylcholine

The acetyl moiety of acetylcholine (ACh) is normally derived from glucose (via pyruvate) as shown schematically in Fig. 4. Therefore, impaired glucose (pyruvate) oxidation might be expected to cause impaired synthesis of ACh. Such a possibility was originally discounted on the grounds that even maximal synthesis of ACh would require less than 1% of the glucose oxidized by brain (McIlwain and Bachelard, 1971). However, studies have shown that a number of substances that inhibited conversion of [1-^{14}C]pyruvate or [2-^{14}C]pyruvate into $^{14}CO_2$ were associated with a corresponding decrease in the conversion of [2-^{14}C]pyruvate into ACh. Furthermore, the amount of ACh (in mass terms) produced from glucose by rat brain minces was similarly decreased (Gibson *et al.*, 1975). The inhibitory substances tested included the irreversible PDHC inhibitor 3-bromopyruvate and the competitive PDHC inhibitor 2-oxobutyrate.

In another series of experiments in brain slices and synaptosomal preparations, production of $^{14}CO_2$ from [3,4-^{14}C]glucose (an index of glucose decarboxylation by PDHC) was examined under conditions of low oxygen tension. Conversion of [^{14}C]glucose to ACh was also measured. Both ACh synthesis and decarboxylation of [3,4-^{14}C]glucose were significantly reduced and these decreases were linearly related (r = 0.74) (Ksiezak and Gibson, 1981). Other supportive evidence for a close link between impaired cerebral PDHC activity

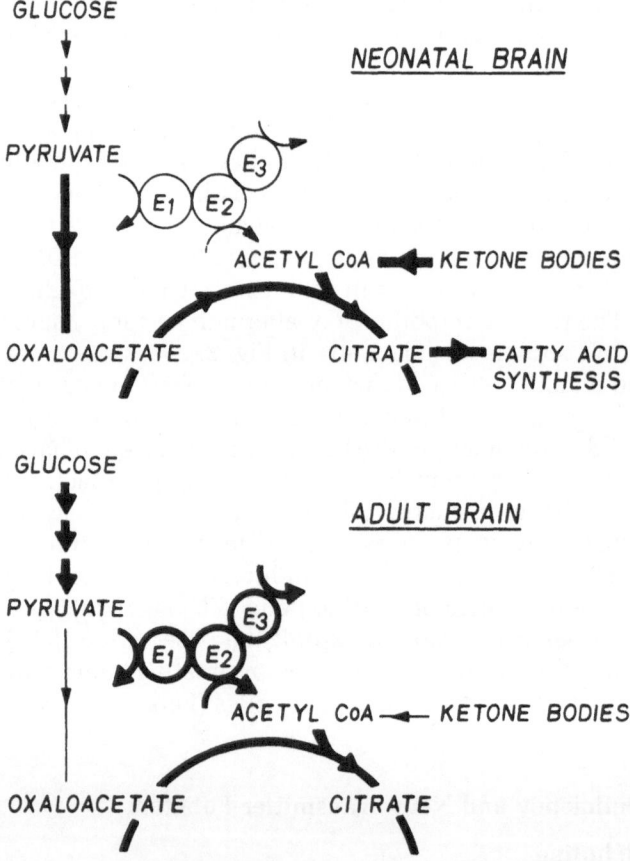

FIGURE 3. Schematic representation of alternative enery sources and role of PDHC in neonatal vs. adult rat brain. Relatively important pathways are shown in bold type.

and ACh metabolism has been provided by studies on thiamine deficiency encephalopathy. In severe experimental thiamine deficiency, cerebral PDHC activity is decreased by up to 70% (Dreyfus and Hauser, 1965; Butterworth, 1982b); acetyl CoA levels have been found to be decreased (Heinrich et al., 1973) and ACh turnover as measured by incorporation of either [^{14}C]glucose or [^{2}H$_4$]choline into ACh has been found to be significantly diminished (Gibson et al., 1981).

4.4.2. Amino Acid Neurotransmitters

There is now a convincing body of evidence to support the contention that the amino acids glutamate, aspartate, and GABA are important neurotransmitters in the mammalian CNS. Several lines of evidence suggest that the neurotransmitter pools of these amino acids are labeled by glucose via pyruvate (Fig. 5).

For example, incorporation of glucose into these amino acids is sensitive to changes in physiological function, whereas incorporation from other precursors is not; anesthetic doses of sodium pentobarbitone cause a decreased incorporation of [^{14}C]glucose into amino acids, but have no effect on incorporation of [^{14}C]acetate or [^{14}C]butyrate (Cremer and Lucas, 1971). In addition, electrical stimulation of guinea pig cortex slices evokes a selective release of [^{14}C]amino acids newly synthesized from [^{14}C]glucose, and this release is calcium dependent and subject to inhibition by the sodium-channel-blocking agent tetrodotoxin (Potashner, 1978).

4.4.3. Neurotransmitter Release

Considerable evidence suggests that mitochondrial metabolism can influence neurotransmitter release by regulatory calcium levels in nerve terminals

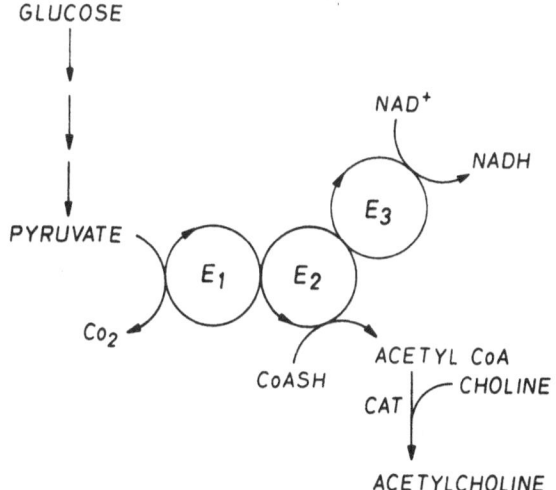

FIGURE 4. Role of PDHC in acetylcholine synthesis in brain.

(Alnaes and Rahaminoff, 1975), and it has been postulated that posttetanic facilitation of transmission is caused by a disturbance of this mitochondrial function (Rahaminoff et al., 1978; Browning et al., 1981a). There is evidence to suggest that calcium accumulation by brain mitochondria may be tightly linked to PDHC activity, the process being mediated by the phosphorylation of the enzyme complex. Dichloroacetate (DCA), a potent inhibitor of PDHC kinase, stimulates pyruvate-supported calcium accumulation at concentrations at which it stimulates PDHC activity and inhibits phosphorylation of the α-subunit of PDHC (Browning et al., 1981a). This suggests that changes in the phosphorylation state of PDHC may be accompanied by changes in calcium transport. Previous studies have shown that repetitive high-frequency stimulation produced changes in the endogenous phosphorylation of the α-subunit of PDHC (Browning et al., 1981b), and it was suggested that repetitive stimulation, by altering the phosphorylation of the α-subunit, changes PDHC activity thereby altering calcium sequestration by nerve terminal mitochondria.

4.5. PDHC and Neuronal Plasticity

The phosphorylation states of brain proteins are altered in the rat by pharmacological and electrophysiological treatments and by behavioral manipulations such as the induction of avoidance responses. It has recently been shown that the phosphorylation in vitro of a 41,000-dalton phosphopeptide, band F-2, was elevated in the frontal cortex in subjects trained on a step-down avoidance task, and furthermore that band F-2 is the α-subunit of PDHC (Mor-

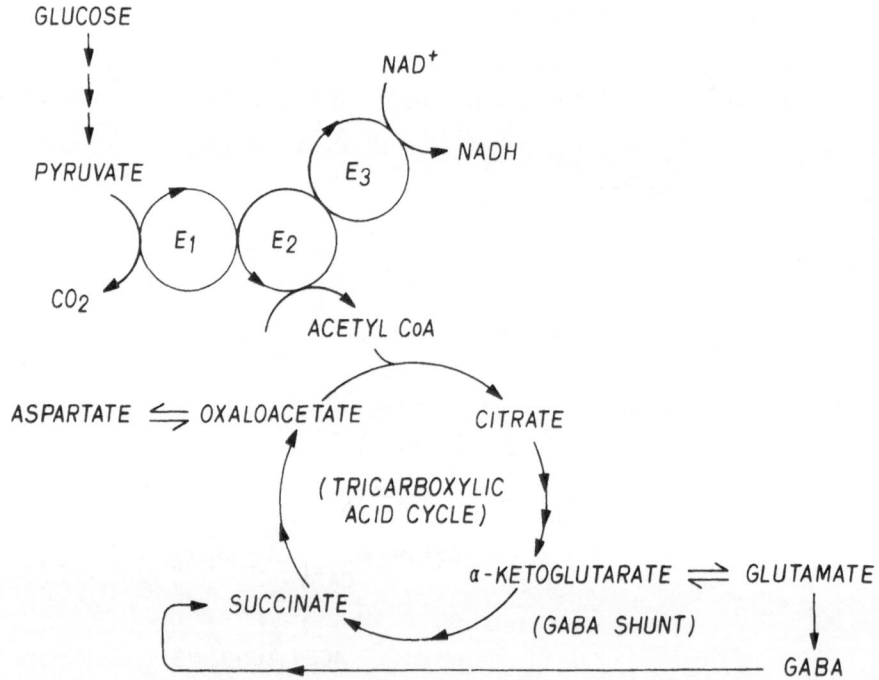

FIGURE 5. Role of PDHC in synthesis of putative amino acid neurotransmitters in brain.

gan and Routtenberg, 1981). Training-induced alterations in cerebral PDHC may thus regulate synaptic function; this could involve neurotransmitter systems such as acetylcholine or the amino acids (GABA, glutamate). Thus cerebral PDHC appears to be sensitive, through the phosphorylation-dephosphorylation state of the complex, to manipulations of cerebral activity, and it has been suggested that this property may participate in the biochemical response of the cells involved in neuronal plasticity of learning and memory (Morgan and Routtenberg, 1981).

5. Treatment of PDHC Deficiency Disorders

Patients with disorders of PDHC, similar to others with chronic metabolic diseases, respond poorly to intercurrent stress, showing rapid deterioration following infection or elective surgery (Blass, 1979). However, several therapeutic approaches yielded positive results, albeit on limited numbers of patients with PDHC deficiency.

5.1. Thiamine

Large doses of thiamine have frequently been used to treat PDHC deficiency disorders, more often than not with disappointing results. However, in one case, a 20-month-old patient with congenital lactic acidosis associated with an inherited deficiency of E_1 was treated with thiamine (1.8 g/day). The treatment successfully corrected the muscular hypotonia and ameliorated the ataxia in this patient. In addition, the biochemical deficit of E_1 in cultured skin fibroblasts from this patient was reversed by incubation with thiamine (Wick et al., 1977). Enzyme induction was ruled out as a mechanism of action of thiamine since an inhibitor of protein synthesis had no effect of the enhancement of activity by thiamine. One possible mechanism for the beneficial effect of thiamine in treatment of PDHC deficiency might involve PDHC regulatory mechanisms. Thiamine pyrophosphate (TPP) inhibits phosphorylation of PDHC, thereby maintaining the enzyme complex in its active form (Roche and Reed, 1972), and administration of thiamine to rats results in a decreased sensitivity of PDHC to inactivation by ATP (Hommes et al., 1973).

5.2. Ketogenic Diets

Since the mammalian brain can effectively use ketone bodies as an alternative source of acetyl CoA, high-fat diets have been tried, with some success, in the treatment of inherited PDHC deficiencies. In the most fully described study, two brothers with an inherited deficiency of E_1 (12% and 20% residual activity) were switched from a standard diet containing 40% of the energy as carbohydrate, 40% fat, and 20% protein to one containing 50% fat, 30% carbohydrate, and 20% protein (Falk et al., 1976). Blood pyruvate levels fell in both patients and this was accompanied by a decrease in the frequency and severity of the episodes of neurological deterioration. There are other reports of definite, if limited, therapeutic value of ketogenic diets in patients with PDHC disorders (Farrell, 1975).

Ketogenic diets do not appear to be useful in the treatment of Leigh's disease associated with PDHC activation defects (De Vivo *et al.*, 1979; Butterworth and Melancon, unpublished findings 1980) or in cases of PDHC phosphate phosphatase deficiency (Robinson *et al.*, 1977b). If these findings are found to be generally applicable, this will reinforce the need for accurate reproducible biochemical measurements in appropriate cells from patients with different types of PDHC deficiency.

5.3. Dichloroacetate

Dichloroacetate (DCA) was found in experimental studies to inhibit PDHC kinase with concomitant activation of PDHC (Alberti and Holloway, 1977). In clinical studies, DCA lowers pyruvate and lactate levels in the serum of diabetic patients (Stacpoole *et al.*, 1978). However, in a trial on a patient with congenital lactic acidosis of infancy, although DCA was found to be effective in decreasing blood pyruvate and lactate, no clinical improvement was observed during a six-month trial period (Coude *et al.*, 1978). Later experimental studies suggested that DCA may not readily cross the blood–brain barrier (Evans, 1981b) and this finding, coupled with the reports of significant toxic side effects of the drug, may render it of limited therapeutic value in the treatment of PDHC deficiency disorders.

5.4. Other Therapeutic Approaches

Steroids have been shown to reduce the severity of attacks of intermittent ataxia in a patient with PDHC deficiency (Blass *et al.*, 1971), but the mechanism of action remains unknown. Administration of citrate to a similar patient, in whom low pyruvate oxidation was demonstrated in intact white blood cells, caused definite amelioration of symptoms (Oka *et al.*, 1976). No confirmation of this interesting finding, however, has been reported to date. A case of Leigh's disease was successfully treated with lipoic acid (Clayton *et al.*, 1967). Accompanying the general clinical improvement in this patient, evidence of a biochemical response to lipoic acid was suggested by the slow but continuous decrease in blood pyruvate.

There is, to this author's knowledge, only one report of attempted prenatal diagnosis of inherited PDHC. In this report, PDHC activity had been found to be 5% of control values in liver and cultured skin fibroblasts from a child born previously. In the subsequent pregnancy, normal pyruvate degradation in cultured amniotic fluid cells was revealed and a healthy child was born (Wendel *et al.*, 1978).

6. Summary

PDHC is a multienzyme complex responsible for the irreversible transformation of pyruvate into acetyl CoA. Inherited neurological diseases involving genetic mutations of each of the constituent enzymes of PDHC have now been described. In addition, there is a growing body of evidence to suggest that an inherited defect of PDHC activation may be associated with Leigh's subacute necrotizing encephalopathy. Acquired disorders of PDHC include heavy metal intoxication and the encephalopathy associated with chronic thiamine deficiency.

Cerebral PDHC increases in activity during the postnatal period reflecting the changeover of cerebral energy substrate from ketone bodies to glucose. PDHC develops in brain in a region-selective manner, in parallel with the establishment of adult patterns of cerebral metabolism and of maturation of cerebral function.

In spite of the possibility that PDHC may be rate limiting, mild impairments of the enzyme complex do not appear to be associated with a cerebral energy deficit. However, neurotransmitter metabolism is impaired. In particular, cholinergic function as well as that mediated by the glucose-derived amino acid neurotransmitters appear to be selectively compromised. Cerebellum appears to be particularly vulnerable to acquired and inherited deficiencies of PDHC.

A recent report suggests that regulation of PDHC phosphorylation may play a key role in the modulation of synaptic plasticity and may be involved in the biochemical response of cells involved in the neuronal plasticity of learning and memory.

Treatment of PDHC deficiencies remains a challenge. Some success has been reported using cofactors of PDHC (thiamine and lipoic acid) or alternative cerebral energy substrates (ketone bodies).

ACKNOWLEDGMENTS. The author wishes to express his thanks to The Medical Research Council of Canada and C.I.L., Inc. for financial assistance and to Ms. Sylvie de Bellefeuille for her assistance with the preparation of this manuscript.

References

Alberti, K. G. M. M., and Holloway, P. A. H., 1977, Dichloroacetate and phenformin-induced lactic acidosis, *Diabetes* **26**:377.

Alnaes, E., and Rahaminoff, R., 1975, On the role of mitochondria in transmitter release from motor nerve terminals, *J. Physiol.* **248**:285–306.

Barbeau, A., Melancon, S. B., Butterworth, R. F., Filla, A., Izumi, K., and Ngo, T. T., 1978, Pyruvate dehydrogenase complex in Friedreich's Ataxia, in: *The Inherited Ataxias*, Volume 21, *Advances in Neurology*, Raven Press, New York, pp. 203–217.

Beaudry, M., Gall, C., Kessler, M., Alapour, H., and Lynch, G., 1983, Denervation-induced decrease in mitochondrial calcium transport in rat hippocampus, *J. Neurosci.* **3**:252–259.

Benjamins, J. A., and McKhann, G. M., 1972, Neurochemistry of development, in: *Basic Neurochemistry* (R. W. Albers, G. J. Siegel, R. Katzman, B. W. Agranoff, eds.), Little, Brown and Co., Boston, pp. 269–298.

Berrera, C. R., Namihira, G., Hamilton, L., Munk, P., Eley, M. H., Linn, T. C., and Reed, L. J., 1972, α-Ketoacid dehydrogenase complexes from bovine kidney and heart, *Arch. Biochem. Biophys.* **148**:343–358.

Bertagnolio, B., Uziel, G., Bottacchi, E., Crenna, G., D'Angelo, A., and Di Donato, S., 1980, Friedreich's Ataxia. II. Biochemical studies in cultured cells, *Ital. J. Neurol. Sci.* **1**:239–243.

Blass, J. P., 1980, Pyruvate dehydrogenase deficiencies, in: *Inherited Disorders of Carbohydrate Metabolism* (D. Burman, J. B. Holton, and C. A. Pennock, eds.), MTP Press, Lancaster, United Kingdom, pp. 239–268.

Blass, J. P., 1981, Hereditary ataxias, in: *Current Neurology*, Volume III (S. H. Appel, ed.), John Wiley and Sons, New York, pp. 66–91.

Blass, J. P., and Gibson, G. E., 1978, Studies of the pathophysiology of pyruvate dehydrogenase deficiency, in: *The Inherited Ataxias*, Volume 21, *Advances in Neurology*, Raven Press, New York, pp. 181–194.

Blass, J. P., and Lewis, C. A., 1973, Kinetic properties of the partially purified pyruvate dehydrogenase complex of ox brain, *Biochem. J.* **131**:31–37.

Blass, J. P., Kark, R. A. P., and Engel, W. K., 1971, Clinical studies of a patient with pyruvate decarboxylase deficiency, *Arch. Neurol.* **25**:449–460.

Blass, J. P., Schulman, J. D., Young, D. S., and Hom, E., 1972, An inherited defect affecting the tricarboxylic acid cycle in a patient with congenital lactic acidosis, *J. Clin. Invest.* **51**:1845–1851.

Blass, J. P., Kark, R. A. P., and Menon, N. K., 1976a, Low activities of the pyruvate and oxoglutarate dehydrogenase complexes in five patients with Friedreich's Ataxia, *N. Engl. J. Med.* **295**:62–67.

Blass, J. P., Cederbaum, S. D., and Dunn, H. G., 1976b, Biochemical abnormalities in Leigh's disease, *Lancet* **1**:1237–1238.

Blass, J. P., Cederbaum, S. D., and Gibson, G. E., 1976c, Clinical and metabolic abnormalities accompanying deficiencies in pyruvate oxidation, in: *Normal and Pathological Development of Energy Metabolism* (F. A. Hommes and C. J. Van den Berg, eds.), Academic Press, New York pp. 193–210.

Booth, R. F. G., Patel, T. B., and Clark, J. B., 1980, The development of enzymes of energy metabolism in the brain of a precocial (guinea pig) and non-precocial (rat) species, *J. Neurochem.* **34**:17–25.

Browning, M., Baudry, M., Bennett, W. F., and Lynch, G., 1981a, Phosphorylation mediated changes in pyruvate dehydrogenase activity influence pyruvate-supported calcium accumulation by brain mitochondria, *J. Neurochem.* **36**:1932–1940.

Browning, M., Bennett, W. F., Kelly, P., and Lynch, G., 1981b, Evidence that the 40,000 MW phosphoprotein influenced by high frequency synaptic stimulation is the α-subunit of pyruvate dehydrogenase, *Brain Res.* **218**:255–266.

Butterworth, R. F., 1982a, Regional amino acid neurotransmitter distribution in thiamine deficiency, *Ann. N. Y. Acad. Sci.* **378**:464–465.

Butterworth, R. F., 1982b, Neurotransmitter function in thiamine-deficiency encephalopathy, *Neurochem. Int.* **4**:449–464.

Butterworth, R. F., and Giguère, J. F., 1984, Pyruvate dehydrogenase activity in regions of the rat brain during post-natal development *J. Neurochem.* **43**:280–282.

Cederbaum, S. D., Blass, J. P., Minkoff, N., Brown, W. J., Cotton, M. E., and Harris, S. H., 1976, Sensitivity to carbohydrate in a patient with familial intermittent lactic acidosis and pyruvate dehydrogenase deficiency, *Pediatr. Res.* **10**:713–720.

Clayton, B. E., Dobbs, R. H., and Patrick, A. D., 1967, Leigh's subacute necrotising encephalopathy: Clinical and biochemical study with special reference to therapy with lipoate, *Arch. Dis. Child.* **42**:467–478.

Coude, F. X., Saudubray, J. M., Demaugre, F., Marsac, C., Leroux, J. P., and Charpentier, C., 1978, Dichloroacetate as treatment for congenital lactic acidosis, *Lancet* **299**:1365–1366.

Cremer, J. E., 1962, The action of triethyltin, triethyl lead, ethyl mercury and other inhibitors on the metabolism of brain and kidney slices in vitro using substrates labeled with ^{14}C, *J. Neurochem.* **9**:289–298.

Cremer, J. E., 1970, Selective inhibition of glucose oxidation by triethyltin in rat brain in vivo, *Biochem. J.* **119**:95–102.

Cremer, J., and Lucas, H. M., 1971, Sodium pentobarbitone and metabolic compartments in rat brain, *Brain Res.* **35**:619–621.

Cremer, J. E., and Teal, H. M., 1974, The activity of pyruvate dehydrogenase in rat brain during postnatal development, *Febs Lett.* **39**:17–20.

De Vivo, D. C., Haymond, M. W., Obert, K. A., Nelson, J. S., and Pagliara, A. S., 1979, Defective activation of the pyruvate dehydrogenase complex in subacute necrotising encephalomyelopathy (Leigh disease), *Ann. Neurol.* **6**:483–494.

Dreyfus, P. M., and Hauser, G., 1975, The effect of thiamine deficiency on the pyruvate decarboxylase system of the central nervous system, *Biochim. Biophys. Acta* **104**:78–84.

Evans, O. B., 1981a, Pyruvate decarboxylase deficiency in subacute necrotising encephalomyelopathy, *Arch. Neurol.* **38**:515–519.

Evans, O. B., 1981b, The effects of dichloroacetate on brain tissue, *Neurology* **31**:87.

Falk, R. E., Cederbaum, S. D., Blass, J. P., Gibson, G. E., Kark, R. A. P., and Carrel, R. E., 1976, Ketonic diet in the management of pyruvate dehydrogenase deficiency, *Pediatrics* **58**:713–721.

Farmer, T. W., Veath, L., Miller, A. L., O'Brien, J. S., and Rosenberg, R. N., 1973, Pyruvate decarboxylase deficiency in a patient with subacute necrotising encephalomyelopathy, *Neurology* **23**:429.

Farrell, D. F., Clark, A. F., Scott, C. R., and Wennberg, R. P., 1975, Absence of pyruvate decarboxylase activity in man: A cause of congenital lactic acidosis, *Science* **187**:1082–1084.

Fowler, B. A., Woods, J. S., and Schiller, C. M., 1977, The ultrastructural and biochemical effects of prolonged oral arsenic exposure on liver mitochondria of rats, *Environ. Health Perspect.* **19**:197–201.

Gibson, G. E., Jope, R., and Blass, J. P., 1975, Decreased synthesis of acetylcholine accompanying impaired oxidation of pyruvic acid in rat brain minces, *Biochem. J.* **148**:17–23.

Gibson, G. E., Peterson, C., and Sansone, J., 1981, Decreases in amino acid and acetylcholine metabolism during hypoxia, *J. Neurochem.* **37**:192–201.

Gremek, D. E., McCafferty, M. R., O'Neill, K. J., Melamed, B. R., and O'Neill, J. J., 1981, Effect of inorganic lead on rat brain mitochondrial respiration and energy production, *J. Neurochem.* **36**:1109–1113.

Hawkins, R. A., Miller, A. L., Cremer, J. E., and Veech, R. L., 1974, Measurement of the rate of glucose utilization by rat brain in vivo, *J. Neurochem.* **23**:917–923.

Haworth, J. C., Perry, T. L., Blass, J. P., Hansen, S., and Urquhart, N., 1976, Lactic acidosis in three sibs due to defects in both pyruvate dehydrogenase and α-ketoglutarate dehydrogenase complexes, *Pediatrics* **58**:564–572.

Heinrich, C. P., Stadler, H., and Weiser, H., 1973, The effect of thiamine-deficiency on the acetylcoenzyme A and acetylcholine levels in rat brain, *J. Neurochem.* **21**:1273–1281.

Hommes, F. A., Berger, R., and Luit-De-Haan, G., 1973, The effect of thiamine treatment on the activity of pyruvate dehydrogenase: Relation to the treatment of Leigh's encephalomyelopathy, *Pediatr. Res.* **7**:616–619.

Jope, R., and Blass, J. P., 1976, The regulation of pyruvate dehydrogenase in brain in vivo, *J. Neurochem.* **26**:709–714.

Kark, R. A. P., Blass, J. P., and Engel, W. K., 1974, Pyruvate oxidation in neuromuscular diseases, *Neurology* **24**:964–971.

Kohlschutter, A., Kraus-Ruppert, R., Rohrer, T., and Herschkowitz, N. N., 1978, Myelin studies in a case of subacute necrotising encephalomyelopathy, *J. Neuropathol. Exp. Neurol.* **317**:155–164.

Ksiezak, H. J., and Gibson, G. E., 1981, Acetylcholine synthesis and CO_2 production from variously labeled glucose in rat brain slices and synaptosomes, *J. Neurochem.* **37**:88–94.

Ksiezak-Reding, H., Blass, J. P., and Gibson, G. E., 1982, Studies on the pyruvate dehydrogenase complex in brain with the arylamine acetyltransferase coupled assay, *J. Neurochem.* **38**:1627–1636.

Land, J. M., and Clark, J. B., 1976, The changing pattern of brain mitochondrial substrate utilization during development, in: *Normal and Pathological Development of Energy Metabolism* (F. A. Hommes and C. J. Van den Berg, eds), Academic Press, New York, pp. 155–167.

Land, J. M., Booth, R. F. G., Berger, R., and Clark, J. B., 1977, Development of mitochondrial energy metabolism in rat brain, *Biochem. J.* **164**:339–348.

Leigh, D., 1951, Subacute necrotising encephalomyelopathy in an infant, *J. Neurol. Neurosurg. Psychiatry,* **14**:216–221.

Livingstone, I. R., Mastaglia, F. L., and Pennington, R. J. T., 1980, An investigation of pyruvate metabolism in patients with cerebellar and spinocerebellar degeneration, *J. Neurol. Sci.* **48**:123–132.

Lonsdale, D., Faulkner, W. R., Price, J. W., and Smeby, R. R., 1969, Intermittent cerebellar ataxia associated with hyperpyruvic acidemia, hyperalanemia, and hyperalaninuria, *Pediatrics* **43**:1025–1034.

McCandless, D. W., and Schenker, S., 1968, Encephalopathy of thiamine deficiency: Studies of intracerebral mechanisms, *J. Clin. Invest.* **47**:2268–2280.

McIlwain, H., and Bachelard, H. S., 1971, *Biochemistry and the Central Nervous System*, Churchill-Livingstone, London, pp. 1–412.

Melancon, S. B., Potier, M., Dallaire, L., Rollin, P., Fontaine, G., and Grenier, B., 1979, Pyruvate dehydrogenase, lipoamide dehydrogenase and citrate synthetase activity in fibroblasts from patients with Friedreich's and Charlevoix–Saguenay ataxia, *Can. J. Neurol. Sci.* **6**:241–242.

Menon, N. K., and Kark, R. A. P., 1976, Inhibition of oxidation in chronic alkyl mercury poisoning, *Trans. Am. Soc. Neurochem.* **7**:151.

Morgan, D. G., and Routtenberg, A., 1981, Brain pyruvate dehydrogenase: Phosphorylation and enzyme activity altered by a training experience, *Science* **214**:470–471.

Oka, Y., Matsuda, I., Arashima, S., Anakura, M., Mitsuyama, T., and Nagamatsu, I., 1976, Citrate treatment in a patient with pyruvate decarboxylase deficiency, *Tohoku J. Exp. Med.* **118**:131–135.

Patel, A., and Koenig, H., 1971, Some neurochemical aspects of fluorocitrate intoxication, *J. Neurochem.* **18**:621–628.

Plaitakis, A., Nicklas, W. J., and Desnick, R. J., 1980, Glutamate dehydrogenase deficiency in three patients with spinocerebellar syndrome, *Ann. Neurol.* **7**:297–303.

Potashner, S. J., 1978, Effects of tetrodotoxin, calcium and magnesium on the release of amino acids from slices of guinea pig cerebral cortex, *J. Neurochem.* **31**:187–195.

Prick, M., Gabreels, F., Reiner, W., Trijbels, F., Jaspar, H., Lamers, K., and Kok, J., 1981, Pyruvate dehydrogenase deficiency restricted to brain, *Neurology* **31**:398–404.

Rahaminoff, R., Erulkar, S. D., Lev-Tov, A., and Meiri, H., 1978, Intracellular and extracellular calcium ions in transmitter release at the neuromuscular synapse *Ann. N.Y. Acad. Sci.* **307**:583–587.

Reynolds, S. F., and Blass, J. P., 1976, A possible mechanism for selective cerebellar damage in partial pyruvate dehydrogenase deficiency, *Neurology* **26**:625–628.

Robinson, B. H., Gall, D. G., and Cutz, E., 1977a, Deficient activity of hepatic pyruvate carboxylase in Reye's syndrome, *Pediatr. Res.* **11**:279.

Robinson, B. H., Taylor, J., and Sherwood, W. G., 1977b, Dihydrolipoyl dehydrogenase deficiency— A cause of congenital lactic acidosis, *Pediatr. Res.* **11**:520.

Robinson, B. H., Taylor, J., and Sherwood, W. G., 1980, The genetic heterogeneity of lactic acidosis: Occurrence of recognizable inborn errors of metabolism in a pediatric population with lactic acidosis, *Pediatr. Res.* **14**:956–962.

Rose, M. S., and Aldrige, W. N., 1966, Triethyltin and the incorporation of (^{32}P) phosphate into rat brain phospholipids, *J. Neurochem.* **13**:103–108.

Schiller, C. M., Fowler, B. A., and Woods, J. S., 1977, Effects of arsenic on pyruvate dehydrogenase activation, *Environ. Health Perspect.* **19**:205–207.

Seiss, E., Wittmann, J., and Wieland, O., 1971, Interconversion and kinetic properties of pyruvate dehydrogenase from brain, *Z. Physiol. Chem.* **352**:447–452.

Sheu, K. F. R., and Blass, J. P., 1983, Normal PDH$_b$ phosphatase activity in pyruvate dehydrogenase complex deficient Leigh fibroblasts, *Clin. Res.* **31**(2):402A.

Sheu, K. F. R., Hu, C. W. C., and Utter, M. F., 1981, Pyruvate dehydrogenase complex activity in normal and deficient fibroblasts, *J. Clin. Invest.* **67**:1463–1471.

Siesjo, B. K., 1978, Hypoglycaemia, in: *Brain Energy Metabolism* (B. K. Siesjo, ed.), John Wiley and Sons, New York, pp. 380–397.

Sorbi, S., and Blass, J. P., 1982, Abnormal activation of pyruvate dehydrogenase in Leigh disease fibroblasts, *Neurology* **32**:555–558.

Sorbi, S., Bird, E. D., and Blass, J. P., 1981, Low activity of the pyruvate dehydrogenase complex as well as of choline acetyltransferase in Huntington caudate and putamen, *Neurosci. Abs.* **8**:494.

Stacpoole, P., Moore, G., and Kornhauser, D., 1978, Metabolic effects of dichloroacetate in patients with diabetes mellitus and hyperlipoproteinemia, *N. Engl. J. Med.* **298**:526–530.

Sterling, G. H., O'Neill, K. J., McCafferty, M. R., and O'Neill, J. J., 1982, Effect of chronic lead ingestion by rats on glucose metabolism and acetylcholine synthesis in cerebral cortex slices, *J. Neurochem.* **39**:592–596.

Stromme, J. H., Borud, O., and Moe, P. J., 1976, Fatal lactic acidosis in a newborn attributable to a congenital defect of pyruvate dehydrogenase, *Pediatr. Res.* **10**:60–66.

Stumpf, B., and Kraus, H., 1979, Regulatory aspects of glucose and ketone body metabolism in infant rat brain, *Pediatr. Res.* **13**:585–590.

Stumpf, D. A., and Parks, J. K., 1979, Friedreich Ataxia. II. Normal kinetics of lipoamide dehydrogenase, *Neurology* **29**:820–826.

Stumpf, D. A., Parks, J. K., Eguren, L. A., and Haas, R., 1982, Friedreich Ataxia. III. Mitochondrial
malic enzyme deficiency, *Neurology* **32**:221–227.

Toshima, K., 1980, Enzymic diagnosis of pyruvate dehydrogenase complex deficiency using blood
platelets, *Nippon Shonika Gakkai Zasshi* **84**:469–476.

Wendel, U., Przyrembel, H., Becker, K., Walther, B., Berger, R., and Bremer, H. J., 1978, Pyruvate-
dehydrogenase deficiency. Lethal course of the disease during infancy, *Monatsschr. Kinder-
heilkd.* **126**:140–147.

Wick, H., Schweizer, K., and Baumgartner, R., 1977, Thiamine dependency in a patient with congenital lactic acidaemia due to pyruvate dehydrogenase deficiency, *Agents Actions* **7**:405–410.

Willems, J. L., Monnens, L. A. H., Trijbells, J. M. F., Sengers, R. A. C., and Veerkamp, J. H., 1974,
Pyruvate decarboxylase deficiency in liver, *N. Engl. J. Med.* **290**:406–411.

6

Carbon Dioxide Narcosis

ALEXANDER L. MILLER

1. Introduction

Carbon dioxide shares the property with several other agents considered in this volume of being an end product of metabolism that must be constantly eliminated from the body in order to keep it from accumulating to toxic levels. Unlike other end products such as urea and bilirubin, carbon dioxide plays many important physiological and biochemical roles in the body which require that it be maintained at an optimal rather than just a low level. Also unlike other endogenous metabolic end products (except ammonia), carbon dioxide is present in multiple chemical forms in the body (dissolved CO_2, H_2CO_3, HCO_3^-). Both the absolute concentrations and the relative proportions of these forms are of great significance to a wide variety of tissue functions.

Given the above, it is clear that abnormal CO_2 levels can be of a variety of types, which are encompassed by the clinical terms respiratory acidosis, respiratory alkalosis, metabolic acidosis, and metabolic alkalosis. All of these can be uncompensated, partially compensated, or fully compensated (where compensation refers to the extent to which pH has returned toward normal from its maximum deviation). This chapter focuses on the cerebral metabolic effects of respiratory (i.e., hypercapnic) acidosis. In order to put this condition into context, the normal physiology and biochemistry of carbon dioxide in the body will first be briefly reviewed, followed by a discussion of the mental and neurophysiological effects of hypercapnic acidosis. This information is intended to emphasize two points. First, the effects of increased carbon dioxide levels are ubiquitous throughout the body, i.e., CO_2 is not simply a neurotoxin, although the central nervous system is especially sensitive to toxic levels of CO_2. Second, the effects of CO_2 on CNS energy metabolism should be viewed in conjunction with its effects on neurophysiological and mental functions if we are to gain an understanding of the links between metabolism and function.

ALEXANDER L. MILLER • Department of Psychiatry, The University of Texas Health Science Center at San Antonio, San Antonio, Texas 78284.

2. Normal Physiology of CO_2

This section is not intended to be an exhaustive review of carbon dioxide physiology, nor does it attempt to shed new light on areas of controversy. It is important for an understanding of subsequent material, however, that the reader who may have forgotten some of the fundamentals of acid-base chemistry and of CO_2 physiology be reminded of the most salient points. Lengthier discussion may be found in texts on the subject (Davenport, 1974; Filley, 1971).

2.1. Production and Elimination of CO_2

Most of the CO_2 produced in the body is from the dehydrogenase reactions of the Krebs cycle [pyruvate dehydrogenase, isocitrate dehydrogenase (NAD), and α-ketoglutarate dehydrogenase]. Other decarboxylating reactions (e.g., dopa decarboxylase, 6-phosphogluconate dehydrogenase, glutamic acid decarboxylase) are of great functional importance, but make only minor contributions to total CO_2 production. Since the Krebs cycle rapidly ceases to operate if the NADH produced by the dehydrogenases is not oxidized by the respiratory chain, it follows that CO_2 production is largely dependent on oxygen availability and that increases in oxidative phosphorylation (as in aerobic exercise) cause increases in CO_2 production. The exact stoichiometry between O_2 consumed and CO_2 produced depends on the fuel being consumed. The respiratory quotient (CO_2 produced/O_2 consumed) is 1.0 for carbohydrates and about 0.7 for lipids.

Most tissues can fix CO_2, with the most active reaction usually being pyruvate carboxylase. Such reactions are typically energy requiring, however, and the products (e.g., oxaloacetate from pyruvate + CO_2) are intermediates, not end products, of metabolism. In all aerobic organisms, elimination of CO_2 is ultimately by diffusion into the surrounding medium. In animals with lungs, this is accomplished by (1) diffusion of CO_2 from tissue into blood, (2) transport in blood to the lungs, (3) diffusion into the alveoli, and (4) exhalation. In principle, interference with any of these four processes could result in excessive CO_2 retention. In clinical reality, however, diffusion is not rate limiting to CO_2 elimination. It is, then, inadequate amount of exhalation of CO_2 or inadequate delivery of CO_2 in blood to the lungs that causes CO_2 retention in patients. Some of the clinical conditions in which CO_2 retention occurs are discussed in a subsequent section.

2.2. Acid–Base Chemistry

CO_2 can dissolve in aqueous solutions and be hydrated to form carbonic acid, H_2CO_3. Carbonic acid in turn dissociates into H^+ + HCO_3^-. The interrelationships among these molecules are expressed in the Henderson-Hasselbach equation:

$$pH = pK + \log \frac{[HCO_3^-]}{apCO_2}$$

where the constant, a, times the partial pressure of CO_2 gas in the solution equals the concentration of dissolved CO_2 in the solution (plus a very small amount of undissociated carbonic acid). In human plasma at 37 °C, pK is 6.1 and a is 0.03. The usual ratio of $[HCO_3^-]$ to $apCO_2$ is about 20/1.

The whole system is well suited to buffer acute increases in H^+ due to production of organic acids within the body or to ingestion of acids. Thus the added H^+ reacts with HCO_3^- to form CO_2 and water and the extra CO_2 is exhaled. By increasing the volume and/or frequency of respiration (hyperventilation), the pCO_2 can be lowered below its normal level (40 mm Hg) to the point where the ratio $[HCO_3^-]/apCO_2$ is again 20/1. In clinical parlance, this constitutes complete respiratory compensation for the initial metabolic acidosis. Such compensation is limited, however, by the amount of available HCO_3^-, which would be completely utilized if excess acid kept being produced or ingested. Thus, the role of the kidneys is to excrete H^+ and regenerate HCO_3^-, restoring it to its normal concentration. Renal compensation occurs in a time frame of hours to days, whereas large changes in pCO_2 and, hence, in pH can occur with a few minutes of hyperventilation or hypoventilation.

It is essential to the proper functioning of the above system that pulmonary respiration be linked to pH and/or pCO_2. Respiration is under the control of the CNS, where changes in blood pH and pCO_2 are detected and respiratory rate and volume adjusted accordingly. The details of these processes are well beyond the scope of this chapter. For present purposes the important point is that one cause of hypercapnic acidosis is a decrease in the sensitivity of the CNS sensors of pH and pCO_2, which results in diminished ventilation and accumulation of CO_2 until a new steady state is achieved. (The state is steady with regard to ventilation. The kidneys will gradually compensate until the pH returns to near normal levels.).

2.3. CO_2 Metabolism in the Brain

Brain is not particularly different from other tissues in metabolism of CO_2, but some aspects merit emphasis. First, brain is aerobic and metabolically active. From its normal rate of oxygen consumption of 3.4 ml/100 g per min (Kety and Schmidt, 1948) and weight of 1500 g, it can be calculated that daily human brain CO_2 production is about 70 liters. Second, dissolved CO_2 is freely permeable across all membranes, including the blood–brain barrier. Thus, changes in plasma pCO_2 immediately result in changes in intracellular pCO_2 and pH. Third, bicarbonate is only slowly permeable across the blood–brain barrier. In fact, when bicarbonate infusions into blood are used to correct an acidosis, pCO_2 in cerebrospinal fluid may rise more rapidly than $[HCO_3^-]$ thus leading to a "paradoxical acidosis." Fourth, as in other tissues, both CO_2 (either as CO_2 or HCO_3^-) and H^+ are participants in many enzymatic reactions in brain, and the activities of many enzymes are sensitive to pH changes in the physiological range, even though H^+ is not a reactant. Thus, there are a very large number of enzymatic steps that can potentially be affected by changes in pCO_2 and pH.

3. Clinical, Psychological, and Neurophysiological Aspects

3.1. Clinical Manifestations

As will be discussed in the next section, medical illnesses and conditions that result in CO_2 retention do so either by causing alveolar hypoventilation or by causing inadequate perfusion of ventilated lung. Thus, both oxygen and carbon dioxide exchange are affected and there is a lowering of arterial pO_2 as well as an increase in pCO_2. Because the pO_2 can fall considerably without a significant effect on oxygen delivery to tissues, however, it is frequently true that the changes in mental status observed clinically are attributable mainly to hypercapnia and not to the associated hypoxemia.

Humans are quite sensitive to acute increases in pCO_2, with subtle alterations in mentation being observable after rises of 10 mm Hg or less. The clinical picture of hypercapnic encephalopathy at CO_2 levels below those which produce coma ($pCO_2 < 70$–80 mm Hg) can include headaches (bilateral), dullness, forgetfulness, drowsiness, confusion, asterixis, increased intracranial pressure, slowing of the EEG, papilledema, and tremor (Victor and Adams, 1980). These effects are all reversible with correction of the blood gases and pH to normal. Because of its coma-producing properties, CO_2 was tried as a general anesthetic for surgery. Muscle twitches and jerks occur at unpredictable intervals, however, so that it did not prove to be a feasible anesthetic.

The typical patient with lung disease who becomes hypercapnic shows some or all of the above signs and symptoms, as CO_2 rises over the course of many minutes to hours. Patients who become acutely hypercapnic as the result of an overdose of a CNS depressant have a clinical picture that is dominated by the drug effect (usually deeply comatose). A different set of reactions are observed in patients who breathe high concentrations of CO_2 and raise their pCO_2 many-fold over a minute or two. The opportunity to study these effects in thousands of patients came about because of the fairly widespread use of CO_2 therapy in the 1950s for the psychoneuroses. (The roster of the Carbon Dioxide Research Association listed 127 members in 1958.) The treatments consisted of inhaling 30% CO_2 ($+20\%$ O_2, 50% N_2) to the point of unconsciousness on multiple occasions. Meduna, who had pioneered the use of pentylenetetrazol-induced seizures for the treatment of schizophrenia in the 1930s, was the most ardent proponent of and prolific writer about CO_2 treatment. His description is of rather dramatic motor excitement and hallucinatory phenomena in the moment preceding loss of consciousness, and of tonic extensions and withdrawal seizures following this (Meduna, 1958). The physical manifestations are similar to those noted in mice and rats placed in atmospheres containing 30–50% CO_2 (Woodbury et al., 1958). It seems, then, that the effects of very acute, extreme hypercapnia differ from those of the more moderate, more slowly developing hypercapnia typical of medical conditions. CO_2 treatment has largely disappeared from use, as judged by an absence of reports in the recent psychiatric literature. Whether it benefitted a group of patients not helped by presently available treatments was never tested in adequately controlled studies.

3.2. Medical and Nonmedical Causes of Hypercapnic Acidosis

Chronic obstructive lung disease is the single medical illness most frequently associated with episodes of symptomatic hypercapnia (often superimposed on chronic, mild, assymptomatic hypercapnia). Many of these patients have respiratory centers that have lost their sensitivity to CO_2 and their breathing is maintained by hypoxic stimuli. If they are given a gas mixture to breathe that is too high in O_2, this stimulus is lost, respirations slow, and CO_2 can accumulate to dangerous, narcotic levels (West, 1980). Other causes of hypoventilation include drug-induced depression of the respiratory centers (especially by opiates and barbiturates), diseases of the brain stem, diseases or injuries affecting the motor nerve output to the muscles of respiration (e.g., polio), diseases of the respiratory muscles (e.g., muscular dystrophy), injuries to or abnormalities of the thoracic cage, obstruction of the upper airways, and extreme obesity (West, 1980). Drug-induced hypoventilation is particularly significant at birth, when a mild hypercapnic acidosis normally occurs as the task of oxygenating the newborn's blood shifts from the placenta to the lungs. Failure of the respiratory centers to adequately respond to this stimulus requires that the child be given ventilatory assistance until the drug effects have abated, if hypoxic brain damage is to be avoided.

The conditions just described cause the body to retain excessive amounts of endogenously produced CO_2. Hypercapnic acidosis can also be caused by breathing atmospheres containing high amounts of CO_2. Carbon dioxide therapy has already been mentioned. Its use now is very rare. Persons in submarines, spacecraft, or any other container that has limited contact with the atmosphere, are potentially at risk for CO_2 accumulation. Most have very efficient CO_2-trapping mechanisms, and problems with CO_2 intoxication are rare. No gross deficits in motor or problem-solving abilities were found in young men continuously breathing an atmosphere of 4% CO_2 for two weeks (Storm and Giannetta, 1974).

3.3. Amnestic Properties of CO_2

The clinical descriptions of CO_2 therapy emphasize the vivid dreams and abreactive experiences of patients as they go into coma (e.g., Meduna, 1958). There is an extensive literature on the amnestic effects of high concentrations of CO_2 in animals. Although much of the work has used rodents placed in a 100% CO_2 atmosphere until respirations cease (e.g., Leukel and Quinton, 1964), it has been shown that anoxia is not the cause of the amnesia (Taber and Banuazizi, 1966). The retrograde amnesia produced by CO_2 in mice and goldfish is most evident if the exposure to CO_2 occurs within a few minutes of the test situation and is not detectable if the interval is greater than 60–120 min (Taber and Banuazizi, 1966; Riege and Cherkin, 1973). The lowest level of CO_2 found to produce retrograde amnesia in rats was 30% (Quinton, 1966). At this level, but not with 50% CO_2, increasing the duration of exposure increased the degree of retrograde amnesia (Quinton, 1966). Although CO_2 effects on human memory do not seem to have been systematically studied, memory problems during hypercapnia are clinically evident (Victor and Adams, 1980), and there is no reason to doubt that CO_2 can cause retrograde amnesia in humans. Whether the amnestic effects of CO_2 are directly or indirectly related to its effects on brain energy metabolism is an open question. A number of agents have been reported to lessen CO_2-induced amnesia, including ACTH and vasopressin analogues, α-methyl-p-tyrosine, propranolol, flurothyl, and the enkephalins (Rigter, 1973; Leonard and Rigter, 1973; Riege and Cherkin, 1976; Rigter, 1978).

3.4. Neurophysiological Effects of Hypercapnic Acidosis

As might be predicted from the huge numbers of biochemical reactions affected by H^+ and CO_2, the functioning of most body organs is influenced by hypercapnic acidosis. The scope of such effects is indicated in an entire issue of *Anesthesiology* (Volume 21, Number 6) in 1960 and the published proceedings of a 1971 symposium (Nahas and Schaefer, 1974). On the basis of present knowledge, however, it seems to be a fair generalization that systemic and peripheral effects are not the causes of most of the physiological and biochemical changes that occur in the CNS during acute hypercapnic acidosis. Rather, CO_2 and H^+ have direct effects on nervous tissue independent of their effects on the pulmonary, cardiovascular, and other systems.

There are two possible exceptions to this generalization. First, hypercapnia disturbs temperature regulation in mammals. The early effect is to inhibit both heat-losing and heat-retaining mechanisms, therefore making the animals more susceptible to both hypothermia and hyperthermia, depending on the environmental temperature (Stupfel, 1974; Schaefer *et al.*, 1975). Since hypothermia and hyperthermia affect cerebral glucose utilization, oxygen consumption, and metabolite levels (Hagerdal *et al.*, 1975a,b; Carlsson *et al.*, 1976; McCulloch *et al.*, 1982), CO_2 can indirectly affect cerebral metabolism in this way if normal body temperature is not maintained. Second, acute hypercapnic acidosis causes increased adrenal release of epinephrine (Schaefer *et al.*, 1968). There is some evidence that the adrenals play a role in increasing cerebral oxygen consumption of rats during immobilization stress (Carlsson *et al.*, 1977), although this did not seem to be a factor in hypercapnia (Berntman *et al.*, 1979).

One of the best-studied and most dramatic effects of hypercapnic acidosis is to increase cerebral blood flow (CBF). An early application of the nitrous oxide method for determining CBF was to the study of hypercapnia. An increase in arterial pCO_2 from 43 to 52 mm Hg caused a 75% increase in CBF in normal young men (Kety and Schmidt, 1948). Changes in CBF are approximately linear with changes in pCO_2 up to about 80 mm Hg in man, monkey, dog, goat, and rat, but above this level they asymptotically approach peak flow rates, and may even decline with 40% CO_2 in the rat (Reivich, 1964; Smith *et al.*, 1971; Grubb *et al.*, 1974; Berntman *et al.*, 1979; Artru and Michenfelder, 1980). The increased blood flow is associated with increased cerebral blood volume (Smith *et al.*, 1971; Grubb *et al.*, 1974), increased permeability of the blood–brain barrier to water (Preskorn *et al.*, 1981) and sucrose (Cameron *et al.*, 1969), and increased intracranial pressure (Victor and Adams, 1980). The mechanism of the cerebral vasodilatation with CO_2 is not known, but several studies suggest that prostaglandins are involved, since the prostaglandin inhibitor, indomethecin, prevents or greatly attenuates the increase in CBF with CO_2 (Pickard and MacKenzie, 1973; Sakabe and Siesjo, 1979; Pickard *et al.*, 1980; Dahlgren *et al.*, 1981).

The effects of CO_2 on the electrical activity and properties of nervous tissue are complex and no attempt will be made to review this literature. Some of the observations on intact animals are worth noting here, because the metabolic findings to be discussed next have been obtained under similar conditions.

In man, monkeys, mice, and rats, the seizure threshold (to electroshock) is raised by CO_2 (Pollock *et al.*, 1949; Stein and Pollock, 1949; Woodbury *et al.*, 1958). In mice and rats, this effect was seen at CO_2 concentrations of 5–20% (Woodbury *et al.*, 1958). The seizure threshold was below normal in mice in 40% CO_2 and not determinable in rats because of spontaneous seizures in 30–40% CO_2 (Woodbury *et al.*, 1958). CO_2 above 40% was anesthetic in rats (Woodbury *et al.*, 1958). In monkeys and man it took several minutes of breathing 15–20% CO_2 for the full effect on seizure threshold to be seen, whereas the full effect was seen after 30 sec of 30% CO_2 (Stein and Pollock, 1949; Pollock *et al.*, 1949). Lennox *et al.* (1956) found that CO_2 suppressed petit mal seizures and normalized the EEG under mildly hypoxic conditions. All of the above observations suggest a membrane-stabilizing effect of CO_2, which is in accord with the finding of increased membrane potential and increased threshold to electrical stimulation of isolated frog nerve in high CO_2 media (Lorente de No, 1947). The fact that spontaneous tonic and clonic movements occur with high concentrations of CO_2 probably reflects the differential sensitivity of different neurons to the CO_2 effects. Neurons in sensorimotor cortex, for instance, are much more sensitive to CO_2 than spinal motor neurons (Carpenter *et al.*, 1974). Thus, the seizurelike movements with high CO_2 may be due to release of lower motor neurons from tonic inhibitory influences. As will be seen in the subsequent section on metabolic effects of CO_2, cortex and forebrain show none of the changes in metabolite levels associated with epileptic seizures, at any level of hypercapnia.

4. Metabolic Aspects

The literature on the effects of CO_2 on cerebral energy metabolism from the past 35 years illustrates how new techniques have allowed investigators to obtain vital new information that has contributed to an increasingly coherent and consistent picture. It also illustrates the significant methodological gaps that continue to exist and keep several areas of controversy in an unresolved state.

4.1. Rates of Oxygen Consumption and Glucose Utilization

As noted above, an early application of the nitrous oxide technique for measuring CBF was to determine the effects of mild hypercapnia in man (Kety and Schmidt, 1948). The increase in CBF that had been qualitatively evident from preceding work was quantitatively confirmed, and it was shown that the arteriovenous difference of oxygen across the brain decreased in proportion to the increase in CBF, leaving the cerebral metabolic rate of oxygen consumption (CMR_{O_2}) unchanged. Subsequent studies in man found essentially the same result (Novack *et al.*, 1953; Cohen *et al.*, 1964; Fujishima *et al.*, 1971). In most of these studies the cerebral metabolic rate of glucose utilization (CMR_G) was not determined. In contrast to oxygen, the arteriovenous difference of glucose across the brain is normally small (e.g., 7% of the arterial level in anesthetized, normocapnic man) (Cohen *et al.*, 1964). It is, therefore, more difficult to accurately detect changes in CMR_G than in CMR_{O_2} when these rates are determined from the product of CBF and the arteriovenous differences of glucose and oxygen. Nevertheless, Cohen *et al.* (1964) did note that the ratio of oxygen to glucose use increased slightly during mild hypercapnia in man and hypothesized that, in addition to glucose, other substances might have been oxidized by brain under this condition. In a later study of canine brain, a significant decrease in CMR_G without any change in CMR_{O_2} during mild hypercapnia was reported (Xanalatos and James, 1972).

It is probably fair to say that the impressive evidence favoring glucose as the sole metabolic fuel for brain, which had been accumulated in the middle third of this century, made the finding of a discrepancy between oxygen consumption and glucose utilization harder to accept and interpret, especially given the problems in determining CMR_G from arteriovenous differences of glucose across the brain. With the discovery that uptake and metabolism of ketone bodies could account for over 50% of human CMR_{O_2} during prolonged starvation (Owen *et al.*, 1967), the idea that nonglucose fuels might contribute significantly to brain energy metabolism under some conditions gained new credence. The development of radioisotopic methods for quantitating the CMR_G of laboratory animals (Hawkins *et al.*, 1974; Sokoloff *et al.*, 1977) made *in vivo* measurements of CMR_G during hypercapnia simpler and more accurate. It is now well established by these methods that hypercapnic acidosis does inhibit CMR_G in adult rats (Miller *et al.*, 1975; Borgstrom *et al.*, 1976; Des Rosiers *et al.*, 1976). It has recently been found that CO_2 also inhibits the CMR_G of 10- and 20-day-old developing rats and that, for a given degree of hypercapnia, the percent inhibition is approximately the same at all ages, despite widely different baseline CMR_Gs (Miller and Corddry, 1981).

Though the existence of an inhibitory effect of CO_2 on CMR_G seems now to be widely agreed upon, the relationship of the extent of this inhibition to the degree and duration of hypercapnia is more controversial. In the report by Miller *et al.* (1975), it was found that during the first 5 min after onset of hypercapnia, the CMR_G was inversely related to pCO_2 in animals breathing 10, 20, or 30% CO_2. In their abstract, however, Des Rosiers *et al.* (1976) noted that in rats breathing 5 to 10% CO_2 for 40 to 90 min, there was little difference between the two CO_2 groups in their CMR_Gs over the last 30 min of treatment. Using deoxyglucose autoradiography, they found that the decreases in CMR_G were uniformly distributed throughout brain regions. The reports relating pCO_2 to CMR_G are summarized in Table I. The discrepancy between the results of the two studies is unresolved. The data on changes in brain metabolite concentrations during hypercapnia (discussed below) are consistent with the conclusion that increasing the degree of hypercapnic acidosis decreases the CMR_G.

TABLE I. Relationship between CMR_G and pCO_2 in the Rat.

Study	pCO_2 (mm Hg)	% Decrease in CMR_G
Miller *et al.* (1975)	81	13
	152	33
	223	43
Des Rosiers *et al.* (1976)	49	35
	72	40

A second question is whether the duration of hypercapnia influences the CMR_G. Over intervals of 2.5 to 60 min after onset of hypercapnia (20% CO_2 atmosphere), Miller et al. (1975) found CMR_G to range from 50 to 70% of the control rate. Given the nature of the approximations used in the calculations, these results indicate a rather constant effect of CO_2 on CMR_G over the course of time. Des Rosiers et al. (1976) also noted no effect of duration of CO_2 treatment (40–90 min) on CMR_G, and in developing rats, Miller and Corddry (1981) reported that CMR_G was depressed to the same level throughout 15 min treatment with 20% CO_2. Borgstrom et al. (1976), however, found the CMR_G during the first 2 min of hypercapnia to be 29% of control, whereas it was 69% of control after 14–16 min of exposure to 10% CO_2. Again, the issue is unresolved. It is not at all certain that the different observations are due to methodological differences, as Miller et al. (1982b), using essentially the same method as Borgstrom et al. (1976), found only about a 15% decrease in CMR_G during the 2- to 4-min interval after the onset of respiration with 10% CO_2.

Several studies in the past few years have re-examined the effects of CO_2 on CMR_{O_2}. Earlier workers had reported no effect of moderate hypercapnia (pCO_2 50–80 mm Hg) on CMR_{O_2} of man, dog, or rat (Kety and Schmidt, 1948; Novack et al., 1963; Cohen et al., 1964; Fujishima et al., 1971; Xanalatos and James, 1972; Eklof et al., 1973; Gjedde et al., 1975; Alberti et al., 1975; Nilsson and Siesjo, 1976; Gregoire et al., 1978). The question has been raised, however, whether hypercapnia at the upper end of this range and beyond may not depress CMR_{O_2} (Kliefoth et al., 1979; Artru and Michenfelder, 1980), and even whether mild hypercapnia in the rat considerably depresses both CMR_G and CMR_{O_2} (Des Rosiers et al., 1978). This debate is clearly relevant to the possible etiology of CO_2 narcosis, since depression of CMR_{O_2} is a characteristic finding with many general anesthetics (see Siesjo, 1978, for review).

The results of 18 studies of CMR_{O_2} during hypercapnia are summarized in Table II. The work of Kogure et al. (1975) is also listed, since the rate of utilization of high-energy phosphates and the CMR_{O_2} are presumed to be highly correlated. In six studies, the CMR_G was also measured and it is shown in parentheses. As mentioned earlier, the bulk of the evidence indicates a decrease in CMR_G during hypercapnia. In each case, except for the study of Alberti et al. (1975), the CMR_G was lower than would have been predicted from the CMR_{O_2} if glucose were the sole fuel for cerebral oxidative metabolism. A discussion of the oxidation of endogenous metabolites in brain during hypercapnia is contained in the section on changes in metabolite concentrations.

The range of reported CMR_{O_2}s during hypercapnia is from 135% of control in the rat at a pCO_2 of 81 (Hemmingsen et al., 1979) to 70% of control in the rat at a pCO_2 of 49 (Des Rosiers et al., 1978). The latter report was in abstract form only and it was not specified that body temperature was monitored or controlled, so that the possibility that hypothermia occurred exists. Nevertheless, the CMR_{O_2} in monkeys of 71% of control at a pCO_2 of 70 (Kliefoth et al., 1979) represents a very different result from the value of 115% of the control rate in baboons with a pCO_2 of 60 found by Pickard and MacKenzie (1973), and the range of values reported for rats and dogs is also large. In reviewing these results and attempting to reconcile them, Siesjo (1980) examined the possible influence of (1) anesthetic conditions, (2) species differences, and (3) methods for determining CBF and CMR_{O_2}. Although some of the variation between studies is almost certainly attributable to the first two factors, it seems that the role of methodology is vital. Furthermore refinements of existing techniques for measuring CBF may help, but in hypercapnia, as in a number of other conditions, the development of a method for determining CMR_{O_2} in small laboratory animals that does not depend on knowledge of CMF (and yields regional values) would be a signal advance.

Finally, mention should be made of experiments in which hypercapnia has been combined with various pharmacological agents and CMR_{O_2} determined. As noted earlier, indomethacin, an inhibitor of prostaglandin synthesis, greatly attenuates the increase in CBF with hypercapnia. This drug causes only slight decreases in CMR_{O_2} in normocapnic and hypercapnic baboons and rats (Pickard and MacKenzie, 1973; Dahlgren et al., 1981). Thus, the effects of CO_2 on CBF can be modified over a wide range, with essentially no change in CMR_{O_2}. On the other hand, propranolol and diazepam decreased both CBF and CMR_{O_2} in hypercapnic rats, even though in the doses used, they did not greatly affect CMR_{O_2} in normocapnic animals (Berntman et al., 1979; Hemmingsen et al., 1979). Thus it has been argued that CO_2 has a catecholamine-stimulating effect on brain, which enhances CMR_{O_2}, and when this is blocked (or at very high CO_2 concentrations), the CNS depressant properties of CO_2 become evi-

TABLE II. Effect of CO_2 on CMR_{O_2}

Study	Species	pCO_2 (mm Hg)	% Control CMR_{O_2}[a]
Alberti et al. (1975)	Dog	65	100 (101)
Artru and Michenfelder (1980)	Dog	81	93
	Dog	101	89[b]
Berntman et al. (1979)	Rat	81	122[b]
	Rat	156	115
	Rat	296	79[b]
Cohen et al. (1964)	Man	51	88 (77)
Dahlgren et al. (1981)	Rat	84	116
Des Rosiers et al. (1978)	Rat	49	70 (46)
Eklof et al. (1973)	Rat	80	95
Fujishima et al. (1971)	Man	45	100
Gjedde et al. (1975)	Rat	75	104
Gregoire et al. (1978)	Neonatal dog	58	95 (85)
Hemmingsen et al. (1979)	Rat	81	135[b]
Kety and Schmidt (1948)	Man	52	103
Kliefoth et al. (1979)	Monkey	70	71[b]
Kogure et al. (1975)	Rat	61	83[b,c]
Miller et al. (1975)	Rat	152	134 (67)
Nilsson and Siesjo (1976)	Rat	72	104
Novack et al. (1953)	Man	50	100
Pickard and MacKenzie (1973)	Baboon	60	115
Xanalatos and James (1972)	Dog	62	95 (70)

[a] Figures in parentheses indicate CMR_G as a % of control rate in studies in which both CMR_{O_2} and CMR_G were measured.
[b] Indicates that the authors report these results to be different from control rates by a measure of statistical significance.
[c] The determination was of the rate of high-energy phosphate use by the "closed head" technique (Lowry et al., 1964).

dent (Berntman et al., 1979; Siesjo et al., 1980). This interesting hypothesis requires further confirmation, but has the virtue of explaining results that seem otherwise contradictory.

4.2. Brain Metabolite Concentrations

The preceding section summarized reports which found that hypercapnic acidosis (1) decreases CMR_G and (2) does not lower CMR_{O_2} to the same extent (and may even increase it). Changes in metabolite concentrations in brain during hypercapnia are quite pronounced and in combination with knowledge of changes in flux provide important insights into the mechanism of inhibition of glucose utilization and into the use of nonglucose fuels by brain.

A number of investigators, including the author (Miller et al., 1972), had offered a variety of explanations for the many changes in cerebral concentrations of glycolytic and Krebs cycle intermediates and associated amino acids found in hypercapnic animals. These were predicated on the assumption that when CMR_{O_2} is unchanged, CMR_G must also be constant, since glucose is the sole oxidative fuel for brain. The development of methodologies for measuring the CMR_G (Hawkins et al., 1974) and for rapidly freezing the brain (Veech et al., 1973; Quistorff, 1975) permitted experiments to be done which showed that the CMR_G was decreased during hypercapnia and which indicated the site of the inhibition of glycolysis.

Folbergrova et al. (1975) and Miller et al. (1975) showed that in acute hypercapnic acidosis, cerebral concentrations of glycolytic intermediates above the phosphofructokinase step (fructose 6-phosphate, glucose 6-phosphate, glucose) increased, whereas levels of glycolytic intermediates below this reaction decreased. This effect was evident as early as 10 sec after placing the rat in an atmosphere containing 20% CO_2 (Folbergrova et al., 1975). It became apparent, therefore, that the initial effect of CO_2 on glucose utilization was via an inhibition of phosphofructokinase. The mechanism of this effect has not been definitively characterized, but the changes in concentrations of most of the effectors of phosphofructokinase are in directions opposite to those which are inhibitory. The exception is H^+, which increases and is inhibitory (Trivedi and Danforth, 1966). The problem with the conclusion that inhibition of phosphofructokinase is the mechanism of the reduction of glycolytic rate during hypercapnic acidosis is that the concentrations of its reactants gradually return to control levels between 2.5 and 60 min after the onset of exposure to 20% CO_2, even though the acidosis persists during this period and CMR_G remains low (Miller et al., 1975). One possible explanation is that the animals became hypothermic (body temperature was not monitored), which would decrease CMR_G (Michenfelder and Theye, 1968; McCulloch et al., 1982) and CMR_{O_2} (Hagerdal et al., 1975a) without increasing concentrations of glucose and glucose 6-phosphate (Hagerdal et al., 1975b). Thus, it is not known with certainty that the depression in CMR_G of hypercapnic animals persists for prolonged periods in normothermic animals. This is an important question with regard to use of nonglucose substrates by brain.

During acute hypercapnic acidosis there are no changes in brain adenine nucleotide concentrations, but there are decreases in concentrations of (1) glycolytic intermediates below the phosphofructokinase step, (2) Krebs cycle intermediates (except succinate), and (3) glutamate, and it has been argued that there is a net oxidation of these compounds allowing a higher CMR_{O_2} to be maintained than could be sustained by oxidation of glucose alone (Folbergrova et al., 1975; Miller et al., 1975). The evidence for this conclusion and the possible metabolic routes for oxidation of endogenous amino acids in brain are discussed below.

The conceivable dispositions of the intermediary metabolites that progressively decrease in concentration during acute hypercapnic acidosis include: (1) oxidation, (2) leakage from brain into blood, and (3) incorporation into other compounds in brain. The latter possibility is excluded for two reasons. (1) There are no synthetic reactions in brain that proceed at a rate approaching the rate of diminution of the metabolite pools. (2) In rats first injected with [14C]glucose and later made hypercapnic, there is no increased radioactivity in the pellets of perchloric acid extracts of brain and the amount of radioactivity in the soluble fractions is less than in brains from animals given [14C]glucose at the same time but not subsequently made hypercapnic (A. L. Miller and D. H. Corddry, unpublished data). Loss of metabolites into blood, however, could occur at a rapid rate and in animals given [14C]glucose, would lead to more rapid loss of 14C from brain and a lowering of the CMR_G estimated by the method of Hawkins et al. (1974). Three observations mitigate against this possibility and in favor of the conclusion that the "lost" metabolites are oxidized in brain. First, the decline in estimated CMR_G during hypercapnic acidosis is as great or greater when [3H]deoxyglucose is used as tracer as when [2-14C]glucose is used (Miller and Corddry, 1981; Miller et al., 1982a,b). If intermediary metabolites were leaking from brain into blood, this would cause increased loss of 14C and relatively lower estimates of CMR_G with [2-14C]glucose than with [3H]deoxyglucose as tracer—the opposite of what is actually observed. Second, the net decreases in metabolite concentrations are in direct proportion to the decreases in CMR_G (Miller et al., 1975; Miller and Corddry, 1981). That is, the more the absolute change in CMR_G, the more rapid the rate of depletion of endogenous metabolite pools. This is the expected result if these pools are being used to supplement glucose as oxidative fuel for brain, although it does not prove that this is their fate. Third, the concentrations of some intermediary metabolites and related amino acids increase in brain during hypercapnia (e.g., glucose 6-phosphate, fructose 6-phosphate, succinate, and aspartate) (Folbergrova et al., 1975; Miller et al., 1975; Miller and Corddry, 1981). Thus, there is not a generalized increase in leakiness across the blood–brain barrier, and any hypothesis postulating loss of metabolites by efflux into blood would have to explain the pattern of increases as well as the decreases.

Given the weight of evidence favoring the conclusion that endogenous metabolite pools make a net contribution to brain oxidative metabolism during acute hypercapnia, it is appropriate to discuss the relative contributions of different pools and the reactions by which they can be oxidized. In both developing and adult rat brain the metabolites that contribute by far the most, in terms of the change in size of the total carbon pool, are glutamate and lactate (Miller et al., 1975; Miller and Corddry, 1981). Figure 1 illustrates their contribution, relative to that of glucose, to the CMR_{O_2} of adult and developing rats breathing 20% CO_2 for 10 min. The assumption that all glucose taken up by brain is oxidized is quite accurate in the adult rat, but less so in developing rats in which there may be considerable net release of lactate from brain into blood in association with rapid use of ketone bodies (e.g., Hawkins et al., 1971), although Miller et al. (1982a) found the major effect of ketonemia in 20-day-old rats was to depress CMR_G. In any event, the error from assuming complete oxidation of glucose in developing brain in estimating CMR_{O_2} is much smaller than the rate of lactate loss, because lactate loss is associated with ketone body oxidation which contributes to the CMR_{O_2}.

It is apparent from inspection of Fig. 1 that glutamate is the major non-glucose fuel used by brain under these aerobic conditions. (Blood concentrations of ketone bodies and lactate fall during hypercapnia, so that their contribution to CMR_{O_2} will, if anything, decrease.) Since the glutamate concentration of brain is very high (10–12 μmole/g in the brain of adult rats) (e.g., Van Den Berg, 1970), it is well suited to be a temporary oxidative fuel when glucose utilization is impaired. There are several ways in which glutamate may be oxidized in brain, and the changes in metabolite levels in brain indicate which paths are operative. Thus, in the first few minutes after the onset of hypercapnia, brain aspartate concentration increases to about the same extent as glutamate concentration decreases. This observation implies that glu-

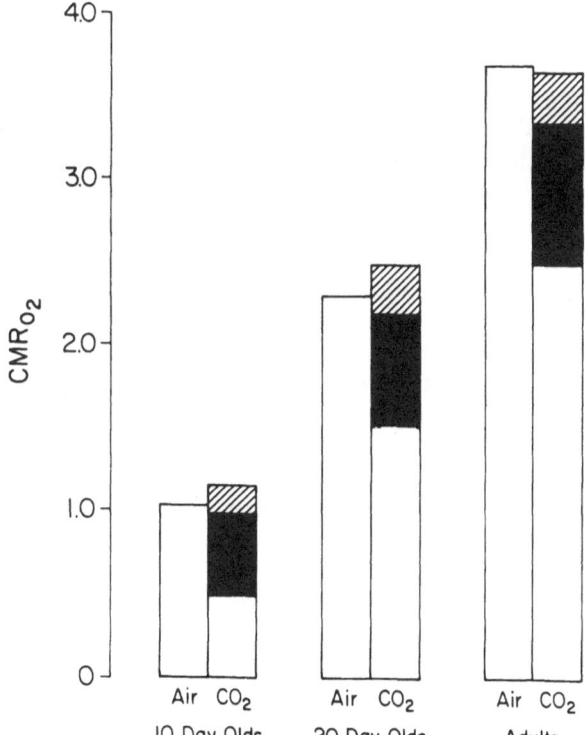

FIGURE 1. Use of cerebral pools of glutamate and lacate by developing and adult rat brain during acute hypercapnia. CMR_{O_2} (μmol/g per min) was calculated from $CMR_G \times 6$ in control animals. For CO_2-treated rats (20% CO_2:21% O_2:59% N_2 for 10 min), CMR_{O_2} was calculated from (a) $CMR_G \times 6$, (b) (Δ lactate × 3)/10, and (c) [(Δ glutamate × 4.5) − (Δ aspartate × 3)]/10. The contributions of lactate and glutamate (less correction for net formation of aspartate) are shown individually. Oxidation of glucose (□), oxidation of glutamate (■), and oxidation of lactate (▨).

tamate is entering the Krebs cycle via transamination with oxaloacetate to form α-ketoglutarate and aspartate. Subsequent oxidation of the α-ketoglutarate replaces the oxaloacetate, so that the series of reactions may be written as follows:

1. Glutamate + oxaloacetate → α-ketoglutarate + aspartate
2. α-Ketoglutarate + NAD^+ + CoA → succinyl CoA + CO_2 + NADH + H^+
3. Succinyl CoA + GDP → succinate + GTP + CoA
4. Succinate + FAD^+ → fumarate + FADH + H^+
5. Fumarate → malate
6. Malate + NAD^+ → oxaloacetate + NADH + H^+

Net: Glutamate + $2NAD^+$ + FAD^+ + GDP → aspartate + CO_2 + 2 NADH + FADH + $3H^+$ + GTP

When hypercapnia persists beyond 2–3 min, however, increases in brain aspartate levels are not as great as the decreases in glutamate levels, and after about 10–15 min, aspartate levels begin to decrease while ammonia and glutamine levels increase (Folbergrova et al., 1975; Miller et al., 1975). There are at least two pathways that lead to production of ammonia and Krebs cycle intermediates from amino acids. One is direct oxidation of glutamate to α-ketoglutarate and ammonia in the glutamate dehydrogenase reaction. The other is via the purine nucleotide cycle (Lowenstein, 1972) which consists of three reactions the net of which is:

$$\text{Aspartate} + \text{GTP} + H_2O \rightarrow \text{fumarate} + \text{GDP} + Pi + NH_4^+$$

In either case, the ammonia is detoxified by glutamine formation.

When flux through glutamate dehydrogenase is in the direction of α-ketoglutarate production, aspartate can be metabolized by first transaminating with α-ketoglutarate to produce oxaloacetate and glutamate. The oxidative deamination of glutamate by glutamate dehydrogenase regenerates the α-ketoglutarate. If the purine nucleotide cycle is operative, glutamate is metabolized by transaminating with oxaloacetate to yield aspartate and α-ketoglutarate. Oxidation of the latter in the Krebs cycle regenerates the oxaloacetate. No evidence has been presented to date as to which of these pathways accounts for the production of ammonia during hypercapnia of more than a few minutes' duration. For aspartate and glutamate to be fully oxidized in the Krebs cycle, it should be noted that acetyl CoA production must keep pace with the rate of oxaloacetate formation. There is a deficit of acetyl CoA production from glucose when the CMR_G falls and CMR_{O_2} is constant. Additional acetyl CoA could come from oxidation of fatty acids or from carboxylation of oxaloacetate to phosphoenolpyruvate (phosphoenolpyruvate carboxykinase), which would be converted first to pyruvate (pyruvate kinase) and then to acetyl CoA (pyruvate dehydrogenase). Whether one or both of these pathways are more active during prolonged hypercapnia is an important but unanswered question. The necessity for nonglucose fuels will persist if the CMR_G stays at rates that cannot sustain the CMR_{O_2}. As mentioned above, it needs to be established whether or not this is the case during prolonged hypercapnia.

5. Overview

This final section of the chapter will discuss the relationship between brain energy metabolism and CO_2 narcosis. That is, what role do the metabolic changes detailed above play in the clinical picture of CO_2 intoxication?

Several plausible arguments can be made, each with some support from experimental observations. First, there is some basis for the viewpoint that decreased level of consciousness and degree of anesthesia correlate with reductions in CMR_{O_2}. Thus, rats are not anesthetized until CO_2 levels are 40% or more in the inhaled gas (Woodbury et al., 1958), which is also the level at which one group (Berntman et al., 1979) found a significant decrease in CMR_{O_2}. Moreover, in dogs, the first evidence of anesthesia with CO_2 occurs when pCO_2 exceeds 95 mm Hg (Eisele et al., 1967), which is the pCO_2 above which a significant decrease in CMR_{O_2} has been found (Artru and Michenfelder, 1980) (Table II). The effects of CO_2 on CMR_{O_2} are, however, controversial (see above), and even if we accept the premise that CO_2 reduces CMR_{O_2}, we are still left with the question whether the effect on CMR_{O_2} is a cause of, consequence of, or unrelated to an altered level of consciousness.

Second, one could postulate that it is the decrease in CMR_G that accounts for the behavioral effects of hypercapnic acidosis. This line of reasoning holds that the same basic mechanisms are involved in producing CO_2 narcosis and hypoglycemic coma. Indeed, many of the biochemical findings are similar, including (1) a large decrease in CMR_G with lesser or no changes in CMR_{O_2}, (2) depletion of the pools of glycolytic and Krebs cycle intermediates, (3) decreases in brain glutamate levels in association with increases in aspartate levels, and (4) increased ammonia levels (Kety et al., 1948; Della Porta et al., 1964; Pappenheimer and Setchell, 1973; Lewis et al., 1974; Feise et al., 1976; Gorell et al., 1976, 1977; Norberg and Siesjo, 1976; Agardh et al., 1978; Hernandez et al., 1980; Butterworth et al., 1982; Ghajar et al., 1982). Given these conditions, the neurological findings could be attributed to decreased synthesis of acetylcholine because of the low levels of pyruvate (Gibson and Blass, 1976), to low concentrations of the excitatory amino acid glutamate (Cooper et al., 1978), to ammonia intoxication (Ghajar et al., 1982), or to other consequences of altered metabolite levels and/or use of nonglucose endogenous substrates (e.g., lipids) (Pappenheimer and Setchell, 1973; Agardh et al., 1981). These possible contributors to the behavioral effects of hypercapnia have not been intensively investigated, but seem very worth of study. It should be noted that there is not exact concordance between the effects of hypoglycemia and those of hypercapnia: (1) glucose, glucose 6-phosphate, and fructose 6-phosphate levels are elevated in hypercapnia (Folbergrova et al., 1975; Miller et al., 1975), which could lead, for example, to altered flux through the hexose monophosphate shunt, (2) aspartate levels decrease after their initial increase in hypercapnia, and (3) the levels of ammonia found in brains of hypercapnic rats (Folbergrova et al., 1975; Miller et al., 1975) are not nearly as high as those found in brains of severely hypoglycemic rats (Agardh et al., 1978; Ghajar et al., 1982). Moreover, acetylcholine levels were found to be unchanged in hypoglycemic mice (Gorell et al., 1981), although the possibility of decreased cholinergic transmissive activity is not excluded by this observation.

The roles of changes in H^+ and CO_2 in the etiology of both the cerebral metabolic and the behavioral effects of hypercapnic acidosis must be considered. This has been a difficult problem to examine experimentally *in vivo* because, although CO_2 is readily permeable across cell membranes, bicarbonate movement is slow. It is not possible, therefore, to keep brain pH constant during acute hypercapnic acidosis by administering bicarbonate along with CO_2. Nevertheless, there are data from animals and man that indicate the importance of pH effects, independent of pCO_2. Thus encephalopathy correlates fairly well with spinal fluid pH in patients with respiratory and metabolic acidoses (Posner *et al.*, 1965; Posner and Plum, 1967). Moreover, during chronic hypercapnia of rats and guinea pigs (i.e., 24 hr or longer), many of the acute metabolic changes observed in blood and brain (e.g., reduced lactate levels) are reversed toward their normal values, in parallel with the return of pH toward normal (Weyne *et al.*, 1970; Messeter and Siesjo, 1971; Schaefer, 1974). Again, however, proof that changes in pH are primary and that the effects on intermediary metabolism and behavior are causally related is lacking. As noted above, an initial pH effect on phosphofructokinase activity is evident from metabolite data, but it is less clear what happens to the CMR_G in normothermic animals during hypercapnia of more than 15 min duration. Given the number of reactions in which CO_2 or HCO_3^- is a participant, it seems certain that flux through some reactions is directly affected by changes in CO_2 and HCO_3^-. On the other hand, Krebs cycle intermediates decrease in concentration, indicating that any increase in the rate of CO_2 fixation is minor by comparison with the rate of the cycle.

Lastly, the focus of this chapter and of this book is on the toxic, encephalopathic effects of various agents. In the case of CO_2, however, possible beneficial effects of hypercapnia should be at least briefly mentioned. In hypoxic or oligemic rats and monkeys, hypercapnic acidosis reduces the accumulation of lactic acid in brain and improves brain tissue oxygenation (Gottesfeld and Miller, 1969; Metzger *et al.*, 1971; MacMillan and Siesjo, 1972; Michenfelder and Sundt, 1973). During seizures as well, hypercapnia (10% CO_2) slows the rate of production of lactate and, in fact, halves the CMR_G compared with the rate found in normocapnic, seizing rats (Miller *et al.*, 1982b). Thus, in conditions that are characterized by rapid rates of glycolysis and development of lactic acidosis in brain, CO_2 has a moderating effect on the biochemical disturbances. Whether it also modifies the neuronal damage and/or behavioral disturbances caused by these conditions is an unanswered question.

References

Agardh, C.-D., Folbergrova, J., and Siesjo, B. K., 1978, Cerebral metabolic changes in profound, insulin-induced hypoglycemia, and in the recovery period following glucose administration, *J. Neurochem.* **31**:1135–1142.

Agardh, C.-D., Chapman, A. G., Nilsson, B., and Siesjo, B. K., 1981, Endogenous substrates utilized by rat brain in severe insulin-induced hypoglycemia, *J. Neurochem.* **36**:490–500.

Alberti, E., Hoyer, S., Hamer, J., Stoeckel, H., Packschiess, P., and Weinhardt, F., 1975, The effect of carbon dioxide on cerebral blood flow and cerebral metabolism in dogs, *Br. J. Anaesth.* **47**:941–946.

Artru, A. A., and Michenfelder, J. D., 1980, Effects of hypercarbia on canine cerebral metabolism and blood flows with simultaneous direct and indirect measurement of blood flow, *Anesthesiology* **52**:466–469.

Berntman, L., Dahlgren, N., and Siesjo, B. K., 1979, Cerebral blood flow and oxygen consumption in the rat brain during extreme hypercarbia, *Anesthesiology* **50**:299–305.

Borgstrom, L., Norberg, K., and Siesjo, B. K., 1976, Glucose consumption in rat cerebral cortex in normoxia, hypoxia, and hypercapnia, *Acta Physiol. Scand.* **96**:569–574.

Butterworth, R. F., Merkel, A. D., and Landreville, F., 1982, Regional amino acid distribution in relation to function in insulin hypoglycaemia, *J. Neurochem.* **38**:1483–1489.

Cameron, I. R., Davson, H., and Segal, M. B., 1969, The effect of hypercapnia on the blood–brain barrier to sucrose in the rabbit, *Yale J. Biol. Med.* **42**:241–247.

Carlsson, C., Hagerdal, M., and Siesjo, B. K., 1976, The effect of hyperthermia upon oxygen consumption and upon organic phosphates, glycolytic intermediates, citric acid cycle intermediates and associated amino acids in rat cerebral cortex, *J. Neurochem.* **26**:1001–1006.

Carlsson, C., Hagerdal, M., Kaasik, A. E., and Siesjo, B. K., 1977, A catecholamine-mediated increase in cerebral oxygen uptake during immobilization stress in rats, *Brain Res.* **119**:223–231.

Carpenter, D. O., Hubbard, J. H., Humphrey, D. R., Thompson, H. K., and Marshall, W. H., 1974, Carbon dioxide effects on nerve cell function, in: *Carbon Dioxide and Metabolic Regulations* (G. Nahas and K. E. Schaefer, eds.), Springer-Verlag, New York, pp. 49–62.

Cohen, P. J., Wollman, H., Alexander, S. C., Chase, P. E., and Behar, M. G., 1964, Cerebral carbohydrate metabolism in man during halothane anesthesia, *Anesthesiology* **25**:185–191.

Cooper, J. R., Bloom, F. E., and Roth, R. H., 1978, *The Biochemical Basis of Neuropharmacology*, Oxford University Press, New York, pp. 223–258.

Dahlgren, N., Nilsson, B., Sakabe, T., and Siesjo, B. K., 1981, The effect of indomethacin on cerebral blood flow and oxygen consumption in the rat at normal and increased carbon dioxide tensions, *Acta Physiol. Scand.* **111**:475–485.

Davenport, H. W., 1974, *The ABC of Acid-Base Chemistry*, University of Chicago Press, Chicago.

Della Porta, P., Maiolo, A. T., Negri, V. U., and Rosella, E., 1964, Cerebral blood flow and metabolism in therapeutic insulin coma, *Metabolism* **13**:131–140.

Des Rosiers, M. H., Kennedy, C., Shinohara, M., and Sokoloff, L., 1976, Effects of CO_2 on local cerebral glucose utilization in the conscious rat, *Neurology* **26**:346.

Des Rosiers, M. H., Kennedy, C., Sakurado, O., Shinohara, M., and Sokoloff, L., 1978, Effects of hypercapnia on cerebral oxygen and glucose consumption in the conscious rat, *Stroke* **9**:98.

Eisele, J. H., Eger, E. I., and Muallem, M., 1967, Narcotic properties of carbon dioxide in the dog, *Anesthesiology* **28**:856–865.

Eklof, B., Lassen, N. A., Nilsson, L., Norberg, K., and Siesjo, B. K., 1973, Blood flow and metabolic rate for oxygen in the cerebral cortex of the rat, *Acta Physiol. Scand.* **88**:587–589.

Feise, G., Kogure, K., Busto, R., Scheinberg, P., and Reinmuth, O. M., 1976, Effect of insulin hypoglycemia upon cerebral energy metabolism and EEG activity in the rat, *Brain Res.* **126**:263–280.

Filley, G. F., 1971, *Acid-Base and Blood Gas Regulation*, Lea and Febiger, Philadelphia.

Folbergrova, J., Norberg, K., Quistorff, B., and Siesjo, B. K., 1975, Carbohydrate and amino acid metabolism in rat cerebral cortex in moderate and extreme hypercapnia, *J. Neurochem.* **25**:457–462.

Fujishima, M., Scheinberg, P., Busto, R., and Reinmuth, O. M., 1971, The relation between cerebral oxygen consumption and cerebral vascular reactivity to carbon dioxide, *Stroke* **2**:251–257.

Ghajar, J. B., Plum, F., and Duffy, T. E., 1982, Cerebral oxidative metabolism and blood flow during acute hypoglycemia and recovery in unanesthetized rats, *J. Neurochem.* **38**:397–409.

Gibson, G. E., and Blass, J. P., 1976, Impaired synthesis of acetylcholine in brain accompanying mild hypoxia and hypoglycemia, *J. Neurochem.* **27**:37–42.

Gjedde, A., Caronna, J. J., Hindfelt, B., and Plum, F., 1975, Whole-brain blood flow and oxygen metabolism in the rat during nitrous oxide anesthesia, *Am. J. Physiol.* **229**:113–118.

Gorell, J. M., Dolkhart, P. H., and Ferrendelli, J. A., 1976, Regional levels of glucose, amino acids, high energy phosphates, and cyclic nucleotides in the central nervous system during hypoglycemia stupor and behavioral recovery, *J. Neurochem.* **27**:1043–1049.

Gorell, J. M., Law, M. M., Lowry, O. H., and Ferrendelli, J. A., 1977, Levels of cerebral cortical glycolytic and citric acid cycle metabolites during hypoglycemic stupor and its reversal, *J. Neurochem.* **29**:187–191.

Gorell, J. M., Navarro, C. P., and Schwendner, S. P. W., 1981, Regional CNS levels of acetylcholine and choline during hypoglycemic stupor and recovery, *J. Neurochem.* **36:**321–324.

Gottesfeld, Z., and Miller, A. T., Jr., 1969, Metabolic response of rat brain to acute hypoxia: Influence of polycythemia and hypercapnia, *Am. J. Physiol.* **216:**1374–1379.

Gregoire, N. M., Gjedde, A., Plum, F., and Duffy, T. E., 1978, Cerebral blood flow and cerebral metabolic rates for oxygen, glucose, and ketone bodies in newborn dogs, *J. Neurochem.* **30:**63–69.

Grubb, R. L., Raichle, M. E., Eichling, J. O., and Ter-Pogossian, M. M., 1974, The effects of changes in P_aCO_2 on cerebral blood volume, blood flow, and vascular mean transit time, *Stroke* **5:**630–639.

Hagerdal, M., Harp, J., Nilsson, L., and Siesjo, B. K., 1975a, The effect of induced hypothermia upon oxygen consumption in the rat brain, *J. Neurochem.* **24:**311–316.

Hagerdal, M., Harp, J., and Siesjo, B. K., 1975b, Effect of hypothermia upon organic phosphates, glycolytic metabolites, citric acid cycle intermediates and associated amino acids in rat cerebral cortex, *J. Neurochem.* **24:**743–748.

Hawkins, R. A., Williamson, D. H., and Krebs, H. A., 1971, Ketone-body utilization by adult and suckling rat brain *in vivo, Biochem. J.* **122:**13–18.

Hawkins, R. A., Miller, A. L., Cremer, J. E., and Veech, R. L., 1974, Measurement of the rate of glucose utilization by rat brain *in vivo, J. Neurochem.* **23:**917–923.

Hemmingsen, R., Hertz, M. M., and Barry, D. E., 1979, The effect of propranol on cerebral oxygen consumption and blood flow in the rat: Measurements during normocapnia and hypercapnia, *Acta Physiol. Scand.* **105:**274–281.

Hernandez, M. J., Vannucci, R. C., Salcedo, A., and Brennan, R. W., 1980, Cerebral blood flow and metabolism during hypoglycemia in newborn dogs, *J. Neurochem.* **35:**622–628.

Kety, S. S., and Schmidt, C. F., 1948, The effects of altered arterial tensions of carbon dioxide and oxygen on cerebral blood flow and cerebral oxygen consumption of normal young men, *J. Clin. Invest.* **27:**484–492.

Kety, S. S., Woodford, E. B., Harmel, M. H., Freyhan, F. A., Appel, K. E., and Schmidt, C. F., 1948, Cerebral blood flow and metabolism in schizophrenia. The effects of barbiturate semi-narcosis, insulin coma and electroshock, *Am. J. Psychiatry* **104:**765–770.

Kliefoth, A. B., Grubb, R. L., Jr., and Raichle, M. E., 1979, Depression of cerebral oxygen utilization by hypercapnia in the rhesus monkey, *J. Neurochem.* **32:**661–663.

Kogure, K., Busto, R., Scheinberg, P., and Reinmuth, O., 1975, Dynamics of cerebral metabolism during moderate hypercapnia, *J. Neurochem.* **24:**471–478.

Lennox, W. G., Gibbs, E. L., Gibbs, F. A., Hurwitz, R., and Chase, H., 1956, The relationship between brain metabolism and brain function and compensation for oxygen lack by changes in other blood constituents, *Physiol. Rev.* **36**(suppl. 2):66.

Leonard, B. E., and Rigter, H., 1973, The effect of α-methyl-p-tyrosine, p-chlorophenylalanine, methysergide and propranolol in CO_2-induced amnesia in rats, *Br. J. Pharmacol.* **49:**177P.

Leukel, F., and Quinton, E., 1964, Carbon dioxide effects on acquisition and extinction of avoidance behavior, *J. Comp. Physiol. Psychol.* **57:**267–270.

Lewis, L. D., Ljunggren, B., Norberg, K., and Siesjo, B. K., 1974, Changes in carbohydrate substrates, amino acids and ammonia in the brain during insulin-induced hypoglycemia, *J. Neurochem.* **23:**659–671.

Lorente de No, R., 1947, A Study of nerve physiology, *Stud. Rockefeller Inst.* **131:**148–194.

Lowenstein, J. M., 1972, Ammonia production in muscle and other tissues: The purine nucleotide cycle, *Physiol. Rev.* **52:**382–414.

MacMillan, V., and Siesjo, B. K., 1972, The effect of hypercapnia upon the energy metabolism of the brain during arterial hypoxemia, *Scand. J. Clin. Lab. Invest.* **30:**237–244.

McCulloch, J., Savaki, H. E., Jehle, J., and Sokoloff, L., 1982, Local cerebral glucose utilization in hypothermic and hyperthermic rats, *J. Neurochem.* **39:**255–258.

Meduna, L. J., 1958, The effect of carbon dioxide upon the function of the human brain, in: *Carbon Dioxide Therapy* (L. J. Meduna, ed.), Charles C Thomas, Springfield, Illinois. pp. 35–56.

Messeter, K., and Siesjo, B. K., 1971, The effect of acute and chronic hypercapnia upon the lactate, pyruvate, α-ketoglutarate, glutamate and phosphocreatine contents of the rat brain, *Acta Physiol. Scand.* **83:**344–351.

Metzger, H., Erdmann, W., and Thews, G., 1971, Effect of short periods of hypoxia, hyperoxia, and hypercapnia on brain O_2 supply, *J. Appl. Physiol.* **31:**751–759.

Michenfelder, J. D., and Sundt, T. M., 1973, The effect of $PaCO_2$ on the metabolism of ischemic brain in squirrel monkeys, *Anesthesiology* **38**:445–453.

Michenfelder, J. D., and Theye, R. A., 1968, Hypothermia: Effect on canine brain and whole-body metabolism, *Anesthesiology* **29**:1107–1112.

Miller, A. L., and Corddry, D. H., 1981, Brain carbohydrate metabolism in developing rats during hypercapnia, *J. Neurochem.* **36**:1202–1210.

Miller, A. L., Hawkins, R. A., Harris, R. L., and Veech, R. L., 1972, The effects of acute and chronic morphine treatment and of morphine withdrawal on rat brain *in vivo*, *Biochem. J.* **129**:463–469.

Miller, A. L., Hawkins, R. A., and Veech, R. L., 1975, Decreased rate of glucose utilization by rat brain *in vivo* after exposure to atmospheres containing high concentrations of CO_2, *J. Neurochem.* **25**:553–558.

Miller, A. L., Kiney, C. A., Corddry, D. H., and Staton, D. M., 1982a, Interactions between glucose and ketone body use by developing brain, *Dev. Brain Res.* **256**:443–450.

Miller, A. L., Shambam, A. T., Corddy, D. H., and Kiney, C. A., 1982b, Cerebral metabolic responses to electroconvulsive shock and their modification by hypercapnia, *J. Neurochem.* **38**:916–924.

Nahas, G., and Schaefer, K. E. (eds.), 1974, *Carbon Dioxide and Metabolic Regulations*, Springer-Verlag, New York.

Nilsson, B., and Siesjo, B. K., 1976, A method for determining blood flow and oxygen consumption in the rat brain, *Acta Physiol. Scand.* **96**:72–82.

Norberg, K., and Siesjo, B. K., 1976, Oxidative metabolism of the cerebral cortex of the rat in severe insulin-induced hypoglycaemia, *J. Neurochem.* **26**:345–352.

Novack, P., Shenkin, H. A., Bortin, L., Goluboff, B., and Soffe, A. M., 1953, The effects of carbon dioxide inhalation upon the cerebral blood flow and cerebral oxygen consumption in vascular disease, *J. Clin. Invest.* **32**:696–702.

Owen, O. E., Morgan, A. P., Kemp, H. G., Sullivan, J. M., Herrera, M. G., and Cahill, G. F., Jr., 1967, Brain metabolism during fasting, *J. Clin. Invest.* **46**:1589–1595.

Pappenheimer, J. R., and Setchell, B. P., 1973, Cerebral glucose transport and oxygen consumption in sheep and rabbits, *J. Physiol.* **233**:529–551.

Pickard, J. D., and MacKenzie, E. T., 1973, Inhibition of prostaglandin synthesis and the response of baboon cerebral circulation to carbon dioxide, *Nature (New Biology)* **245**:187–188.

Pickard, J., Tamura, A., Stewart, M., McGeorge, A., and Fitch, W., 1980, Prostacyclin, indomethacin and the cerebral circulation, *Brain Res.* **197**:425–431.

Pollock, G. H., Stein, S. N., and Gyarfas, K., 1949, Central inhibitory effects of carbon dioxide, III. Man, *Proc. Soc. Exp. Biol. Med.* **71**:291–292.

Posner, J. B., and Plum, F., 1967, Spinal fluid pH and neurologic symptoms in systemic acidosis, *N. Engl. J. Med.* **277**:605–613.

Posner, J. B., Swanson, A. G., and Plum, F., 1965, Acid-base balance in cerebrospinal fluid, *Arch. Neurol.* **12**:479–496.

Preskorn, S. H., Irwin, G. H., Simpson, S., Friesen, D., Rinne, J., and Jerkovich, G., 1981, Medical therapies for mood disorders alter the blood–brain barrier, *Science* **213**:469–471.

Quinton, E. E., 1966, Retrograde amnesia induced by carbon dioxide inhalation, *Psychonom. Sci.* **5**:417–418.

Quistorff, B., 1975, A mechanical device for the rapid removal and freezing of liver or brain tissue from unanesthetized and nonparalyzed rats, *Anal. Biochem.* **68**:102–118.

Reivich, M., 1964, Arterial pCO_2 and cerebral hemodynamics, *Am. J. Physiol.* **206**:25–35.

Riege, W. H., and Cherkin, A., 1973, Retroactive facilitation of memory in goldfish by flurothyl, *Psychopharmacologia* **30**:195–204.

Riege, W. H., and Cherkin, A., 1976, Memory performance after flurothyl treatment in rainbow trout, *Psychopharmacologia* **46**:31–35.

Rigter, H., 1973, Pharmacological influences on carbon-dioxide-induced amnesia, *Arch. Int. Pharmacodyn.* **206**:397–398.

Rigter, H., 1978, Attenuation of amnesia in rats by systemically administered enkephalins, *Science* **200**:83–85.

Sakabe, T., and Siesjo, B. K., 1979, The effect of indomethacin on the blood flow-metabolism couple in the brain under normal, hypercapnic and hypoxic conditions, *Acta Physiol. Scand.* **107**:283–284.

Schaefer, K. E., 1974, Metabolic aspects of adaptation to carbon dioxide, in: *Carbon Dioxide and*

Metabolic Regulations (G. Nahas and K. E. Schaefer, eds.), Springer-Verlag, New York, pp. 253–265.

Schaefer, K. E., McCabe, N., and Withers, J., 1968, Stress response in chronic hypercapnia, *Am. J. Physiol.* **214:**543–548.

Schaefer, K. E., Messier, A. A., Morgan, C., and Baker, G. T. III., 1975, Effect of chronic hypercapnia on body temperature regulation, *J. Appl. Physiol.* **38:**900–906.

Siesjo, B. K., 1978, Anesthesia, analgesia, and sedation, in: *Brain Energy Metabolism*, Wiley, New York, pp. 233–265.

Siesjo, B. K., 1980, Cerebral metabolic rate in hypercarbia—A controversy, *Anesthesiology* **52:**461–465.

Siesjo, B. K., Berntman, L., and Nilsson, B., 1980, Regulation of microcirculation in the brain, *Microvasc. Res.* **19:**158–170.

Smith, A. L., Neufeld, G. R., Ominsky, A. J., and Wollman, H., 1971, Effect of arterial CO_2 tension on cerebral blood flow, mean transit time, and vascular volume, *J. Appl. Physiol.* **31:**701–707.

Sokoloff, L., Reivich, M., Kennedy, C., Des Rosiers, M. H., Patlak, C. S., Pettigrew, K. D.,Sakurada, O., and Shinohara, M., 1977, The [^{14}C]deoxyglucose method for the measurement of local cerebral glucose utilization: Theory, procedure and normal values in the conscious and anesthetized albino rat, *J. Neurochem.* **28:**897–916.

Stein, S. N., and Pollock, G. H., 1949, Central inhibitory effects of carbon dioxide, II. Macacus Rhesus, *Proc. Soc. Exp. Biol. Med.* **71:**290–291.

Storm, W. F., and Gianetta, C. L., 1974, Effects of hypercapnia and bedrest on psychomotor performance, *Aerospace Med.* **45:**431–433.

Stupfel, M., 1974, Carbon dioxide and temperature regulation of homeothermic mammals, in: *Carbon Dioxide and Metabolic Regulations* (G. Nahas and K. E. Schaefer, eds.), Springer-Verlag, New York, pp. 163–186.

Taber, R. I., and Banuazizi, A., 1966, CO_2-induced retrograde amnesia in a one-trial learning situation, *Psychopharmacologia* **9:**382–391.

Trivedi, B., and Danforth, W. H., 1966, Effect of pH on the kinetics of frog muscle phosphofructokinase, *J. Biol. Chem.* **241:**4110–4114.

Van Den Verg, C. J., 1970, Glutamate and glutamine, in: *Handbook of Neurochemistry*, Volume III (A. Lajtha, ed.), Plenum Press, New York, pp. 355–379.

Veech, R. L., Harris, R. L., Veloso, D., and Veech, E. H., 1973, Freeze-blowing: A new technique for the study of brain *in vivo*, *J. Neurochem.* **20:**183–188.

Victor, M., and Adams, R. D., 1980, Hypercapnic encephalopathy, in: *Principles of Internal Medicine* (K. J. Isselbacher, R. D. Adams, E. Braunwald, R. G. Petersdorf, and J. D. Wilson, eds.), McGraw-Hill, New York, p. 1979.

West, J. B., 1980, Disorders of regulation of respiration, in: *Principles of Internal Medicine* (K. J. Isselbacher, R. D. Adams, E. Braunwald, R. G. Petersdorf, and J. D. Wilson, eds.), McGraw-Hill, New York, pp. 1271–1276.

Weyne, J., Demeester, G., and Leusen, I., 1970, Effects of carbon dioxide, bicarbonate, and pH on lactate and pyruvate in the brain of rats, *Pflug. Arch. Ges. Physiol.* **314:**292–311.

Woodbury, D. M., Rollins, L. T., Gardner, M. D., Hirschi, W. E., Hogan, J. R., Rallison, M. L., Tanner, G. S., and Brodie, D. A., 1958, Effects of carbon dioxide on brain excitability and electrolytes, *Am. J. Physiol.* **192:**79–90.

Xanalatos, C., and James, I. M., 1972, Effect of arterial CO_2 pressure on the response of cerebral and hind-limb blood flow and metabolism to isoprenaline infusion in the dog, *Clin. Sci.* **42:**63–68.

Encephalopathy Due to Short- and Medium-Chain Fatty Acids

LESLIE ZIEVE

1. Introduction

1.1. Definition, Properties, Source, and Disposition

Fatty acids are hydrocarbon chains with a terminal carboxylate group. Carbon chain length varies from one (formic acid) to twenty-six (ximenic acid). Most are saturated straight-chain acids. Some are unsaturated, some branched-chain, and some hydroxylated. Fatty acids are in general hydrophobic and lipophilic. The very-short-chain fatty acids ($\leqq 5$ carbons) are volatile and soluble in water. The medium-chain fatty acids (C_6–C_{12}) are slightly soluble in water except for hexanoic and lauric (dodecanoic) acids which are insoluble. The long-chain fatty acids ($\geqq 14$ carbons) are insoluble in water. They also have melting points greater than 50 °C. The term "short-chain" fatty acids has been used loosely in many published papers to refer to non-long-chain fatty acids, encompassing the truly short- and some medium-chain fatty acids such as octanoic acid. Short- and medium-chain fatty acids, as defined herein, have been shown to have encephalopathic properties that will be discussed later.

The chemical properties of fatty acids are dependent on their chain length and degree of unsaturation. Shorter chain length and unsaturation enhance the fluidity of fatty acids. The fluidity of biological membranes is determined by the fluidity of their fatty acids. Fatty acids are present largely as esters in triglycerides, phospholipids, glycolipids, and cholesterol. They function as the building blocks of these compounds and as a source of fuel.

The sources of fatty acids are the diet and lipogenesis from acetyl CoA. The fatty acids of the diet vary greatly in chain length and degree of unsaturation depending on the food eaten. The short- and most of the medium-chain

---

LESLIE ZIEVE • Hennepin County Medical Center, University of Minnesota, Minneapolis, Minnesota 55415.

fatty acids pass directly to the liver via the portal vein after absorption. The long-chain fatty acids enter the intestinal lymphatics and pass through the thoracic duct and the lung before they enter the general circulation and reach the liver. Lipogenesis is primarily in the liver from carbohydrate and certain amino acids. Synthesis takes place in the cytosol by the action of a multienzyme complex called fatty acid synthetase. The reductant is NADPH. Chain length is increased by two carbon units derived from acetyl CoA. Intermediates are linked to acyl carrier proteins. Fatty acids of varying chain length result. The liver shortens the carbon chain of fatty acids as well, again two carbons at a time. It also saturates and unsaturates fatty acids. Short- and medium-chain fatty acids are thus also derived from long-chain fatty acids, which are oxidized in mitochondria. Acylcarnitine carries long-chain fatty acids into the mitochondria where they are degraded by sequential removal of two carbon units. The fatty acids are linked to CoA before they are oxidized. The end product of even-chain fatty acid breakdown is acetyl CoA and of odd-chain fatty acid breakdown, acetyl CoA and propionyl CoA. At these points they enter the Krebs cycle.

1.2. Clinical Evidence of an Encephalopathic Role for Fatty Acids

There is no direct evidence that fatty acids have induced encephalopathy in humans. However, abnormalities of short- and medium-chain fatty acids particularly have been described in association with liver failure. Muto (1966) first described elevation of serum valeric and hexanoic acids in hepatic coma, and we have shown that the short-chain fatty acids increase progressively in plasma and red cells after total hepatectomy in the dog (Zieve and Nicoloff, 1976). Rabinowitz and co-workers (1978) found that the serum octanoate rises in liver failure associated with cirrhosis, and that much of the octanoate is derived from incomplete beta oxidation of long-chain fatty acids. In fulminant hepatic failure, no association could be found between plasma levels of short-

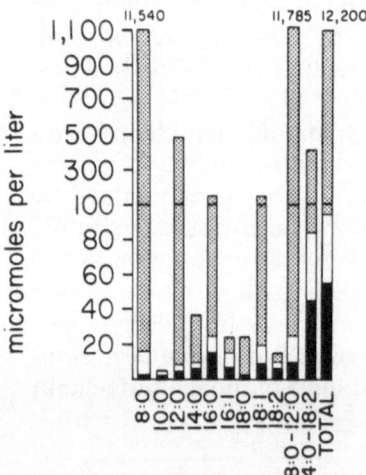

FIGURE 1. Serum-free fatty acids in an infant with Reye's syndrome. Note change in scale on ordinate above 100. The various fatty acids are indicated on abscissa by carbon number. Tops of black and white bars represent mean and highest values for controls. Tops of stippled bars give values of the various fatty acids in the patient on admission to the hospital. (From Mamunes et al. (1975).

chain fatty acids and the presence or absence of coma (Lai et al., 1977). In Reye's syndrome, short- and medium-chain fatty acids characteristically increase (Mamunes et al., 1975). In one patient the serum octanoate rose to levels several times higher than that which induced coma in rats (Fig. 1). There is at least circumstantial evidence that the encephalopathy of Reye's syndrome may be due to the increase in circulating fatty acids and ammonia.

165

ENCEPHA-
LOPATHY
DUE TO
SHORT- AND
MEDIUM-CHAIN
FATTY ACIDS

Long-chain fatty acids increase in the plasma in the presence of hepatic failure (Mortiaux and Dawson, 1961; Wilcox et al., 1978). However, no correlation with the presence or absence of encephalopathy could be demonstrated. The interpretation of these results is confounded by methodological difficulties. The method (Dole) for fatty acids measures other acids besides fatty acids in liver failure, and phthalates interfere with the identification of individual fatty acids separated gas-chromatographically (Geisler et al., 1979).

2. Biochemical Effects

Fatty acids are excellent respiratory substrates for mitochondria of most tissues (Wojtczak, 1976). They increase the permeability of mitochondria to monovalent cations and inhibit adenine nucleotide translocation. At high concentrations they cause uncoupling of oxidative phosphorylation. Increases in the concentration of short- and medium-chain fatty acids to levels of 5–10 mM are required to cause uncoupling of oxidative phosphorylation, loss of respiratory control, and inhibition of oxygen consumption in mitochondria in vitro (Hird and Weidemann, 1966). The longer the fatty acid chain, the more potent as an uncoupler. Carnitine and serum albumin protect the mitochondria from these effects of the fatty acids. Long-chain fatty acids such as oleate have similar effects in vitro (Davis and Gibson, 1969), however, oleate was shown to have no uncoupling effect in vivo (Williamson et al., 1969).

Normal human serum to which 10 mM octanoate was added induced alterations in rat brain mitochondrial respiratory activity (Ansevin, 1980). The state 4 respiratory rate was increased about 90% compared with an increase of about 30% with control serum, and the respiratory control ratio (state 3/state 4) was decreased by about one third (Fig. 2). The increase in ATPase activity in state 4 was about five times that of control serum. Serum from patients with Reye's syndrome showed similar alterations when added to rat brain mitochondria. The extent of the alterations was related to the severity of the encephalopathy. When these serums were pretreated with fatty acid-free bovine serum albumin, the alterations were prevented.

In tissue slices and in the isolated perfused liver, gluconeogenesis was enhanced by fatty acids (Ruderman et al., 1969). In vivo, similar results were observed in carbohydrate-fed rats after a single injection of octanoate (Friedmann et al., 1967). However, in normal, fasted dogs infused intravenously with octanoate, hypoglycemia resulted (Sanbar et al., 1965). Glucose production and utilization were reduced by 30% and 24%, respectively. In fasted dogs with portacaval shunts, acute elevation of long-chain fatty acids resulted in reduced glucose production of 37% and peripheral utilization of 30%. These changes were associated with and believed due to a significant increase (33%) in insulin and a decrease (40%) in glucagon secretion (Seyffert and Madison, 1967). We have found that hypoglycemia, occasionally severe, results from several intraperitoneal injections of octanoic acid in nonfasted normal rats. A single intraperitoneal coma-inducing dose of octanoate caused a fall in the blood sugar in 1 hr from approximately 90 to 45 mg/dl, with recovery in 30–45 min. (Zieve et al., 1983). Certain unsaturated short-chain fatty acids, such as 4-pentenoic acid, given acutely or chronically in the fasted or nonfasted state, injured mitochondria and interfered with both in vivo and in vitro mitochondrial reactions (Sherratt, 1969). Hypoglycemia was characteristic. Gluconeogenesis was strongly inhibited, while there was little effect on glycolysis. Fatty acid oxidation was blocked, and pyruvate oxidation and endogenous respiration inhibited. Methylenecyclopropylacetic acid, the active compound in hypoglycin which causes Jamaican vomiting sickness, had similar effects.

The key glycolytic enzymes, glucokinase, hexokinase, phosphofructokinase, and pyruvate kinase, were inhibited in rat liver homogenates by octanoate and longer-chain fatty acids (Weber et al. 1966). The longer the chain, the greater the effect for a given concentration of fatty acid. Glucose-6-phosphate and 6-phosphogluconate dehydrogenases were also markedly inhibited. In rat brain homogenates, octanoate was similarly shown to inhibit α-glycerophosphate dehydrogenase activity (Schwark et al., 1970). Other dehydrogenases such as glutamate dehydrogenase are also known to be inhibited (Derr and Zieve, 1976). It is likely that other enzyme systems are similarly affected.

Short- and medium-chain fatty acids inhibited rat brain microsomal (Na$^+$, K$^+$)ATPase in vitro at concentrations that corresponded to the blood levels causing coma in vivo (Dahl, 1968). The degree of inhibition was determined by the logarithm of the concentration of the fatty acid, and the inhibitory ca-

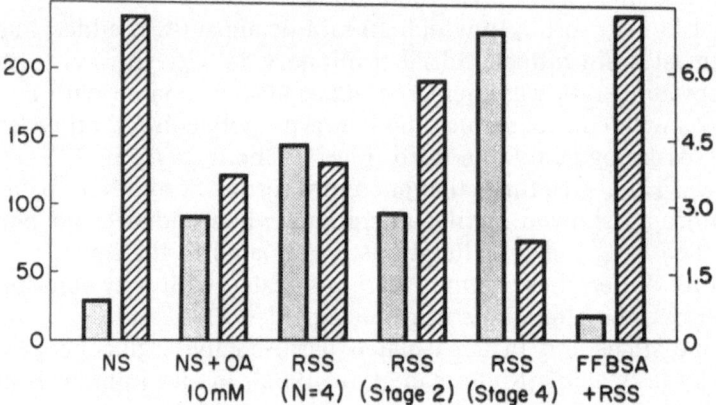

FIGURE 2. Alterations in rat brain mitochondrial function induced by octanoate and by serum from patients with Reye's syndrome. Protective effect of fat-free bovine serum albumin. Left ordinate is percent increase in state 4 respiratory rate. Right ordinate is respiratory control ratio. Respiratory control ratio (▨) = state 3/state 4 after addition of serum; state 3 = active state of respiration (ATP synthesized), state 4 (▨) = resting state of respiration (no ATP synthesized); NS = normal serum; OA = octanoic acid; RSS = Reye's syndrome serum; FFBSA = fat-free bovine serum albumin. From data of Ansevin (1980).

pacity increased by a factor of 2.3 for each carbon atom added to the fatty acid chain. Long-chain fatty acids such as myristate, palmitate, stearate, and oleate were also inhibitory at low concentrations (Ahmed and Thomas, 1971). Unsaturation increased the inhibitory effect, and the inhibition was freely reversible.

167

ENCEPHA-
LOPATHY
DUE TO
SHORT- AND
MEDIUM-CHAIN
FATTY ACIDS

3. Pharmacologic Effects

3.1. Coma Induction—Energy Metabolism

Samson et al. (1956) first demonstrated that short- and medium-chain fatty acids induced a state of unconsciousness when injected into animals intravenously or intraperitoneally. The comatose state was reversible. For a given fatty acid, the coma-inducing dose depended on the concentration and the site of injection. The potency of the fatty acids increased with increase in chain length (Fig. 3). These observations have been verified many times. Infusion of subcoma amounts of octanoate into rabbits resulted in elevated intracranial pressure and hyperventilation (Trauner and Huttenlocher, 1978; Trauner and Adams, 1981). In a study of butyrate, evidence was presented that the fatty acid adsorbs onto membrane lecithin in the process of coma induction (Rizzoli and Galzigna, 1970).

Coma induction with butyrate, valerate, and octanoate in rats was not associated with changes in cortical or brain stem ATP or phosphocreatine (Walker et al., 1970). Energy utilization rates were normal. The adenylate energy charge was normal in brains of 200-g rats in coma induced with a variety of agents including octanoate (Derr and Zieve, 1973). In these coma states, brain mito-

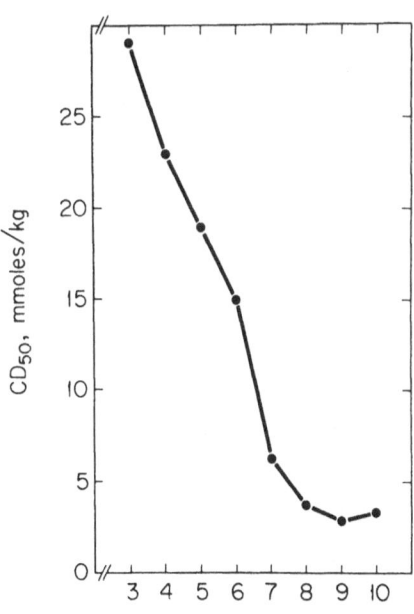

FIGURE 3. Relationship of coma-inducing dose of fatty acid to length of carbon chain in rats. CD_{50} = dose of fatty acid causing coma in 50% of rats. From Samson, Jr., et al. (1956).

chondrial oxidative phosphorylation was normal. In those rats injected with octanoate, blood levels were about 2–3 mM.

3.2. Ammonia Disposition

Fatty acids affect the metabolism of ammonia. In liver homogenates, citrulline and urea synthesis were inhibited by medium- and long-chain fatty acids, the long-chain fatty acids being much more potent than the shorter ones (Derr and Zieve, 1976). A short unsaturated fatty acid such as 4-pentenoate was also much more potent in this effect than octanoate. The *in vitro* liver enzyme assays indicated that the inhibition in the urea cycle was largely due to the depression of carbamyl phosphate synthetase. In isolated normal rat liver perfusions, proprionate and octanoate reduced urea synthesis from NH_4^+ (Fig. 4), but not from carbamyl phosphate (Fig. 5), suggesting that in this system also these fatty acids had their inhibitory effect at the carbamyl phosphate synthetase step of urea synthesis (Derr *et al.*, 1982). Octanoate was more potent than propionate, and 4-pentenoate was much more potent than octanoate. In extracts of brain and liver, octanoate inhibited glutamate dehydrogenase (but not glutamine synthetase) (Fig. 6). Thus the pathways for processing of ammonia to either urea or glutamine were compromised.

3.3. Electrical Activity

In rabbits, short- and medium-chain fatty acids induced coma that was associated with an EEG slow-wave sleep pattern similar to that produced by sodium pentobarbital anesthesia (White and Samson, 1956). The fatty acids—propionate, butyrate, valerate, and octanoate—were infused intravenously at

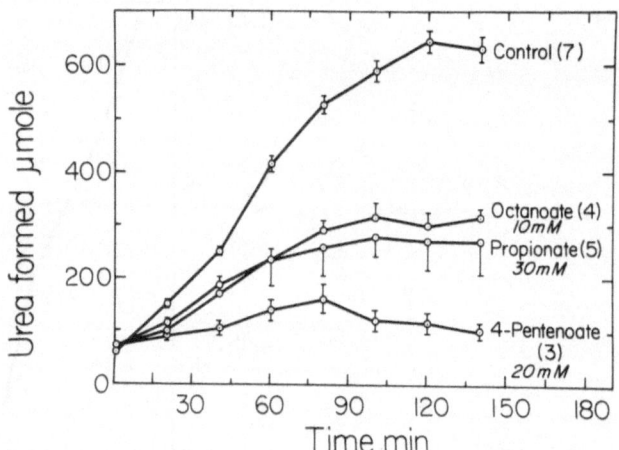

FIGURE 4. Effect of fatty acids on urea formation from ammonia by the isolated perfused rat liver. NH_4Cl (1500 μmole) and ornithine (400 μmole) were added at time zero after the liver had been perfused with medium for 20 min. The fatty acids shown were also added at time zero to give the concentrations indicated. Number of livers perfused in each case is given in parentheses. Vertical lines about plotted points give SEMs. All points but one (without SEM designated) were significantly depressed, statistically. From Derr and Zieve (1982).

169

ENCEPHA-
LOPATHY
DUE TO
SHORT- AND
MEDIUM-CHAIN
FATTY ACIDS

a dose of 4 mmole/kg. The longer the chain, the more potent the fatty acid. The finding of synchrony with slow waves in the EEG following the injection of butyrate and octanoate was verified in cats in whom simultaneous cortical vasodilatation with a pronounced increase in blood flow was demonstrated (Holmquist and Ingvar, 1957). Evidence has been presented that the main effect of short-chain fatty acids in such cats is suppression of the midbrain neocortical reticular activating systems (Muto et al.,1964).

In rats, fatty acids from butyrate to decanoate produced progressive changes in the EEG with the following sequence: desynchronization, intermittent hypersynchrony, continuous hypersynchrony without or with spiking, and finally spiking with intermittent periods of silence that were followed by death (Marcus et al., 1967) (Fig. 7). Prior to the occurrence of spiking with periods of silence, the process was reversible. When the state of 1–2 cycles/sec hypersynchrony and spiking occurred, the righting reflex was lost and there was no behavioral or electrical arousal. There was an inverse relationship between the length of the carbon chain and the dose required for hypersynchrony or the loss of consciousness.

The five-carbon fatty acids—valeric, isovaleric, β-methylcrotonic, tiglic, α-ketoisovaleric, and α-hydroxyisovaleric acids—were studied in rabbits and found to be encephalopathic with slow-wave electrical activity on the EEG (Teychenne et al., 1976). The encephalopathic effect was less with the last four acids.

The effect of octanoic acid on the visual-evoked potential (VEP) of rats has been compared with that observed in experimental hepatic coma. The VEP is an average of the cortical wave patterns generated by the succession of neuronal membrane potential changes that follow stimulation with a flash of light. Normally there are four components occurring with a latency period of 240 msec (Coob and Dawson, 1960). These have been labeled P_1, N_1, P_2, and N_2. The first and third are positive deflections with amplitudes of about 10 and 30 μV, respectively. The second and fourth are negative deflections with amplitudes of about −50 and −15 μV, respectively. The last deflection is highly variable

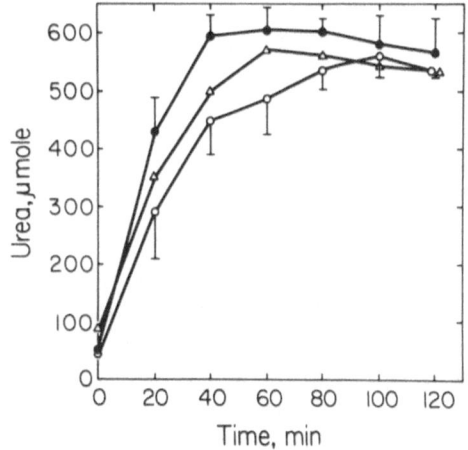

FIGURE 5. Effect of octanoate (○) (10 mM, 3 subjects) and propionate (△) (30 mM, 4 subjects) on urea synthesis from carbamyl phosphate (1500 μmole) and ornithine (400 μmole) in the isolated perfused rat liver [controls (●), 4 subjects]. Vertical lines about points give SEMs. None of the individual points were significantly depressed, statistically. From Derr and Zieve (1982).

and not as reproducible as the first three. The VEP following coma induction with intraperitoneal octanoate was characterized by about 40% reduction in the amplitude of N_1, a disappearance of N_2 in a few instances, and a slight reduction in latency (Zeneroli *et al.*, 1982). Coma induced with NH_4Cl gave a similar VEP. In contrast, hepatic coma induced by injection of galactosamine showed an extreme reduction in N_1 (90%), about 50% reduction in P_2, and about 30% reduction in the overall latency. A combination of reduced doses of octanoate, NH_4^+, and dimethyl disulfide ($\rightarrow$ methanethiol), that caused coma after several injections, resulted in VEPs that were similar to those observed in hepatic coma. N_1 was reduced by about 80%, P_2 by 40%, and the latency by 40% (Zeneroli *et al.*, 1982).

Medium- and long-chain fatty acids have produced severe arrhythmias including ventricular fibrillation in normal isolated rat hearts (Willebrands *et al.*, 1973). A study of the transmembrane potentials of electrically stimulated calf Purkinje's fibers found that 2.4 mM octanoate markedly shortened the duration of the action potential and decreased the rate of depolarization and the conductivity of the fibers (Borbola *et al.*, 1974). In perfused guinea pig hearts, octanoate, palmitate, and linoleate potentiated the intracellular action potential shortening due to mild ischemia (Cowan and Vaughan Williams, 1980).

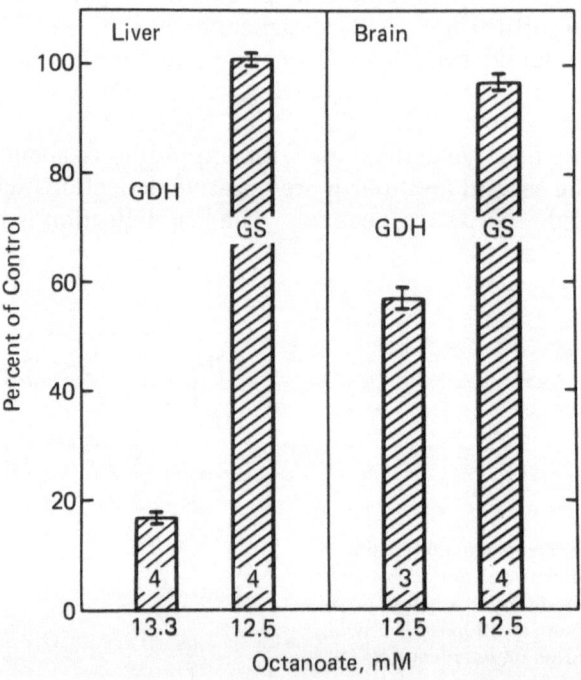

FIGURE 6. Effect of octanoate on activities of glutamate dehydrogenase (GDH) and glutamine synthetase (GS) in extracts from liver and brain. From Derr and Zieve (1976).

3.4. Endocrine Effects

In vitro studies of isolated islets of Langerhans from rats and guinea pigs have suggested that short- and medium-chain fatty acids may be important regulators of glucose-6-phosphate levels in pancreatic islet cells and in this way influence insulin secretion (Montague and Taylor, 1968). Butyrate and octanoate at 5 mM concentrations stimulated the release of insulin from isolated islets of Langerhans. The increased insulin secretion was parallel to the increased intracellular concentration of glucose-6-phosphate. Octanoate and β-hydroxybutyrate also markedly inhibited the release of glucagon (Edwards et al., 1969). In man, in vivo elevation of plasma free fatty acids (mostly long chain) after administration of heparin caused a 50% decrease in plasma glucagon (Gerich et al., 1974). We have already seen that elevation of free fatty acids in the portacaval-shunted dog resulted in increased insulin and decreased glucagon secretion (Seyffert and Madison, 1967).

3.5. Synergism with Other Encephalopathic Agents

Octanoic acid, as a representative fatty acid, interacts with most coma-inducing substances. The dose of NH_4^+ inducing coma in 50% of animals

171

ENCEPHA-
LOPATHY
DUE TO
SHORT- AND
MEDIUM-CHAIN
FATTY ACIDS

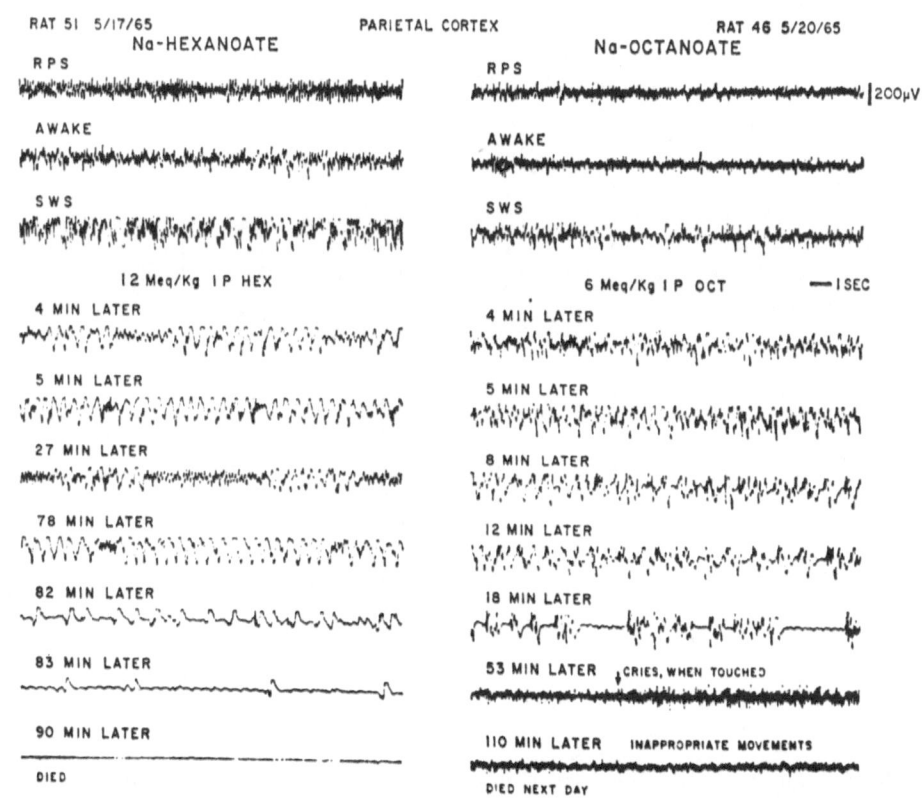

FIGURE 7. EEG in the rat before and after sodium hexanoate and sodium octanoate. RPS = rhombencephalic phase of sleep. From Marcus et al. (1967).

(CD$_{50}$) was reduced by 52% when a subcoma dose (0.4 mmole) of octanoate was given simultaneously to 300-g rats (F. J. Zieve *et al.*, 1974). When the maximum subcoma doses of both these substances were injected simultaneously into these animals, the incidence of unconsciousness rose from 0% to 100%. The converse effect of a subcoma dose of NH$_4^+$ on the dose–response curve to octanoate was similar. Subcoma doses of octanoate also influenced the blood level of ammonia resulting from injection of the NH$_4^+$. The blood ammonia rose twofold to threefold when subcoma doses of both substances were injected. This might have been predicted from the studies, described previously, showing inhibitory effects of octanoate on liver carbamyl phosphate synthetase activity and glutamate dehydrogenase activity (Derr and Zieve, 1976). This effect of a fatty acid on blood ammonia was similarly demonstrated with hexanoate, decanoate, and oleate (F. J. Zieve *et al.*, 1974) (Fig. 8). The dose–response curve for brain ammonia was likewise shifted downward by the simultaneous injection of a subcoma dose of octanoate. Although the brain concentration of ammonia associated with coma in 100% of animals remained approximately the same in those receiving and in those not receiving octanoate, the dose of NH$_4^+$ required for this effect was significantly lower in the presence of octanoate (Fig. 9).

The synergistic interaction of octanoate with mercaptans in the induction of coma was even more striking than that with NH$_4^+$ (L. Zieve *et al.*, 1974). The dose–response to octanoate was reduced by 66–100% in the presence of subcoma doses of methanethiol or dimethyl sulfide (Fig. 10). Octanoate influ-

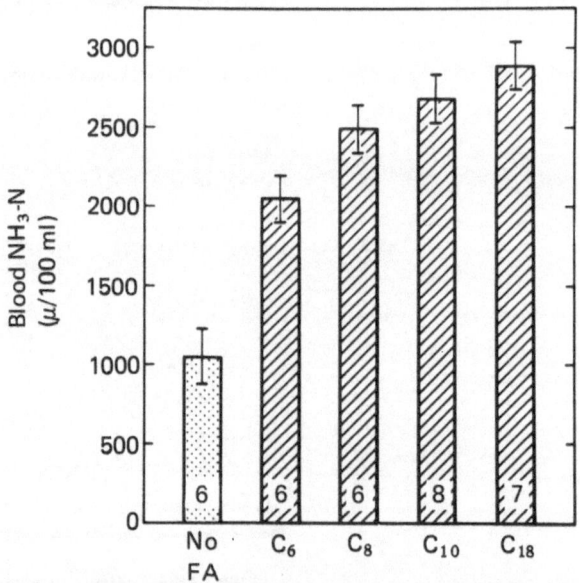

FIGURE 8. Effect [no coma (▨), coma (▨)] of subcoma doses of medium- and long-chain fatty acids on blood ammonia after intraperitoneal (IP) injection of 1.0 mmole NH$_4$Cl. Fatty acids injected IP just prior to injection of NH$_4$Cl. Bars give mean ± SEM. Number of animals given at base of each bar. Doses of fatty acids were 1.2 mmole for C$_6$ (hexanoate) and 0.4 mmole for C$_8$ (octanoate), C$_{10}$ (decanoate), or C$_{18}$ (oleate). From F. J. Zieve *et al.* (1974).

173

ENCEPHA-
LOPATHY
DUE TO
SHORT- AND
MEDIUM-CHAIN
FATTY ACIDS

enced markedly the blood level of methanethiol resulting from a given dose of the mercaptan. Small amounts of octanoate seemed to modulate the encephalopathic effects of ammonia and methanethiol. In the presence of a small amount of octanoate, a combined coma-inducing dose of NH_4^+ and methanethiol was found for normal rats which gave brain levels of these substances that closely approximated those found in experimental hepatic coma (Zieve and Doizaki, 1980).

Phenols are other metabolic products that accumulate during hepatic failure and are encephalopathic in experimental animals. As with ammonia and mercaptans, octanoate enhances phenol's encephalopathic effects and vice versa (Windus-Podehl et al., 1982). The CD_{50} of phenol in rats was reduced by 17% with a subcoma dose of octanoate, and the CD_{50} of octanoate was reduced by 19% with a subcoma dose of phenol. The synergistic interaction of octanoate with phenol was approximately the same as that of NH_4^+, but significantly less than that of dimethyl disulfide ($\rightarrow$ methanethiol).

The encephalopathic effects of octanoate were enhanced by the presence of hypoglycemia and by the injection of a subcoma dose of a barbiturate. Rats made hypoglycemic with insulin (blood sugars 25–45 mg/dl) were more susceptible to coma induction with octanoate (Zieve, 1980). The CD_{50} and CD_{100} for octanoate were reduced by 30%. This corresponded to the effect of hypoglycemia on the encephalopathic effect of NH_4^+. The interaction of octanoate with pentobarbital was as striking as with any other agent (Fig. 11). The maximum subcoma dose of pentobarbital reduced the CD_{50} and CD_{100} of octanoate by 60%. Conversely, one would anticipate that octanoate would influence the sedative action of pentobarbital. This has not been studied. However, as a case in point, oleate has been shown to potentiate thiopental anesthesia (Ohmiya, 1971).

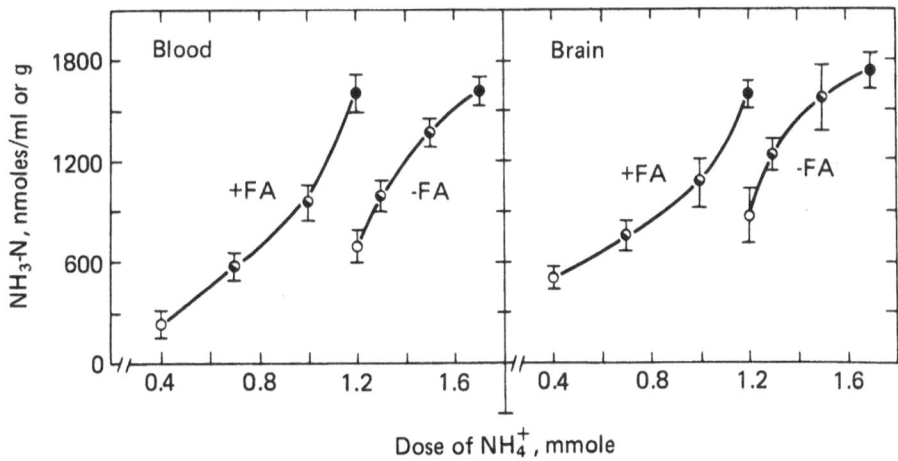

FIGURE 9. Relationships of blood and brain ammonia concentrations to IP dose of NH_4^+, presence or absence of a subcoma IP dose (0.4 mmole) of octanoate (FA), and presence or absence of coma in 300 ± 30 g. male rats. Means ± SEMs are plotted. Coma, all (●); coma, ¼–½ (◐); alert (○).

4. Pathophysiologic Role of Fatty Acids in Disease

Fatty acids probably play a role in glucose homeostasis (Ruderman *et al.*, 1969). As they accumulate excessively, they interfere with glycolysis and with gluconeogenesis. Insulin secretion is increased and glucagon secretion decreased. They also enhance membrane permeability. The full consequences of these effects in diseases is not clear. In diseases with selective mitochondrial injury, medium-chain fatty acid accumulations in association with excess ammonia appear to have toxic effects on the nervous system. This may be the case in Reye's syndrome and in Jamaican vomiting sickness. In liver diseases causing hepatic failure and ultimately encephalopathy and coma, fatty acids accumulate, but their role in pathogenesis of the encephalopathy is uncertain. There is no direct evidence that they play a role at all. However, their intrinsic encephalopathic properties, their synergistic interaction experimentally with all substances studied so far that have encephalopathic effects, and their profound effects on membranes provide circumstantial evidence of a possible role. Reliable measurements of fatty acids in CNS tissues would enable correlations to be made experimentally that could help resolve the question. From the experimental data on synergistic interactions, it seems likely that fatty acids have at least a modulating effect on the pathogenetic actions of toxins such as ammonia and methanethiol.

5. Summary

Fatty acids are derived from the diet or lipogenesis in the liver from carbohydrate and certain amino acids. Their chemical properties are dependent

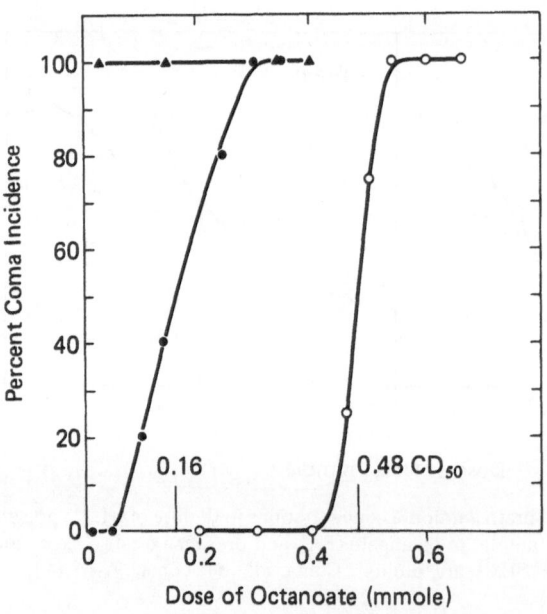

FIGURE 10. Dose-response curves for induction of coma with IP octanoate in the presence and absence of subcoma inhalation doses of methanethiol (MT, 0.12%) and dimethyl sulfide (DMS, 6.36%). [+MT (●), +DMS (▲), −MT or DMS (○)]. Male rats, 300 ± 30 g. From L. Zieve *et al.* (1974).

175

*ENCEPHA-
LOPATHY
DUE TO
SHORT- AND
MEDIUM-CHAIN
FATTY ACIDS*

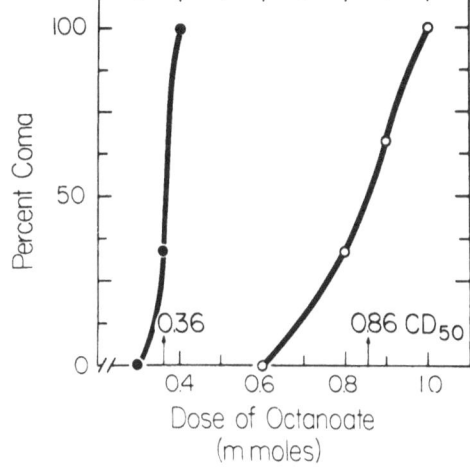

FIGURE 11. Dose-response curves for induction of coma with IP octanoate in the presence and absence of a subcoma IP dose of pentobarbital (PB, 10 mg/kg). [+PB (●), −PB (○)]. Male rats, 300 ± 30 g.

on their chain length and degree of unsaturation. The liver shortens or lengthens the carbon chain two carbons at a time. It also saturates or unsaturates the fatty acids. Abnormalities of fatty acids have been described in association with liver failure. However, there is no direct evidence that fatty acids have induced encephalopathy in humans. Excess of fatty acids interferes with mitochondrial respiratory activity. Certain unsaturated fatty acids injure mitochondria and interfere with their reactions. An excess of fatty acids also inhibits key glycolytic enzymes, reduces glucose production, and inhibits peripheral glucose utilization. Insulin secretion is increased and glucagon secretion decreased. Fatty acids also inhibit microsomal (Na^+, K^+)ATPase.

Short- and medium-chain fatty acids induce coma in normal experimental animals. The state of unconsciousness is reversible. The potency of the fatty acids for this effect increases with increased chain length. Energy utilization by the brain is unaffected. The metabolism of ammonia is disturbed, the formation of urea and glutamine being depressed because of inhibition of carbamyl phosphate synthetase and glutamate dehydrogenase. The electrical activity of the brain is affected by fatty acids. On the EEG, synchrony with slow waves are observed. Visual-evoked potentials are modified slightly by fatty acid alone, but in combination with NH_4^+ and mercaptan, marked alterations similar to those of experimental hepatic coma are produced. Fatty acids can also cause cardiac arrhythmias and shorten the action potential of isolated muscle fibers.

Fatty acids interact synergistically with most coma-inducing substances. Their coma-inducing dose is strikingly reduced by simultaneous injection of subcoma amounts of NH_4^+ and mercaptans, and vice versa. Similar though less-striking synergisms have been demonstrated with phenol, pentobarbital, and hypoglycemia. Fatty acids appear to have a pathogenetic role in diseases, such as Reye's syndrome, with selective mitochondrial injury. From the experimental data on synergistic interactions, it seems likely that in liver diseases causing hepatic failure in general they have at least a modulating effect on the pathogenetic actions of accumulating toxins such as ammonia and mercaptans.

References

Ahmed, K., and Thomas, B. S., 1971, The effects of long chain fatty acids on sodium plus potassium adenosine triphosphatase of rat brain, *J. Biol. Chem.* **246**:103–109.

Ansevin, C. F., 1980, Reye syndrome: Serum-induced alterations in brain mitochondrial function are blocked by fatty-acid-free albumin, *Neurology* **30**:160–166.

Borbola, Jr., J., Papp, J. G., and Szekeres, L., 1974, Effects of octanoate on the electrical activity of Purkinje fibres, *Experientia* **30**:262–264.

Coob, W. A., and Dawson, G. D., 1960, The latency and form in many of the occipital potentials evoked by bright flashes, *J. Physiol.* **152**:108–121.

Cowan, J. C., and Vaughan Williams, E. M., 1980, The effects of various fatty acids on action potential shortening during sequential periods of ischemia and reperfusion, *J. Mol. Cell. Cardiol.* **12**:347–369.

Dahl, D. R., 1968, Short chain fatty acid inhibition of rat brain Na-K adenosine triphosphatase, *J. Neurochem.* **15**:815–820.

Davis, E. J., and Gibson, D. M., 1969, Regulation of the metabolism of rabbit liver mitochondria by long chain fatty acids and other uncouplers of oxidative phosphorylation, *J. Biol. Chem.* **244**:161–170.

Derr, R. F., and Zieve, L., 1973, Decreased cerebral uptake of oxygen in coma—A consequence of decreased utilization of ATP, *J. Neurochem.* **21**:1555–1557.

Derr, R. F., and Zieve, L., 1976, Effect of fatty acids on the disposition of ammonia, *J. Pharmacol. Exp. Ther.* **197**:675–680.

Derr, R. F., and Zieve, L., 1982, Methanethiol and fatty acids depress urea synthesis by the isolated perfused rat liver, *J. Lab. Clin. Med.* **100**:585–592.

Edwards, J. C., Howell, S. L., and Taylor, K. W., 1969, Fatty acids as regulators of glucagon secretion, *Nature* **224**:808–809.

Friedmann, B., Goodman, Jr., E. H., and Weinhouse, S., 1967, Effects of insulin and fatty acids on gluconeogenesis in the rat, *J. Biol. Chem.* **242**:3620–3627.

Geisler, C., Swanson, A. R., Zieve, L., and Anders, M. W., 1979, Phthalate interference in gaschromatographic determination of long-chain fatty acids in plasma, *Clin. Chem.* **25**:308–310.

Gerich, J. E., Langlois, M., Schneider, V., Karam, J. H., and Noacco, C., 1974, Effects of alterations of plasma free fatty acid levels on pancreatic glucagon secretion in man, *J. Clin. Invest.* **53**:1284–1289.

Hird, F. G. R., and Weidemann, M. J., 1966, Oxidative phosphorylation accompanying oxidation of short-chain fatty acids by rat liver mitochondria, *Biochem. J.* **98**:378–388.

Holmquist, B., and Ingvar, D. H., 1957, Effects of short chain fatty acid anions upon cortical blood flow and EEG in cats, *Experientia* **13**:331–333.

Lai, J. C. K., Silk, D. B. A., and Williams, R., 1977, Plasma short-chain fatty acids in fulminant hepatic failure, *Clin. Chim. Acta* **78**:305–310.

Mamunes, P., DeVries, G. H., Miller, C. D., and David, R. B., 1975, Fatty acid quantitation in Reye's syndrome, in: *Reye's Syndrome* (J. D. Pollack, ed.), Grune and Stratton, New York, pp. 245–254.

Marcus, R. J., Winters, W. D., Mori, K., and Spooner, C. E., 1967, EEG and behavioral comparison of the effects of gamma-hydroxybutyrate, gamma-butyrolactone and short chain fatty acids in the rat, *Int. J. Neuropharmacol.* **6**:175–185.

Montague, W., and Taylor, K. W., 1968, Regulation of insulin secretion by short chain fatty acids, *Nature* **217**:853.

Mortiaux, A., and Dawson, A. M., 1961, Plasma free fatty acid in liver disease, *Gut* **2**:304–309.

Muto, Y. I., 1966, Clinical study on the relationship of short chain fatty acids and hepatic encephalopathy, *Jpn. J. Gastroenterol.* **63**:19–32.

Muto, Y. I., Takahashi, Y., and Kawamura, H., 1964, Effects of short-chain fatty acids on the electrical activity of the neo-, paleo-, and archicortical system, *Brain Nerve (Tokyo)* **16**:601–608.

Ohmiya, Y., 1971, Potentiation of thiopental anesthesia by fatty acids, *Folia Pharmacol. Jpn.* **67**:202–208.

Rabinowitz, J. L., Staeffen, J., Blanquet, P., Vincent, J. D., Terme, R., Series, C., and Myerson, R. M., 1978, Sources of serum ^{14}C-octanoate in cirrhosis of the liver and hepatic encephalopathy, *J. Lab. Clin. Med.* **91**:223–227.

Rizzoli, A. A., and Galzigna, L., 1970, Molecular mechanisms of unconscious state induced by butyrate, *Biochem. Pharmacol.* **19**:2727–2736.

177

ENCEPHA-
LOPATHY
DUE TO
SHORT- AND
MEDIUM-CHAIN
FATTY ACIDS

Ruderman, N. B., Toews, C. J., and Shafrir, E., 1969, Role of free fatty acids in glucose homeostasis, *Arch. Intern. Med.* **123**:299–313.

Samson, Jr., F. E. Dahl, N., and Dahl, D. R., 1956, A study on the narcotic action of the short chain fatty acids, *J. Clin. Invest.* **35**:1291–1298.

Sanbar, S. S., Hetenyi, Jr., G., Forbath, N., and Evans, J. K., 1965, Effects of infusion of octanoate on glucose concentration in plasma and the rates of glucose production and utilization in dogs, *Metabolism* **14**:1311–1323.

Schwark, W. S., Singhal, R. L., and Ling, G. M., 1970, Free fatty acid inhibition of α-glycerophosphate dehydrogenase activity in rat brain, *J. Pharm. Pharmacol.* **22**:458–460.

Seyffert, Jr., W. A., and Madison, L. L., 1967, Physiologic effects of metabolic fuels on carbohydrate metabolism. I. Acute effect of elevation of plasma free fatty acids on hepatic glucose output, peripheral glucose utilization, serum insulin, and plasma glucagon levels, *Diabetes* **16**:765–776.

Sherratt, H. S. A., 1969, Hypoglycin and related hypoglycemic compounds, *Br. Med. Bull.* **25**:250–255.

Teychenne, P. F., Walters, I., Claveria, L. E., Calne, D. B., Price, J., Macgillivary, B. B., and Gempertz, D., 1976, The encephalopathic action of five-carbon-atom fatty acids in the rabbit, *Clin. Sci. Mol. Med.* **50**:463–472.

Trauner, D. A., and Adams, H., 1981, Intracranial pressure elevations during octanoate infusion in rabbits: An experimental model of Reye's syndrome, *Pediatr. Res.* **15**:1097–1099.

Trauner, D. A., and Huttenlocher, P. R., 1978, Short chain fatty acid-induced central hyperventilation in rabbits, *Neurology* **28**:940–944.

Walker, C. O., McCandless, D. W., McGarry, J. D., and Schenker, S., 1970, Cerebral energy metabolism in short-chain fatty acid-induced coma, *J. Lab. Clin. Med.* **76**:569–583.

Weber, G., Hird Convery, H. J., Lea, M. A., and Stamm, N. B., 1966, Feedback inhibition of key glycolytic enzymes in liver: Action of free fatty acids, *Science* **154**:1357–1360.

White, R. P., and Samson, Jr., F. E., 1956, Effects of fatty acid anions on the electroencephalogram of unanesthetized rabbits, *Am. J. Physiol.* **186**:271–274.

Wilcox, H. G., Dunn, C. D., and Schenker, S., 1978, Plasma long chain fatty acids and esterified lipids in cirrhosis and hepatic encephalopathy, *Am. J. Med. Sci.* **276**:293–303.

Willebrands, A. F., Ter Welle, H. F., and Tasseron, S. A. J., 1973, The effect of a high molar FFA/albumin ratio in the perfusion medium on rhythm and contractility of the isolated rat heart, *J. Mol. Cell. Cardiol.* **5**:259–273.

Williamson, J. R., Scholz, R., and Browning, E. T., 1969, Control mechanisms of gluconeogenesis and ketogenesis. II. Interactions between fatty acid oxidation and the citric acid cycle in perfused rat liver, *J. Biol. Chem.* **244**:4617–4627.

Windus-Podehl, G., Lyftogt, C., Zieve, L., and Brunner, G., 1983, Encephalopathic effect of phenol in rats, *J. Lab. Clin. Med.* **101**:586–592.

Wojtczak, L., 1976, Effect of long-chain fatty acids and acyl-CoA on mitochondrial permeability, transport, and energy-coupling processes, *J. Bioenerg. Biomembranes* **8**:293–311.

Zeneroli, M. L., Ventura, E., Baraldi, M., Penne, A., Messori, E., and Zieve, L. 1982, Visual evoked potentials in encephalopathy induced by galactosamine, ammonia, dimethyldisulfide, and octanoic acid, *Hepatology* **2**:532–538.

Zieve, F. J., Zieve, L., Doizaki, W. M., and Gilsdorf, R. B., 1974, Synergism between ammonia and fatty acids in the production of coma: Implications for hepatic coma. *J. Pharmacol. Exp. Ther.* **191**:10–16.

Zieve, L., 1980, Coma production with NH_4^+: Synergistic factors, *Gastroenterology* **78**:1327.

Zieve, L., and Doizaki, W., 1980, Brain and blood methanethiol and ammonia concentrations in experimental hepatic coma and coma due to injections of various combinations of these substances, *Gastroenterology* **79**:1070.

Zieve, L., Lyftogt, C., and Draves, K., 1983, Toxicity of a fatty acid and ammonia: Interactions with hypoglycemia and Krebs cycle inhibition, *J. Lab. Clin. Med.* **101**:930–939.

Zieve, L., and Nicoloff, D., 1976, Alterations in volatile free fatty acids of blood after hepatectomy, *Surgery* **80**:554–557.

Zieve, L., Doizaki, W. M., and Zieve, F. J., 1974, Synergism between mercaptans and ammonia or fatty acids in the production of coma: A possible role for mercaptans in the pathogenesis of hepatic coma, *J. Lab. Clin. Med.* **83**:16–28.

8

Encephalopathy Due to Mercaptans and Phenols

LESLIE ZIEVE and GORIG BRUNNER

MERCAPTANS*

1. Introduction

Mercaptans are thio alcohols. Most are acyclic hydrocarbon derivatives, the alkane thiols. Methanethiol, CH_3SH, is the prototype. It is one of the most potent mercaptans and of greatest importance medically. The lower molecular weight alkane thiols occur naturally in coal, tar, and petroleum distillates and are also found in foods and vegetables. They are intermediates in the synthesis of pharmaceuticals and agricultural chemicals such as herbicides and insecticides. More than 180,000 employees are potentially exposed to these compounds in the United States (NIOSH, 1978). Mercaptans are noted for their obnoxious odor, and methyl and ethyl derivatives have been put in piped natural gas in very low concentrations (0.02 parts per billion) to serve as warning agents for gas leaks.

Inhalation is the main route of exposure to mercaptans. Methanethiol boils at 6.2°C and ethanethiol at 35°C. There have been few reports of the consequences of toxic exposure in man. A recent publication by the National Institute for Occupational Safety and Health reviewed the subject comprehensively, and set exposure limits (NIOSH, 1978). The first published report of a fatal case of methanethiol poisoning appeared in 1970 (Schults et al., 1970). The most prominent feature was coma that persisted for 28 days until death from a massive

---

* This section was written by L. Z.

LESLIE ZIEVE • Hennepin County Medical Center, University of Minnesota, Minneapolis, Minnesota 55415. *GORIG BRUNNER* • Medizinische Hochschule Hannover, Hannover, West Germany.

pulmonary embolus. The subject was a 53-year-old laborer who had emptied several tanks containing residual methanethiol. He was found unconscious and reeked of methanethiol. During hospitalization he had random myoclonic jerks and rigidity of all four extremities. For a few days he had an acute severe hemolytic anemia and methemoglobinemia. The serum bilirubin rose to 3.5 mg/dl. The brain at autopsy showed no gross or microscopic abnormality. The liver was not mentioned and was presumably also normal.

The first published evidence of the clinical significance of methanethiol appeared in 1955. Challenger and Walshe (1955) isolated mercury dimethyl mercaptide, $Hg(SCH_3)_2$, from the urine of a woman with acute fulminant hepatic failure by passing a stream of nitrogen through the urine into aqueous mercuric cyanide. The woman had an intense fetor hepaticus that was less fetid although similar to the odor of her urine. From the isolated urine mercaptide, they could detect the odor of methanethiol by acidification and of dimethyl sulfide by heating. The boiling point of dimethyl sulfide (CH_3SCH_3) is 37.3°C and of dimethyl disulfide ($CH_3S\text{-}SCH_3$) 109.7°C. They suggested that the breath odor was likely due to a mixture of CH_3SH, CH_3SCH_3, and $CH_3S\text{-}SCH_3$.

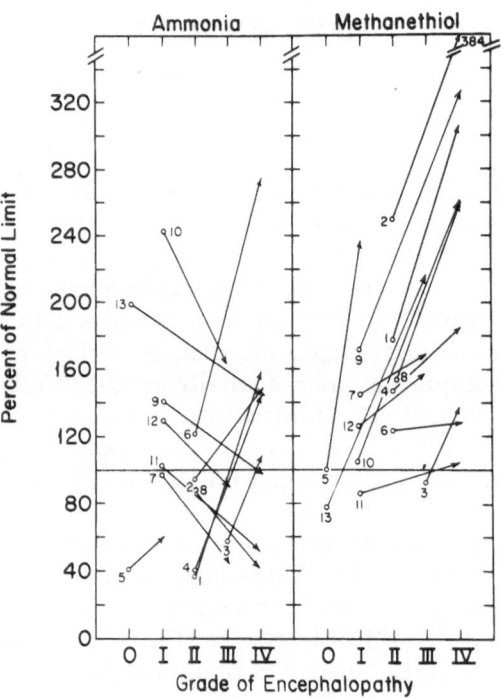

FIGURE 1. Change in blood ammonia and methanethiol values in relation to change in clinical grade of hepatic encephalopathy in 13 cirrhotic patients with progressive disease who eventually died. The arrows indicate the direction of change and the values when the encephalopathy was at its worst. The circles indicate values when the encephalopathy was least. The numbers by the lines indicate the individual patients. The grade of encephalopathy is given on the abscissa. Grade 0 meant absence of encephalopathy, and grade IV presence of frank coma. Values of ammonia and methanethiol are expressed as a percentage of their upper limits of normal (90 nmole/ml and 500 pmole/ml, respectively). From McClain et al. (1980).

Methanethiol is a product of methionine breakdown in the intestine, much, if not most, resulting from bacterial action. It is largely removed by the liver and converted to dimethyl sulfide in the process of detoxification. Utilizing gas chromatography it has been possible to demonstrate small amounts of methanethiol normally in the blood (less than 0.5 nmole/ml) (Doizaki and Zieve, 1977) and breath (less than 0.03 nmole/liter) (Chen et al., 1970). Circulating methanethiol is bound to red blood cells, and a strong reducing agent in an acid medium is necessary to release it into a gas phase. Metallic zinc, relatively free of hydrogen sulfide, has been used successfully as the reducing agent, but care must be exercised not to release methanethiol from circulating methionine in the process, or spuriously high values will be found.

In patients with liver failure, increased amounts of methanethiol are found in the blood and breath (Chen et al., 1970). In rats in hepatic coma after acute ischemic necrosis of the liver, the brain methanethiol increases fivefold (Zieve and Doizaki, 1980). In patients with cirrhosis without overt encephalopathy, the blood methanethiol increased 1.5-fold. In those with overt hepatic encephalopathy, blood levels rose 2.5-fold (McClain et al., 1980). Values as high as 2.1 μM have been found in comatose patients and as high as 2.4 μM in comatose rats. A predictable relationship has been observed between severity of encephalopathy and degree of elevation of blood methanethiol, when serial measurements have been made in a given patient (Fig. 1). Others have also found increased blood levels in hepatic failure, although the absolute blood values are not comparable methodologically (Brunner and Scharff, 1978; Mar-

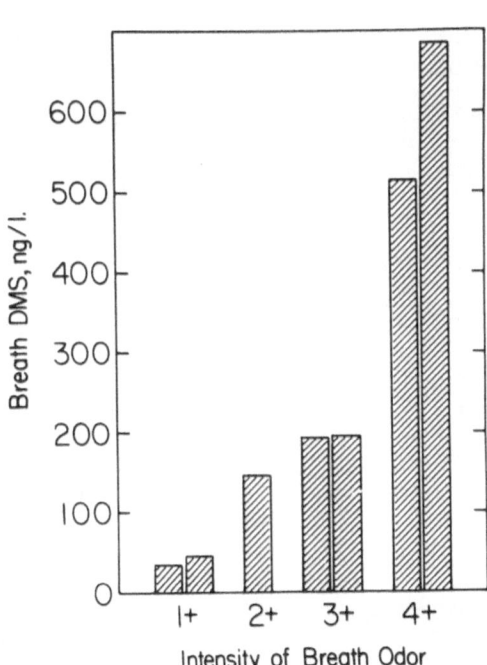

FIGURE 2. Relationship of breath dimethyl sulfide to intensity of the breath odor in cirrhotics fed methionine. Data from seven different patients. From Chen et al. (1970).

dini *et al.*, 1981). Cirrhotics who had been fed methionine became somnolent and developed a breath odor that was directly related in intensity to the amount of dimethyl sulfide in the breath (Fig. 2) (Chen *et al.*, 1970). An association between the ingestion of methionine and the occurrence of coma in patients with cirrhosis was first recognized in 1949 (Watson, 1949; Kinsell *et al.*, 1949), and the prevention of the neuropsychiatric effects with a broad-spectrum antibiotic was demonstrated in 1956 (Phear *et al.*, 1956).

2. Metabolism—Biochemical Effects

Few studies of the metabolism of methanethiol and its derivatives have been reported. Methanethiol and hydrogen sulfide are normally formed in the process of tissue metabolism of methionine as well as bacterial action on methionine in the colon (Fig. 3). Acidification of the stool reduces the formation of these substances. The colon contains more thiol S-methyltransferase than any other tissue including liver (Weisiger *et al.*, 1980). This enzyme converts H_2S to CH_3SH and CH_3SH to CH_3S-CH_3. Thus much of the hydrogen sulfide and methanethiol formed in the colon is probably converted to dimethyl sulfide by the colon. The unconverted methanethiol and the dimethyl sulfide are largely removed and processed by the liver (Canellakis and Tarver, 1953; Bremer and Greenberg, 1961; Weisiger *et al.*, 1980). The methanethiol is rapidly methylated (thiol S-methyltransferase) by the liver to form dimethyl sulfide and also oxidized along with the dimethyl sulfide to carbon dioxide and sulfate. The process of methylation of hydrogen sulfide and methanethiol are important steps in detoxification, since dimethyl sulfide is many-fold less toxic than either of its precursors.

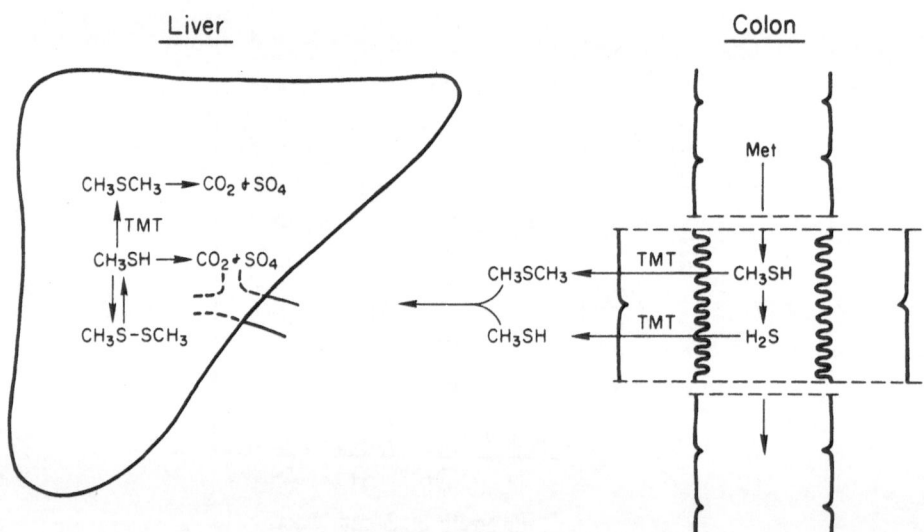

FIGURE 3. Metabolism of hydrogen sulfide and methanethiol. Met, methionine; TMT, thiol S-methyltransferase.

As little as 1 μM methanethiol inhibited rat liver mitochondrial glutamate oxidation (Waller, 1977). In more elaborate studies of mitochondrial function, Meijer and co-workers (Vahlkamp et al., 1979) found that methanethiol and ethanethiol were effective inhibitors of mitochondrial electron transfer in both liver and brain (Fig. 4). The relative degrees of inhibition by methanethiol, ethanethiol, and dimethyl sulfide corresponded with their differences in potency for producing coma in rats in vivo. Methanethiol and ethanethiol inhibited cytochrome oxidase. Ethanethiol was also shown to inhibit gluconeogenesis and ureagenesis from suitable substrates in rat hepatocytes, depressing cellular ATP and increasing the mitochondrial reduction state. These effects

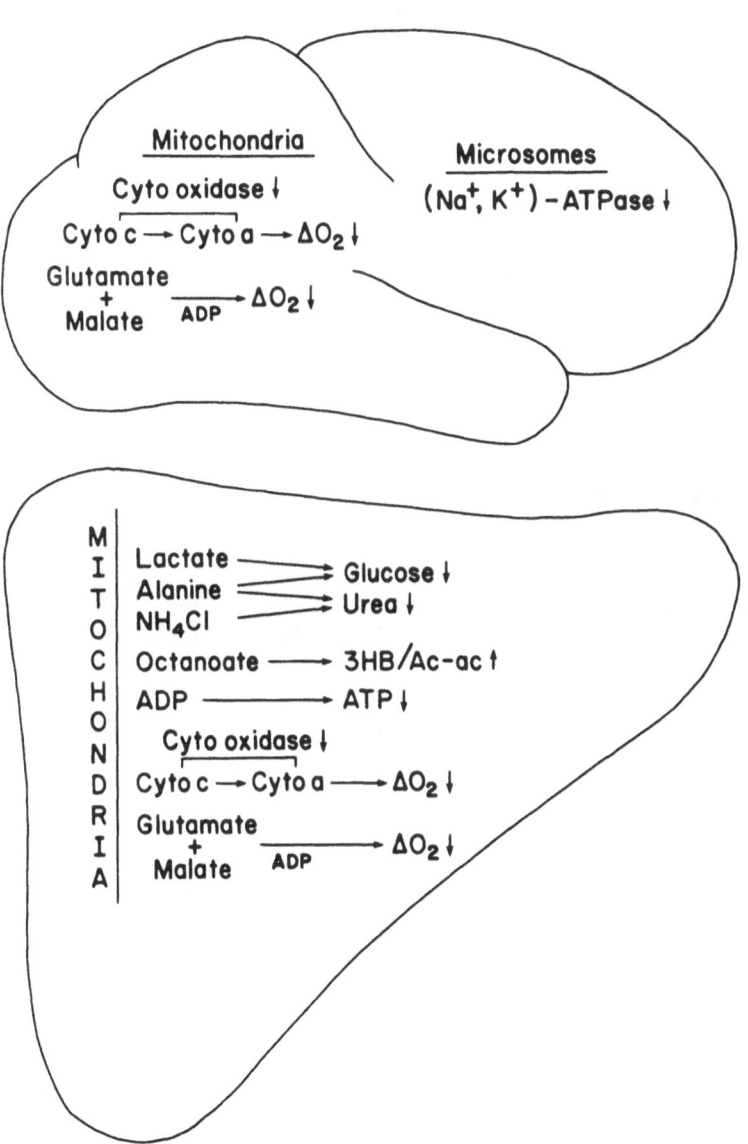

FIGURE 4. Known biochemical toxicity of methanethiol and ethanethiol.

were probably the result of the electron-transfer inhibition. The authors speculated that the inhibition of mitochondrial electron transfer by methanethiol in the brain might be relevant to the mechanism by which energy production in the brain is depressed during hepatic coma.

Methanethiol also inhibits brain microsomal (Na^+, K^+)ATPase (Quarfoth et al., 1976). The compound apparently acts at several sites on the enzyme system. Although the ATPase was inhibited, there was concurrent stimulation of the associated K^+-dependent phosphatase. Methanethiol inhibits the phosphoenzyme formation in the (Na^+, K^+)ATPase. These effects of methanethiol are fully reversible. The concentrations at which the toxin produces 50% inhibition of the brain (Na^+, K^+)ATPase are of the same order as those that give a 50% protection against hemolysis of red blood cells.

Other enzyme systems are probably inhibited by mercaptans. One that has been studied is erythrocyte carbonic anhydrase (Schwimmer, 1969), and it is possible that methanethiol has a role in intestinal carbon dioxide transport inhibition. The observation of Vahlkamp et al. (1979) that urea synthesis was inhibited by ethanethiol in isolated hepatocytes was supported by studies of urea synthesis in the isolated perfused rat liver cited below.

The biochemical studies of the effects of methanethiol on brain microsomal and mitochondrial components offer possible explanations for the encephalopathic effects of mercaptans, particularly when their synergistic interaction with other substances (ammonia, fatty acids) that also affect these components is taken into account.

3. Pharmacological Effects

3.1. Toxicity Studies (Fig. 5)

Little experimental data is available. In 1943 Ljunggren and Norberg studied the relative toxicity of H_2S, CH_3SH, dimethyl sulfide (DMS), and dimethyl disulfide (DMDS) in a few (number not specified) female rats weighing 90–130 g. Being volatile, the substances were inhaled in a closed chamber. The fatal concentration of H_2S was 0.1% by volume. The fatal concentration of DMDS was about five times greater, CH_3SH ten times greater, and DMS 54 times greater.

In 1958, Fairchild and Stokinger reported the most systematic study of acute and subacute toxicity of thiols. They studied the 2-, 3-, 4-, 6-, and 7-carbon aliphatic thiols and two aromatic thiols. The main experimental animals were male rats weighing approximately 200 g. Mice and a few rabbits were also studied. The routes of administration studied were inhalation, intraperitoneal (IP) injection, oral gavage, and in a few instances, eye and skin exposure. Sedation approaching anesthesia was characteristic of acute toxicity. This was reversible. With lethal doses the sequence was CNS depression, respiratory paralysis, and death. Even-number aliphatic thiols were more toxic than odd. Longer-chain compounds were more toxic after IP administration. The range of acute toxicity was greater with inhalation than with the IP or oral routes. Intraperitoneal administration was more toxic than oral, the LD_{50} dose for ethanethiol and propanethiol after IP injection being about one half that of oral gavage. Pulmonary elimination of thiols was an important factor in determining their toxicity.

These two studies were the most informative reports on toxicity. Additional information and commentary on the general subject of toxicity will be found in the monograph published by NIOSH (1978).

3.2. Coma Induction Studies

Because of the clinical observations of increased pulmonary and urinary excretion of methanethiol (MT), DMDS, and DMS in human hepatic failure (Challenger and Walshe, 1955; Chen et al., 1970), we studied in more detail the coma-producing properties of these substances, as well as ethanethiol (ET), in male Sprague-Dawley rats weighing approximately 300 g (Zieve et al., 1974). We focused primarily on MT, because that is the parent compound that is generated in liver failure. Fifty percent of 300 g rats became deeply comatose (CD_{50}) when they were exposed for a few minutes to 0.16% by volume of MT in a gas chamber. Complete recovery ensued if the rats were removed from the chamber as soon as coma occurred. Similar results were observed with 3.3%

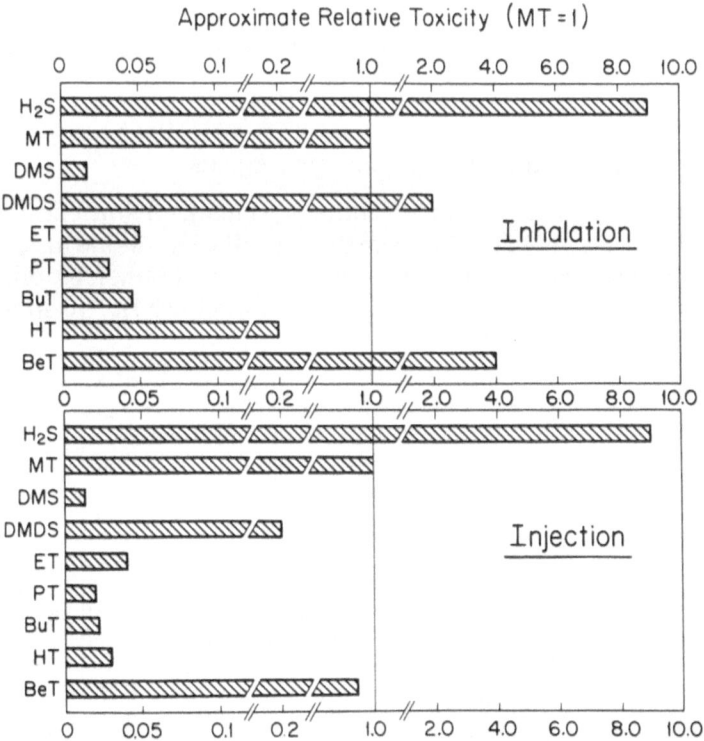

FIGURE 5. Approximate relative toxicity of various mercaptans in rats based upon data from Ljunggren and Norberg (1943), Fairchild and Stokinger (1958), and Zieve et al. (1974). MT, methanethiol; DMS, dimethyl sulfide; DMDS, dimethyl disulfide; ET, ethanethiol; PT, propanethiol; BuT, butanethiol; HT, hexanethiol; BeT, benzenethiol.

ET and 9.6% DMS. Unconsciousness did not occur in any animal until the concentration of MT exceeded 0.12%, ET exceeded 3%, or DMS exceeded 7%.

The IP dose of MT causing deep coma in 100% of such rats was 80–100 μmole/kg (Zieve and Doizaki, 1980). Up to 75 μmole/kg could be given without causing unconsciousness in any of the rats. Using our procedures for measuring blood and tissue concentrations of MT, the normal average rat blood and brain MT concentrations were 182 ± 13 pmole/ml and 218 ± 15 pmole/g (mean ± SEM), respectively. Comatose rats that received 80–100 μmole/kg IP had blood levels of 7400 ± 900 pmole/ml and brain levels of 5100 ± 1100 pmole/g (Fig. 6). Noncomatose rats that received 60–75 μmole/kg IP had blood levels of 1800 ± 200 pmole/ml and brain levels of 1275 ± 125 pmole/g. In experimental hepatic coma due to acute massive ischemic necrosis of the liver, the brain level of MT was 1135 ± 144 pmole/g, indicating that MT alone could not account for the comatose state.

Although two molecules of MT are produced from one of DMDS, much more DMDS is required to cause coma if it is injected IP, probably because the compound, which is a liquid at body temperature, adsorbs to the peritoneal surface and is not promptly broken down to form MT, and after absorption much of it is excreted unchanged. The IP dose of DMDS causing deep coma in 100% of 300-g rats was 560 μmole/kg (Fig. 7). Up to approximately 400 μmole/kg could be given without causing unconsciousness in any of the rats. In mice, IP injection of DMDS was followed by the appearance of DMDS as well as much smaller amounts of MT and DMS in the breath (Susman *et al.*, 1978).

3.3. Synergism with Other Encephalopathic Agents

An inhaled subcoma dose of MT influenced markedly the encephalopathic effect of NH_4^+ (Zieve *et al.*, 1974). The CD_{50} for IP NH_4^+ alone in 300-g normal rats was approximately 5 mmole/kg. Addition of 0.12% MT by inhalation simultaneously, a subcoma dose, reduced the CD_{50} of the NH_4^+ to approximately

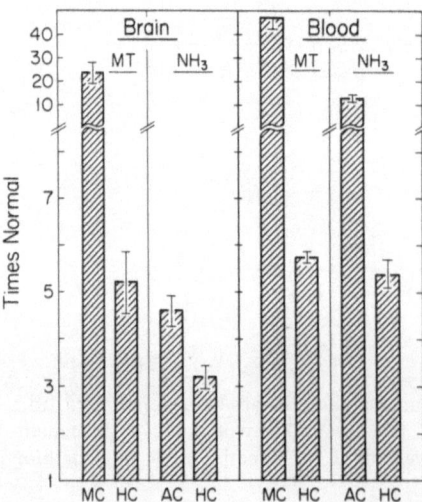

FIGURE 6. Relative elevations in 300-g rats of brain and blood methanethiol (MT) and ammonia in coma induced with each of these substances and in hepatic coma (HC). MC, methanethiol-induced coma; AC, ammonia-induced coma. Normal values based on 8–10 rats were: brain MT, 218 ± 15 pmole/g (mean ± SEM); brain NH_3-N, 421 ± 21 nmole/g; blood MT, 182 ± 13 pmole/ml; blood NH_3-N, 134 ± 5 nmol/ml. From Zieve *et al.* (1984).

1.5 mmole/kg, a reduction of about two thirds (Fig. 8). The incidence of coma could be raised from 0 to 100% by simultaneous administration of subcoma doses of both substances. DMS had a similar but less pronounced effect. Methanethiol and DMS also affected the blood level of ammonia resulting from a given IP dose of NH_4^+, causing it to increase by 60–80%. As little as one sixth the coma dose of MT in combination with one fifth the coma dose of octanoic acid caused the blood ammonia resulting from a subcoma IP dose of NH_4^+ to increase by 80%, raising it from a noncoma to a coma-inducing level (Fig. 9). Subcoma doses of NH_4^+ and octanoate had a similar converse effect on the blood MT concentration resulting from a subcoma dose of MT.

Methanethiol and DMS by inhalation had similar effects on the encephalopathy caused by octanoic acid (Zieve et al., 1974). The CD_{50} for IP octanoate alone in normal 300-g rats was approximately 1.65 mmole/kg. Addition of 0.12% MT or 5.62% DMS by inhalation simultaneously, subcoma doses, reduced the CD_{50} of octanoate by approximately 70%. Phenol also causes deep coma in rats of this size at doses in the range of 1.6–2.0 mmole/kg (Windus-Podehl et al., 1982). The CD_{50} was 1.8 mmole/kg. Dimethyl disulfide was shown to interact synergistically with phenol, a subcoma dose of 400 μmole/kg causing a reduction of 28% in the CD_{50} of phenol. Conversely, phenol enhanced the encephalopathic properties of DMDS. The CD_{50} for DMDS was reduced by approximately 25% with simultaneous injection of 1.5 mmole/kg phenol. The CD_{50} for DMDS was similarly reduced by 38% with simultaneous injection of 3.3 mmole/kg of NH_4^+, a subcoma dose by itself. Studies with octanoate are unavailable.

When normal rats are injected simultaneously with NH_4^+, MT, and octanoic acid, much less of each substance is needed to induce coma (Zieve et al., 1974). By trial of a variety of dose combinations we found that simultaneous IP injection of 2.5 mmole/kg NH_4^+, 0.5 mmole/kg octanoate, and 60 μmole/kg

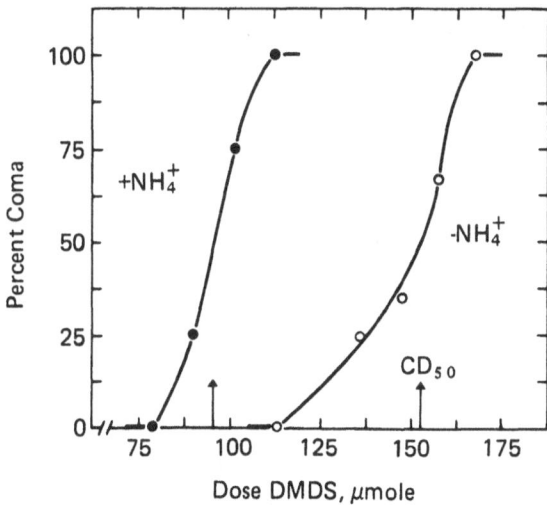

FIGURE 7. Coma induction dose–response curves for IP dimethyl disulfide in the presence and absence of a subcoma dose (1.0 nmole per 300-g rat) of NH_4^+. CD_{50}, dose inducing coma in 50% of rats.

MT caused deep coma in normal 300-g male rats at brain levels of ammonia (1303 ± 108 nmole/g, mean ± SEM) and MT (1079 ± 64 pmole/g) that corresponded to those (1370 ± 101 nmole/g and 1135 ± 144 pmole/g, respectively) observed in rats with hepatic coma following acute massive ischemic necrosis of the liver (Zieve and Doizaki, 1980).

3.4. Visual-Evoked Potential (VEP) Studies (Fig. 10)

The VEP is an average of the cortical wave patterns generated by the succession of neuronal membrane potential changes that follow stimulation with a flash of light. Normally there are four components occurring within a latency period of 240 msec (Coob and Dawson, 1960). These have been labeled P_1, N_1, P_2, and N_2. The first and third are positive deflections with amplitudes of about 10 and 30 μV, respectively. The second and fourth are negative deflections with amplitudes of about −50 and −15 μV, respectively. The last deflection is highly variable and not as reproducible as the first three. The VEP following coma induction with IP DMDS is characterized by about 35% reduction in the absolute amplitude of N_1, a disappearance of N_2, and a marked reduction in latency (Zeneroli et al., 1982). Coma induced with octanoic acid or NH₄Cl gives similar VEPs but without the shortened latency and the loss of N_2 in most instances. In contrast, hepatic coma induced by injection of gal-

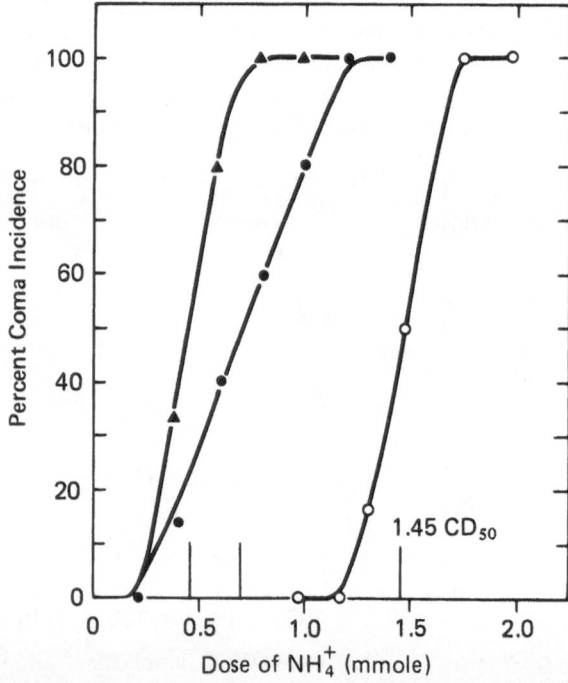

FIGURE 8. Coma induction dose–response curves for NH_4^+ [NH_4^+ only (○)] in the presence and absence of an inhaled subcoma dose of MT (0.12% by volume) [+MT (▲)] or DMS (6.36% by volume) [+ DMS (●)]. From Zieve et al. (1974).

actosamine shows an extreme reduction in N_1 (90%), about 50% reduction in P_2, and about 30% reduction in overall latency. A combination of reduced doses of DMDS, NH_4^+, and octanoate that causes coma after five injections in 24 hr results in VEPs that approximate those observed in hepatic coma. N_1 is reduced by about 80%, P_2 by 40%, and the latency by 40%.

From these observations and the previous ones on brain concentrations of MT, it is apparent that coma induced with mercaptan alone differs from that observed after hepatic failure in the amount of mercaptan required in the brain to induce coma and in the resulting electrical activity of the cortex. A combination of reduced doses of mercaptan, NH_4^+, and octanoate has been found that causes coma in normal animals that is similar to hepatic coma in the electrical response of the cortex, as reflected by the VEP, and in the concentrations of MT and of ammonia in the brain.

3.5. Urea Formation

The observations, cited previously, that the blood ammonia rises disproportionately when a subcoma dose of MT or DMS is given simultaneously with a subcoma dose of NH_4^+, was indirect *in vivo* evidence of an effect of mercaptans on urea synthesis, particularly in view of Vahlkamp's *et al.* (1979) *in vitro* observations that they inhibited mitochondrial electron transfer. The

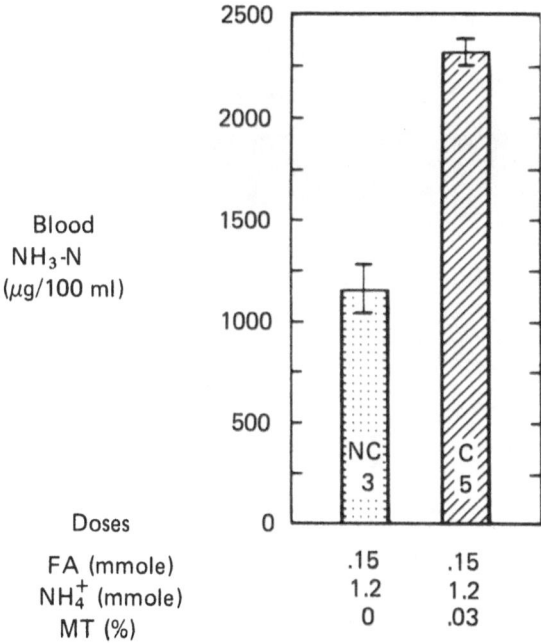

FIGURE 9. Blood NH_3 concentrations in 300-g rats after IP injections of the highest subcoma dose of NH_4Cl simultaneously with very small subcoma doses of octanoic acid (FA) and methanethiol (MT). FA given IP. MT given by inhalation (% by volume). Number of rats indicated at base of each bar. NC, no coma, C, coma induction. From Zieve *et al.* (1974).

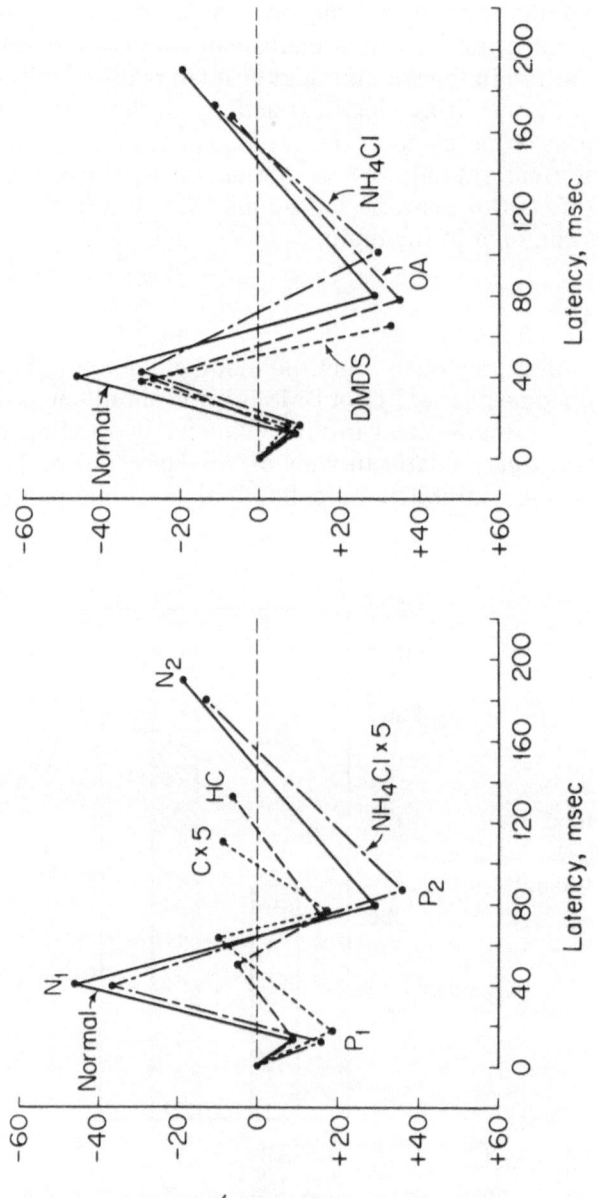

FIGURE 10. Visual-evoked potentials in 125-g rats after coma induction with IP injections of various toxins. DMDS, single injection of 500 μmole dimethyl disulfide; OA, single injection of 0.5 mmole octanoic acid; NH$_4$Cl, single injection of 1.7 mmole; NH$_4$Cl × 5, five injections within 24 hr, four of 1.0 mmole, and the fifth of 1.5 mmole; C × 5, five injections of combination of (NH$_4$Cl + OA + DMDS), four of (0.23 mmole + .05 mmole + 23 μmole) and a fifth of (0.7 mmole + 0.15 mmole + 70 μmol); HC, hepatic coma induced with galactosamine. From Zeneroli *et al.* (1982).

question has been studied directly in the isolated perfused normal rat liver (Derr *et al.*, 1982). Concentrations of 0.5–5.0 μM MT in the perfusing medium reduced urea synthesis from NH_4^+ (+ ornithine) by approximately 60% over 2 hr (Fig. 11). When carbamyl phosphate was used in place of NH_4^+, urea synthesis was not different from controls at the end of 1 or 2 hr, although the rate of synthesis of urea during the first 40 min was slightly slower than controls. Thus the depressant effect of MT appears to be at the carbamyl phosphate synthetase step of urea synthesis. These results are comparable to those observed with 10 mM octanoate in the same perfusion system. Thus both mercaptans and fatty acids interfere with the first step in urea synthesis, providing at least a partial explanation for the disproportionate rise in blood ammonia when they are injected along with NH_4^+. *In vitro* studies indicate that the effect of these two agents is on mitochondrial function in general (Vahlkamp *et al.*, 1979), suggesting that other *in vivo* derangements will be found if looked for.

4. Pathophysiologic Role of Mercaptans in Disease

Methanethiol and its derivative DMS are products of metabolism normally efficiently but not entirely removed by the liver. They are readily identifiable in the breath. In human hepatic failure they accumulate in the blood and breath. This has been verified in experimental hepatic failure in animals in whom elevated brain levels have been demonstrated. Pharmacologic doses of these substances, particularly MT which is most potent, cause a reversible encephalopathy including deep coma in animals. The cellular effects of MT at these dosage levels is profound, affecting mitochondrial respiration and membrane ion transfer. The encephalopathy produced with MT alone differs from that observed in severe hepatic failure. However, when the synergistic interactions of MT with other encephalopathic agents accumulating during hepatic failure

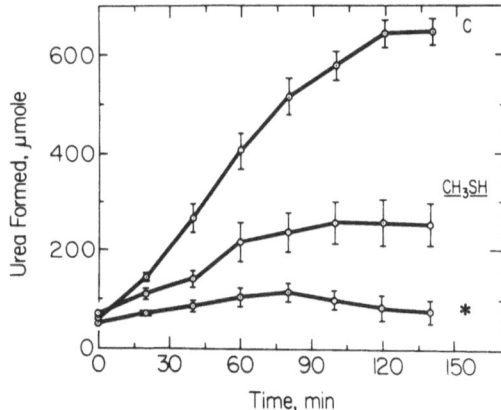

FIGURE 11. Effect of methanethiol on synthesis of urea from ammonia and ornithine in the isolated perfused normal rat liver. C, control (n = 14). CH_3SH, methanethiol added to perfusate at intervals to maintain concentration at 0.5–5.0 μM (n = 6). *, addition of 14 μmole methanethiol twice, at 0 time and at 10 minutes (n = 5). From Derr *et al.* (1982).

are taken into account experimentally, the encephalopathy that results mimics that of hepatic coma fairly closely. The amount of brain MT required for these results is only about one fifth that required for coma with MT alone. In human hepatic encephalopathy, MT probably plays a pathogenetic role while interacting with the abnormal accumulations of ammonia, fatty acids, and phenols.

5. Summary

Methyl mercaptan (methanethiol) is a very toxic odoriferous product of methionine breakdown in the gut. It and a less toxic derivative, dimethyl sulfide, are normally efficiently metabolized in the liver. Increased amounts of methanethiol and dimethyl sulfide are found in the blood, breath, and urine of patients with liver failure. Methanethiol has profound biochemical effects on mitochondrial electron transfer and on microsomal (Na^+, K^+)ATPase. It also depresses the activity of various other enzymes including carbamyl phosphate synthetase of liver that results in reduced urea synthesis.

Sedation approaching anesthesia is characteristic of acute toxicity with mercaptans. The range of toxicity is greater with inhalation than with injection of the mercaptans. Pulmonary elimination is important in reducing toxicity. In experimental animals, a controlled reversible state of coma can be induced with methanethiol, the prototype mercaptan. The brain level of methanethiol in comatose rats injected with methanethiol alone is about 25 times normal and the blood level about 45 times normal. Mercaptans interact synergistically with NH_4^+, fatty acids, and phenol to induce coma with much smaller amounts of each of these substances. In normal rats, a combination of reduced doses of mercaptan, NH_4^+, and octanoate was found which induced coma that was similar to experimental hepatic coma in the electrical response of the brain, as reflected by visual-evoked potentials, and in the concentrations of methanethiol and ammonia in the brain and blood. Methanethiol probably plays a pathogenetic role in human hepatic encephalopathy while interacting with abnormal accumulations of ammonia, fatty acids, and phenols.

PHENOLS*

6. Introduction

Phenols are aromatic compounds with one or more hydroxyl groups replacing the hydrogen atoms of the benzene moiety. Despite detailed knowledge of their appearance and properties, they have long been neglected in medicine. Many phenolic substances in free and conjugated form are found in the urine of man (Schmidt, 1949; Armstrong *et al.*, 1956; Hartmann, 1961; Hartmann and Ruge, 1962). Of the many phenols in the urine of healthy persons, phenol, p-cresol (4-methyl phenol) and 49 more phenolic acids have been identified. Among the other phenols identified are p-ethylphenol, catechol, quinol, heterocyclic

---

* This section was written by G. B.

phenols, and phenolic hormones such as adrenaline, noradrenaline, thyroxine, and the phenolic steroids estrone and estradiol. The phenolic amino acid tyrosine is also found in urine (Frerichs, 1854; Ruge, 1971). These phenols are derived from food materials and from breakdown of endogenous compounds. Daily urinary excretion of phenol amounts to 5–15 mg/day and p-cresol 65–120 mg/day; the total daily output of phenols varies between 0.2 and 0.5 g (Williams, 1959). Phenol and p-cresol are not only derived from the degradation of tyrosine, but also from estrone. The amount of free tyrosine excreted in the urine is about 35mg/day, while another 30 mg are excreted as the sulfate ester (Williams, 1959).

Phenols can be measured by a variety of methods. Total phenolic compounds including genuine phenols, phenolic acids, phenolic amines, and tyrosine can be measured quickly and easily by colorimetric methods (Folin and Ciocalteu, 1927; Bray *et al.*, 1952; Muting *et al.*, 1970). Qualitative determination of phenolic compounds is carried out by paper chromatography (Hartmann, 1961; Scott, 1968). Accurate quantitative analysis of most specific phenols and phenolic acids requires extraction, derivatization, and gas chromatography (Williams and Leonard, 1963) or liquid high-pressure chromatography.

7. Metabolism

7.1. Origin

Phenols originate mainly from the metabolism of phenylalanine and tyrosine (Dalgliesh, 1955; Knox, 1955; Ruge, 1971). One major source of phenols is the gut where both amino acids are subject to oxidative deamination by bacteria (Knox, 1955; Levine and Conn, 1967; Muting *et al.*, 1968). In tissues, tyrosine and phenylalanine have a number of possible metabolic pathways (Dalgliesh, 1955). Most is normally metabolized after deamination to p-hydroxyphenyl pyruvic acid, homogentisic acid, and finally fumaric and acetoacetic acids, which enter the citric acid cycle (Knox, 1955). If this pathway is blocked by an inborn error of metabolism or by a disturbance of liver function, a variety of phenols and phenolic acids are found in serum, urine, and cerebrospinal fluid (Becher *et al.*, 1926; Becher, 1930; Muting and Reikowski, 1965; Prader, 1966; Jervis and Drejza, 1966; Levine and Conn, 1967; Muting *et al.*, 1968; Frimpter, 1973). In renal failure, phenolic acids also pile up in serum and urine (Hicks *et al.*, 1964; Muting, 1965).

7.2. Disposition

In the metabolism of phenolic compounds, one must distinguish between the genuine phenols, which are lipophilic, more toxic, and cannot be excreted unchanged in the urine, and the phenolic acids, which are less lipophilic, less toxic, can more easily be excreted into the urine without conjugation reactions, and can be broken down into metabolites that finally enter the citric acid cycle.

Phenols undergo three different reactions: conjugation, hydroxylation, and methylation. The main reaction is conjugation of the hydroxyl group with glucuronic acid to form glucuronides and with sulfuric acid to form the ethereal sulfates or monoesters of sulfuric acid (Fig. 12). Conjugation with glucuronic acid or sulfate decreases the toxic properties of phenols, as will be shown later. Some phenols are hydroxylated so that an additional hydroxyl group may be added to the aromatic system, but this reaction takes place only to a minor extent. Methylation occurs to an even lesser extent. It has been shown that conjugation of phenols with glucuronic acid and sulfuric acid is influenced by a number of factors including the phenol concentration, temperature, metabolic state, and diet. The rate of glucuronic acid conjugation of phenols is proportional to the body level of phenol, whereas the rate of sulfate conjugation is independent of the phenol level, but is dependent on the availability of sulfate (Williams, 1959, 1964).

8. Pharmacologic Effects

8.1. Toxicity

Louis Pasteur was the first to observe that phenol kills bacteria, and he used this substance to produce vaccines. This observation led Lister to develop the first antiseptic wound treatment as early as 1867. By spraying infected open leg fractures with a 3% aqueous phenol solution (carbolic acid), he could lower the mortality from such wounds from 60% to 10%. As a consequence of these results, "carbolic acid" was used over decades for the disinfection of hospitals and operating theaters, causing the typical unpleasant hospital smell of the time.

Phenol accidentally taken by mouth causes burns in the esophagus and stomach, generalized cramps, and coma. Ten grams of phenol are usually fatal. Inhalation of phenols leads to nausea, giddiness, and agitation (von Oettingen,

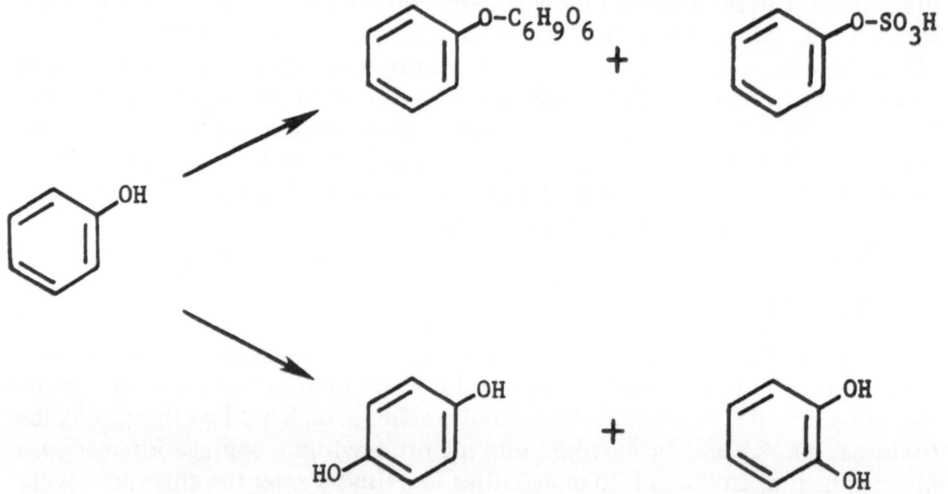

FIGURE 12. Metabolic pathways of phenols. Glucuronide and sulfate conjugates are the major metabolites. The dihydroxy derivatives, quinol and catechol, are minor metabolites.

1949). Phenolic compounds not only cause encephalopathy as observed in acute phenol poisoning, but also inhibit brain development leading to mental retardation, as is well known from cases of phenylketonuria with phenylalanine hydroxylase deficiency, where large amounts of phenolic acids (phenylpyruvic acid, OH-phenylacetic acid) appear in blood and spinal fluid and accumulate in brain tissue (Sandler and Close, 1959; Gjessing, 1966: Prader, 1966). This increase in phenols leads to defects in intelligence, general hyperkinesis, muscular hypertonicity, and tremors.

8.2. Cerebral Effects

The effect of various phenols on cerebral functions has been studied by a number of investigators. Hanson (1958) and Tashion (1961) reported that metabolites of phenylalanine inhibit the cerebral enzyme glutamic acid decarboxylase. The increase in urine phenolic acids in uremia prompted Hicks *et al.* (1964) to investigate the effect of these hydrophilic phenolic compounds on a variety of brain enzymes in the rat. They found that many phenolic acids accumulating in uremia strongly inhibited several enzymatic functions of the brain. On the basis of the abnormalities in phenolic compounds, Tholen and Bigler (1962) postulated that the pathogenesis of uremic and hepatic coma have elements in common.

8.3. Abnormality in Liver Disease

In 1930 Becher reported that phenols are increased in the serum of patients with liver cirrhosis. Muting (1965) pointed out that phenols pile up in the blood and spinal fluid of patients with severe kidney and liver disease. Stott (1968) also reported on an increase of phenolic compounds in blood of patients with liver disease. By paper chromatography, he identified several phenolic acids. These findings were confirmed by other investigators (Hartmann and Ruge, 1962; Ruge, 1971; Brunner, 1974). The concentrations of free phenols in different forms of liver disease are listed in Table I.

The effect of increased phenolic compounds in severe liver disease was elucidated by investigations on enzymatic functions of liver and brain tissue (Brunner, 1974; Brunner *et al.*, 1981). When phenols, extracted from serums

TABLE I. Free Phenols in Serum of Patients with Liver Disease (Mean ± SD)

	N	Free phenols μmole/ml
Healthy persons	10	0.65 ± 0.09
Acute hepatitis	18	1.38 ± 0.40
Chronic persistent hepatitis	15	0.78 ± 0.22
Chronic aggressive hepatitis	20	0.81 ± 0.25
Cirrhosis	12	1.12 ± 0.53
Coma hepaticum serum	14	4.20 ± 1.62
spinal fluid	4	3.74 ± 1.46

of patients in hepatic failure, were added to rat liver and brain homogenates and to isolated liver mitochondria in the same concentrations as were found in the patient's serums, significant inhibition in the activity of key enzymes of both organs were found (Table II). Mitochondrial respiration was also severely inhibited. That the inhibition was actually caused by phenols could be demonstrated by incubation of the extracted phenols with immobilized UDP-glucuronyltransferase. After incubation of the extracted phenols with this enzyme in the presence of UDPGA, the toxic effect on the enzymes was almost completely abolished and could be recovered after further treatment with β-glucuronidase (Table III). Enzyme inhibition by phenol was also demonstrated by Wardle (1978) with $(Na^+, K^+)ATPase$.

8.4. Coma Induction and Synergism with Other Toxins

In rabbits, the coma-inducing capacity of free phenols was much greater than that with phenolic acids (Brunner *et al.*, 1981). Less phenol than NH_4^+ was required to induce coma in normal rabbits after intravenous infusions. The coma dose of phenol was 1.1 ± 0.1 mmole/kg and of NH_4^+ 2.62 ± 0.56 mmole/kg. Seven times as much hexanoic acid as phenol was needed to induce coma. Simultaneous infusion of phenol with NH_4^+ or hexanoate resulted in a 33% reduction of the coma-inducing dose of NH_4^+ and 21% reduction in the coma-inducing dose of hexanoate. The fatty acid similarly reduced the coma-inducing dose of phenol by 33%.

The coma-inducing effect of phenol was also studied in normal 300 ± 50 g Sprague–Dawley rats (Windus-Podehl *et al.*, 1982). Dose–response curves were developed that showed that one half the animals became deeply comatose with 1.8 mmole/kg of intraperitoneal (IP) phenol, and 100% with 2.0 mmole/kg. Five stages of encephalopathy were readily distinguished. In stage 1, spontaneous activity was noticeably decreased, posture was upright, back-leg control normal, muscle tonus slightly increased, and responses to stimuli normal.

TABLE II. Inhibition of Rat Liver Enzymes by Phenols Extracted from Serum of Patients in Fulminant Hepatic Failure

Enzyme (μmole/mg per min)	Extract from normal serum (%)	Extract from coma serum (%)	Extract from glucuronidated coma serum (%)
Monoamine oxidase (0.12)	12	95	22
Succinic dehydrogenase (0.16)	23	51	30
Proline oxidase (0.022)	14	63	25
NADPH-cytochrome *c*-reductase (0.054)	12	54	26
Glucuronyl transferase (0.013)	0	9	0
Aspartate aminotransferase (1121.2)	0	10	0

In stage 3, spontaneous activity was markedly decreased, rats were on their side or back and could not get upright, back-leg control was present, muscle tone was not increased, and responses to stimuli were decreased and slow. In stage 5, activity was gone, the rats were deeply unconscious, righting reflex and back-leg control were gone, muscles showed no resistance to passive movement, and there were no responses to stimuli. With a coma-inducing IP dose of phenol, spontaneous activity decreased rapidly. After 1 min, a body tremor developed interspersed with unpredictable jumping, and all four limbs began to shake. Leg control and then the righting reflex were lost by 2 min. Within 5 min, rats were deeply unconscious and muscles showed no resistance, while shaking of limbs continued until recovery 20–60 min later or death. With 1.5 mmole/kg or less, none of the rats became comatose, but the shaking of limbs was present. The coma-inducing dose of phenol was reduced by 10–20% with simultaneous injection of subcoma doses of NH_4^+ or octanoate and 20–30% with simultaneous dimethyl disulfide (DMDS → 2 methanethiol). Conversely, a subcoma dose of phenol reduced the coma-inducing doses of NH_4^+, octanoate, and DMDS by approximately 20–25%. Thus phenol induces coma by itself in rats and acts synergistically with other hepatic failure toxins.

9. Pathophysiologic Role of Phenols in Disease

As with mercaptans, phenol and its derivatives are products of metabolism normally largely removed by the liver or excreted in the urine. In human hepatic and renal failure, they accumulate in the blood and tissues. In animals, pharmacologic doses of phenol cause a reversible encephalopathy including deep coma. Phenols extracted from serums of patients with acute fulminant hepatic failure inhibit a variety of enzymes when added to rat liver or brain homogenates. In normal rats, phenols augment the toxicity of ammonia, mercaptans, and fatty acids in vivo. From this circumstantial evidence, it appears likely

TABLE III. Inhibition of Rat Cerebral Enzymes by Phenols Extracted from Serum of Patients in Fulminant Hepatic Failure

Enzyme (μmole/mg per min)	Extract from normal serum (%)	Extract from coma serum (%)	Extract from glucuronidated coma serum (%)
Monoamine oxidase (0.09)	14	91	28
Succinic dehydrogenase (0.072)	20	70	28
NADPH cytochrome c-reductase (0.026)	12	51	14
Lactate dehydrogenase (1712.7)	0	68	12
Malate dehydrogenase (72.8)	5	60	18
Leucineaminopeptidase (12.1)	18	51	29

that phenols may have a pathogenic role in human hepatic encephalopathy while interacting with the abnormal accumulations of ammonia, mercaptan, and fatty acids.

10. Summary

Many phenolic substances in free and conjugated form are found in the urine of man. Phenols are derived from phenylalanine and tyrosine. They are removed in the liver by the processes of conjugation with glucuronide or sulfate, hydroxylation, or methylation. Glucuronidation is the main method of detoxifying and removing phenols. Phenol poisoning has resulted in encephalopathy, and in children phenols inhibit brain development as in phenylketonuria. They inhibit several cerebral enzymes. In liver disease and in renal failure, phenols accumulate in the blood, urine, and spinal fluid. Phenols extracted from serums of patients with fulminant hepatic failure inhibited several liver and cerebral enzymes such as monoamineoxidase, several dehydrogenases and transferases, and cytochrome c reductase, among others. After incubation of the extracted phenols with glucuronyl transferase, the inhibition was eliminated. In normal rabbits and rats, phenol induced encephalopathy and coma that were reversible. The coma-inducing dose of phenol in rats caused a generalized tremor that persisted while spontaneous body movements and body control decreased. Within 5 min, the animals were deeply unconscious, showed no muscle resistance to passive movement, and did not respond to painful stimuli. The coma-inducing dose was reduced by 10–30% with simultaneous injection of subcoma doses of NH_4^+, octanoate, or dimethyl disulfide. In human hepatic encephalopathy, the accumulated phenols may play a role while interacting synergistically with abnormal accumulations of ammonia, mercaptan, and fatty acids.

References

Mercaptans

Bremer, J., and Greenberg, D. M., 1961, Enzymic methylation of foreign sulfhydryl compounds, *Biochim. Biophys. Acta* **46**:217–224.

Brunner, G., and Scharff, P., 1978, Untersuchungen uber den diagnostischen Wert der Bestimmung von Mercaptenen im Serum bei Leberkrankungen, *Dtsch. Med. Wochenschr.* **103**:1796–1800.

Canellakis, E. S., and Tarver, H., 1953, The metabolism of methyl mercaptan in the intact animal, *Arch. Biochem. Biophys.* **42**:446–455.

Challenger, F., and Walshe, J. M., 1955, Methyl mercaptan in relation to foetor hepaticus, *Biochem. J.* **59**:372–375.

Chen, S., Zieve, L., and Mahadevan, V., 1970, Mercaptans and dimethyl sulfide in the breath of patients with cirrhosis of the liver, *J. Lab. Clin. Med.* **75**:628–635.

Coob, W. A., and Dawson, G. D., 1960, The latency and form in many of the occipital potentials evoked by bright flashes, *J. Physiol.* **152**:108–121.

Derr, R. F., and Zieve, L., 1982, Methanethiol and fatty acids depress urea synthesis by the isolated perfused rat liver, *J. Lab. Clin. Med.* **100**:585–592.

Doizaki, W. M., and Zieve, L., 1977, An improved method for measuring blood mercaptans, *J. Lab. Clin. Med.* **90**:849–855.

Fairchild, E. J., and Stokinger, H. E., 1958, Toxicologic studies on organic sulfur compounds. I. Acute toxicity of some aliphatic and aromatic thiols (mercaptans), *Amer. Ind. Hyg. Assoc. J.* **3**:171–189.

Kinsell, L. W., Harper, H. A., Giese, G. K., Morgen, S., McCallie, D. P., and Hess, J. R., 1949, Studies in methionine metabolism. II. Fasting plasma methionine levels in normal and hepatopathic individuals in response to daily methionine ingestion, *J. Clin. Invest.* **28**:1439–1450.

Ljunggren, G., and Norberg, B., 1943, On the effect and toxicity of dimethyl sulfide, dimethyl disulfide and methyl mercaptan, *Acta Physiol. Scand.* **5**:248–255.

Mardini, H. A., Bartlett, K., and Record, C. O., 1981, An improved gas chromatographic method for the detection and quantitation of mercaptans in blood, *Clin. Chim. Acta* **113**:35–41.

McClain, C. J., Zieve, L., Doizaki, W. M., Gilberstadt, S., and Onstad, G. R., 1980, Blood methanethiol in alcoholic liver disease with and without hepatic encephalopathy, *Gut* **21**:318–323.

NIOSH, 1978, Occupational exposure to n-alkane mono thiols, cyclohexanethiol, and benzenethiol, Department of Health, Education, and Welfare, (NIOSH) Publication No. 78–213, Washington, D.C.

Phear, E. A., Reubner, B., Sherlock, S., and Summerskill, W. H. J., 1956, Methionine toxicity in liver disease and its prevention by chlortetracycline, *Clin. Sci.* **15**:93–117.

Quarfoth, G., Ahmed, K., Foster, D., and Zieve, L., 1976, Action of methanethiol on membrane (Na^+, K^+)-ATPase of rat brain, *Biochem. Pharmacol.* **25**:1039–1044.

Schwimmer, S., 1969, Inhibition of carbonic anhydrase by mercaptans, *Enzymologia* **37**:163–173.

Shults, W. T., Fountain, E. N., and Lynch, E. C., 1970, Methanethiol poisoning. Irreversible coma and hemolytic anemia following inhalation, *J. Am. Med. Assoc.* **211**:2153–2154.

Susman, J. L., Hornig, J. F., Thomae, S. C., and Smith, R. P., 1978, Pulmonary excretion of hydrogen sulfide, methanethiol, dimethyl sulfide and dimethyl disulfide in mice, *Drug Chem. Toxicol.* **1**:327–338.

Vahlkamp, T., Meijer, A. J., Wilms, J., and Chamuleau, R. A. F. M., 1979, Inhibition of mitochondrial electron transfer in rats by ethanethiol and methanethiol, *Clin. Sci.* **56**:147–156.

Waller, R. L., 1977, Methanethiol inhibition of mitochondrial respiration, *Toxicol. Appl. Pharmacol.* **42**:111–117.

Watson, C. J., 1949, The prognosis and treatment of hepatic insufficiency. *Ann. Intern. Med.* **31**:405–423.

Weisiger, R. A., Pinkus, L. M., and Jakoby, W. B., 1980, Thiol S-methyltransferase: Suggested role in detoxication of intestinal hydrogen sulfide, *Biochem. Pharmacol.* **29**:2885–2887.

Windus-Podehl, G., Lyftogt, C., Zieve, L., and Brunner, G., 1982, Encephalopathic effect of phenol in rats, *J. Lab. Clin. Med.* **101**:586–592.

Zeneroli, M. L., Ventura, E., Baraldi, M., Penne, A., Messori, E., and Zieve, L., 1982, Visual evoked potentials in encephalopathy induced by galactosamine, ammonia, dimethyldisulfide, and octanoic acid, *Hepatology* **2**:532–538.

Zieve, L., and Doizaki, W. M., 1980, Brain and blood methanethiol and ammonia concentrations in experimental hepatic coma and coma due to injections of various combinations of these substances, *Gastroenterology* **79**:1070.

Zieve, L., Doizaki, W. M., Lyftogt, C., 1984, Brain methanethiol and ammonia concentrations in experimental hepatic coma and coma due to injections of various combinations of these substances, *J. Lab. Clin. Med.* **104**.

Zieve, L., Doizaki, W. M., and Zieve, F. J., 1974, Synergism between mercaptans and ammonia or fatty acids in the production of coma: A possible role for mercaptans in the pathogenesis of hepatic coma, *J. Lab. Clin. Med.* **83**:16–28.

Phenols

Armstrong, M. D., Silaw, K., and Wall, P., 1956, The phenolic acids of human urine, *J. Biol. Chem.* **218**:293–303.

Becher, E., 1930, Uber Steigerungen des Blutwertes von Phenol und Phenol-derivaten und uber das Auftreten von freiem Phenol im Blute bei Leberzirrhose, *Munch. Med. Wochenschr.* **77**:751–752.

Becher, E., Doenecke, F., and Litzner, S., 1926, Quantitative Studien uber die Fraktion der aromatischen Oxysauren im Blut bei Krankheiten, *Z. Klin. Med.* **104**:29–43.

Bray, H. G., Humphris, B., Thorpe, W., White, K., and Wood, P., 1952, Kinetic studies of metabolism of foreign organic compounds; conjugation of phenols with glucuronic acid, *Biochem. J.* **52**:416–419.

Brunner, G., 1974, Experimentelle und klinische Untersuchungen zur Entwicklung einer extra-corporalen Entgiftung beim Leberversagen unter besonderer Berucksichtigung der freien Serum Phenole, Habilitation thesis, University of Gottingen.

Brunner, G., Windus, G., and Losgen, H., 1981, On the role of free phenols in the blood of patients in hepatic failure, in: *Artifical Liver Support* (G. Brunner and F. W. Schmidt, eds.), Springer-Verlag, Heidelberg, pp. 25–31.

Dalgliesh, C. E., 1955, Metabolism of aromatic amino acids, in *Advances in Protein Chemistry*, Volume 10 (N. L. Anson, K. Bailey, and J. T. Edsall, eds.), Academic Press, New York, pp. 31–150.

Folin, O., and Ciocalteu, V., 1927, On tyrosine and tryptophane determinations in proteins, *J. Biol. Chem.* **73**:627–649.

Frerichs, F. Th., 1854, Offenes Schreiben an den Herrn Hofrat Dr. Oppholzer in Wien, *Wien. Med. Wochenschr.* **4**:465–466.

Frimpter, G. W., 1973, Aminoacidurias due to inherited disorders of metabolism, *N. Engl. J. Med.* **289**:835–895.

Gjessing, L. R., 1966, *Symposium on Tyrosinosis*, Oslo University Press, Oslo.

Hanson, A., 1958, Inhibition of brain glutamic acid decarboxylase by phenylalanine metabolites, *Naturwissenschaften* **45**:423.

Hartmann, F., 1961, Das Phenolsauremuster im Urin Leberkranker, *Klin. Wochenschr.* **39**:273–280.

Hartmann, F., and Ruge, W., 1962, Untersuchungen uber das Phenolsauremuster bei Leberkranken, *Dtsch. Arch. Klin. Med.* **208**:298–322.

Hicks, J. M., Wootton, I. D. P., and Young, D. S., 1964, The effect of uraemic blood constituents on certain cerebral enzymes, *Clin. Chim. Acta* **9**:228–235.

Jervis, G. A., and Drejza, E. J., 1966, Phenylketonuria: Blood levels of phenylpyruvic and ortho-hydroxy-phenylacetic acids, *Clin. Chim. Acta* **13**:435–441.

Knox, W. E., 1955, The metabolism of phenylalanine and tyrosine, in: *A Symposium on Aminoacid Metabolism* (W. D. McElroy and H. B. Glass, eds.), Johns Hopkins Press, Baltimore, pp. 836–866.

Levine, R. L., and Conn, H. O., 1967, Tyrosine metabolism in patients with liver disease, *J. Clin. Invest.* **46**:2012–2020.

Muting, D., 1965, Studies on the pathogenesis of uremia. Comparative determinations of glucuronic acid, indican, free and bound phenols in the serum; cerebrospinal fluid and urine of renal diseases with and without uremia, *Clin. Chim. Acta* **12**:551–554.

Muting, D., and Reikowski, H., 1965, Protein metabolism in liver disease, in: *Progress In Liver Diseases*, Volume II (H. Popper and F. Schaffner, eds.) Grune and Stratton, New York, pp. 84–94.

Muting, D., Reikowski, J., and Reikowski, H., 1968, Uber die Wirkung von Neomycin auf Blutammoniak, freie Aminosauren und freie Phenole im Serum und Urin bei Leberzirrhose, *Med. Welt.* **19**:1235–1239.

Muting, D., Keller, H. E., and Kraus, W., 1970, Quantitative colorimetric determination of free phenols in serum and urine of healthy adults using modified diazo reactions, *Clin. Chim. Acta* **27**:177–180.

Prader, A., 1966, Hyperaminoacidurien, *Schweiz. Med. Wochenschr.* **96**:53–56.

Ruge, W., 1971, Veranderungen im Phenylalanin und Tyrosinstoffwechsel bei Leberinsuffizienz als Zeichen gestorter Entgiftung, in: *Akute Hepatitis* (H. Wannagat, ed.), Thieme, Stuttgart, pp. 281–285.

Sandler, M., and Close, H. G., 1959, Biochemical effect of phenylacetic acid in a patient with 5-hydroxytryptophan—Secreting carcinoid tumor, *Lancet* **2**:316–318.

Schmidt, E. G., 1949, Urinary phenols. IV. The simultaneous determination of phenol and p-cresol in urine, *J. Biol. Chem.* **179**:211–215.

Stott, A. W., 1968, Paper chromatography of serum phenolic acids, *Clin. Chim. Acta* **20**:181–184.

Tashian, R. E., 1961, Inhibition of brain glutamic acid decarboxylase by phenylalanine, valine, and leucine derivates: A suggestion concerning the etiology of the neurological defect in phenylketonuria and branched-chain ketonuria, *Metabolism* **10**:393–402.

Tholen, H., and Bigler, F., 1962, Pathogenetische Beziehungen zwischen uramischem und hepa-
    tischem Koma, *Dtsch. Med. Wochenschr.* **87**:1188–1192.
von Oettingen, W. F., 1949, Phenol toxicity, *Nat. Inst. Health Bull.* **190**:1–408.
Wardle, E. N., 1978, Phenols, phenolic acids and sodium-potassium ATPases, *J. Mol. Med.* **3**:319–
    327.
Williams, C. M., and Leonard, R. H., 1963, Microanalytical determination of dihydroxy aromatic
    acids by gas chromatography, *Ann. Biochem.* **5**:362–368.
Williams, R. T., 1959, *Detoxication Mechanism*, 2nd ed., Chapman and Hall Ltd., London.
Williams, R. T., 1964, *Metabolism of Phenolic Compounds*, Academic Press Inc., New York-London.
Windus-Podehl, G., Lyftogt, C., Zieve, L., and Brunner, G., 1983, Encephalopathic effect of phenols
    in rats, *J. Lab. Clin. Med.* **101**:586–592.

Ammonia-Induced Encephalopathy

ALAN H. LOCKWOOD

1. Introduction

Ammonia is one of the principal end products of the digestion of nitrogenous compounds. It is generally recognized to be a normal constituent of the blood and, when present in excessive concentrations, to produce encephalopathy. This has not always been the case. Ammonia is present in very low concentrations in blood and is formed rapidly when blood is allowed to stand for more than a few minutes without being treated to prevent enzymatic production of ammonia. Early investigators thus concluded that normal blood did not contain ammonia. Improved methods for the measurement of ammonia led to the acceptance of ammonia as a normal chemical constituent of blood. The concept that ammonia is toxic to the nervous system is also one that evolved slowly. Early observations by European and Scandinavian investigators revealed a clear association between the ingestion of compounds containing ammonia and the evolution of a syndrome that was identical to hepatic encephalopathy (Van Coulaert *et al.*, 1933; Fuld, 1933; Kirk, 1936). This information was not generally appreciated in the United States until 1952. At that time, Gabuzda and his colleagues (1952) reported their attempts to reduce ascites by treating patients with resins that exchanged sodium in the blood for ammonium ion bound to the resin. Although this produced a prompt and effective diuresis that reduced the volume of the ascitic fluid, the patients quickly became encephalopathic. Further studies by this group correlated the change in the mental status with elevated blood ammonia levels and produced an admonition against the administration of ammonia-containing compounds to patients with liver disease (Phillips *et al.*, 1952). This relatively weak association between hyperammonemia and neurological dysfunction has since been greatly strengthened by a large number of studies that all link hyperammonemia and encephalopathy, especially in patients with liver disease (Phear *et al.*, 1955;

ALAN H. LOCKWOOD • Department of Neurology, University of Texas Health Science Center at Houston, Houston, Texas 77025.

Sherlock, 1958; Walker and Schenker, 1970; Lockwood *et al.*, 1979; Hoyumpa *et al.*, 1979). As experimental methods have improved and more encephalopathic states have been scrutinized by the biochemist, a growing body of evidence has begun to link a variety of seemingly unrelated encephalopathic states with elevated levels of blood or brain ammonia. Results of biochemical and physiological studies have begun to suggest mechanisms by which ammonia may induce encephalopathy.

2. Conditions Associated with Hyperammonemia

Along with a greater awareness of the encephalopathic effects of ammonia, there is an increasing awareness that high brain levels of ammonia are associated with many conditions that are characterized by neurological dysfunction. Although a high brain ammonia level may not be the primary metabolic problem, it may contribute to the development of the disorder. In this section, clinical conditions associated with hyperammonemia will be discussed. These disorders will be divided into discussions of primary and secondary disorders.

2.1. Primary Disturbances of Ammonia Metabolism

The Krebs-Henseleit cycle is the primary metabolic pathway for the elimination of ammonia from the body. The enzymes for the complete cycle are all found in the liver and result in the conversion of ammonia into urea. Inherited enzyme deficiencies in this metabolic pathway produce serious illness in the perinatal period or in childhood (Bruton *et al.*, 1970; Shih, 1976). The children affected by this group of disorders, including carbamylphosphate synthetase deficiency, ornithine transcarbamylase deficiency, arginosuccinate deficiency (citrullinemia), arginosuccinase deficiency (arginosuccinic aciduria), and arginase deficiency (hyperargininemia), are usually normal at birth, but become affected after eating protein, including milk. The symptoms include irritability, lethargy, poor feeding, coma, seizures, and often death. Hyperammonemia is associated with the periods of neurological symptomatology, and treatment consists of reducing dietary protein, thus reducing the amount of ammonia produced, peritoneal dialysis, exchange transfusion, and more recently, by keto acid analogue therapy (Batshaw *et al.*, 1975; Ballard *et al.*, 1978). The diagnosis of these disorders cannot be made on clinical criteria alone; testing for specific enzyme deficiencies is required.

2.2. Secondary Disturbances of Ammonia Metabolism

2.2.1. Hepatic Encephalopathy

Hepatic coma, or hepatic encephalopathy (HE), is probably the most important disorder associated with hyperammonemia. Hepatic encephalopathy becomes an increasingly important cause of morbidity and mortality as patients with cirrhosis of the liver, the seventh ranked cause of death in Americans between the ages of 25–65 (Public Health Service, 1979), enter the terminal phases of their illness. Patients with acute yellow atrophy of the liver invariably exhibit neurological signs and symptoms that may be the first indication of the illness. Although the association between liver disease and neurological dysfunction was noted by early physicians such as Hippocrates and Galen, the first clearly described reports were published by Frerichs in 1860 (as quoted in translation by Adams and Foley, 1953). Frerichs described alterations in behavior and consciousness that invariably occurred in patients who went on to die of acute yellow atrophy of the liver. The link between encephalopathy and cirrhosis has been most clearly described by Adams and Foley (1953). Their treatise presented a description of the syndrome, neuropathological observations, and speculations with regard to its pathogenesis.

The hallmark of HE is an alteration in the mental status. This usually includes an alteration in the level of consciousness and the content of consciousness, as discussed by Plum and Posner (1972). Hepatic encephalopathy is far from being a homogeneous stereotyped disorder. The onset is frequently insidious, beginning with a blunting of consciousness, with a variable degree of progression to deep coma in the most severely affected individuals. Occasionally, the patient will be agitated at the onset of the condition. This is more common in patients with acute yellow atrophy of the liver. Usually there is a history of antecedent liver disease, but occasionally patients without known liver disease will develop encephalopathy as the initial manifestation of a hepatic disorder.

The cranial nerve examination fails to reveal evidence of structural disease; in extreme cases, the extraocular movements or oculocephalic reflex may be absent, but ice-water caloric testing almost always elicits a complete, tonic deviation of the eyes. On motor examination, asterixis is often present. This sign is elicited by asking the patient to extend the hand, and manifested by a sudden lapse of the posture. Electromyographic studies have shown that there is electrical silence in the muscle at the time of the postural lapse (Adams and Foley, 1953). This sign was originally thought to be pathognomonic for HE, but is relatively nonspecific, and has been observed in association with a variety of metabolic and structural diseases of the brain. Muscle tone is often increased. Brisk deep tendon reflexes are usually easily elicited (unless the patient has a peripheral neuropathy) and there is often an extensor plantar sign. Epileptic seizures are uncommon. When they are observed, an alternate cause should be sought, such as alcohol withdrawl or head injury.

A clear association between hyperammonemia and the development of HE has only emerged in the past 30 years. In the early 1950s, Gabuzda and associates (1952) attempted to treat ascites with ion-exchange resins that adsorbed sodium and released ammonium. Although this was very effective in reducing the amount of ascites, many patients developed a syndrome that was indistinguishable from HE. Further studies showed that feeding similar patients certain nitrogenous compounds produced a similar result (Phillips et al., 1952). Subsequent to these observations, Sherlock and her colleagues (Phear et al., 1955; Sherlock, 1958) measured the arterial ammonia concentration in a large group of patients with HE and found a high association between hyperammonemia and HE. Glutamine, and its transamination product alpha-ketoglutaramate, have been found to be present in higher than normal concentrations in the cerebrospinal fluid of patients with HE (see Fig. 1) (Steigmann *et al.*, 1963; Vergara *et al.*, 1974).

2.2.2. Perinatal Hyperammonemia

Infants with perinatal hyperammonemia due to disorders of the urea cycle are rare. However, an increase in the awareness of this group of disorders has led to more widespread measurement of blood ammonia levels in infants with relatively nonspecific symptoms. As a consequence, less severe forms of hyperammonemia have been recognized. Goldberg *et al.* (1979), described twelve hyperammonemic infants that were recognized during the course of one year who presented with severe fetal bradycardia or who required prolonged resuscitation. In the seven survivors, five were left with severe neurological deficits. In another group of preterm infants, transient hyperammonemia produced an overwhelming illness that responded promptly to aggressive treatment that included exchange transfusions and peritoneal dialysis (Ballard *et al.*, 1978). The activity of enzymes of the urea cycle was not reduced, and the etiology of the hyperammonemia was not discovered. Survivors in this group developed normally. Both groups of authors suggest that neonatal hyperammonemia may be much more common than generally appreciated, and stress the need for prompt diagnosis and aggressive therapy.

2.2.3. Reye's Syndrome

In 1963, Reye *et al.* described a syndrome that was characterized by encephalopathy and fatty degeneration of the viscera. Later reports described

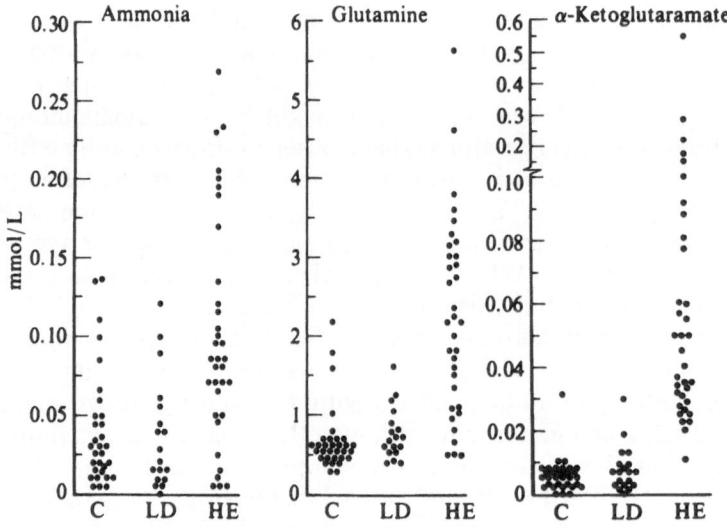

FIGURE 1. Cerebrospinal fluid concentrations of ammonia, glutamine, and alpha-ketoglutaramate in comatose patients without liver disease, C; alert patients with liver disease, LD; and patients with hepatic encephalopathy, HE. The highest levels of each compound are found in the HE group, and the best differentiation between C and HE is seen with alpha-ketoglutaramate and glutamine measurements. Reproduced by copyright permission of Little, Brown and Company, from Duffy and Plum (1981).

hyperammonemia, often severe, that is associated with the syndrome (Hutten-locher *et al.*, 1969). Further investigations of these children have shown that there are transient reductions in carbamyl phosphate synthetase and ornithine transcarbamylase (Snodgrass and DeLong, 1976; Brown *et al.*, 1976). These are mitochondrial enzymes and are probably affected as a part of a generalized syndrome of mitochondrial destruction. Mitochondrial injury in the liver (Bove *et al.*, 1975) and neurons (Partin *et al.*, 1975) are pathological hallmarks of this disorder, and it is probable that mitochondrial injury is primarily responsible for the development of the syndrome. Although hyperammonemia is present and severe, astrocyte swelling, myelin blebs, and universal injury to neuronal mitochondria (Partin *et al.*, 1975) serve to separate this disorder from other conditions characterized by hyperammonemia alone, where neuropathological changes are confined to astrocytes. (See Section 6 for a discussion of neuro-pathological changes due to hyperammonemia.)

2.2.4. Miscellaneous Disorders Associated with Hyperammonemia

Hyperammonemic encephalopathy has been reported in several patients who have had ureterosigmoidostomies and coexisting liver disease (Silberman, 1958; Egense and Schwartz, 1970; Mortensen *et al.*, 1972). The ammonia is formed by urease-containing bacteria in the colon and the liver disease causes a failure to detoxify this additional ammonia load, as in patients with hepatic encephalopathy. McDermott (1957) has reported a single patient who had an ilial pouch created who became hyperammonemic during an infection by urea-splitting bacteria. This patient, too, had mild liver disease. More recently, a patient with a neurogenic bladder was reported to develop hyperammonemia associated with a bladder infection caused by urease-containing diphtheroids, even though liver function tests were normal (Drayna *et al.*, 1981). In this patient, obstruction of the urinary bladder outlet and an increase in the surface area of the bladder were implicated as predisposing factors.

Hyperammonemic encephalopathy associated with drug-induced liver failure from salicylates and valproic acid has been described (Makela *et al.*, 1980).

2.2.5. Other Encephalopathies Associated with Elevated Brain Ammonia Levels

Brain ammonia levels are increased in a variety of seemingly unrelated clinical conditions characterized by encephalopathy, such as ischemia (Thorn and Heimann, 1958), hypercapnic acidosis (Kazemi *et al.*, 1973), hypoglycemia (Tews, *et al.*, 1965; Lewis *et al.*, 1974), and hypoxia (Seisjo and Nilsson, 1971), and after fluoroacetate poisoning (Tews and Stone, 1965). In some cases, the increase in ammonia is perhaps the consequence of intracellular acidosis, since ammonia enters the cells as a gas and then forms the ammonium ion. In each of these conditions, the role of the increase in brain ammonia concentration and the pathogenesis of altered brain function is unknown. However, it is pos-sible that the abnormally high level of ammonia contributes to the development of neurological dysfunction, either directly or synergistically.

3. Ammonia Metabolism by Noncerebral Tissues

Ammonia metabolism in man is extremely complicated and is dependent on a variety of metabolic reactions carried out in various organs. Although this chapter is primarily concerned with the effects of ammonia on cerebral function, it is essential to have a broader understanding of ammonia metabolism in the whole organism, since there is an interdependence of organs that determines the eventual metabolic fate of ammonia and the probability of developing hyperammonemic symptomatology.

One of the first comprehensive studies of ammonia metabolism in man was published by McDermott et al. (1954). They measured the level of ammonia in the hepatic portal vein, the inferior venacava, and the renal artery and vein in patients, during the course of intra-abdominal operations. They found ammonia concentrations in the portal vein that were about five times those in the peripheral blood (renal artery). Ammonia levels in the renal vein were about twice those measured in the artery. Their study clearly identified the gastrointestinal tract as the most important site for ammonia production in the body. Their work has been expanded by others, and it is now known that most of the ammonia in the portal vein is derived from the colon (Summerskill and Wolpert, 1970), where ammonia is believed to be produced by bacteria that contain urease or deaminating enzymes (Evans et al., 1966; Sabbaj et al., 1970). Urease may be of primary importance, especially in uremic patients, since some 15–20% of all urea is enzymatically degraded in the body, presumably to ammonia and carbon dioxide, and resynthesized into urea (Walser and Bordenlos, 1959).

Skeletal muscle is also involved in ammonia metabolism. Active exercise causes the release of ammonia into the blood. This is presumed to be the result of the purine nucleotide cycle, which has also been implicated in renal ammonia production (for a review, see Lowenstein, 1972). Under resting conditions, Bessman and Bradley (1955) showed that the venous ammonia concentration was about 40% of that measured in arterial blood from the same subject. Ganda and Ruderman (1976) similarly showed that there was a net extraction of ammonia by the forearm and that muscle wasting reduced the efficiency of the extraction.

Since most body organs contain enzymatic systems that are capable of both ammonia consumption and production, studies that have used chemical methods to examine ammonia metabolism have produced results that only describe the net uptake or production. To examine ammonia metabolism more closely, tracer techniques must be used. Although ammonia can be labeled with ^{15}N and used for this purpose, appreciable amounts of ammonia must be administered and steady-state conditions cannot be maintained. Ammonia labeled with cyclotron-produced ^{13}N is ideal for tracer studies. It has a very high specific activity and is radioactive. It decays by positron emission and is thus well-suited to investigations that are based on positron emission tomographic (PET) scanning techniques. (For a review of PET scanning, see Phelps et al., 1982.) Lockwood et al. (1979) used this isotope to investigate ammonia metabolism in man. Their study showed that the whole-body ammonia clearance rate (exclusive of ammonia in the important hepatic portal system) was about 0.66

mole/day in normal subjects, and about 1 mole/day in five subjects with end-to-side portacaval shunts on diets that contained only 40 g of protein per day. The brain metabolized about 7% of all of the ammonia in the systemic circulation and the brain, liver, blood, and urinary bladder together accounted for about 30% of all of the ammonia injected. Their data suggested that large amounts of ammonia are metabolized by skeletal muscle, largely due to the

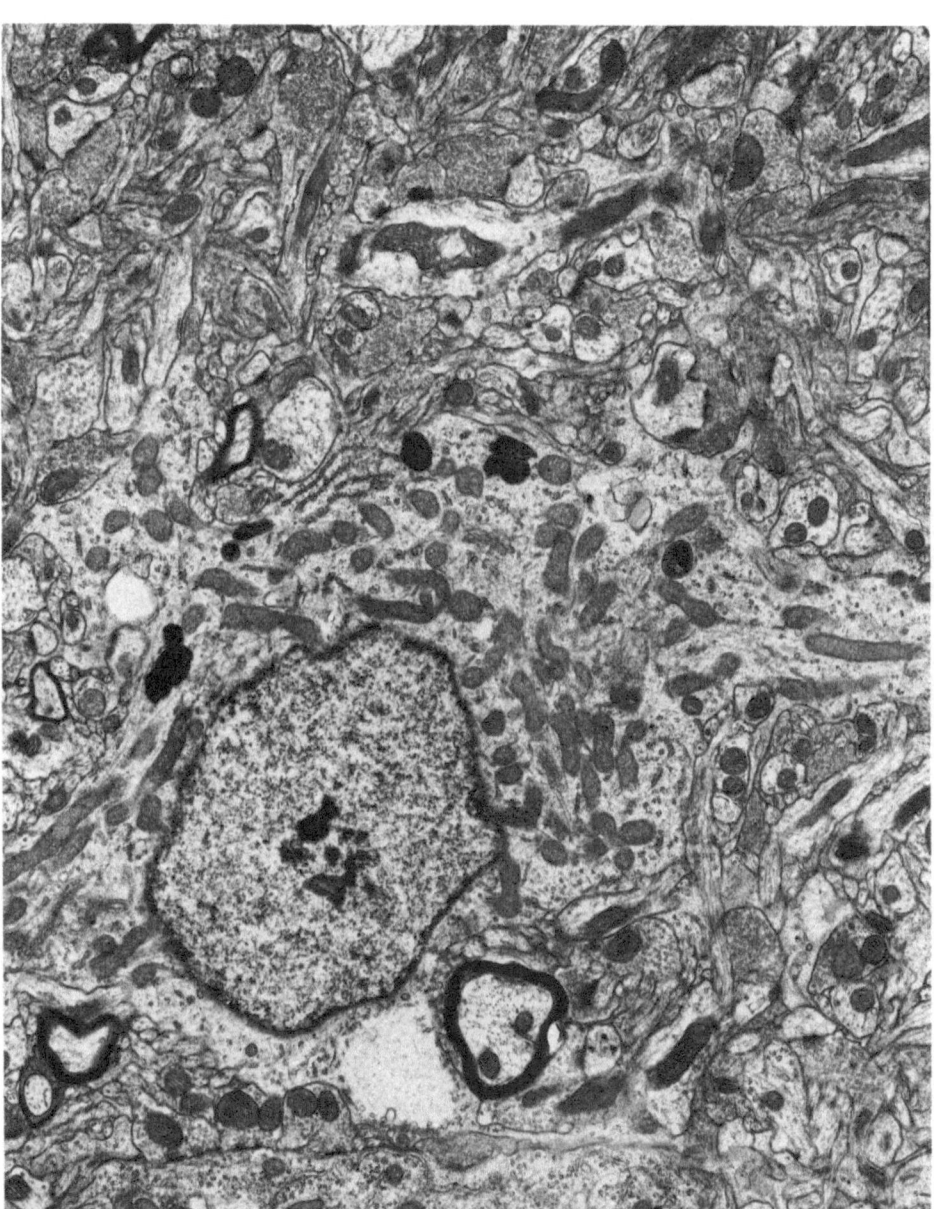

FIGURE 2. This electron micrograph of an astrocyte from a rat with a portacaval shunt fed an ammoniated resin shows a proliferation of mitochondria and cellular organelles that suggests that this is a metabolically activated cell. Photograph courtesy of Michael D. Norenberg.

fact that it accounts for about 40% of the normal body weight and emphasized the importance of earlier studies (Bessman and Bradley, 1955; Ganda and Ruderman, 1976). This observation led to the speculation that skeletal muscle may become the most important organ in ammonia homeostasis in patients that have end-to-side or side-to-side portacaval shunts.

Under normal conditions, ammonia, along with other toxins that cause encephalopathy, is carried from the gastrointestinal tract to the liver, where metabolic detoxification reactions take place. In patients with liver disease and cirrhosis with portal hypertension or in individuals without liver disease, but who have required portacaval shunt construction, ammonia-rich blood is diverted from the liver and enters the systemic circulation where it can be delivered to the brain. Figure 2 shows this relationship. In a series of patients that were fed exogenous ammonia, a secondary rise in the peripheral blood ammonia level, indicating abnormal ammonia tolerance, was more closely linked with the existence of portacaval shunting than with the degree of hepatocellular impairment (Conn, 1960). Conn thus proposed test feedings of small amounts of ammonia followed by serial measurements of arterial blood ammonia levels as a simple, safe, and reliable test for the presence of shunts between the portal and systemic circulatory systems.

Although portacaval shunting has been clearly shown to reduce the frequency of bleeding from esophageal varices, the frequency of postoperative encephalopathy prompted a re-evaluation of the pathophysiological and anatomical factors involved in the selection of a suitable surgical procedure. The result, the distal splenorenal shunt, appears to be the operation of choice in suitable candidates (Zeppa, 1972). The procedure avoids shunting toxin-laden blood into the systemic circulation, reducing the incidence of encephalopathy, yet pressure in esophageal varices is relieved and bleeding is prevented.

These studies of ammonia metabolism in the complete organism have formed the rational basis for the most effective therapies for hyperammonemia and encephalopathy associated with liver disease. Protein restriction decreases the amount of amino acids presented to the deaminating enzymes in the colon. Nonabsorbed antibiotics exert their therapeutic effect by reducing the numbers of bacteria in the colon. Similar reasoning led to a trial of surgical exclusion of the colon, which was later abandoned because of the high operative mortality (Resnick et al., 1968). The synthetic disaccharide, lactulose, has been shown to be an effective drug for treating hyperammonemic encephalopathy (Elkington et al. 1969; Kersh and Rifkin, 1973). Although this drug is of proven utility, the mechanism of action, originally thought to be due to a lowering of colonic pH with a reduction of colonic ammonia uptake, is not known with certainty (Conn, 1978). Although other strategies have been developed, such as urease inhibition (Thomson and Holmes, 1967), they are not a part of the usual treatment of hyperammonemia associated with liver disease.

4. Neuropathology of Hyperammonemia and Related Disorders

The present ability of the neuropathologist to attribute specific neuropathological changes to persistent hyperammonemia is the result of many painstaking clinical and experimental examinations. Before 1924, most descriptions of nervous system tissue obtained from patients with liver disease showed nonspecific changes related to the agonal state, pre-existing nervous system disease, and jaundice. The protoplasmic astrocytic changes, now recognized as the neuropathological hallmark of hepatic coma, were reported by Insabato and shortly thereafter by Pollack (see Adams and Foley, 1953, for an English language summary of their findings). In their comprehensive report on the neurological disorder associated with liver disease, Adams and Foley (1953) described a proliferation and enlargement of protoplasmic astrocytes in the cerebral cortex, lenticular, lateral thalamic, dentate, and red nuclei. Although their study did not use contemporary methods for correcting cell counts for the effects of nuclear size variations, groupings of two or more protoplasmic astrocytes lent credence to their report of astrocyte proliferation. Although the astrocytic changes were most severe in patients with prolonged periods of coma, they did not regard this correlation as being definitive.

The etiology of this change was a source of controversy and not clearly attributable to any one toxic agent until experimental studies were conducted. In a study that used monkeys infused with ammonia to produce blood ammonia levels comparable to those observed in patients with hepatic coma, Cole et al. (1972) produced a clinical syndrome and neuropathological changes that were indistinguishable from liver coma. Their findings have been supported by those of other investigators who have produced hyperammonemia after constructing portacaval shunts in rats, by feeding shunted rats ion-exchange resins that release ammonia into the blood, by treating animals with methionine sulfoximine, and by administering urease extracted from jack beans (Cavanaugh and Kyu, 1971; Zamora et al., 1973; Norenberg and Lapham, 1974; Norenberg, 1977; Gibson et al., 1974; Rizzuto and Gonatas, 1974).

Experimental studies, largely performed in rats with portacaval shunts, with and without ammoniated resin feedings, have helped clarify the nature of the glial response to hyperammonemia. In shunted rats, Cavanaugh and Kyu (1971) produced changes in the nuclei of astrocytes that became evident after the fifth postoperative week. The nuclei developed lobulation associated with unusual shapes and an increase in size. The size change was at least in part attributed to immersion fixation. In a later study, electron microscopic studies showed an increase in the amount of smooth intracellular membranes after the fourth week (Zamora et al., 1973). This was associated with an increase in the number of mitochondria and ribosomes. Norenberg (1977), using both light and electron microscopic methodology, was able to correlate astrocyte morphological changes with the clinical state in animals with portacaval shunts that were fed ammoniated resins. Before the development of coma, there was an increase in the amount of astrocytic protoplasm associated with a proliferation of mitochondria and endoplasmic reticulum. Figure 3 shows an astrocyte with ammonia-induced changes. This has been interpreted as being evidence of an increase in the metabolic activity of these cells. The more typical Alzheimer type II cell change was observed after the development of coma. These cells exhibited additional hydropic changes and evidence of mitochondrial and nuclear degeneration. Norenberg suggests that the clinical manifestations of hyperammonemia may be in part due to mitochondrial dysfunction, and that the irreversible clinical course of the resin-fed animals was due to the failure of astrocytes to function normally once degenerative changes were established.

The anatomical evidence linking astrocytes with hyperammonemia is strengthened by histochemical studies. Martinez-Hernandez *et al.* (1977), using a horseradish peroxidase immunohistochemical method, localized glutamine synthetase to astrocytes. It is glutamine synthetase that catalyzes the ATP requiring enzymatic conversion of glutamate and ammonia to glutamine. Furthermore, that study showed an increase in the immunologically active enzyme in the brains of resin-fed portacaval-shunted rats, thus strengthening the hypothesis that these astrocytes are more active than normal and are engaged in ammonia detoxification. Norenberg (1976) has also successfully shown that glutamate dehydrogenase activity is increased in the astrocytes of similarly treated animals.

Although the primary glial response to portacaval shunting is astrocytic, oligodendroglia have been shown to be affected as well (Cavanaugh *et al.*, 1971; Zamora *et al.*, 1973). In their initial studies, these investigators observed an

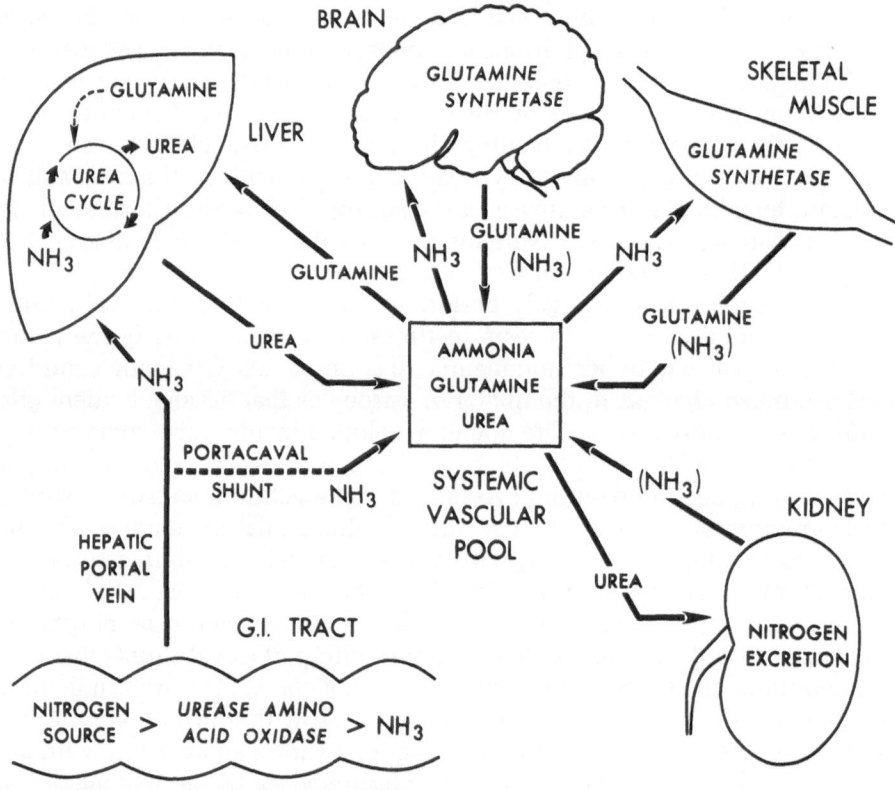

FIGURE 3. The whole-body metabolic scheme for ammonia. Ammonia, produced in the colon by the action of bacteria, is transported to the liver in normal humans where enzymes of the urea cycle convert it to urea. In patients with portal-systemic shunts, ammonia enters the systemic circulation and is carried to the brain and other organs. Under these circumstances skeletal muscle may gain importance in ammonia detoxification. Reproduced by copyright permission of The American Society for Clinical Investigation, from Lockwood *et al.* (1979).

increase in fibers in oligodendroglial cells in shunted animals. Since the proteins that are in these fibers contain large amounts of glutamic acid believed to be involved in the ammonia detoxification process, it was suggested that hyperammonemia produced a colchicinelike effect by interfering with tubular function. In later studies, however, other tubular functions in the brain were not observed to be abnormal and that hypothesis was revised. More recent suggestions attribute this response to cellular swelling in the shunted animals (Zamora *et al.*, 1973).

5. Physicochemical Considerations

A compound must gain entry to the brain before it can affect brain function. Since ammonia is a weak base (measured K'a values range between 8.89 and 9.49×10^{-10} mole/liter) (Bates and Pinching, 1949; Jacquez *et al.*, 1959), ammonia exists in two molecular forms, ammonia gas, NH_3, and the ammonium ion, NH_4^+. The concentration of each species is determined by the blood ammonia concentration and the pH of the blood. Within the extremes of the pH range observed in human subjects (7.0 to 7.6) between 0.9 and 3.4% of the total amount of ammonia in the blood exists as the gas. Since most ammonia is in the ionic form and the blood–brain barrier is generally regarded as being impermeable or having a very low permeability to ions, many shared the belief that ammonia did not cross the blood–brain barrier to enter the brain. Early experimental studies in man, where arterial and cerebral venous blood ammonia concentrations were measured chemically, showed small arteriovenous concentration differences, although there were some exceptions (Fazekas *et al.*, 1956; Webster and Gabuzda, 1958; Posner and Plum, 1960).

The availability of ^{13}N has made it possible to label ammonia with a radioactive atom and to apply tracer techniques to study the behavior of ammonia. Studies of the distribution of ^{13}N after the intravenous injection of $[^{13}N]$ammonia have shown appreciable amounts of the isotope in the brain. In a study of normal subjects and patients with liver disease, Lockwood *et al.* (1979) found that about 7.0% of all ammonia in the systemic circulation is taken up by the brain. Their calculations also showed that about 49% of all ammonia presented to the normal brain is extracted and metabolized by the brain. This observation indicated that the blood–brain barrier was in fact quite permeable to ammonia, but that the barrier is not freely permeable to ammonia; i.e., the entry of ammonia into the brain is diffusion limited, otherwise all of the ammonia would be extracted. This result is supported by the observations of others in nonhuman species (Phelps *et al.*, 1977; Lockwood *et al.*, 1980; Raichle and Larson, 1981).

The recognition that there is a limitation to the diffusion of ammonia at the blood–brain barrier leads to the prediction that ammonia extraction should vary inversely with cerebral blood flow. In other words, at low rates of blood flow, the contact time for a given volume of blood with the blood–brain barrier increases, and the probability of a given ammonia molecule entering the brain increases. This result has been observed in several studies (Phelps *et al.*, 1977; Raichle and Larson, 1981).

In order for nearly half of all arterial ammonia to be extracted by the brain at blood pH values where about 1% of the ammonia is in the nonionic gas form, one of two things must be true: either the barrier must be permeable to the ammonium ion or the supply of ammonia gas must be constantly replaced from the ionic reserves with which it is in equilibrium. Measurements of rate constant values indicate that the equilibration between NH_3 and NH_4^+ is an extraordinarily rapid process (on the order of 3.4×10^{10} M/sec) (Eigen and DeMaeyer, 1963). In other words, virtually all NH_4^+ may exist as NH_3 at some time during the period of blood–brain barrier contact. More recently, Raichle and Larson (1981) have been able to measure the permeability of the blood–brain barrier to ammonia gas and the ammonium ion. Their data show, as expected, that the gas crosses the barrier almost 200 times more easily than the ion. Nevertheless, since the ion is so much more abundant than the gas, they conclude that as much as 25% of all blood-borne ammonia that is trapped by the brain enters as the ion.

Ammonia is a weak base, and the theory that predicts the distribution of compounds across a membrane that separates compartments of differing pH states that the total ammonia concentration should be higher on the side of the membrane with the lower pH, if the membrane is permeable to the unionized, but not ionized form (Shore et al., 1957). Attempts to demonstrate this have met with varied results. The studies of Lockwood et al. (1980) and more recently Raichle and Larson (1981) clearly show that the blood–brain pH gradient does have an important, determining effect on ammonia extraction by the brain. Restated, as blood becomes more alkalotic, relative to brain, the fraction of the total ammonia existing as the diffusible gas increases, and thus ammonia influx into the brain is facilitated. Lockwood et al. (1980) have shown that the brain uptake index for ammonia increased as the per cent of the total ammonia, existing as a gas, increased, as shown in Fig. 4. This increase was nonlinear and interpreted in the light of the more recent measurements of ionic and nonionic permeability constants for ammonia lends support to their data and the conclusion that ionized ammonia does cross the blood–brain barrier.

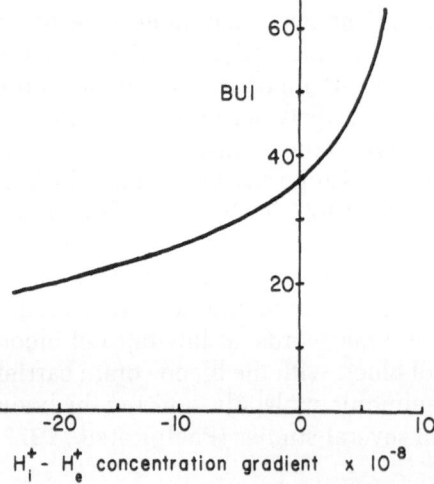

FIGURE 4. The ammonia brain uptake index, BUI, is shown as a function of the hydrogen ion gradient between the blood and the brain. At high extracellular hydrogen ion concentrations, H_e^+, the extraction is low, and rises as the hydrogen ion concentration in blood falls, demonstrating the importance of hydrogen ion gradients and ammonia uptake by the brain. Reproduced by copyright permission of the Elsevier Biomedical Press from Lockwood et al. (1980).

Using equations derived from the Henderson-Hasselbach equation, the calculated ratio of brain:blood ammonia is 1.96:1.00, assuming a brain pH of 7.10 and a blood pH of 7.40 (Lockwood *et al.*, 1980). The measured ratio in rats is 1.25:1.00 (Hindfelt *et al.*, 1975). The discrepancy between the predicted and the measured ratios indicates that the ammonia concentration is not in equilibrium, but is consistent with the hypothesis that pH effects steady-state concentration ratios. It seems probable that rapid metabolism of ammonia entering the brain from the blood prevents the ammonia concentration from reaching an equilibrium concentration in brain parenchyma (see Section 7.2 for data on metabolic compartmentation and trapping).

6. Neurophysiological Aspects of Hyperammonemia

Since epileptic seizure activity is a feature of ammonia intoxication, neurophysiologists have evaluated the effects of ammonium ions on electrophysiological aspects of nerve function. In a series of voltage clamp experiments using giant squid axons, Binstock and Lecar (1969) showed that ammonium ions will substitute for sodium and for potassium ions. Ammonium ions carried the early transient current with a permeability that was 30% of that associated with sodium ions and carried the delayed current with 30% the permeability of potassium ions. Tetrodotoxin and tetraethyl ammonium chloride blocked the early and late components of the currents when both were carried by ammonium ions.

Additional studies from a number of laboratories identified more specific abnormalities that are attributable to ammonium ions. In a study of motor neurons in the cat spinal cord, Lux *et al.* (1970) found that ammonium ions produced a reversible depolarizing shift of the inhibitory postsynaptic equilibrium potential, E_{IPSP}, toward the resting membrane potential. The shift was observed when ammonium ions were given intravenously or by intracellular or extracellular injections. There was a higher sensitivity and shorter duration of the potential shift when ammonium was applied extracellularly or given systemically compared with intracellular injections. This suggested that the effect of intracellular injections was the consequence of ammonium ions existing from the cell and exerting an effect in the extracellular space. They hypothesized that the phenomenon of the reversible E_{IPSP} shift was attributable to interference with chloride ion extrusion. In a later set of experiments, Lux (1971) injected chloride ions, iontophoretically, into motor neurons of the cat spinal cord before and after intravenous injections of ammonium acetate. He found that ammonium ions altered the magnitude and the time course of the E_{IPSP} in a fashion that was consistent with an interference with the active extrusion of chloride. In a study of an isolated neuronal system, the crayfish stretch receptor neuron, Meyer and Lux (1974) again produced results that suggested that ammonium ions blocked the extrusion of chloride.

Electrophysiological studies of the effects of ammonium ions have also been conducted using more complex systems: the pyramidal tract neurons of cat cortex (Raabe and Gumnit, 1975). These investigators stimulated the pyramidal tract in the brain stem to produce antidromic firing of pyramidal tract neurons. Using both intracellular and extracellular recording techniques, they found that the postsynaptic inhibition that was normally produced by stimulus trains administered antidromically was abolished by ammonium salts that were administered intravenously. In this same study, measurements of the postsynaptic membrane resistance suggested that ammonium ions did not act at the postsynaptic membrane or on presynaptic mechanisms involved with neurotransmitter release. Thus these studies were consistent with an action of ammonium ions on ionic mechanisms such as the chloride pump.

Although ammonium ions do substitute for other ions such as sodium and potassium, this mechanism is probably not as important in the pathogenesis of hyperammonemic encephalopathy as chloride pump inhibition. At the concentrations of ammonium that are found in patients with hyperammonemia, the substitution of ammonium for sodium or potassium would be expected to have negligible effects. However, the effects on the extrusion of chloride have been observed after intravenous administration of ammonium salts and the production of blood ammonia levels that are in the range encountered in encephalopathic patients (Raabe and Gumnit, 1975). The resulting loss of inhibitory neurophysiological inhibitory mechanisms undoubtedly contributes to some of the manifestations of hyperammonemia.

7. Pathophysiological Aspects of Ammonia Intoxication

7.1. Cerebral Blood Flow, Oxygen and Glucose Metabolism

Since cerebral blood flow (CBF) and oxygen metabolism are linked with overall brain function, these variables have been studied extensively in patients with hyperammonemia due to liver disease and in animals with chronic portacaval shunt-induced hyperammonemia. Studies of whole brain CBF have generally shown normal, or slightly reduced values for flow in patients with liver disease, but with no evidence of encephalopathy (Fazekas et al., 1956; Posner and Plum, 1960). Oxygen metabolism was usually reduced in these same subjects. The reduction in oxygen metabolism was an unexpected finding in these studies, since signs of mental impairment were not detected in these subjects. Fazekas et al. (1956) suggested that there may have been minimal impairment that could have been detected with more extensive psychological testing. In patients with overt encephalopathy, these same authors, as well as others (James and Garassini, 1971; Morgan et al., 1980), have found significant reductions in CBF. The transient and reversible nature of the CBF deficit can be seen in patients studied before and after drug treatment and improvement in the degree of encephalopathy (James and Garassini, 1971; Morgan et al., 1980). All authors point out the problems associated with interpreting these studies, namely, it is not certain whether the decrease in blood flow follows a change in metabolism or whether there are primary factors that reduce flow with encephalopathy developing secondarily.

Blood flow and oxygen metabolism have been studied in a variety of experimental animal models of hyperammonemia. The rat with chronic portacaval-shunt induced hyperammonemia is the most exhaustively investigated and has produced the most consistent data, since arterial partial pressures of carbon dioxide can be easily controlled, thereby reducing the effects of respiratory alkalosis on CBF. Eklof *et al.* (1974) reported that CBF was unchanged from control values, when animals were studied one week after end-to-side portacaval shunts were constructed. The arteriovenous oxygen content was reduced, indicating a reduced cerebral metabolic rate for oxygen. Gjedde *et al.* (1978) measured blood flow and oxygen metabolism four and eight weeks after portacaval shunt construction and found that oxygen metabolism was unaffected by the procedure, but that there was a significant increase in flow observed in the eight week but not the four week group. This change in CBF is similar to that measured in humans undergoing portacaval shunts (Posner and Plum, 1960; Bianchi Porro *et al.*, 1969).

The whole brain metabolic response to a superimposed ammonia load was investigated by Gjedde *et al.* (1978). Control animals and animals with portacaval shunts constructed four or eight weeks prior to the study were given an ammonia load sufficient to produce an altered mental status in the shunted animals. Arterial ammonia levels rose only slightly in the control animals, consistent with the absence of portal-systemic blood shunts. In normal animals, there was an initial fall in CBF, then flow values rose back to control levels and the metabolic rate for oxygen increased, perhaps due to ammonia-induced cerebral activation. In the four-week shunt group, the ammonia load more than doubled the arterial ammonia concentration, but CBF and oxygen metabolism were unaffected. In the eight-week shunt group, resting CBF was above normal with normal oxygen metabolism. In response to the ammonia load, both CBF and oxygen metabolism were significantly reduced. This increase in the sensitivity to ammonia in the eight-week group was associated with a significant net uptake of ammonia and the production of glutamine by the brain. In control and four-week animals, there was no measurable net uptake of ammonia or net release of glutamine. These data were interpreted as being the physiological counterparts to "enhanced toxin sensitivity" described by Shenker *et al.* (1974). This idea has been advanced to explain the clinical observations that indicate that patients with long-standing liver disease or repeated episodes of coma are more likely to be affected by minor stresses than patients without a history of prolonged exposure of the brain to toxins. Whether or not this corresponds to the astrocyte changes that are evident in shunted animals and the implied alterations in cerebral function suggested by Norenberg (1977) (see also Section 4), cannot be determined from the data.

In spite of the central role of glucose in cerebral metabolism, there is a considerable amount of controversy in the literature that describes cerebral glucose metabolism during hyperammonemia. In most patients with encephalopathy, blood flow and glucose metabolism are reported as being lower than normal. However, James *et al.* (1972) reported that the cerebral glucose metabolic rate, along with blood flow and oxygen metabolism, rose immediately after the construction of a portacaval shunt. He also reported an increase in glucose metabolism in dogs given intravenous infusions of ammonia that was accompanied by a lightening of anesthesia (James *et al.*, 1974). Hawkins *et al.* (1973) also studied the acute hyperammonemia and measured a 31% increase in the cerebral glucose metabolic rate. More recently, Morgan *et al.* (1980), in agreement with most others, reported glucose metabolic rates that were below normal in encephalopathic patients and that increased after bromocriptine therapy and clinical improvement. Lockwood *et al.* (1982) have found that intracarotid ammonia infusions that produced EEG slowing produced large increases in the glucose metabolic rate, measured with [^{14}C]deoxyglucose, that were confined primarily to the deep grey structures of the brain, most prominently the substantia nigra. Thus it seems likely that different regions of the brain differ in their sensitivity to ammonia.

7.2. Metabolic Fate of Ammonia in Brain

Isotopic experiments have clearly shown that ammonia enters the brain from the blood (see Section 5). Lockwood *et al.* (1979) have used [^{13}N]ammonia to measure the rate of transfer of ammonia from blood to brain in a group of normal subjects and patients with liver disease. Their study showed that the rate of ammonia uptake was a linear function of the arterial ammonia concentration, and that in normal subjects nearly half of the arterial ammonia was taken up by the brain and metabolically trapped. Since this extraction ratio is considerably higher than that reported by others, who used chemical techniques to measure net brain ammonia uptake (Bessman and Bessman, 1955; Fazekas *et al.*, 1956; Gjedde *et al.*, 1978), Lockwood *et al.* (1979) concluded that the normal brain continually produces and consumes ammonia. In this study, the rate of ammonia metabolism by the brain was higher in the group with encephalopathy than the group with normal mentation. These data were interpreted as being further evidence for the importance of hyperammonemia in the pathogenesis of hepatic encephalopathy.

The metabolic fate of ammonia in the brain was investigated by Berl *et al.* in 1962. They infused [^{15}N]ammonia into cats, removed their brains, separated and measured the specific activity of various amino acids. They reported evidence for two separate pools of metabolic activity. A small glutamate pool, with rapid turnover, synthesized the most glutamine, whereas a larger glutamate pool led to the slower synthesis of glutamine. As with most investigations that use [^{15}N]ammonia, the steady state was not preserved and the experimental animals developed electroencephalographic abnormalities that were attributable to hyperammonemia. The study of Berl *et al.* (1962) was repeated by Cooper *et al.* (1979), using high-specific-activity [^{13}N]ammonia. The relative specific activities of glutamate to the alpha-amino group of glutamine and to the amide group of glutamine were 1:5:400. The preferential labeling of the amide group of glutamine is indicative of a two-pool system, as reported by Berl *et al.* (1962). Additional experiments showed that 60% of the injected label had been converted to glutamine within 5 sec of an intracarotid [^{13}N]ammonia injection (Cooper *et al.*, 1979). From this data they calculated that the half-life (t½) for the conversion of glutamate to glutamine in the small fast pool was probably less than 3 sec. The data of Lockwood *et al.* (1979) suggested that there were at least two ammonia pools in the human brain. When glutamine synthetase was inhibited by methionine sulfoximine, the isotopic labeling pattern after [^{13}N]ammonia injections gave evidence for persistence of the large slow pool and abolition of the small fast pool (Cooper *et al.*, 1979).

Isotopic-labeling experiments do not provide information about the anatomical location of metabolic pools; however, anatomical and immunohistochemical data, reviewed in Section 4 and as discussed by Cooper et al. (1979), suggest that the rapid pool for glutamine synthesis is located in astrocytes.

In order to be able to synthesize glutamine, there must be a continuous supply of glutamate in the small pool. There are a number of likely sources for the carbon skeletons required. Isotopic-labeling studies have shown that glucose (Cremer et al., 1975) and leucine (Cremer et al., 1977) both contribute to the glutamate pool. Other amino acids such as phenylalanine (Van den Berg and Van den Velden, 1970) also give rise to glutamate in the small glutamate pool, presumably by amino acid transamination reactions.

Although a number of studies have shown that there is an increase in the glutamine concentration in the brains of hypoammonemic animals, the mechanism for the elevation is not known. From the results of isotopic-labeling studies using [^{14}C]glucose as a glutamine precursor, Cremer et al. (1975) concluded that the overall rate of glutamine synthesis was not increased and the increased concentration of glutamine was the consequence of a reduction in the rate of glutamine use or egress from the brain. In studies of acute marked hyperammonemia, Hawkins et al. (1973) calculated an excess glutamine synthesis rate of 0.33 mmole/min per g of tissue. Gjedde et al. (1978) found that the ammonia production rate by the brains of animals that had had portacaval shunts constructed eight weeks earlier was significantly higher than the glutamine production rates of sham-operated control animals (rates were products of arteriovenous concentration differences and CBF). Thus it seems likely that the rate of glutamine production in the brains of hyperammonemic animals is increased. The increase is most likely a part of a metabolic defense mechanism that seeks to protect the brain against hyperammonemia, a hypothesis consistent with neuropathological observations (Norenberg, 1977).

Although glutamine occupies a central position in many metabolic pathways due to its importance as an amide group donor, the potential for disruption of these pathways due to hyperammonemia has not been the subject of many studies. It seems likely that a great deal of the glutamine formed in the brain leaves the brain (Gjedde et al., 1978; Phelps et al., 1977). Glutamine also undergoes transamination in the brain to form alpha-ketoglutaramate. The finding of excess alpha-ketoglutaramate in the CSF of patients with hepatic encephalopathy prompted speculation that this compound was a neurotoxin that was important in the pathogenesis of hyperammonemic nervous system dysfunction (Vergara et al., 1974). However, additional investigations failed to find evidence that this compound affected cerebral metabolism, and it is probably formed as a consequence of the law of mass action and low omega-amidase levels in the brain (Duffy et al., 1974a,b; Lockwood and Duffy, 1977). Thus hyperammonemia appears to lead to secondary increases in glutamine levels and tertiary increases in alpha-ketoglutaramate (Duffy et al., 1974a). Measurement of levels of glutamine and alpha-ketoglutaramate are useful in the diagnosis of hepatic or hyperammonemic encephalopathy (Duffy et al., 1974b) (Fig. 1).

7.3. Effects of Ammonia on Brain Energy Metabolism

The finding of a net ammonia extraction by the brain led to the hypothesis by Bessman and Bessman (1955) that hyperammonemia depleted metabolites related to energy production in the brain, and has led to a number of studies that have focused on this question. Schenker *et al.* (1967) reported that ATP and phosphocreatine levels were depleted in the caudal midbrain and pontomedullary areas of animals that were made hyperammonemic. These high-energy compounds were not reduced in cortical regions. Shorey *et al.* (1967) continued this study and found that ammonia intoxication did not deplete alpha-ketoglutarate and that levels of this metabolite did not correlate with ATP depletion, which was confirmed in their study. The effects of ammonia intoxication on energy metabolism were reinvestigated by Hindfelt and Siesjo (1971) because of the possibility that the tissue fixation methods employed by Schenker *et al.* (1967) and Shorey *et al.* (1967) were not adequate to maintain labile metabolites at levels that are present during life. Their study failed to show any effect of ammonia on brain stem ATP levels. In a later study, Hindfelt *et al.* (1977) investigated the time course of energy metabolite levels in relationship to clinical symptoms in rats with portacaval shunts that were given an acute ammonia load. This investigation showed that ATP levels in the brain stem were still normal at the time that consciousness first became impaired, and that after deep coma was established, ATP levels were indeed reduced. Finally, McCandless and Schenker (1981) have employed microneurochemical methods to measure metabolite levels in the reticular formation of normal mice that were given intraperitoneal injections of ammonia. They found that ATP, phosphocreatine, and glucose levels were significantly reduced in the early coma period, with preservation of these metabolites at normal levels in the nearby posterior colliculus. Lactate levels in both the reticular formation and colliculus were normal, suggesting that the reduction in ATP was not a fixation artifact. They hypothesized that a reduction in energy metabolism in the reticular formation may lead to a generalized reduction of brain metabolism because of the known importance of the reticular formation in the control of consciousness.

Hindfelt *et al.* (1977) suggested that hyperammonemia may lead to coma production by inhibiting the malate–aspartate shuttle. They measured alterations in the concentrations of the components of the shuttle that suggested that the cytoplasmic $NADH:NAD^+$ ratio was increased, while the ratio within the mitochondria was reduced. This implied that reducing equivalents failed to enter the mitochondria and led to an impairment of oxidative metabolism and encephalopathy.

7.4. Amino Acids and Biogenic Amines

Hepatic coma is associated with altered blood and brain amino acid levels (Iber *et al.*, 1957; Fischer *et al.*, 1974; Hindfelt *et al.*, 1977). The association of hepatic coma with abnormal levels of biogenic amine precursors led Fischer and Baldessarini (1971) to propose that false neurotransmitters were formed during periods of liver failure. These compounds then competed with dopamine and norepinephrine at postsynaptic receptors and, because of their low potency, led to selective neurotransmitter blockade and coma. Octopamine is the compound that received the most attention. Fischer and his group (for review, see Fischer, 1974) demonstrated that brain, serum, and heart octopamine levels were elevated during acute hepatic coma or after portacaval shunt construction. In later studies (Fischer et al., 1974), the valine plus leucine plus isoleucine to phenylalanine plus tyrosine ratio was seen as a critical factor in determining whether neurotransmitter derangement would occur and cause coma.

The most recent studies from that group have suggested that hyperammonemia may be the most important factor in the alteration of brain amino acid transport and amino acid levels. Ammonia itself may directly stimulate the uptake of some amino acids by the brain (Cangiano et al., 1980), and the carrier-mediated efflux of glutamine, formed as a consequence of hyperammonemia, may cause an influx of amino acids that then alter neurotransmitter synthesis (James et al., 1980).

The false neurotransmitter hypothesis suffers from several weaknesses. Zieve and Olsen (1977) have shown that intracerebral injections of octopamine that were sufficient to increase levels to 20,000 times those measured experimentally were without effect on the level of consciousness. This, coupled with the failure of others to find neurotransmitter precursor amino acid imbalance in patients with coma (Cascino et al., 1978) or to produce coma by amino acid infusion, led Baker (1979) to urge a reconsideration of the altered neurotransmitter hypothesis.

Hindfelt and his colleagues (1977) have suggested additional mechanisms by which ammonia-induced cerebral amino acid imbalance might lead to the production of encephalopathy. They found a reduced level of the putative neurotransmitters glutamate and aspartate. Although the reduction in glutamate was modest, they suggested that if it was confined to a single compartment, such as synaptosomes, the effect could be marked. This hypothesis is strengthened by the finding of Benjamin (1981) that ammonium ions are competitive inhibitors of glutaminase activity and the proposal that hyperammonemia interferes with the conversion of glutamine, formed by astrocytes, to glutamate in nerve terminals.

Finally, hyperammonemia and portacaval shunting have been found to interfere with the rate of amino acid incorporation into brain proteins. Wasterlain et al. (1978) reported that portacaval shunting in rats reduced the rate of lysine incorporation into protein by about 50% and that an acute ammonia load caused further reductions in protein synthesis rates. Brun et al. (1977) and Hindfelt et al. (1979) have reported reductions in soluble protein concentrations in patients that died from hepatic failure and in brains of rats after portacaval shunt construction. Although any explanation that seeks to link coma production with reduced protein synthesis must be tentative, selection actions on critical proteins could lead to coma production or could prolong coma beyond the period of hyperammonemia.

7.5. Toxin Synergism

Although hyperammonemia by itself is a well-recognized cause of coma, ammonemia acts synergistically with other toxins to produce encephalopathy. Mercaptans and short-chain fatty acids have both been implicated in the pathogenesis of hepatic coma. Subcoma doses of both of these classes of compounds produce coma if the concentration of ammonia is elevated (Zieve *et al.*, 1974a,b). (For a more complete discussion of a mercaptan and short-chain fatty acid-induced coma, see Chapters 7 and 8).

8. Summary

The metabolism, and hence the action, of ammonia within the nervous system is very complex and is governed by a variety of chemical and physical considerations. Since ammonia is a weak base, it exists in ionized and unionized forms that exhibit different behaviors and are responsible for different phenomena. The gas, or unionized form, crosses the blood–brain barrier, but is diffusion-limited. The amount of ammonia entering the brain depends on the arterial concentration and the hydrogen ion gradient between the blood and the brain. In the brain, the ion substitutes for sodium and potassium, thus affecting membrane electrical properties, and inhibits the chloride pump, affecting postsynaptic inhibitory mechanisms. Within the brain, ammonia is metabolized to form glutamine in a highly compartmentalized system. The consequences of hyperammonemia appear to exert different effects on different brain regions with the reticular formation being perhaps more sensitive to hyperammonemia than other nearby locations. The metabolic consequences of ammonia detoxification in the brain are extensive, leading to changes in astrocyte morphology and metabolic capacity and alterations in the concentrations of a variety of metabolites, each with their own potential to disrupt normal metabolic processes and to contribute to the development of encephalopathy. There is hardly any aspect of cerebral metabolism that is unaffected by hyperammonemia.

It seems unlikely that there is a single mechanism by which hyperammonemia causes encephalopathy. Rather, the evidence suggests that hyperammonemic encephalopathy is the consequence of multiple abnormalities that follow any increase in the blood or the brain ammonia concentration.

ACKNOWLEDGMENTS. Portions of the work described in this chapter were supported by National Institutes of Health research grants NS 14996 and NS 15639.

References

Adams, R. D., and Foley, J. M., 1953, The neurological disorder associated with liver disease, in: *Metabolic and Toxic Diseases of the Nervous System*, Volume 32, *Proceedings of the Association for Research in Nervous and Mental Disease*, William and Wilkins, Baltimore, pp. 198–237.

Baker, A. L., 1979, Amino acids in liver disease: A cause of hepatic encephalopathy?, *J. Am. Med. Assoc.* **242**:355–356.

Ballard, R. A., Vincour, B., Reynolds, J. W., Wennberg, R. P., Merritt, A., Sweetman, L., and Nyhan, W. L., 1978, Transient hyperammonemia of the preterm infant, *N. Engl. J. Med.* **299**:920–925.

Bates, R. G., and Pinching, G. D., 1949, Acidic dissociation constant of ammonium ion at 0° to 50° and the base strength of ammonia, *J. Res. Natl. Bureau Standards* **42**:419–430.

Batshaw, M., Brusilow, S., and Walser, M., 1975, Treatment of carbamylphosphate synthetase deficiency with keto-analogues of essential amino acids, *N. Engl. J. Med.* **292**:1085–1090.

Benjamin, A. M., 1981, Control of glutaminase activity in rat brain cortex *in vitro*: Influence of glutamate, phosphate, ammonium, calcium and hydrogen ions, *Brain Res.* **208**:363–377.

Berl, S., Takagaki, G., Clarke, D. D., and Waelsch, H., 1962, Metabolic compartments *in vivo*. Ammonia and glutamic acid metabolism in brain and liver, *J. Biol. Chem.* **237**:2562–2569.

Bessman, S. P., and Bradley, J. E., 1955, Uptake of ammonia by muscle: Its implications in ammoniagenic coma, *N. Engl. J. Med.* **253**:1143–1147.

Bianchi Porro, G., Maiolo, A. T., and Della Porta, P., 1969, Cerebral blood flow and metabolism in hepatic cirrhosis before and after portacaval shunt operation, *Gut* **10**:894–897.

Binstock, L., and Lecar, H., 1969, Ammonium ion currents in the squid giant axon, *J. Gen. Physiol.* **53**:342–360.

Bove, K. E., McAdams, A. J., and Partin, J. C., The hepatic lesion in Reye's Syndrome, *Gastroenterology* **69**:685–697.

Brown, T., Hug, G., Lansky, L., Bove, K., Scheve, A., Ryan, M., Brown, H., Schubert, W. K., Partin, J. C., and Lloyd-Still, J., 1976, Transiently reduced activity of carbamyl phosphate synthetase and ornithine transcarbamylase in liver of children with Reye's syndrome, *N. Engl. J. Med.* **294**:861–867.

Brun, A., Dawiskiba, S., Hindfelt, B., and Olsson, J-E., 1977, Brain proteins in hepatic encephalopathy, *Acta Neurol. Scand.* **55**:213–225.

Bruton, C. J., Corsellis, J. A. N., and Russell, A., 1970, Hereditary hyperammonemia, *Brain* **93**:423–434.

Cangiano, C., Cardelli-Cangiano, P., James, J. H., and Fischer, J. E., 1980, Ammonia stimulates amino acid uptake by isolated bovine brain microvessels, *Gastroenterology* **78**:1302.

Cascino, A., Cangiano, C., Calcaterra, V., Rossi-Fanelli, F., and Capocaccia, L., 1978, Plasma amino acid imbalance in patients with liver disease, *Am. J. Dig. Dis.* **23**:591–598.

Cavanaugh, J. B., and Kyu, M. H., 1971, Type II Alzheimer change experimentally produced in astrocytes in the rat, *J. Neurol. Sci.* **12**:63–75.

Cavanaugh, J. B., Blakemore, W. F., and Kyu, M. H., 1971, Fibrillary accumulations in oligodendroglial processes of rats subjected to portocaval anastomosis, *J. Neurol. Sci.* **14**:143–152.

Cole, M., Rutherford, R. B., and Smith, F. O., 1972, Experimental ammonia encephalopathy in the primate, *Arch. Neurol.* **26**:130–136.

Conn, H. O., 1960, Ammonia tolerance in liver disease, *J. Lab. Clin. Med.* **55**:855–871.

Conn, H. O., 1978, Lactulose: A drug in search of a modus operandi, *Gastroenterology* **74**:624–626.

Cooper, A. J. L., McDonald, J. M., Gelbard, A. S., Gledhill, R. S., and Duffy, T. E., 1979, The metabolic fate of ^{13}N-labeled ammonia in rat brain, *J. Biol. Chem.* **254**:4982–4992.

Cremer, J. E., Heath, D. F., Teal, H. M., Woods, M. S., and Cavanaugh, J. B., 1975, Some dynamic aspects of brain metabolism in rats given a portocaval anastomosis, *Neuropathol. Appl. Neurobiol.* **3**:293–311.

Cremer, J. E., Teal, H. M., Heath, D. F., and Cavanaugh, J. B., 1977, The influence of portocaval anastomosis on the metabolism of labelled octanoate, butyrate and leucine in rat brain, *J. Neurochem.* **28**:215–222.

Drayna, C. J., Titcomb, C. P., Varma, R. R., and Soergel, K. H., 1981, Hyperammonemic encephalopathy caused by infection in a neurogenic bladder, *N. Engl. J. Med.* **304**:766–768.

Duffy, T. E., and Plum, F., 1981, Seizures, coma and major metabolic encephalopathies, in: *Basic Neurochemistry* (G. J. Siegel, R. W. Albers, B. W. Agranoff, and R. Katzman, eds.), Little, Brown and Company, Boston Massachusetts, pp. 693–718.

Duffy, T. E., Cooper, A. J. L., and Meister, A., 1974a, Identification of alpha-ketoglutaramate in rat liver, kidney and brain: Relationship to glutamine transaminase and omega-amidase activities, *J. Biol. Chem.* **249**:7603–7606.

Duffy, T. E., Vergara, F., and Plum, F., 1974b, Alpha-ketoglutaramate in hepatic encephalopathy, in: *Brain Dysfunction in Metabolic Disorders* (F. Plum, ed.), *Res. Publ. Assoc. Nerv. Ment. Dis.*, Volume 53, Raven Press, New York, pp 39–52.

Egense, J., and Schwartz, M., 1970, Recurrent hepatic coma following ureterosigmoidostomy, *Scand. J. Gastroenterol.* **5**(suppl. 7):149–152.

Eigen, M., and DeMaeyer, 1963, Relaxation methods, in: *Investigations of Rates and Mechanisms of Reactions*, second ed. (S. L. Friess, E. S. Lewis, and A. Weissberger, eds.), Interscience Publishers, New York, p. 1034.

Eklof, B., Holmin, T., Johannsson, H., and Siesjo, B. K., 1974, Cerebral blood flow and cerebral metabolic rate for oxygen in rats with a porta-caval anastomosis, *Acta Physiol. Scand.* **90**:337–344.

Elkington, S. G., Floch, M. H., and Conn, H. O., 1969, Lactulose in the treatment of chronic portal-systemic encephalopathy: A double-blind clinical trial, *N. Engl. J. Med.* **281**:408–412.

Evans, W. B., Aoyagi, T., and Summerskill, W. H. I., 1966, Part II: Urea hydrolysis and ammonia absorption in upper and lower gut lumen and effect of neomycin, *Gut* **7**:635–639.

Fazekas, J. F., Ticktin, H. E., Ehrmantraut, W. R., and Alman, R. W., 1956, Cerebral metabolism in hepatic encephalopathy, *Am. J. Med.* **21**:843–849.

Fischer, J. E., 1974, False neurotransmitters and hepatic coma, in: *Brain Dysfunction in Metabolic Disorders* (F. Plum, ed.), *Res. Publ. Assoc. Nerv. Ment. Dis.*, Volume 53, Raven Press, New York, pp. 53–73.

Fischer, J. E. and Baldessarini, R. J., 1971, False neurotransmitters and hepatic failure, *Lancet* **2**:75–80.

Fischer, J. E., Yoshimura, N., Aguirre, A., James, J. H., Cummings, M. G., Abel, R. M., and Deindoerfer, F., 1974, Plasma amino acids in patients with hepatic encephalopathy. Effects of amino acid infusions, *Am. J. Surg.* **127**:40–47.

Fuld, H., 1933, Uber de Dignostische Ververkbarkeit von Ammoniakbestimmung im Blut, *Klin. Wochenschr.* **12**:1364–1366.

Gabuzda, G., Jr., Phillips, G. B., and Davidson, C. S., 1952, Reversible toxic manifestations in patients with cirrhosis of the liver given cation-exchange resins, *N. Engl. J. Med.* **246**:124–130.

Ganda, O. P., and Ruderman, N. B., 1976, Muscle nitrogen metabolism in chronic hepatic insufficiency, *Metabolism* **25**:427–435.

Gibson, G. E., Zimber, A., Krook, L., and Visek, W. J., 1974, Brain histology and behavior of mice injected with urease, *J. Neuropathol. Exp. Neurol.* **33**:201–211.

Gjedde, A., Lockwood, A. H., Duffy, T. E., and Plum, F., 1978, Cerebral blood flow and metabolism in chronically hyperammonemic rats: Effect of an acute ammonia challenge, *Ann. Neurol.* **3**:325–330.

Goldberg, R. N., Cabal, L. A., Sinatra, F. R., Plajstek, C. E., and Hodgman, J. E., 1979, Hyperammonemia associated with perinatal asphyxia, *Pediatrics* **64**:336–341.

Hawkins, R. A., Miller, A. L., Nielsen, R. C., and Veech, R. L., 1973, The acute action of ammonia on rat brain metabolism *in vivo*, *Biochem. J.* **134**:1001–1008.

Hindfelt, B., 1975, The distribution of ammonia between extracellular and intracellular compartments of the rat brain, *Clin. Sci. Molec. Med.* **48**:33–37.

Hindfelt, B., and Siesjo, B. K., 1971, Cerebral effects of acute ammonia intoxication. The effect upon energy metabolism, *Scand. J. Clin. Lab. Invest.* **28**:365–374.

Hindfelt, B., Plum, F., and Duffy, T. E., 1977, Effect of acute ammonia intoxication on cerebral metabolism in rats with portacaval shunts, *J. Clin. Invest.* **59**:386–396.

Hindfelt, B., Holmin, T., and Olsson, J-E., 1979, Brain proteins in experimental portal-systemic shunting, *Acta Neurol. Scand.* **59**:275–280.

Hoyumpa, A. M., Jr., Desmond, P. V., Avant, G. R., Roberts, R. K., and Schenker, S., 1979, Hepatic encephalopathy, *Gastroenterology* **76**:184–195.

Huttenlocher, P. R., Schwartz, A. D., and Klatskin, G., 1969, Reye's syndrome: Ammonia intoxication as a possible factor in the encephalopathy, *Pediatrics* **43**:443–454.

Iber, F. L., Rosen, H., Levenson, S. M., and Chalmers, T. C., 1957, The plasma amino acids in patients with liver failure, *J. Lab. Clin. Med.* **50**:417–425.

Jacquez, J. A., Poppell, J. W., and Jeltsch, R., 1959, Solubility of ammonia in human plasma, *J. Appl. Physiol.* **14**:255–258.

James, I. M., and Garassini, M., 1971, Effect of lactulose on cerebral metabolism in patients with chronic portosystemic encephalopathy, *Gut* **12**:702–704.

James, I. M., MacDonnell, L., and Xanalatos, C., 1972, The effect of acute portacaval shunting in dogs on cerebral and peripheral blood flow and metabolism, *Clin. Sci.* **42**:769–774.

James, I. M., MacDonell, L., and Xanalatos, C., 1974, Effect of ammonium salts on brain metabolism, *J. Neurol. Neurosurg. Psychiatry* **37**:948–953.

James, I. M., Cangiano, L., Cardelli-Cangiano, P., and Fischer, J. E., 1980, Glutamine links hyperammonemia and neurotransmitter derangements in portal systemic shunting, *Gastroenterology* **78**:1308.

Kazemi, H., Shore, N. S., Shih, V. E., and Shannon, D. C., 1973, Brain organic buffers in respiratory acidosis and alkalosis, *J. Appl. Physiol.* **34**:478–482.

Kersh, E. S., and Rifkin, H., 1973, Lactulose enemas, *Ann. Intern. Med.* **78**:81–84.

Kirk, E., 1936, Amino acid and ammonia metabolism in liver disease, *Acta Med. Scand.* (suppl.)**77**:1–147.

Lewis, L. D., Ljunggren, B., Norberg, K., and Siesjo, B. K., 1974, Changes in carbohydrate substrates, amino acids and ammonia in the brain during insulin-induced hypoglycemia, *J. Neurochem.* **23**:659–671.

Lockwood, A. H., and Duffy, T. E., 1977, Glutamine transaminase and omega-amidase: Species variations in brain activity and effect of portacaval shunting, *J. Neurochem.* **28**:673–675.

Lockwood, A. H., McDonald, J. M., Reiman, R. E., Gelbard, A. S., Laughlin, J. S., Duffy, T. E., and Plum, F., 1979, The dynamics of ammonia metabolism in man: Effects of liver disease and hyperammonemia, *J. Clin. Invest.* **63**:449–460.

Lockwood, A. H., Finn, R. D., Campbell, J. A., and Richman, T. B., 1980, Factors that affect the uptake of ammonia by the brain: The blood-brain pH gradient, *Brain Res.* **181**:259–266.

Lockwood, A. H., Ginsberg, M. D., Butler, C. M., and Gutierrez, M. T., 1982, Selective effects of ammonia on regional brain glucose metabolism, *Ann. Neurol.* **12**:114.

Lowenstein, J. M., 1972, Ammonia production in muscle and other tissues: The purine nucleotide cycle, *Physiol. Rev.* **52**:382–413.

Lux, H. D., 1971, Ammonium and chloride extrusion: Hyperpolarizing synaptic inhibition in spinal motoneurons, *Science* **173**:555–557.

Lux, H. D., Loracher, C., and Neher, E., 1970, The action of ammonium on postsynaptic inhibition of cat spinal motoneurons, *Brain Res.* **11**:431–447.

Makela, A-L., Lang, H., and Korpela, P., 1980, Toxic encephalopathy with hyperammonemia during high-dose salicylate therapy, *Acta Neurol. Scand.* **61**:146–151.

Martinez-Hernandez, A., Bell, K. P., and Norenberg, M. D., 1977, Glutamine synthetase: Glial localization in brain, *Science* **195**:1356–1358.

McCandless, D. W., and Schenker, S., 1981, Effect of acute ammonia intoxication on energy stores in the cerebral reticular activating system, *Exp. Brain Res.* **44**:325–330.

McDermott, W. V., Jr., Adams, R. D., and Riddell, A. G., 1954, Ammonia metabolism in man, *Ann. Surg.* **140**:539–556.

McDermott, W. V., Jr., 1957, Diversion of the urine to the intestines as a factor in ammoniagenic coma, *N. Engl. J. Med.* **256**:460–462.

Meyer, H., and Lux, H. D., 1974, Action of ammonium on a chloride pump: Removal of hyperpolarizing inhibition in an isolated neuron, *Pflug. Arch. Eur. J. Physiol.* **350**:185–195.

Morgan, M., Jakobovitz, A. W., James, I. M., and Sherlock, S., 1980, Successful use of bromocriptine in the treatment of chronic hepatic encephalopathy, *Gastroenterology* **78**:663–670.

Mortensen, E., Lyng, G., and Juhl, E., 1972, Ammonia-induced coma after ureterosigmoidostomy, *Lancet* **1**:496.

Norenberg, M. D., 1976, Histochemical studies in experimental portal-systemic encephalopathy, *Arch. Neurol.* **33**:265–269.

Norenberg, M. D., 1977, A light and electron microscopic study of experimental portal-systemic (ammonia) encephalopathy: Progression and reversal of the disorder, *Lab. Invest.* **36**:618–627.

Norenberg, M. D., and Lapham, L. W., 1974, The astrocyte response in experimental portal-systemic encephalopathy: An electron microscopic study, *J. Neuropathol. Exp. Neurol.* **33**:422–435.

Partin, J. C., Partin, J. S., Schubert, W. K., and McLaurin, R. L., 1975, Brain ultrastructure in Reye's syndrome (encephalopathy and fatty alteration of the viscera), *J. Neuropathol. Exp. Neurol.* **34**:425–444.

Phear, E. A., Sherlock, S., and Summerskill, W. H. J., 1955, Blood ammonium levels in liver disease and "hepatic coma", *Lancet* **268**:836–840.

Phelps, M. E., Hoffman, E. J., and Raybard, C., 1977, Factors which affect cerebral uptake and retention of $^{13}NH_3$, *Stroke* **8**:694–702.

Phelps, M. E., Mazziotta, J. C., and Huang, S-C., 1982, Study of cerebral function with positron computed tomography, *J. Cereb. Blood Flow Metab.* **2**:113–162.

Phillips, G. B., Schwartz, R., Gabuzda, G., Jr., and Dawson, C. S., 1952, Syndrome of impending hepatic coma in patients with cirrhosis of the liver given certain nitrogenous substances, *N. Engl. J. Med.* **247**:239–246.

Plum, F., and Posner, J. B., 1972, *The Diagnosis of Stupor and Coma*, 2nd ed., F. A. Davis, Co., Philadelphia.

Posner, J. B., and Plum, F., 1960, The toxic effects of carbon dioxide and acetazolamide in hepatic encephalopathy, *J. Clin. Invest.* **39**:1246–1258.

Public Health Service, 1979, U.S. Department of Health Education and Welfare, *Vital Statistics of the United States 1975*, Volume II, *Mortality*, Part A, Hyattsville, Maryland.

Raabe, W., and Gumnit, R. J., 1975, Disinhibition in cat motor cortex by ammonia, *J. Neurophysiol.* **38**:347–355.

Raichle, M. E., and Larson, K. B., 1981, The significance of the NH_3-HN_4^+ equilibrium on the passage of ^{13}N-ammonia from blood to brain, *Circ. Res.* **48**:913–937.

Resnick, R. H., Ishihara, A., Chalmers, T. C., Schlimmel, E., and The Boston Inter-Hospital Liver Group, 1968, A controlled trial of colon bypass in chronic hepatic encephalopathy, *Gastroenterology* **54**:1057–1069.

Reye, R. D. K., Morgan, G., and Baral, J., 1963, Encephalopathy and fatty degeneration of the liver— A disease entity in childhood, *Lancet* **2**:749–752.

Rizzuto, N., and Gonatas, N. K., 1974, Ultrastructural study of the effect of methionine sulfoximine on developing and adult rat cerebral cortex, *J. Neuropathol. Exp. Neurol.* **33**:237–250.

Sabbaj, J., Sutter, V. L., and Feingold, S. M., 1970, Urease and deaminase activities of fecal bacteria in hepatic coma, *Antimicrob. Agents Chemother.* **1970**:181–185.

Schenker, S., McCandless, D. W., Brophy, E., and Lewis, M. S., 1967, Studies on the intracerebral toxicity of ammonia, *J. Clin. Invest.* **46**:838–848.

Schenker, S., Breen, K. J., and Hoyumpa, A. M., 1974, Hepatic encephalopathy: Current status, *Gastroenterology* **66**:121–151.

Sherlock, S., 1958, Pathogenesis and management of hepatic coma, *Am. J. Med.* **24**:805–813.

Shih, V. E., 1976, Congenital hyperammonemic syndromes, *Clin. Perinatol.* **3**:3–14.

Shore, P. A., Brodie, B. B., and Hogben, C. A. M., 1957, The gastric secretion of drugs: A pH partition hypothesis, *J. Pharmacol. Exp. Ther.* **119**:361–369.

Shorey, J., McCandless, D. W., and Schenker, S., 1967, Cerebral -ketoglutarate in ammonia intoxication, *Gastroenterology* **53**:706–711.

Siesjo, B. K., and Nilsson, L., 1971, The influence of arterial hypoxemia upon labile phosphates and upon extracellular and intracellular lactate and pyruvate concentrations in the rat brain, *Scand. J. Clin. Lab. Invest.* **27**:83–96.

Silberman, R., 1958, Ammonia intoxication following ureterosigmoidostomy in a patient with liver disease, *Lancet* **2**:937–939.

Snodgrass, P. J., and De Long, G. R., 1976, Urea-cycle enzyme deficiencies and increased nitrogen load producing hyperammonemia in Reye's Syndrome, *N. Engl. J. Med.* **294**:855–860.

Steigmann, F., Kazemi, F., Dubin, A., and Kissane, J., 1963, Cerebrospinal fluid glutamine in the diagnosis of hepatic coma, *Am. J. Gastroenterol.* **40**:378–386.

Summerskill, W. H. J., and Wolpert, E., 1970, Ammonia metabolism in the gut, *Am. J. Clin. Nutr.* **23**:633–639.

Tews, J. K., and Stone, W. E., 1965, Free amino acids and related compounds in brain and other tissues: Effects of convulsant drugs, *Prog. Brain Res.* **16**:135–163.

Tews, J. K., Carter, S. H., and Stone, W. E., 1965, Chemical changes in the brain during insulin hypoglycaemia and recovery, *J. Neurochem.* **12**:679–693.

Thomson, A., and Holmes, A. W., 1967, Immune inhibition of urea breakdown in patients with cirrhosis, *Gastroenterology* **52**:14–17.

Thorn, W., and Heimann, J., 1958, The effect of anoxia, ischemia, asphyxia and reduced temperature on the ammonia level in the brain and other organs, *J. Neurochem.* **2**:166–177.

Van den Berg, C. J., and Van den Velden, J., 1970, The effect of methionine sulfoximine on the incorporation of labelled glucose, acetate phenylalanine and proline into glutamate and related amino acids in the brains of mice, *J. Neurochem.* **17**:985–991.

Van Caulaert, C., Deviller, C., and Halff, M., 1933, Troubles provoques par l'ingestion de sels ammoniacaux chez l'homme atteint de cirrhose de Laennec, *C. R. Seances Soc. Biol. Ses Fil.* **111**:739–740.

Vergara, F., Plum, F., and Duffy, T. E., 1974, Alpha-ketoglutaramate: Increased concentrations in the cerebrospinal fluid of patients in hepatic coma, *Science* **183**:81–83.

Walker, C. O., and Schenker, S., 1970, Pathogenesis of hepatic encephalopathy with special reference to the role of ammonia, *Am. J. Clin. Nutr.* **23**:619–632.

Walser, M., and Bordenlos, L. J., 1959, Urea metabolism in man, *J. Clin. Invest.* **38**:1617–1626.

Wasterlain, C. G., Lockwood, A. H., and Conn, M., 1978, Chronic inhibition of brain protein synthesis after portacaval shunting: A possible pathogenic mechanism in chronic hepatic encephalopathy in the rat, *Neurology* **28**:233–238.

Webster, L. T., and Gabuzda, G., Jr., 1958, Ammonium uptake by the extremities and brain in hepatic coma, *J. Clin. Invest.* **37**:414–424.

Zamora, A. J., Cavanaugh, J. B., and Kyu, M. H., 1973, Ultrastructural response of the astrocytes to protacaval anastomosis in the rat, *J. Neurol. Sci.* **18**:25–45.

Zeppa, R., 1972, New surgical approaches to portal hypertension, in: *Progress in Liver Diseases,* Volume 4 (H. Popper and F. Schaffner, eds.), Grune and Stratton, New York, pp. 289–299.

Zieve, L., and Olsen, R. L., 1977, Can hepatic coma be caused by a reduction of brain noradrenaline or dopamine?, *Gut* **18**:688–691.

Zieve, L., Doizaki, W. M., and Zieve, F., 1974a, Synergism between mercaptans and ammonia or fatty acids in the production of coma: A possible role for mercaptans in the pathogenesis of hepatic coma, *J. Lab. Clin. Med.* **83**:16–28.

Zieve, F., Zieve, L., Doizaki, W. M., and Gilsdorf, R. B., 1974b, Synergism between ammonia and fatty acids in the production of coma: Implications for hepatic coma, *J. Pharmacol. Exp. Ther.* **191**:10–16.

10

Bilirubin Encephalopathy

GERARD B. ODELL and HENRY S. SCHUTTA

1. Introduction—Human Bilirubin Encephalopathy

"Kernikterus" is the term coined by Schmorl (1903) to characterize the autopsy findings in six of 120 infants who had died with icterus gravis neonatorum and exhibited yellow staining of selected nuclear centers of the brain. Orth (1875) previously reported an example of similar yellow nuclear staining in a single infant, and Esch (1908) has been credited with the correlation of the antemortem clinical features with the postmortem findings of kernicterus, as it is now known (Gerrard, 1952).

Diamond *et al.* (1932) recognized that icterus gravis neonatorum was but one manifestation of the erythroblastosis fetalis syndrome to which they also included hydrops fetalis and anemia of the newborn within its clinical spectrum. These observations were made before the discovery of the Rh factor as a basis for these varied manifestations for isoimmune hemolytic disease of the newborn (Levine and Stetson, 1939; Landsteiner and Wiener, 1940).

The typical clinical manifestations of kernicterus are seen in deeply jaundiced infants after the first 48 hr of life. Such infants develop "lethargy" recognizable by an incomplete Moro response with extension but absence of flexion of their extremities in response either to abrupt postural changes or to a loud clapping noise. Often, only opisthotonic posturing occurs with nuchal extension of the head in response to such stimuli. Sucking responses to feeding are feeble, and the infant's cry in response to noxious stimuli is high-pitched (cerebral). In lethal instances, thermal instability is found with episodic periods of hypothermia and hyperthermia. Some of these infants have a fixed downward gaze ("setting sun") interrupted by bizarre rotary eye movements (oculogyric crises). Episodic decerebrate posturing with characteristic extension and pronation of the arms with fist clenching and opisthotonus occur in the

GERARD B. ODELL • Department of Pediatrics, University of Wisconsin School of Medicine, Clinical Sciences Center, Madison, Wisconsin 53792.   *HENRY S. SCHUTTA* • Department of Neurology, University of Wisconsin School of Medicine, Clinical Sciences Center, Madison, Wisconsin 53792.

end-stage of the disease. Terminally, pulmonary and gastric hemorrhaging is frequent (Vaughan *et al.*, 1950). At autopsy, the brains of such neurologically impaired infants that die during the first weeks of life while still jaundiced exhibit a diffuse yellow hue of the leptomeninges and surface of the cerebral cortex. The nuclear masses of the brain whose staining persists through the usual formalin fixation are listed in Table I (Claireaux, 1961; Haymaker *et al.*, 1961) and illustrated in Fig. 1. This particular infant died of kernicterus on the fourth day of life, and the brain was preserved by intracarotid perfusion of fixatives within 4 hr of death.

The histopathology ascribed to bilirubin injury antemortem is difficult to dissociate from postmortem changes secondary to the antecedent asphyxia and anoxia prior to death. Despite such limitations, the earliest histologic signs of bilirubin-induced encephalopathy are swelling and fine vacuolation of the cytoplasm of neurons and loss of their Nissl substance. The nucleus becomes eccentric, fuzzy in outline, and often the nucleoli are absent. In more advanced changes, the nuclei become hyperchromatic and later pyknotic, and the neurons eventually disappear (Claireaux, 1961). There is not always a close correlation between the macroscopic yellow staining of the nuclear areas and the severity of the neuronal lesions seen histologically (Vaughan *et al.*, 1950; Claireaux, 1959; Schutta *et al.*, 1970). Deposits of yellow pigment can be seen within neuronal cells and surrounding glial elements, but axons themselves are mostly spared. In contrast to asphyxia and hypoxia, the cerebral cortex does not often show significant histopathology.

This clinical and pathologic disorder was almost exclusively found during the neonatal period in association with hemolytic disease and has been previously ascribed to anemic hypoxia (Meriwether *et al.*, 1955), anoxia (Govan and Scott, 1953), direct injury on cerebral tissue by circulating maternal an-

TABLE I. Frequency of Distribution of Pigmentation in 35 Cases of Kernicterus[a]

Basal ganglia	32
Cerebellum	25
Hippocampus	24
Medulla oblongata	24
Subthalamic nuclei	19
Thalamus	18
Corpus striatum	18
Fourth ventricle	18
Dentate nucleus	17
Olivary nuclei	15
Nuclei in floor of fourth ventricle	15
Lentiform nucleus	14
Midbrain	11
Spinal cord	10
Globus pallidus	9
Ependyma	9
Pons	6
Corpora mamillaria	6
Caudate nucleus	6
Cerebral cortex	5

[a] From Claireaux (1961).

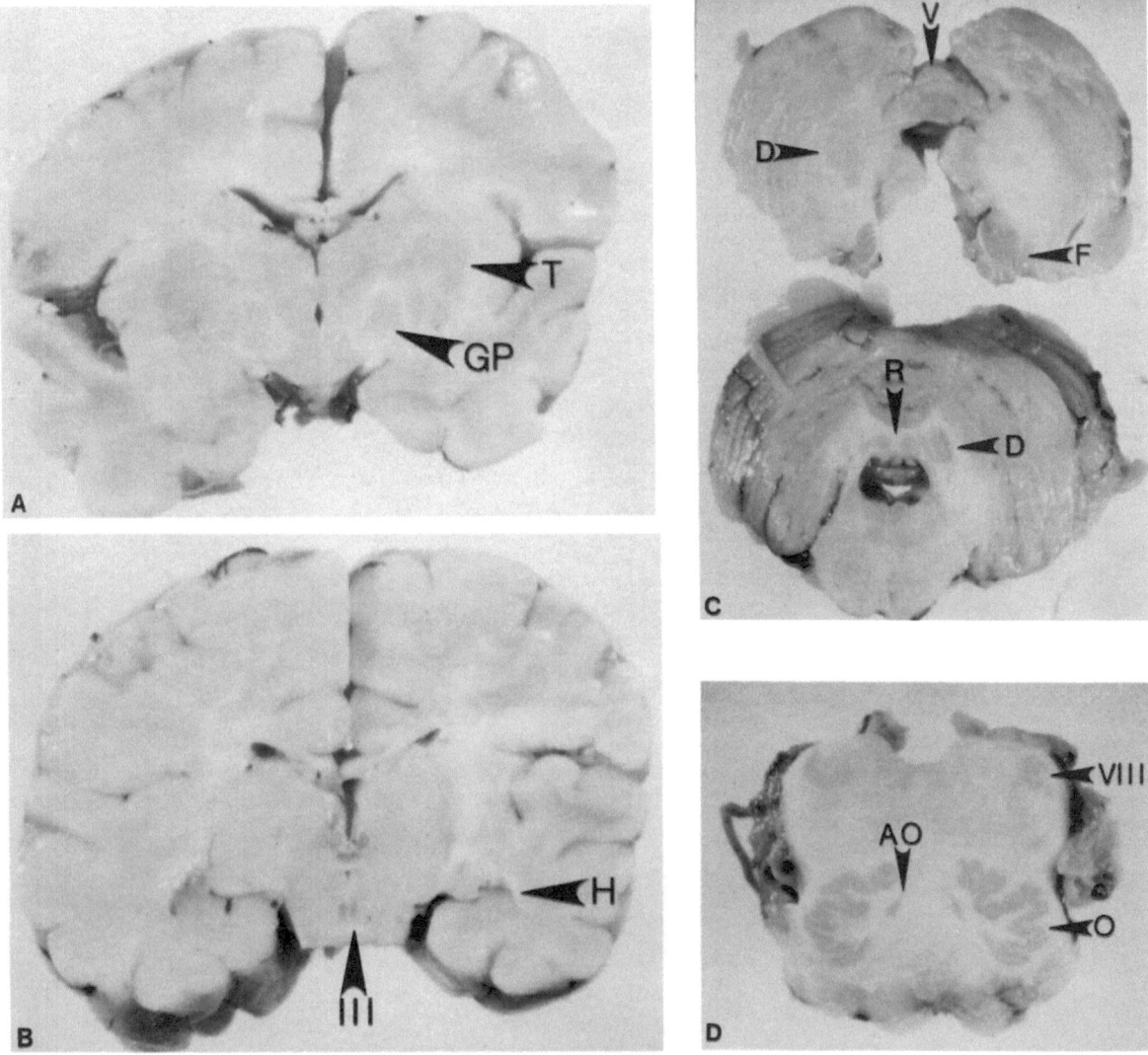

FIGURE 1. Coronal sections of the cerebrum and cerebellum and cross-sections of the brain stem from an untreated infant with Rh isoimmune hemolytic disease who died with hyperbilirubinemia at 45 hours of age. The brain was fixed *in situ* with buffered 10% formalin. **A:** Coronal section at the level of the thalamus, striatum, and pallidum. Yellow staining of the thalamus (T), striatum, and globus pallidus (GP) is evident. **B:** Coronal section of the cerebrum and midbrain. The oculomotor nuclei (III) and hippocampal formation (H) are yellow stained. **C:** Cross-sections through the cerebellum and pons. The flocculonodular lobe (F), vermis (V), dentate (D), and roof nuclei of the cerebellum (R) are intensely stained as well as the tegmentum of the pons. **D:** Cross-section through the medulla oblongata. There is intense yellow staining of the olivary nucleus (O) and the medial accessory olivary nucleus (AO). Staining is also seen in the region of the eighth nerve nuclei (VIII).

tibody (Vaughan, 1946), liver injury (Parsons, 1947), agglutinated cells plugging cerebral vessels (Wiener and Brody, 1946), hypoglycemia (Gerrard, 1952), hyperammonemia (Gorten et al., 1958), porphyrin intermediates of heme catabolism (Abelson and Boggs, 1956), and bacterial toxins from terminal septicemia (Zimmerman and Yannet, 1933; Vaughan, 1946).

The serious consideration that bilirubin itself was the toxic agent responsible for neuronal injury emerged when Claireaux et al. (1953) isolated nonconjugated bilirubin from the yellow-stained nuclear center of the brain of infants that died with kernicterus, and Hsia et al. (1952) as well as Mollison and Cutbush (1951) were able to correlate the level of serum bilirubin concentration and the risk of kernicterus in infants suffering from hemolytic disease despite correction of their anemia by exchange transfusion. Furthermore, the antemortem and postmortem diagnosis of kernicterus during neonatal jaundice in the absence of hemolytic disease was recognized in infants who died secondary to either *Escherichia coli* septicemia (Zimmerman and Yannet, 1933), prematurity (Zuelzer and Mudgett, 1950), or congenital familial nonhemolytic jaundice (Crigler and Najjar, 1952). Bilirubinemia was the only common denominator to account for kernicterus in this latter group of infants.

Those infants that survive the clinical manifestations of kernicterus in early neonatal life show gradual reduction of hypertonia and opisthotonic posturing that is associated with an improvement in their feeding. By 10 to 14 days of age, many infants whose jaundice had receded appear almost normal. However, beginning at 6 weeks to 3 months of age, the increased tone often reappears with varying degrees of athetosis. Many infants exhibit developmental delay with various degrees of mental retardation. Sensoriperceptual deafness has also been frequently noted (Perlstein, 1950; Gerrard, 1952; Byers et al., 1955). More subtle degrees of bilirubin encephalopathy associated with neonatal hyperbilirubinemia have been described in the absence of spasticity and athetosis in later childhood and consist of a lowered I.Q. (Day and Haines, 1954) and disturbances in cognitive functioning (Hyman et al., 1969; Odell et al., 1970a; Johnson and Boggs, 1974).

The autopsy observations on late survivors of the clinical manifestations of neonatal kernicterus are limited, but have shown atrophy with marked diminution of neuronal elements in many nuclear areas that are characteristically stained neonatally, as well as gliosis particularly in the globus pallidus, subthalamic nucleus, the hippocampal formation, the pons, and the tegmentum of the medulla oblongata. The cerebral cortex appears normal, but cerebellar atrophy has been recorded. The atrophic nuclei are no longer stained even in patients with familial nonhemolytic jaundice (Jervis, 1959; Haymaker et al., 1961; Gardner and Konigsmark, 1969). Furthermore, the clinical neurological manifestations of extrapyramidal motor disorders in the late survivors correlated well with the anatomical pathology.

The observations in these late survivors of severe neonatal hyperbilirubinemia due to hemolytic disease as well as the patients with familial nonhemolytic jaundice who manifest bilirubin encephalopathy in later infancy and childhood are pivotal to the recognition that bilirubin itself can be neurotoxic to the central nervous sytem and implies that certain neuronal elements of the CNS are particularly sensitive to bilirubin. During early infancy when such cellular elements are present, their affinity for bilirubin is manifest by the pigment staining, visible both macroscopically and microscopically. The subsequent necrobioses of these cells and reactive glial proliferation are associated with absence of yellow staining of the atrophic nuclear masses of the brain even though the hyperbilirubinemia is persistent in those patients with familial nonhemolytic jaundice whose hyperbilirubinemia is secondary only to their deficiency in hepatic glucuronyl transferase activity for the conjugation of bilirubin (Haymaker *et al.*, 1961). Thus, the presence or absence of yellow staining of the nuclear masses of the brain need not be prima facie evidence of bilirubin encephalopathy. This distinction is important, because recent postmortem observations in jaundiced premature infants with multiple antemortem hypoxic and asphyxial episodes and intracranial hemorrhage have exhibited yellow staining of nuclear masses of the brain and are called kernicterus—literally, yellow staining of brain nuclei (Turkel *et al.*, 1982, and contained references). Such infants do not often exhibit, antemortem, the neurologic signs of bilirubin encephalopathy described above, and the pathologic findings appear different because of the spongy vacuolated appearance of the parenchyma with intact neurons and greater evidence of gliosis and demyelination.

This dilemma in assigning the pathogenesis of the macroscopic observation of yellow-stained nuclear masses of the brain is not recent, for long ago, Beneke (1907) summarized the then-prevailing views concerning the pathogenesis of kernicterus as follows: (1) a peculiar attraction of bile pigments for ganglion cells resulting in their necrosis; (2) primary damage of ganglion cells by substances, such as bile salts, with secondary cellular pigmentation; and (3) ganglion cell damage by ischemia or trauma and subsequent pigmentation of these cells (quoted from Zimmerman and Yannet, 1933). Bilirubin staining can occur in an intact brain (primary), and this represents bilirubin encephalopathy; antecedent damage to the blood–brain barrier can also result in bilirubin staining of the brain (secondary). An early but informative example of this distinction was illustrated by the observations of Day (1947). Experimental rats were X-ray irradiated, but one half of the brain was lead shielded. The animals were subsequently infused with a bilirubin solution intravenously, and at autopsy the unshielded hemicortex of the rats' cerebrum exhibited yellow staining (kernicterus). More recently, comparable observations have been reported (Levine *et al.*, 1982) by temporarily increasing the porosity of the cerebral capillaries by unilateral infusion of hyperoncotic solutes into the carotid artery of experimental animals. Subsequent infusion of bilirubin into animals so treated resulted in diffuse bilirubin staining of the ipsilateral hemicortex normally perfused by the carotid circulation that had been previously infused with the hypertonic solutions. In these examples, the massive leakage of albumin across the cerebral capillary barrier was easily visible because of the bilirubin bound to the albumin. Although such observations have been called kernicterus, they do not represent bilirubin encephalopathy (Odell, 1982).

2. Animal Models of Bilirubin Encephalopathy

The best-known animal model of human kernicterus is the Gunn rat. This rodent is a mutant of the Wistar strain discovered by C. K. Gunn (1938). He described these albino rats (Gunn, 1944) as having lifelong unconjugated hyperbilirubinemia, and documented that their jaundice was inherited as an autosomal recessive characteristic. Carbone and Grodsky (1957) and Lathe and Walker (1958) demonstrated that liver homogenates of jaundiced homozygous animals could not form bilirubin glucuronide when bilirubin was substrate, and Schmid et al. (1958) demonstrated defects in glucuronide formation with multiple substrates both in vivo and in vitro. Those animals heterozygous for the conjugation defect in microsomal UDP-glucuronyl transferase activity are not jaundiced, but do show only 50% of the conjugating activity for bilirubin in vitro (Strebel and Odell, 1971). This juandiced animal is particularly valuable, because it frequently develops bilirubin encephalopathy comparable to humans (Blanc and Johnson, 1959), and its defect in bilirubin conjugation is similar to that seen in humans with familial nonhemolytic jaundice type I (Crigler and Najjar, 1952). The latter disorder has also been demonstrated to be of autosomal recessive inheritance (Childs et al., 1959; Odell and Childs, 1980), and many of these infants died during neonatal life with kernicterus (Haymaker et al., 1961). Johnson et al. (1959, 1961) documented the natural occurrence of kernicterus in this species and reproduced in vivo many of the clinical circumstances that are believed either to promote or decrease the likelihood of kernicterus in human neonates. Of particular importance, these animals provided the opportunity to apply microscopic study to the earliest changes in morphology, as well as biochemical measurements associated with the pathophysiology of kernicterus.

The homozygous animal (jj) is not jaundiced at birth if born of a heterozygous female (Jj), but does become visibly icteric after the first day of life. Their bilirubin concentrations rise to 5 to 10 mg/dl by the seventh day and reach their greatest levels of 12–16 mg/dl during the second and third weeks of life. By four weeks of age, the animal's bilirubin concentration is comparable to adult levels, 6–8 mg/dl.

The earliest clinical signs of neurologic impairment in these animals are seen at 10–12 days of age when they develop weakness of the hindlimbs and a wobbly gait. Such animals usually have been noted to have higher serum bilirubin concentrations than littermates. In some, the ataxia progresses with consequent poor feeding, failure of weight gain, hypotonia mixed with erratic hyperactivity, convulsions, and finally death. The greatest incidence of fatality occurs in the third week of life. Symptomatic animals sacrificed during the second week of life exhibit characteristic bilirubin staining of nuclear centers of the brain. The basal ganglia, thalamus, hippocampus, vermis of the cerebellum, anterior and posterior colliculi, flocculus and nodulus of the cerebellum and dentate, and roof nuclei show intense bilirubin staining. The third and eighth nuclear complexes are also prominently stained. After the third week of life, many animals fail to show macroscopic staining at autopsy despite the neurologic symptoms and continuing hyperbilirubinemia of the animal (Johnson et al., 1961; Schutta and Johnson, 1967). This observation parallels that previously noted in humans with late-onset kernicterus with familial nonhemolytic jaundice (see Section 1).

The earliest histologic signs of injury are most predictively found in the Purkinje cells of the cerebellum which by light microscopy exhibit the accumulation of osmiophilic granules. Electron microscopy reveals the osmiophilic granules to be membranous whorls within the cytoplasm and, in addition, there occur striking alterations in the mitochondria (Schutta and Johnson, 1967; Schutta et al., 1970). The latter organelles become enlarged, bizarre-shaped and contain glycogen-laden vacuoles. Morphologic abnormalities are also found in the neurons of the hippocampus, basal ganglia, the brain stem, and rarely the cerebral cortex. These morphologic changes have been confirmed (Karp et al., 1982) and described in detail in other nuclear centers; e.g., the cochlear nucleus (Jew and Williams, 1977) and the substantia nigra (Batty and Millhouse, 1976). Such changes have been observed in vitro in myelinating cerebellum cultures on exposure to bilirubin (Silberberg et al., 1970a,b).

The early and fully developed Purkinje cell lesions are illustrated in Fig. 2, and the end-stage loss in cerebellar mass of affected jaundiced rats is reproduced in Fig. 3. This striking absence of cerebellar mass has been ascribed by one group of investigators as a consequence of hypoplasia with its onset at 10 days of age (Sawasaki et al., 1976) rather than a degenerative atrophy. It is noteworthy that in the Gunn rat not all nuclear centers that exhibit yellow staining show evidence histologically of neuronal damage; e.g., gasserian ganglia did not show fine structure abnormalities even though significantly discolored with bilirubin.

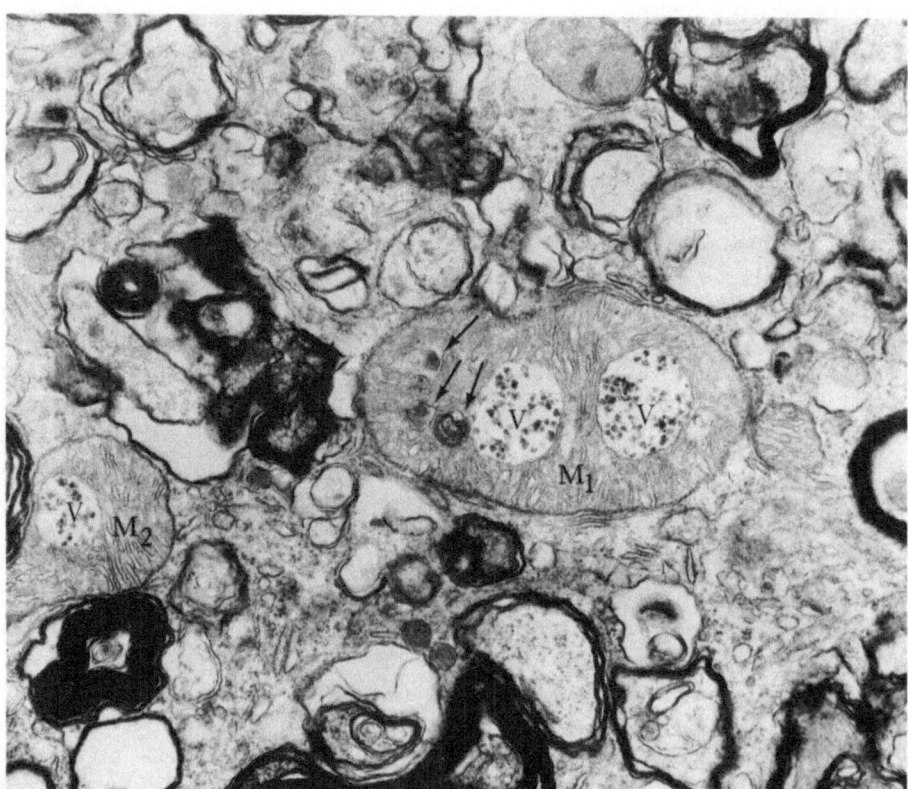

FIGURE 2. Enlarged mitochondria (M_1, M_2) in a Purkinje cell profile that also show cytoplasmic membranous inclusions. The mitochondria also exhibit vesicles (V) filled with glycogen granules and myelin bodies (arrows); × 29,000. Reproduced by permission from Schutta et al. (1970).

Attempts to reproduce kernicterus in the restricted sense that required yellow staining of brain nuclei in the experimental animals have been for the most part disappointing. In order to produce macroscopic bilirubin staining of nuclear centers of the brain characteristic of kernicterus in neonatal life, an additional insult to the animal species in conjunction with systematic infusion of bilirubin has been necessary. Such insults have included osmotic shocking of the blood–encephalon barrier in rats (Levine *et al.*, 1982), severe hypoxia in primate monkeys (Lucey *et al.*, 1964), hypoglycemia and/or hypercapnia in puppies (Rozdilsky and Olszewski, 1961; Lending *et al.*, 1966), and asphyxia in rabbits (Chen *et al.*, 1965). It is only in the newborn kitten that systemic bilirubin infusion per se could produce the characteristic neurologic signs and bilirubin staining of the basal ganglia (Rozdilsky, 1966). However, if one accepts neurologic signs characteristic of bilirubin encephalopathy, such as sluggishness and hyperreactivity to handling as well as opisthotonus and seizures, then many animal models exhibit clinical signs compatible with bilirubin encephalopathy after acute exposure to bilirubin administered systemically. Again, the macroscopic visibility of yellow-stained nuclear centers of the brain (kernicterus) and the neuronal injury caused by bilirubin (encephalopathy) requires distinction and disassociation.

3. The Transport of Bilirubin in the Circulation and Its Mode of Entry into the Central Nervous System

The means by which bilirubin gains access to the CNS in order to produce an encephalopathy requires an understanding of its transport within the systemic circulation. Bilirubin is normally bound to serum albumin which serves

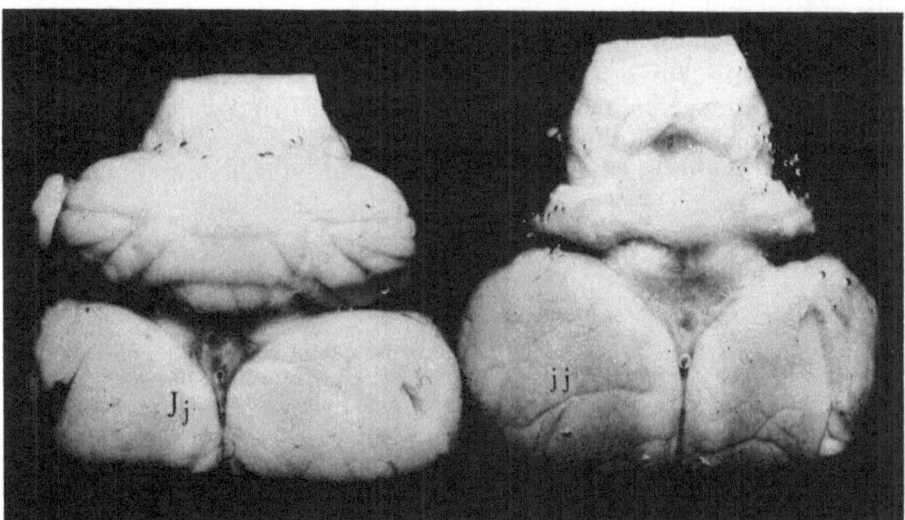

FIGURE 3. Brains from heterozygous (Jj) and homozygous (jj) Gunn rats aged 3 months. The jj animal was severely ataxic and exhibits the marked cerebellar atropy; × 3.6. Reproduced by permission from Schutta and Johnson (1967).

as its vehicle in the translocation of bilirubin from its sites of formation to its organ of excretion, the liver. The binding affinity of human albumin for bilirubin is very high, $K_a = 10^8$ M^{-1} (Jacobsen, 1969). At birth, the average concentration of bilirubin and albumin is 1.0 mg and 3.0 g/dl, respectively (1.7 $\times$ 10^{-5} M bilirubin and 4.4 $\times$ 10^{-4} M albumin). Given the affinity constant, one can calculate that the aqueous phase concentration of bilirubin dissociated from albumin is 5.8 $\times$ 10^{-10} M. Under such circumstances, very little bilirubin exists in the aqueous phases of body fluids and delivery of bilirubin to the liver is almost quantitative (Odell, 1980). However, during neonatal hyperbilirubinemia, at which time the serum bilirubin concentrations frequently reach 20 mg/dl (3.4 $\times$ 10^{-4} M), it becomes equimolar with that of albumin (4.4 $\times$ 10^{-4} M). The corresponding calculated concentration of dissociated bilirubin increases almost 200-fold (i.e., 4.95 $\times$ 10^{-8} M vs. 5.8 $\times$ 10^{-10} M) (Odell, 1980). Thus, the opportunity for bilirubin to filter across capillary beds and be sequestered in extravascular tissues is enhanced. Furthermore, under physiologic circumstances, not every molecule of albumin within the circulation is able to bind bilirubin at its primary affinity site (Odell *et al.*, 1977), and secondary binding sites on albumin of lower affinity ($K_a = 10^6$ M^{-1}) (Jacobsen, 1969) also serve for the transport of bilirubin in the circulation.

The circumstances of bilirubin transport are schematically shown in Fig. 4. As illustrated in this model, bilirubin bound to albumin at its high-affinity site is depicted at the right-hand side of the molecule and is relatively resistant to displacement by other more water-soluble organic anions (A). The high affinity for binding of bilirubin is dependent on constituent histidines (Odell *et al.*, 1970b; Jacobsen, 1972) and lysine residue, 240, in human albumin (Jacobsen, 1978). The binding is independent of tryptophan (Chen, 1974) even though bilirubin, when bound to albumin, quenches the fluorescence of the latter amino acid. Likewise, binding does not involve the binding site of the first 2 moles of fatty acids depicted as the V-notched domains, numbered 1 and 2, at the north and south poles of the albumin molecule in the figure (Berde *et al.*, 1979). When bilirubin is bound at its primary site, it exhibits characteristic circular dichroism and optical rotary dispersion (Blauer *et al.*, 1975) and emission fluorescence (McCluskey *et al.*, 1975; Lee and Gillespie, 1981). Brodersen (1979a) recently published an excellent review of bilirubin binding to which the interested reader is recommended for further discussion.

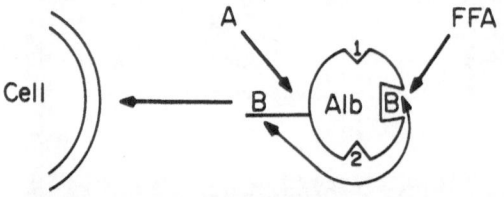

FIGURE 4. The transport of bilirubin in extracellular fluid. The albumin (Alb) molecule contains a single high-affinity binding site for bilirubin (B) indicated as the indented keyhole at the right, and lower affinity sites illustrated as the projecting horizontal bar at the left of the protein molecule. The V-notched areas at the north and south poles of the molecule represent the two (1,2) primary binding sites for free fatty acids (FFA). When these sites are occupied, additional FFA will compete with bilirubin for binding at its primary binding site forcing its transport at secondary sites indicated by the reversible arrow. Bilirubin bound at its secondary sites is subject to competitive displacement by organic anions (A) such as sulfonamides which thereby promote its diffusion into cell membranes shown at the left of the figure.

During neonatal jaundice, the plasma concentrations of bilirubin are increased and the transport function of albumin may require that secondary binding sites of albumin be utilized even at molar concentrations less than 1:1 as illustrated on the left-hand side of the albumin molecule. An example, as indicated in the figure, is the circumstance when the primary sites for binding of free fatty acids (FFA) are filled. Any additional FFA will compete with the albumin binding of bilirubin at its primary binding site. Such circumstances are not unusual, because at serum albumin concentrations of 0.44 mM (3.0 g/dl), FFA concentrations in excess of 1.0 mM are frequently encountered (Odell *et al.*, 1977) during neonatal life. Thus, a considerable fraction of circulating serum bilirubin during neonatal jaundice may be transported by albumin with it bound to secondary binding sites and governed by an affinity constant of 10^6 M^{-1} or less. This implies an even greater concentration of bilirubin (5×10^{-8} M) exists in the aqueous phases of plasma water dissociated from albumin.

As the reciprocal arrows in Fig. 4 indicate, the amount of bilirubin transported by albumin at its primary and secondary sites for any given total bilirubin concentration can be quite variable and depends in part on the FFA concentration. Reduction of elevated FFA concentrations in plasma can reverse the dependency of bilirubin binding at secondary sites (Odell *et al.*, 1977; Ostrea *et al.*, 1983) and thereby reduce the filterable aqueous-phase concentrations of bilirubin.

Importantly, bilirubin molecules bound to these secondary sites are more readily displaced by albumin-bound organic anions such as sulfonamides and

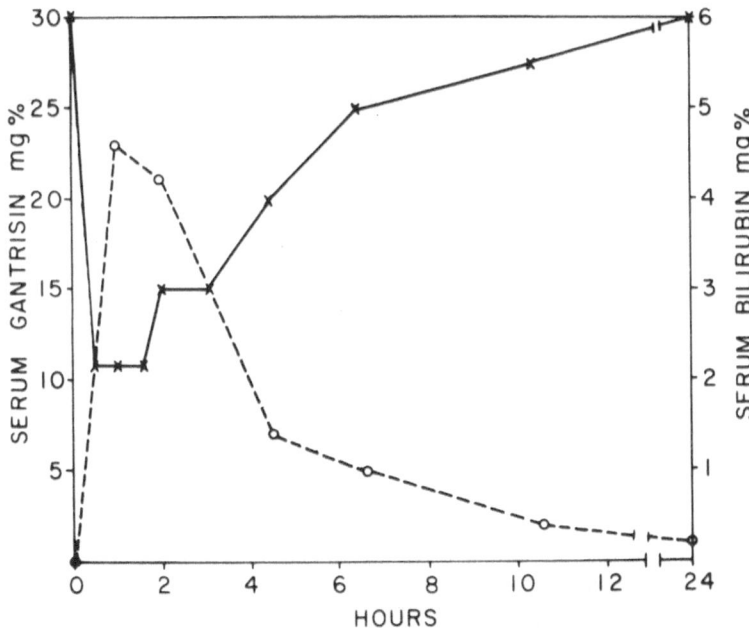

FIGURE 5. The reciprocal concentration relationship between the serum bilirubin (———) and sulfonamide in a homozygous (jj) Gunn rat injected with 200 mg/kg of gantrisin (------) subcutaneously. Reproduced by permission from Johnson *et al.* (1959).

salicylates (A) that would further increase the aqueous-phase concentration of bilirubin available for filtration across capillary beds (Brodersen, 1974).

The application of these principles of albumin binding of bilirubin to bilirubin encephalopathy help explain, clinically, the high incidence of kernicterus seen at postmortem in premature infants who had received prophylactic therapy with the sulfonamide, sulfisoxazole (Silverman *et al.*, 1956). These infants at death actually had lower serum bilirubin concentrations than concurrent control infants who had not received sulfonamides (Harris *et al.*, 1958). Such a paradox was explicable by demonstration of the displacement of bilirubin from albumin and its ultrafilterability *in vitro* (Odell, 1959a,b) and its reproduction in the experimental jaundiced Gunn rat (Johnson *et al.*, 1959). Figure 5 illustrates the reciprocal relationship between the serum bilirubin and

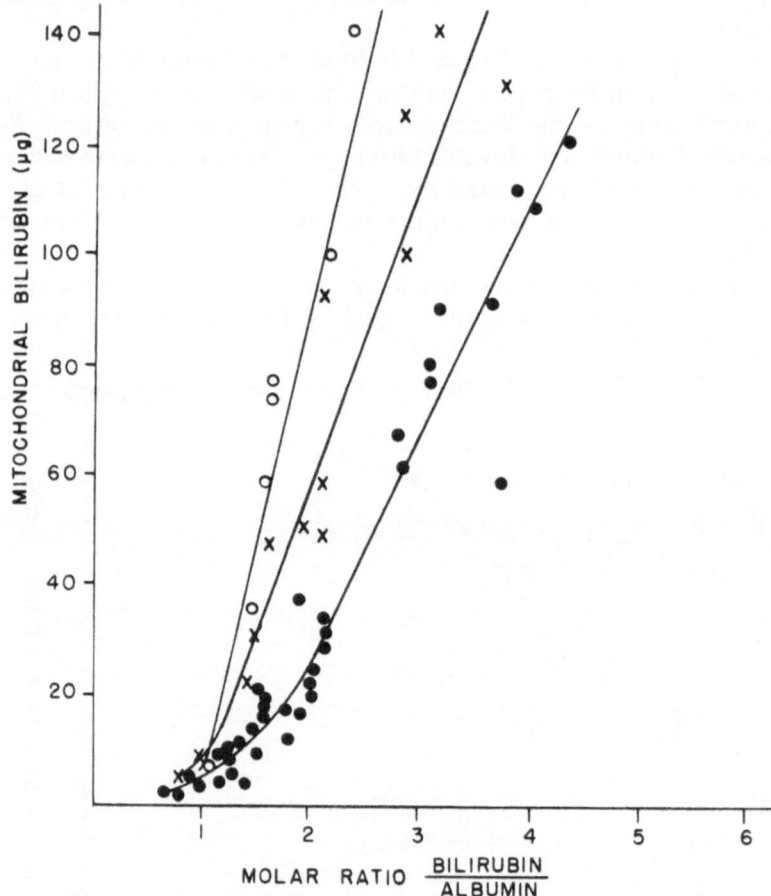

FIGURE 6. The distribution of bilirubin between albumin and mitochondria suspended in a common aqueous media isotonic to plasma to pH 7.37. The bilirubin concentration was 0.1 mM (250 µg) in all suspensions and the albumin concentration ranged between 0.02–0.2 mM to provide the molar ratios depicted on the abcissa. The recoveries of bilirubin in the mitochondrial phase of the suspensions is indicated on the ordinate. (●) represent the recoveries in the absence of added organic anions, (X) and (O) are the recoveries from suspensions which included 0.5 mM salicylate and oleate respectively. Reproduced by permission from Odell (1966).

sulfonamide concentrations in jaundiced rats injected with sulfonamide. Such abrupt displacement of the circulating bilirubin led to its ultrafililtration from the intravascular space and has been repeatedly shown to exacerbate the sub-acute encephalopathy of the Gunn rat and to induce acute neuronal necrosis (Schutta and Johnson, 1969, 1971; Rose and Wisniewski, 1979).

This displacement of bilirubin from albumin by organic anions has been often demonstrated *in vitro*, as illustrated in Fig. 6 in which isolated liver mitochondria were suspended in a common isotonic medium containing bi-lirubin–albumin solutions at various molar ratios. To be noted, relatively little bilirubin was recovered in the mitochondrial organelles until the molar ratio of bilirubin to human albumin exceeded 1.0. Substitution of erythrocytes (Bra-tlid, 1972) or isolated brain capillaries (Katoh-Semba and Kashiwamata, 1980) for the suspended mitochondria yield essentially the same results. Substitution of bovine or rat albumin for human albumin shifted the distribution curves to the left with reference to the abscissa, indicating their affinity for bilirubin is less resistant to displacement by organic anions than that of human albumin (Odell, 1966). Such observations indicate that rodent species—e.g., the Gunn rat—might be more susceptible to the inducement of bilirubin encephalopathy than humans. The important aspect of these considerations is the relative bind-ing efficacy and capacity of albumin for bilirubin will predict the concentration to which one must saturate the circulating albumin before sufficient bilirubin exists within the filterable phases of extracellular fluid reaching the CNS (Odell, 1974; Wells *et al.*, 1982).

4. The Aqueous Solubility and Behavior of Bilirubin

The aqueous solubility of bilirubin is very limited and pH sensitive be-tween pH 7.0 and 8.0 (Overbeek *et al.*, 1955). Brodersen (1979b) found at pH 7.4, ionic strength 0.15 M, that its solubility was 7–10 nM by dissolution from solid-state bilirubin. Thus, when referring, as above, to aqueous-phase con-centrations of bilirubin dissociated from albumin *in vivo* during neonatal jaun-dice which exceed 5 μM, free bilirubin must either exist in a supersaturated state or be sequestered from the aqueous phases of body fluids. Stable aqueous concentrations as high as 86 μM have been observed *in vitro* (Burnstine and Schmid, 1961). Comparable solubility studies with testosterone (Bischoff and Stauffer, 1954) reported its aqueous solubility to be much greater when equi-librated with bovine albumin even though the latter was separated by a dialysis casing. It is also well known that cholesterol often exists in a metastable su-persaturated state *in vivo* (Carey, 1982). Thus, the existence of poorly soluble solutes in body fluids beyond their *in vitro* aqueous solubility is not unique to bilirubin. Brodersen (1972) noted *in vitro* that dimerization and colloidal aggregation of bilirubin do occur and can account for some of the solubilities of bilirubin in aqueous solutions that far exceed 10 nM at pH values less than 7.4.

The limitations of the aqueous solubility and reactivity of bilirubin IXα become more predictable from its various molecular configurations recently discovered. X-ray crystallography by Bonnet *et al.* (1976) demonstrated that bilirubin has a Z configuration at its C4–C5 and C15–C16 meso bonds similar to its precursor heme. Because of this configuration and the saturated carbon at C10, it is able to twist into a conformation that allows intramolecular hydrogen bonding, as illustrated in Fig. 7. Kuenzle *et al.* (1973) deduced similar intramolecular hydrogen bonding in which the polar groups are masked and can account in part for its limited aqueous solubility. However, Fig. 7 illustrates the fully protonated molecule and, as argued by Lee *et al.* (1974) and Hansen *et al.* (1979), at physiologic pH in aqueous media one might expect the propionic acid side chains to be dissociated and the molecule to be more water soluble. The latter investigators were able to demonstrate in nonaqueous media two pKa of about 4.5 in titrations of bilirubin that would correspond to the carboxyl groups of the propionic acid side chains and reaffirmed that bilirubin in alkaline media exists primarily as the dianion (Brodersen, 1979b). Mugnoli *et al.* (1978), by X-ray crystallography, demonstrated the diisopropyl ammonium salt of bilirubin retained the same twisted configuration with hydrogen bonding between the carbonyl oxygens and the spacially opposed dipyrrole nitrogens. This form of the dianion of bilirubin in aqueous media is thought to predominate and be in equilibrium with bilirubin bound to albumin (Brodersen, 1979a).

The titration of bilirubin in alkaline aqueous media is difficult to conduct, because the titration curve of aqueous bilirubin below pH 8.0 is associated with abrupt changes in the aqueous solubility of the bilirubin dianion (Brodersen, 1979b). Its solubility becomes so poor that aggregation occurs without consumption of additional protons, and measurements below pH 7.0 are beyond current laboratory technology. The changes in the molecule at pH 8.0 and below that severely limit its solubility are unknown (Schmid, 1978) and may be essential to the understanding of the toxicity of bilirubin at the cellular level.

The critical importance of environmental hydrogen ion concentration on the distribution of bilirubin between extracellular fluid and cells is illustrated in Fig. 8. These mitochondrial suspensions were in a common aqueous medium at frequently encountered extracellular concentrations of bilirubin and albumin *in vivo*. From such "simple" studies, it is apparent that although the total concentrations of bilirubin within the suspension remain the same, the distribution of bilirubin between the intracellular organelle and the extracellular albumin with which it is in equilibrium are very dependent on the environmental pH. Figure 8A illustrates a significant shift of bilirubin from the albumin to the mitochondrial phase of the equilibration system that does not occur until the molar ratio of bilirubin-to-albumin exceeds 1:1. Substitution of red blood

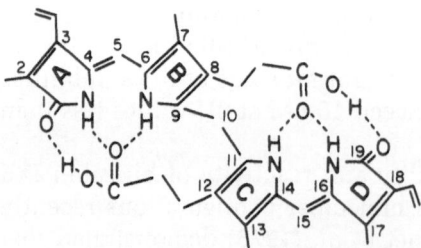

FIGURE 7. The Z-Z structure of bilirubin IXα illustrating the extensive H-bonding (·······) that masks its polar groups and help to account for its poor solubility in aqueous media at pH 7.4.

cells (Bratlid, 1972) or brain capillaries (Katoh-Semba and Kashiwamata, 1980) results in similar distributions of the bilirubin in the suspensions. However, the coexistence of low pH and organic anions that can compete with bilirubin for protein binding to albumin in the suspension media (Fig. 8B) illustrate that the translocation of bilirubin from the extracellular phase of the suspension to the intracellular membranous mitochondrial organelles occurs at higher hydrogen ion concentrations below the presaturation levels of albumin with bilirubin of 1:1 molar ratio. Since hydrogen ion concentration per se within the pH range of 5–9 does not alter the affinity constant of bilirubin dianion with albumin (Jacobsen and Brodersen, 1976), one must infer either the aqueous configuration of bilirubin is altered from its dianionic state that is in equilibrium with albumin or the cellular organelle has a greater affinity for the bilirubin dianion at the higher environmental hydrogen ion concentrations. In either instance, dianionic bilirubin is sequestered from the aqueous phase and

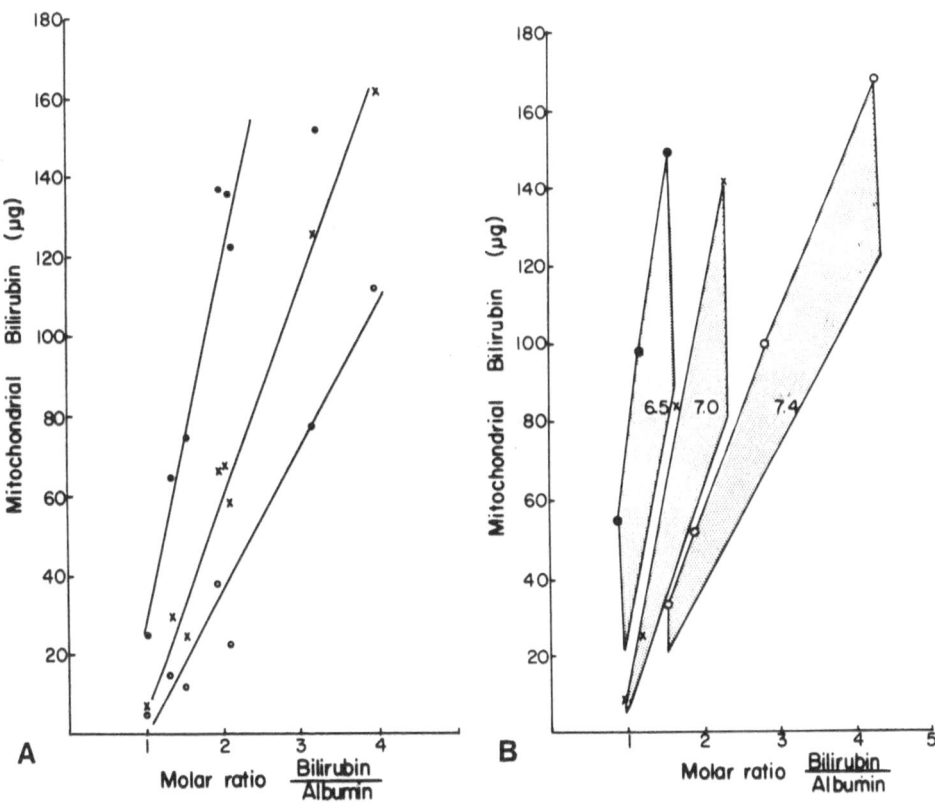

FIGURE 8. The influence of pH on the distribution of bilirubin between albumin and mitochondria suspended in a common aqueous media. A: The compositions of the suspension media are as in Fig. 6 except they were buffered at pH 6.5 (●), 7.0 (X), and 7.4 (○) and did not contain added organic anions. B: Suspensions similar to A, but include the addition of 0.5 mM sodium salicylate. The right-hand base of each quadrilateral are the results from A due to pH alone, whereas the plotted points (●, X, ○) reflect the additive influence or organic anions and hydrogen ion concentration on the distribution of bilirubin in the suspensions. Reproduced by permission from Odell (1965).

more bilirubin dissociates from albumin to restore its equilibrium with the depleted aqueous-phase concentration of the dianion. The redistribution of bilirubin from extracellular albumin to intracellular tissue associated with increased hydrogen ion concentrations of the extracellular fluid was dramatically demonstrated by the toxicity studies of Silberberg *et al.* (1970a) in myelinating cerebellar neurons in tissue culture. They found neuronal toxicity occurred when bilirubin-to-albumin concentration ratios were less than 1:1 if environmental pH was below 7.4. Of equal importance, if the pH of the media was maintained above pH 7.6, toxicity did not occur even though the molar ratio of bilirubin-to-albumin exceeded 1:1. Some investigators have proposed (Nel-

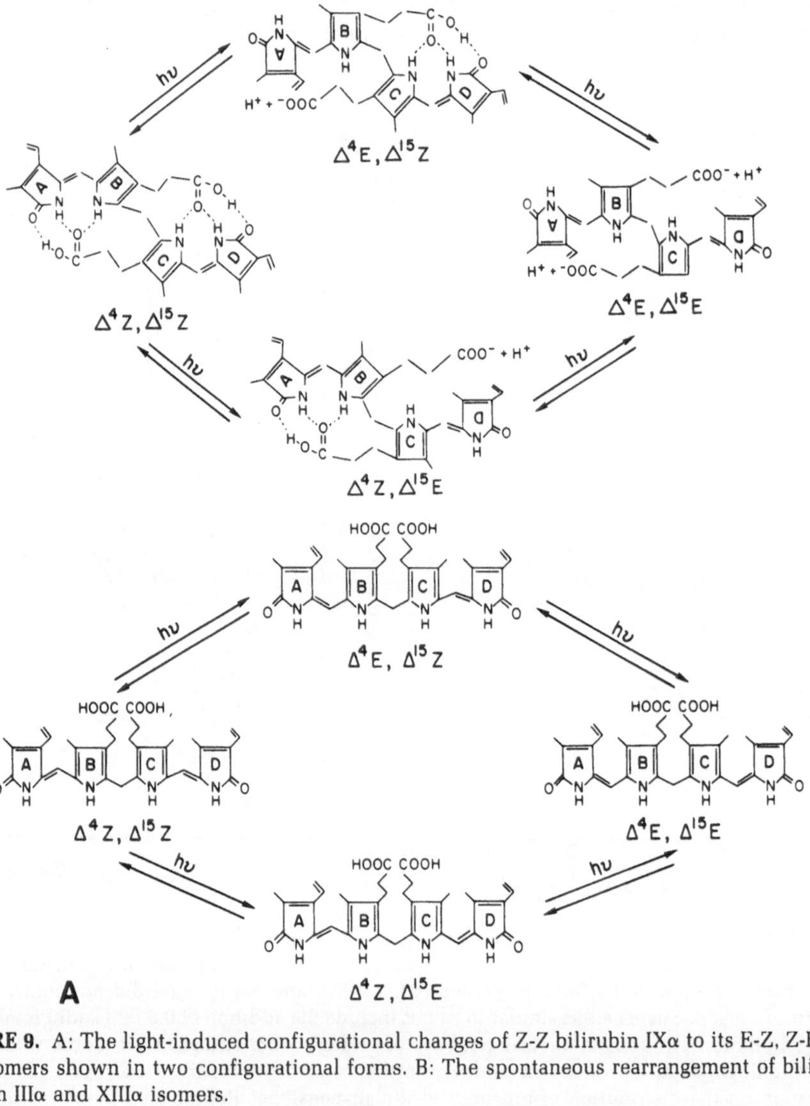

FIGURE 9. A: The light-induced configurational changes of Z-Z bilirubin IXα to its E-Z, Z-E, and E-E isomers shown in two configurational forms. B: The spontaneous rearrangement of bilirubin to form IIIα and XIIIα isomers.

son *et al.*, 1974; Brodersen, 1979a) that cell membranes have a greater affinity for dianionic bilirubin at higher environmental hydrogen ion concentrations and sequester it into their lipid membranes. Associated with such sequestration, the dianionic bilirubin becomes protonated, aggregates and precipitates as the insoluble bilirubin acid (Erikson *et al.*, 1981) and they suggest it is the latter that represents the toxic form of bilirubin.

The recent studies of the effects of light on bilirubin illustrate the complexities of this molecule and its behavior in aqueous solution. Figure 9 indicates that photons of the appropriate energy level can induce a 180° rotation of the A or D rings of bilirubin in its Z–Z configuration and form 4E,15Z or 4Z,15E bilirubin or even 4E,15E bilirubin (Brown and McDonagh, 1980, and included references). Each of these photoisomers, although unstable, are configurational isomers of bilirubin but are more water-soluble, rapidly excreted by the liver without prior conjugation and apparently nontoxic to tissues (Cohen and Ostrow, 1980). These photoisomers bind differently to albumin (Tran and Beddard, 1982). Such photoisomers, because of the induced rotations of the A and D rings of bilirubin, can no longer form hydrogen bonds between the pyrrole nitrogens and the carbonyl oxygens and further expose these hydrophylic groups to the aqueous media. Thus, their increased water solubility as charged anions appears more evident, as well as their lack of lipid solubility. Given these configurations and their separation by chromatographic techniques (McDonagh *et al.*, 1982a,b), one cannot but wonder if some other configurational change occurs in the bilirubin dianion between pH 8.0 and pH 7.0 in aqueous solution that might account for its sudden insolubility in aqueous media. Such speculation and inquiry seem necessary since current data are conflicting as to whether Z–Z bilirubin acid, its dianion, or some other confirmational isomer is the toxic formation responsible for bilirubin encephalopathy.

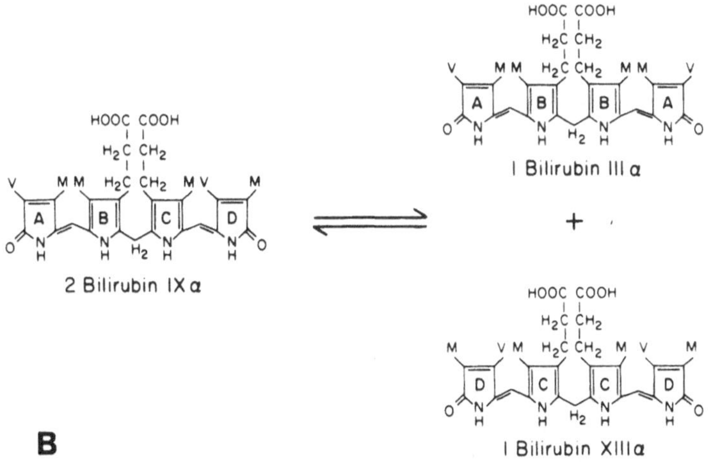

FIGURE 9. (*continued*)

5. The Susceptibility of Neonatal Central Nervous System Tissue to Bilirubin Toxicity

There has been much conjecture concerning the apparent selective susceptibility of the neonatal infant or animal to bilirubin encephalopathy. Early reports ascribed this to an "immaturity" of the blood–brain barrier (Lending et al., 1966; Diamond, 1969; Rapaport, 1976, with included references), but no convincing evidence has even been provided to demonstrate that the capillary–encephalon barrier during neonatal life is more permeable either to bilirubin or albumin-bound bilirubin than in adult animals (Diamond, 1969). The greater oil/water partition coefficient of bilirubin and its inherent molecular flexibility to either expose or mask its polar groups indicates unconjugated Z–Z bilirubin would have little difficulty crossing the lipid membranes of capillary endothelial cells and gain access to the central nervous system. The importance of the studies of Diamond and Schmid (1966) in the experimental rat was the demonstration that bilirubin dissociated from albumin within the systemic circulation could readily enter the CNS independent of the maturity of the animal. Thus, the barrier function of the cerebral capillary does not appear critical for the exclusion of bilirubin from the brain. What may be of greater importance is the rapidity by which bilirubin filtered into the CNS can be cleared from interstitial fluid and adjacent CSF of the CNS. The studies of Davson et al. (1962) and Pappenheimer (1961, 1962) demonstrated the rapid clearance of anions from intraventricular fluid. Organic anions such as phenolsulfonphthalein (PSP) and para-aminohippuric acid (PAH) are rapidly excreted from the CSF and are thereby prevented from accumulating within the extracellular fluid environment of neuronal tissue. In this regard, the studies of Ernster et al. (1957) are of pertinence, for they found that intracisternal instillation of the sulfhydryl poison, p-chloromercuribenzoate, resulted in a much greater concentration of bilirubin within CSF after systemic infusion of bilirubin in young rabbits and was associated with yellow staining of nuclear centers of the brain. The apparent interference either with the formation of CSF and/or clearance function by the choroid plexus after the p-chloromercuribenzoate exposure allowed greater concentrations of bilirubin to accumulate within the central nervous system than normally occurs. Although never demonstrated, one suspects that the choroid plexus participates in the clearance of CSF of bilirubin much as it does with such organic anions as PSP and PAH. The intense bilirubin staining of the choroid plexus observed by Rozdilsky and Olszewski (1960) after systemic infusion of bilirubin in rabbits might as readily be interpreted as net accumulation of bilirubin reabsorbed from the CSF as by bilirubin filtered from the systemic circulation across the choroid vasculature.

Another functional aspect of the susceptibility of brain tissue during neonatal life to bilirubin toxicity is related to the so-called "sink function" of CSF—i.e., the bulk removal of filtered and endogenously formed solutes found in the CSF back into the systemic circulation. Davson et al. (1962) noted that removal of solutes from CSF was not only a function of the rate of formation of CSF by a choroid plexus, but also its rate of removal across the CSF–blood barrier either by solute-selective transport via the choroid plexus or by bulk filtration across the arachnoid villi into the sagittal and lateral sinuses that constitute a part of the venous drainage from the brain.

The extracellular fluid space of the central nervous system in neonatal life is considerably greater than in adulthood (Aprison and Segar, 1963; Ferguson and Woodbury, 1969), and consequently its contained solutes are less rapidly cleared, particularly since the rate of formation of CSF is reduced (Bass and Lundborg, 1973). The clearance of solutes filtered across the capillary endothelium of the brain into the relatively expanded extracellular space of the neonatal central nervous system tissue will be delayed because of the larger fluid volume into which they enter, and the opportunity for their concentration to increase is enhanced because of the inefficient "sink function" during neonatal life (Davson et al., 1962).

An additional, but untested, speculation for the potentially greater susceptibility to bilirubin toxicity during neonatal life is that the target neurons may be less protected after bilirubin filters across the capillary bed of the encephalon. In the mature CNS, bilirubin, in order to reach the target cell, may have to pass through multiple plasma membrane and cytoplasmic barriers because of the restricted extraneuronal fluid space. By contrast in the neonatal brain, the extraneuronal fluid space between the capillary ultrafiltrate and the susceptible target neurons may be in immediate continuity. Such a suggestion derives credence from the observed greater diffusion of trypan blue in the brains of young animals (Rapaport, 1976), even though the site of entry of the dye is at the common nonbarrier sites of the central nervous system. An example of the sensitivity of neonatal infants to bilirubin encephalopathy is illustrated in Table II in which the distribution of bilirubin among albumin-containing body fluids is illustrated in a hypothetical jaundiced 2-kg infant.

The infant is 5 days old, and the plasma bilirubin concentration has reached a level of 25 mg/dl. Since the hyperbilirubinemia has developed over five days duration, one may presume a steady state between the albumin spaces with reference to the distribution of bilirubin. The concentrations of bilirubin in plasma and respective extravascular spaces are illustrated, and the molar

TABLE II. Bilirubin/Albumin Concentrations and Distribution in a 2-kg Infant Aged 5 Days

Concentrations		Plasma (100 ml)	Systemic interstitial fluid (400 ml)	Intracranial extracellular fluid (50 ml)
Bil.	mg %	25.0	5.8	0.58
Alb.	g	3.0	0.7	0.07
	$\dfrac{mM}{mM}$	$\dfrac{0.43}{0.43} = 1.0$	$\dfrac{0.1}{0.1} = 1.0$	$\dfrac{0.01}{0.01} = 1.0$
Injection of Sulfonamide or acute acidemia				
		(25 − 5)	(5.8 + 0.94)	(0.58 + 2.50)
Bil.	mg %	20.0	6.74	3.08
Alb.	g	3.0	0.7	0.07
	$\dfrac{mM}{mM}$	$\dfrac{0.34}{0.43} = 0.8$	$\dfrac{0.115}{0.10} = 1.15$	$\dfrac{0.053}{0.01} = 5.3$

concentration ratios of bilirubin:albumin are given in columns 2, 4, and 6 of the table.

Now suppose that an antibiotic such as sulfonamide is injected intravenously which displaces bilirubin from the circulating albumin or a hypoxic episode with atelectasis occurs associated with a sharp reduction in blood pH and lactic acid acidemia. In either instance, bilirubin becomes dissociated from the circulating plasma albumin and its total plasma concentration will abruptly drop to 20 mg/dl without significant change in the albumin concentration. The dissociated bilirubin (5 mg) will filter across the nearest capillary bed. Since 25% of cardiac output enters the cerebral circulation, 1.25 mg (5 mg $\times$ 0.25) will enter the CNS, whereas the remaining 3.75 mg (5 mg $\times$ 0.75) will be filtered into peripheral tissues. Given these circumstances, the concentrations of bilirubin within the volumes of systemic interstitial fluid (ISF) and intracranial extracellular fluid (ICEF) spaces will be raised by 0.94 mg/dl (3.75 mg/400 ml = 0.94 mg/dl) in the ISF and 2.50 mg/dl (1.25 mg/50 ml = 2.50 mg/dl) in the ICEF. As illustrated in the bottom half of Table II, the molar ratios of bilirubin-to-albumin become seriously perturbed and the concentration ratio within the ICEF has suddenly become acutely elevated (5.3) and less able to prevent bilirubin diffusion to adjacent cells. That such can occur forms the basis for the production of experimental bilirubin encephalopathy by the administration of sulfonamides, salicylate, and acids to the experimentally jaundiced animals (Johnson et al., 1961; Diamond and Schmid, 1966; Schenker et al., 1966; Menken et al., 1966a), and the acidemia may be essential in neonatal animals that require asphyxiation prior to the infusion of bilirubin in order to produce kernicterus (Lucey et al., 1964; Chen et al., 1965).

A further mechanism that permits the high concentrations of bilirubin to persist in the ICEF is due to the autonomy of CSF hydrogen ion concentration that can often maintain its pH at 7.35 in the presence of metabolic acidosis where systemic blood pH is below 7.2 (Posner et al., 1965). Under such circumstances, the solubility of bilirubin is favored in the CSF (Swatzski et al., 1968), and its back diffusion across the CSF–blood barrier as an undissociated or hydrophobic anion is retarded.

The poorer clearance function of the CSF during neonatal life may predispose the infant to bilirubin encephalopathy. But older children and adults have also been documented to exhibit bilirubin encephalopathy clinically and exhibit "kernicterus" at autopsy (Rosenthal et al., 1956; Gardner and Konigsmark, 1969; Blaschke et al., 1974; Wolkoff et al., 1979; Ho et al., 1980). The combination of unconjugated bilirubin concentrations exceeding the molar saturations of circulating albumin in excess of 1:1 is simply exceedingly rare beyond the neonatal period.

6. The Cytotoxicity of Bilirubin

The hypothesis that bilirubin itself might be cytotoxic and responsible for kernicterus was initially tested by Day (1954), who found that inclusion of bilirubin into the media of chopped brain slices depressed their oxygen consumption. Zetterstrom and Ernster (1956) subsequently reported that bilirubin uncoupled oxidative phosphorylation in respiring mitochondria *in vitro*. Both of the latter investigators' studies utilized bilirubin concentrations *in vitro* comparable to that seen *in vivo* in the sera of jaundiced neonatal infants who were at risk to kernicterus—namely, 25 mg/dl or 4×10^{-4} M. To expect such concentrations *in vivo* in the microenvironment of neurons within the central nervous sytem was considered doubtful. The most explicit study of the influence of bilirubin on isolated mitochondrial respiration was provided by Mustafa *et al.* (1969) and illustrated in Fig. 10. These investigators found an initial stimulation of oxygen consumption followed by depression with increasing bilirubin concentrations in the media. They were able to dissociate oxygen consumption and oxidative phosphorylation in respiring mitochondria and clearly showed that the "respiratory control index" was first impaired (Fig. 10B). The concentrations of bilirubin employed were of the order of 10^{-6} M in the absence of albumin within the incubation media. Such concentrations of bilirubin (5–6 mg/dl) were closer to clinical circumstances, but still insufficient because of the long-known albumin binding of bilirubin (Odell, 1959a). Indeed, Mustafa *et al.* (1969) demonstrated that inclusion of albumin in their incubation media equimolar to bilirubin protected the respiring mitochondrial isolates from any of the noxious effects of bilirubin, and human albumin was more effective than bovine albumin. Menken and colleagues (1966b) had previously reported a greater *in vitro* sensitivity to bilirubin of mitochondrial isolates derived from rat brain than from liver and less protection by albumin with bilirubin concentrations of the order of 4×10^{-5} M. They suggested the greater lipid content of brain mitochondria might account for this more selective sensitivity. Mustafa and King (1970) confirmed that inclusion of lipid extracts from mitochondria could protect against the toxic effects of bilirubin to respiring mitochondrial suspensions comparable to that afforded by albumin. By spectrographic techniques, they characterized the interaction of bilirubin with lipid by the hypochromicity and a red shift in the absorption curve of bilirubin with an extinction shoulder at 490 nm. They suggested the cytotoxic effects of bilirubin on mitochondrial isolates were related to the interaction of bilirubin and mitochondrial lipid rather than its protein constituents. Ernster (1961) also suggested that the toxicity of bilirubin to the membrane-linked enzymatic functions of mitochondria might be related to its detergentlike effect on the membrane structure. Cowger *et al.* (1965) compared the influence of bilirubin with classical poisons of mitochondrial respiration *in vitro* as well as its effect on intact cells in culture media. They discerned that although bilirubin was an effective inhibitor of $NADH_2$ oxidase such as amytal *in vitro*, it was a more effective uncoupler of oxidative phosphorylation *in vivo* as reflected by decreased ATP production and an increase O_2 consumption in intact cells. They furthermore noted a markedly decreased incorporation of leucine into cellular protein in the presence of bilirubin at concentrations of 10^{-6} M, which amytal does not effect. Quastel and Bickis (1959) had similarly observed depression of the incorporation of [^{14}C]glycine into cell proteins in Ehrlich ascites carcinoma cells exposed to bilirubin concentrations of 10^{-4} M. Thus, although bilirubin appeared to exhibit many of the characteristics of classical uncouplers of mitochondrial respiration and oxidative phosphorylation, its effects were more devastating to intact cells. It is pertinent that the cell culture

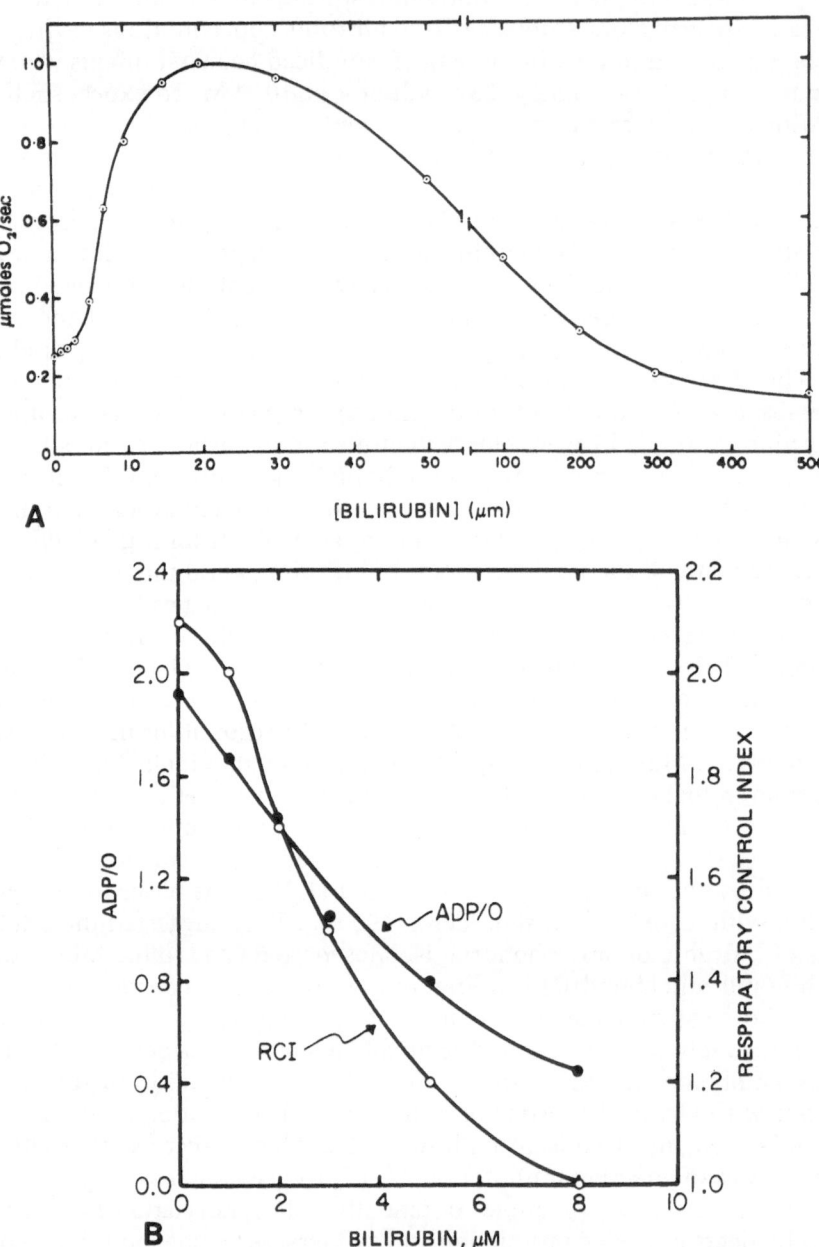

FIGURE 10. The *in vitro* effects of bilirubin on mitochondrial functions. A: The response of respiration (μmoles O₂/sec) in rat liver mitochondria (0.8 mg protein) to progressive increases in bilirubin concentration (abcissa). The suspension contained 100 mM sucrose, 50 mM mannitol, 8 mM MgCl₂, 5 mM potassium phosphate, 5 mM Tris-chloride, 10 mM -hydroxybutyrate, and pH 7.5. From Mustafa *et al.*, 1969. B: The sensitivity of the respiratory control index (RCI) and oxidative phosphorylation (ADP/O) in brain mitochondria (1.5 mg protein) to increasing concentrations of bilirubin (abcissa). The suspensions contained 0.3 mM mannitol, 10 mM potassium chloride, 5 mM potassium phosphate, 10 mM trischloride, 100 mM ADP, 10 mM succinate, 10 mM glycero-phosphate, and pH 7.5. Reproduced by permission, from Mustafa, *et al.* (1969).

system (L-929) studied by Cowger had been adapted to protein-free media and that concentrations of bilirubin as low as 2.5×10^{-6} M exerted a demonstrable influence on the viability of such cells, whereas HeLa cells, similarly adapted to protein-free media, required ten times more bilirubin to affect cell growth. This observation is illustrative of a differential sensitivity of cells to the toxicity of bilirubin. All these toxic effects of bilirubin were entirely reversed if bovine albumin was included within the incubation media of the L-929 cell line if equimolar to the bilirubin concentration. Cowger (1971) recognized that the cytotoxicity of bilirubin could not be adequately explained simply by its uncoupling effects of oxidative phosphorylation on mitochondria. Although bilirubin at concentrations of 5×10^{-6} M caused significant inhibition of $NADH_2$-oxidase activity, similar to amytal in mitochondrial isolates, bilirubin also severely inhibited cell growth in cultured L-929 cell lines and was associated with leakage of cytosol proteins indicative of plasma cell membrane damage uncharacteristic of the classical amytal-like uncouplers of oxidative phosphorylation. Any hypothesis to explain the cytotoxicity of bilirubin must take into account the extreme sensitivity of some neurons to bilirubin (e.g., the Purkinje cells of the Gunn rat) and the apparent resistance to damage of others (e.g., gasserian ganglion cells of the Gunn rat).

A critical observation on the *in vivo* toxicity of bilirubin to mitochondrial function and its relevance to bilirubin encephalopathy was reported by Diamond and Schmid (1967). They infused newborn guinea pigs with bilirubin to a level of clinical neurotoxicity. Yet, mitochondria isolated from the cerebral and cerebellar cortexes of such animals exhibited normal oxidative phosphorylation (P/O ratios > 2.0) *in vitro*. These isolated mitochondria also exhibited the expected uncoupling of oxidative phosphorylation when exposed *in vitro* to bilirubin at concentrations of 6.7×10^{-6} M (i.e., these isolates were not particularly resistant to the expected cytotoxic effects of bilirubin *in vitro*).

Since these early studies on the cytotoxicity of bilirubin, a number of investigators have subsequently described many inhibitory effects of bilirubin on intracellular metabolic pathways that have been recently evaluated by Karp (1979) and are summarized in Table III.

In contrast to the studies on isolated mitochondria (Diamond and Schmid, 1967), other investigators have found a marked reduction of not just ATP in cerebellar isolates (Schenker *et al.*, 1966), but also phosphocreatine within the Purkinje cell layer of the cerebellum of chronically jaundiced Gunn rats in which kernicterus was induced by sulfonamides (McCandless and Abel, 1980). These latter authors demonstrated that only selected neuronal cells of the cerebellum exhibited significant evidence of bilirubin-induced toxicity and therefore results derived from cell isolates of whole organ homogenates might be misleading if only a fraction of the total cell isolates contained evidence of toxicity to bilirubin. Of significance, a much higher glycogen content was also found in the Purkinje cell layer in which ATP and phosphocreatine levels were found to be decreased. Depression of brain cell glycolysis has also been demonstrated in kernicteric jaundiced rats (Katoh-Semba, 1976).

Morphology, as documented by electron microscopy, correlates well with the depletion of ATP and phosphocreatine in the presence of excessive glycogen stores. Purkinje cells exhibit marked accumulations of excessive gly-

TABLE III. Toxicity of Bilirubin on Cellular Metabolism and Specific Enzyme Systems

Activity	System
1. Respiration (Q_{O_2})	Cells in culture
	Mitochondria
2. Oxidative phosphorylation	Cells in culture
	Mitochondria in suspension
a. Mitochondrial enzymes	NADH oxidase
	Succinate oxidase
	Malate dehydrogenase
	Glutamate dehydrogenase
	NAD^+ isocitrate dehydrogenase
3. Glycolysis	Gunn rat brain
	Human red cells
a. Glycolytic enzymes	Brain hexokinase
	Brain phosphofructokinase
	(RBC) glyceraldehyde-3-phosphate dehydrogenase
	(WBC) hexose-monophosate shunt
4. Amino acid incorporated into protein	Glycine in ascites tumor cell
	Leucine in tissue culture cell
	Leucine into serum albumin in intact Gunn rat
	Leucine into neuronal and myelin proteins in:
	a. adult Gunn rat brain
	b. nursing Gunn rat brain
5. Other synthetic processes	
a. Heme biosynthesis	Incorporation of iron into protoporphyrin
b. Phospholipids	Incorporation of P_i into mitochondrial lipid after bilirubin injected into rats
c. Protein phosphorylation	Protein kinase activity of rabbit brain after bilirubin injection

(continued)

TABLE III. (continued)

Effect	Bilirubin concentration	Reference
Increased	5×10^{-6} M	Cowger et al. (1965)
Increased	10^{-5} M	Mustafa et al. (1969)
Decreased	$>10^{-5}$ M	Mustafa et al. (1969)
Uncoupled	2.5×10^{-6} M	Cowger et al. (1965)
Uncoupled	2.5×10^{-6} M	Mustafa et al. (1969)
Inhibition	1.5×10^{-5} M	Cowger et al. (1965)
Inhibition	3.4×10^{-4} M	Cowger et al. (1965)
Inhibition	4×10^{-6} M	Noir et al. (1972)
Inhibition	4×10^{-6} M	Kashiwamata (1975)
Inhibition	7×10^{-5} M	Noir et al. (1972)
Inhibition	$1.0 - 2.5 \times 10^{-5}$ M	Ogasawara et al. (1973)
Decreased	$1.6 - 3.6$ nmoles/g	Katoh-Semba (1976)
Decreased	$7.0 - 8.0 \times 10^{-5}$ M	Petrich et al. (1977)
Inhibition	2.7×10^{-5} M	Katoh-Semba (1976)
Inhibition	10^{-5} M	Katoh-Semba (1976)
Inhibition	$0.1 - 1.0 \times 10^{-3}$ M	Girotti (1976)
Inhibition	1×10^{-5} M	Thong and Rencis (1977)
Inhibition	1.6×10^{-4} M	Quastel and Bickis (1959)
Inhibition	$5.0 - 25.0 \times 10^{-6}$ M	Cowger et al. (1965)
Decreased	1×10^{-4} M	Majumdar et al. (1973)
Decreased	—	Greenfield and Majumdar (1974)
Normal	—	Greenfield and Majumdar (1974)
Inhibition	10^{-3} M	Labbe et al. (1959)
Inhibition	3.3 mg/kg	Youngs and Cornatzer (1963)
	In vivo injection 5–20 mg/ 100 g	Morphis et al. (1982)

cogen both within their mitochondria and the cell cytosol (Schutta and Johnson, 1970, 1971; Jew and Williams, 1977; Jew and Sandquist, 1979; Karp et al., 1982). Apparently, these excessive glycogen stores are unavailable for cell metabolism and glycolysis as an alternative energy source for ATP formation is inoperative. The recent studies of Morphis et al. (1982) may have pertinence to this apparent paradox. They reported that acute systemic infusion of bilirubin into neonatal rabbits produced clinical neurologic symptoms and was associated with a dosage-dependent protein kinase deficiency for protein phosphorylation of brain isolates when using purified histone as substrate. The measured toxicity of bilirubin on protein phosphorylation in the brain isolates was found to be inhibitory to the cyclic-AMP-independent phosphorylations. An obvious speculation applicable to the Gunn rat is the possibility that the activation of glycolytic phosphorylase by phosphorylase kinase may be inhibited by bilirubin and account for the excessive glycogen accumulation.

Since many protein phosphorylation kinases are membrane associated (Williams and Rodnight, 1977; Greengard, 1978), it is conceivable that the numerous toxicities of bilirubin described may be a consequence of the interaction of bilirubin with lipid membranes. The subsequent metabolic consequences may be secondary to the disruption of the lipid constraints and compartmentation of membrane-bound enzymes responsible for cellular homeostatic functions.

The interaction of bilirubin with lipids has been known for many years (Stenhagen and Rideal, 1939) and its penetration believed more strongly influenced by the more hydrophobic portions of the molecule. Thus bilirubin in its Z–Z configuration would have a greater diffusibility in lipid membranes than its more polar photoisomers. The binding of bilirubin by lipids and the associated red shift in its spectral absorption curve have been interpreted as bilirubin being in a hydrophobic environment (Mustafa and King, 1970). The in vitro addition of bilirubin to gangliosides associated with the red shift in absorption indicated a preferential association (Weil and Menkes, 1975), and its isolation and localization in gangliosides from lipid extracts of rat brains made kernicteric by bilirubin perfusion as well as kernicteric brains of newborn infants attest to its preferential sequestration in lipid components of cells (Kahan et al., 1968). More recent studies, in vitro, have indicated that phospholipids have an affinity for bilirubin greater than neutral lipids with reported affinity constants of 10^6 M^{-1} for sphyngomyelin (Nagoka and Cowger, 1978).

Based on bilirubin affinity for phosphatidylcholine dispersions, it has been estimated that 47% of total hepatocyte content of bilirubin may be membrane bound (Tipping et al., 1979). Eriksen et al. (1981) have demonstrated interactions of bilirubin with phosphatidylcholine that indicate disruption of their membrane structure. The pH dependence of the aqueous environment was found to be critical for the interaction.

The methods for measuring bilirubin's interaction with cell membranes are indirect and its consequences on cellular and homeostatic functions may be remote. As an example, the bilirubin nephropathy described in the jaundiced Gunn rat is externally characterized by an inability to concentrate their urine in response to thirsting (Odell et al., 1967). Such a phenomenon was best explained by the inability of medullary epithelia of the kidney (plasma membranes) to preserve the concentration gradients between lumenal fluid and the adjacent medullary interstitium. As illustrated in Fig. 11, the selective permeability of the renal medulla in the jj rats is compromised, for it no longer contained the high concentrations of Na^+, Cl^-, and urea in the papilla necessary for the osmotic reabsorption of solute-free water from distal tubular fluid within collecting ducts during thirsting. The focal accumulation of bilirubin in the papillae that was associated with this disorder of membrane function could be accounted for by the local environment of high ionic strength and hydrogen ion concentrations since both of the latter circumstances are known to increase the dissociation of bilirubin from albumin (Odell, 1966) and foster its localization into adjacent membranous structures.

Pertinent to the purpose of this chapter, when circumstances of cerebral blood flow allow filtration of bilirubin dissociated from albumin into the CNS,

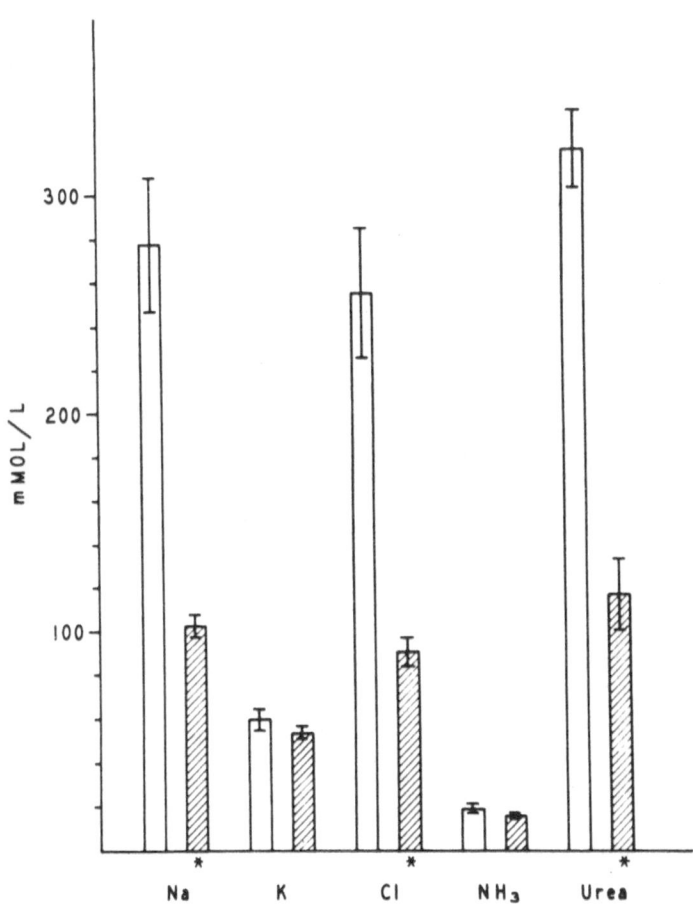

FIGURE 11. Comparison of the composition of the renal papillae in pair-fed and thirsted non-jaundiced heterozygous (Jj) control rats (open columns) and homozygous (jj) rats (cross-hatched columns). The concentrations (ordinate) of the measured solutes (abcissa) are expressed in molar units/liter of tissue water. The * indicates significant differences in tissue concentration and the horizontal bars are SE of the mean values (water content: control rats, 82 ± 0.9% wet wt.; Gunn rats, 85 ± 0.3%). Reproduced by permission from Odell *et al.* (1967).

then its sequestration from the extracellular fluid of the CNS can occur if there are present lipids such as gangliosides and, in particular, phospholipids that have a high affinity for bilirubin, particularly if the environmental pH is low (Silberberg *et al.*, 1970a; Nagoka and Cowger, 1978). Bilirubin once within the membranes undergoes aggregation (Eriksen *et al.*, 1981), and by electronmicroscopy, membranous whorls are seen (Schutta *et al.*, 1970) within the cytoplasm that appear phagocytosed because of a limiting membrane surrounding them. It is conceivable these membranes are derived from bilirubin-damaged plasma membranes that become autophagocytosed and the neuron rapidly reconstitutes its plasma membrane (Fig. 12). This proposed phagocytosis of bilirubin-impregnated membranes results in net transport of large concentrations of bilirubin to within the cytosol of the neuron where, if dissociated from the membrane whorl, it could exert many of the toxic effects of bilirubin cited in Table III. Neurons may be particularly susceptible to bilirubin toxicity, for they lack the anion-binding protein ligandin that is protective of mitochondria to the uncoupling effects of bilirubin *in vitro* (Kamisaki *et al.*, 1975).

The observations of bilirubin's influence on cells in culture (Cowger, 1971) is of relevance since they demonstrated the importance of the sequence of addition of bilirubin and albumin to the cell media. When bilirubin was added even 5 min before the albumin, cell viability was impaired and not all the bilirubin could be recovered in the albumin extracellular phases of the suspension media. Likewise, one can envision circumstances such as in Table II

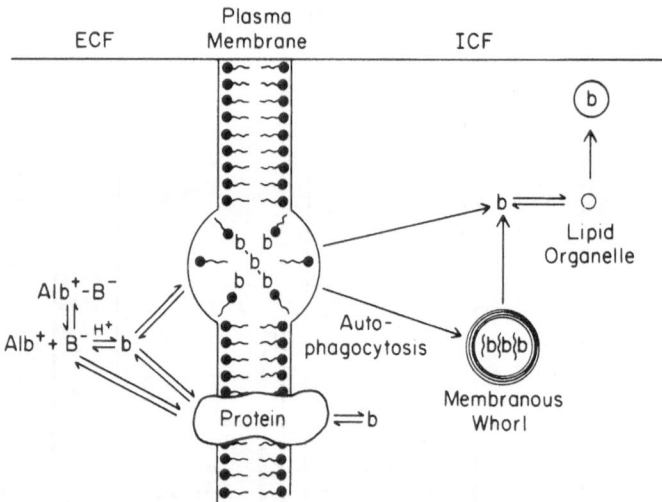

FIGURE 12. Bilirubin transport from the intracranial extracellular fluid (ECF) across the plasma membrane of target neurons into their intracellular fluid (ICF). In this schematic, dianionic bilirubin (B⁻) is in protein-binding equilibrium with albumin (Alb⁺). The dianionic bilirubin dissociated from albumin within extracellular fluid undergoes a hydrogen (H⁺) sensitive configurational change (b) which permits its entry into and aggregation within the plasma membrane. Such membrane fragments become autophagocytosed whereby large concentrations of bilirubin are translocated into the cell cytosol. Once within the cytosol, the bilirubin (B) can either back diffuse to extracellular fluid (reversible bilirubin encephalopathy) or attach other organelles with disruption of their function resulting in the necrobiosis of the cell (irreversible bilirubin encephalopathy).

where bilirubin in excess of albumin gains access to the CNS and can cause neuronal damage that may readily become irreversible even though systemic bilirubin concentrations subsequently favor the back diffusion (clearance) of bilirubin from the intracranial extracellular fluid.

In summary, we propose that bilirubin encephalopathy can occur when sufficient saturation of circulating albumin occurs *in vivo* that permits high aqueous-phase concentrations of bilirubin to exist. Such bilirubin is probably in the IX α Z–Z configuration and is readily filterable across the capillary–encephalon barrier. If there coexists an inefficient "sink function" of CSF such as occurs during neonatal life, the opportunity for bilirubin to accumulate to high concentrations in excess of the available albumin is quite possible.

Within the CNS are neurons whose plasma membranes have a particularly high affinity for bilirubin that permits its accretion to such high concentrations that it is macroscopically visible (kernicterus), and such focal accumulations are sufficient to irreversibly induce the death of such cells (bilirubin encephalopathy). The necrobiosis involves the interruption of glycolysis without interference of glycogen synthesis and there results a loss of ATP and phosphocreatine by either loss of glycolytic mechanisms or direct uncoupling of mitochondrial phosphorylation or both. What appears critical to the pathogenesis of bilirubin encephalopathy is the accretion mechanism by which the plasma membranes of the target neurons sequester high concentrations of bilirubin from their extracellular environs and the membrane-bound bilirubin is subsequently transported to the intraneuronal cytosol.

When sufficient loss of such neurons has occurred, then focal staining is no longer visible since these cells have been replaced by glial elements that do not concentrate bilirubin from ambient aqueous phases of the interstitium of the CNS.

ACKNOWLEDGMENTS. Dr. Odell's work in preparing this chapter was supported in part by USPHS AM 21668.

References

Abelson, N. M., and Boggs, T. R., Jr., 1956, Plasma pigments in erytho-blastosis fetalis. I. Spectrophotometric absorption patterns, *Pediatrics* **17:**452.

Aprison, M. H., and Segar, W. E., 1963, Electrolyte distribution and water content in six discrete areas of the developing mammalian brain, *Recent Adv. Biol. Psychol.* **5:**279.

Bass, N. H., and Lundborg, P., 1973, Postnatal development of bulk flow in the cerebrospinal fluid system of the albino rat: Clearance of carboxyl (14C) insulin after intrathecal infusion, *Brain* **52:**323.

Batty, H. K., and Millhouse, O. E., 1976, Ultrastructure of the Gunn rat substantia nigra, *Acta Neuropathol.* **34:**7.

Beneke, R., 1907, Uber den Kernikterus der Neugeborenen Muchen, 1907, *Med. Wochenschr.* **54:**2023.

Berde, C. B., Hudson, B. S., Simoni, R. D., and Sklar, L. A., 1979, Human serum albumin: Spectroscopic studies of binding and proximity relationships for fatty acids and bilirubin, *J. Biol. Chem.* **254:**391.

Bischoff, F., and Stauffer, R. D., 1954, The dispersion of testosterone in aqueous bovine serum albumin solution, *J. Am. Chem. Soc.* **76:**1962.

Blanc, W. A., and Johnson, L., 1959, Studies in kernicterus. Relationship with sulfonamide intoxication, report of kernicterus in rats with glucuronyl transferase deficiency and review of
pathogenesis, *J. Neuropathol. Exp. Neurol.* **18**:165.

Blaschke, T. F., Berk, P. D., and Scharschmidt, B. F., 1974, Crigler-Najjar Syndrome: an unusual
course with development of neurologic damage at age eighteen, *Pediatr. Res.* **8**:573.

Blauer, G., Harmatz, D., and Naparstek, A., 1975, Circular dischroism of bilirubin-human serum
albumin complexes in aqueous solutions: A comparison among albumins from different species, *Arch. Biochem. Biophys.* **170**:375.

Bonnett, R., Davies, J. E., and Hursthouse, M. B., 1976, Structure of bilirubin, *Nature* **262**:326.

Bratlid, D., 1972, The effect of pH on bilirubin binding by human erythrocytes, *Scand. J. Clin. Lab.
Invest.* **29**:453.

Brodersen, R., 1972, Supersaturation with bilirubin followed by colloid formation and disposition,
with a hypothesis on the etiology of kernicterus, *Scand. J. Lab. Invest.* **29**:447.

Brodersen, R., 1974, Competitive binding of bilirubin and drugs to human serum albumin studied
by enzymatic oxidation, *J. Clin. Invest.* **54**:1353.

Brodersen, R., 1979a, Binding of bilirubin to albumin. Implications for prevention of bilirubin
encephalopathy in the newborn, *CRC Critical Rev. Clin Lab. Sci.* **11**:305.

Brodersen, R., 1979b, Bilirubin solubility and interaction with albumin and phospholipid, *J. Biol.
Chem.* **254**:2364.

Brown, A. K., and McDonagh A. F., 1980, Phototherapy for neonatal hyper-bilirubinemia: Efficacy,
mechanism and toxicity, *Adv. Pediatr.* **27**:341.

Burnstine, R. C., and Schmid, R., 1962, Solubility of bilirubin in aqueous solution, *Proc. Soc. Exp.
Biol. Med.* **109**:356.

Byers, R. K., Paine, R. S., and Crothers, B., 1955, Extrapyramidal cerebral palsy with hearing loss
following erythroblastosis, *Pediatrics* **15**:248.

Carey, M. C., 1982, The enterohepatic circulation in the liver, in: *Biology and Pathobiology.* (I.
Arias, H. Popper, D. Schacter, and D. A. Shafritz, eds.), Raven Press, New York, p. 429.

Carbone, J. V., and Grodsky, G. M., 1957, Constitutional nonhemolytic hyperbilirubinemia in the
rat: Defect of bilirubin conjugation, *Proc. Soc. Exp. Biol. Med.* **94**:461.

Chen, H. C., Lien, I. N., and Lu, T. C., 1965, Kernicterus in newborn rabbits, *Am. J. Pathol.* **46**:331.

Chen, R. F., 1974, Fluorescence stopped-flow study of relaxation processes in the binding of bilirubin to serum albumins, *Arch. Biochem. Biophys.* **160**:105.

Childs, B., Sidbury, J. B., and Migeon, C. J., 1959, Glucuronic acid conjugation by patients with
familial nonhemolytic jaundice and their relatives, *Pediatrics* **23**:903.

Claireaux, A. E., 1959, Hemolytic disease of the newborn. I. A clinical-pathological study of 157
cases. II. Nuclear jaundice (kernicterus), *Arch. Dis. Child.* **25**:61.

Claireaux, A. E., 1961, Pathology of human kernicterus, in: *Kernicterus* (A. Sass-Kortsak, ed.), University of Toronto Press, Toronto, Canada, p. 140.

Claireaux, A. E., Cole, P. G., and Lathe, G. H., 1953, Icterus of the brain in the newborn, *Lancet*
2:1226.

Cohen, A. N., and Ostrow, J. D., 1980, New concepts in phototherapy: Photo-isomerization of
bilirubin IX and potential toxic effects of light, *Pediatrics* **65**:740.

Cowger, M., 1971, Mechanism of bilirubin toxicity on tissue culture cells: Factors that affect toxicity,
reversibility by albumin, and comparison with other respiratory poisons and surfactants,
Biochem. Med. **5**:1.

Cowger, M. L., Igo, R. P., and Labbe, R. F., 1965, The mechanism of bilirubin toxicity studied with
purified respiratory enzyme and tissue culture systems, *Biochemistry* **4**:2763.

Crigler, J. F., and Najjar, V. A., 1952, Congenital familial nonhemolytic jaundice with kernicterus,
Pediatrics **10**:169.

Davson, H., Kleeman, C. R., and Levin, E., 1962, Quantitative studies of the passage of different
substances out of the cerebrospinal fluid, *J. Physiol.* **161**:126.

Day, R., 1947, Kernicterus problem: Experimental in vivo and in vitro staining of brain tissue with
bilirubin, *Am. J. Dis. Child.* **73**:241.

Day, R., 1954, Inhibition of brain respiration *in vitro* by bilirubin. Reversal of inhibition by various
means, *Proc. Soc. Exp. Biol. Med.* **85**:261.

Day, R., and Haines, M. S., 1954, Intelligence quotients of children recovered from erythroblastosis
fetalis since the introduction of exchange transfusion, *Pediatrics* **13**:333.

Diamond, I., 1969, Bilirubin binding and kernicterus, *Adv. Pediatr.* **16**:99.

Diamond, I., and Schmid, R., 1966, Experimental bilirubin encephalopathy. The mode of entry of bilirubin—^{14}C into the central nervous system, *J. Clin. Invest.* **45**:678.

Diamond, I., and Schmid, R., 1967, Oxidative phosphorylation in experimental bilirubin encephalopathy, *Science* **155**:1288.

Diamond, L. K., Blackfan, K. D., and Baty, J. M., 1932, Erythroblastosis fetalis with its association with universal edema of fetus, icterus gravis neonatorum and anemia of the newborn, *J. Pediatr.* **1**:269.

Erikson, E., Danielsen, H., and Brodersen, R., 1981, Bilirubin-liposome interaction, *J. Biol. Chem.* **256**:4269.

Ernster, L., 1961, The mode of action of bilirubin on mitochondria, in: *Kernicterus* (A. Sass-Korsak, ed.), University of Toronto Press, Toronto, Canada, p. 174.

Ernster, L., Herlin, L., and Zetterstrom, R., 1957, Experimental studies on the pathogenesis of kernicterus, *Pediatrics* **20**:647.

Esch, P., 1908, Uber Kernikterus der Neugeborenen, *Ztrbl. Gynak.* **32**:969.

Ferguson, R. K., and Woodbury, D. M., 1969, Penetration of ^{14}C-sucrose into brain, cerebrospinal fluid and skeletal muscle of developing rats, *Exp. Brain Res.* **7**:181.

Gardner, W. A., and Konigsmark, B. W., 1969, Familial nonhemolytic jaundice: Bilirubinosis and encephalopathy, *Pediatrics* **43**:55.

Gerrard, J., 1952, Kernicterus, *Brain* **75**:526.

Girotti, A. W., 1976, Glyceraldehyde-3-phosphate dehydrogenase in the isolated human erythrocyte membrane: Selective displacement by bilirubin, *Arch. Biochem. Biophys.* **173**:210.

Gorten, M. K., Shear, S., Hodson, M., and Bessman, S. P., 1958, Complications of hyperbilirubinemia in the newborn—Possible relation to the metabolism of ammonia, *Pediatrics* **21**:27.

Govan, A. D. T., and Scott, J. M., 1953, Kernicterus and prematurity, *Lancet* **1**:611.

Greenfield, S., and Majumdar, A. P. N., 1974, Bilirubin encephalopathy: Effect of protein synthesis in the brain of the Gunn rat, *J. Neurol. Sci.* **22**:83.

Greengard, P., 1978, Phosphorylated proteins as physiologic effectors, *Science* **199**:146.

Gunn, C. K., 1938, Hereditary acholuric jaundice, *J. Hered.* **29**:137.

Gunn, C. K., 1944, Hereditary acholuric jaundice in the rat, *Can. Med. Assoc. J.* **50**:230.

Hansen, P. E., Thiessen, H., and Brodersen, R., 1979, Bilirubin acidity. Titrimetric and ^{13}CNMR studies, *Acta Chem. Scand.* **33**:281.

Harris, R. C., Lucey, J. F., MacLean, J. R., 1958, Kernicterus in premature infants associated with low concentrations of bilirubin in the plasma. *Pediatrics* **21**:875.

Haymaker, W., Margoles, C., Pentschew, A., 1961, *Pathology of Kernicterus and Posticteric Encephalopathy in Kernicterus and Its Importance in Cerebral Palsy,* Charles C. Thomas, Springfield, Illinois.

Ho, K. C., Hodach, R., Varma, R., Thorsteinson, V., Hess, T., and Dale, D., 1980, Kernicterus and central pontine myelinolysis in a 14-year-old boy with fulminating hepatitis, *Ann. Neurol.* **8**:633.

Hsia, D. Y. Y., Allen, F. H., Jr., Gellis, S. S., and Diamond, L. K., 1952, Erythroblastosis fetalis; studies of serum bilirubin in relation to kernicterus, *N. Engl. J. Med.* **247**:668.

Hyman, C. B., Keaster, J., Hanson, V., Harris, I., Sedgwick, R., Wursten, H., and Wright, A. R., 1969, CNS abnormalities after neonatal hemolytic disease or hyperbilirubinemia; a prospective study of 405 patients, *Am. J. Dis. Child.* **117**:395.

Jacobsen, C., 1972, Chemical modification of the high-affinity bilirubin binding site of human serum albumin, *Eur. J. Biochem.* **27**:513.

Jacobsen, C., 1978, Lysine residue 240 of human serum albumin is involved in the high-affinity binding of bilirubin, *Biochem. J.* **171**:453.

Jacobsen, J., 1969, Binding of bilirubin to human serum albumin; determination of the dissociation constants, *Febs Lett.* **5**:112.

Jacobsen, J., and Brodersen, R., 1976, The effect of pH on albumin-bilirubin binding affinity, *Birth Defects* **12**:175.

Jervis, G. A., 1959, Constitutional nonhemolytic hyperbilirubinemia with findings resembling kernicterus, *AMA Arch. Neurol. Psychiatry* **81**:55.

Jew, J. Y., and Sandquist, D., 1979, CNS changes in hyperbilirubinemia. Functional implications, *Arch. Neurol.* **36**:149.

Jew, J. Y., and Williams, T. H., 1977, Ultra structural aspects of bilirubin encephalopathy in cochlear nuclei of the Gunn rat, *J. Anat.* **124**:599.

Johnson, L., and Boggs, T. R., Jr., 1974, Bilirubin dependent brain damage: Incidence and indications for treatment, in: *Phototherapy in the newborn: An overview* (G. B., Odell, A. P. Simopoulos, and R. Schaffer, eds.), National Academy of Sciences, Washington, D.C., p. 122.

Johnson, L., Sarmiento, F., Blanc, W. A., and Day, R., 1959, Kernicterus in rats with an inherited deficiency of glucuronyl transferase, *Am. J. Dis. Child.* **97**:591.

Johnson, L., Garcia, M. L., Figueroa, E., and Sarmiento, F., 1961, Kernicterus in rat lacking glucuronyl transferase. II. Factors which alter bilirubin concentration and frequency of kernicterus, *Am. J. Dis. Child.* **101**:322.

Kahan, I. L., Timar, M., and Foldi, J., 1968, Bilirubin-binding cerebral lipid, *Acta. Paediatr. Acad. Sci. Hung.* **9**:121.

Kamisaka, K., Gatmaitan, L., Moore, C. L., and Arias, I. M., 1975, Ligandin reverses bilirubin inhibition of liver mitochondrial respiration *in vitro*, *Pediatr. Res.* **9**:903.

Karp, W. B., 1979, Biochemical alterations in neonatal hyperbilirubinemia and bilirubin encephalopathy: A review, *Pediatrics* **64**:361.

Karp, W. B., Moore, P. J., Subramanyam, S. B., and Brown, D. B., 1982, Relationship of plasma total bilirubin, apparent unbound bilirubin and total albumin with cerebellar glycogen and abnormal purkinje cells in Gunn rat, *Biol. Neonate* **41**:294.

Kashiwamata, S., Niva, F., Katoh, R., and Higashida, H., 1975, Malate dehydrogenase of bovine cerebrum: Inhibition by bilirubin, *J. Neurochem.* **24**:189.

Katoh-Semba, R., 1976, Studies on cellular toxicity of bilirubin: Effect on brain glycolysis in the young rat, *Brain Res.* **113**:339.

Katoh-Semba, R., and Kashiwamata, S., 1980, Interaction of bilirubin with brain capillaries and its toxicity, *Biochem. Biophys. Acta* **632**:290.

Kuenzle, C. C., Weibel, M. H., and Pelloni, R. R., 1973, The reaction of bilirubin with diazomethane, *Biochem. J.* **133**:357.

Labbe, R. F., Zaske, M. R., and Aldrich, R. A., 1959, Bilirubin inhibition of heme biosynthesis, *Science* **129**:1741.

Landsteiner, K., and Wiener, A. S., 1940, Agglutinable factor in human blood recognized by immune sera for Rhesus blood, *Proc. Soc. Exp. Biol. N.Y.* **43**:223.

Lathe, G. H., and Walker, M., 1958, The synthesis of bilirubin glucuronide in animal and human liver, *Biochem. J.* **70**:705.

Lee, J. J., and Gillespie, G. D., 1981, The effect of pH on the fluorescence of complexes of serum albumin and bovine serum albumin with bilirubin, *Photochem. Photobiol.* **33**:757.

Lee, J. J., Daley, L. H., and Cowger, M. L., 1974, Bilirubin ion equilibria; their effects on spectra and on conformation, *Res. Commun. Chem. Pathol. Pharmacol.* **9**:763.

Lending, M., Slobody, L. B., and Mestern, J., 1966, The relationship of hypercarbia to the production of kernicterus, *Dev. Med. Child. Neurol.* **9**:145.

Levine, P., and Stetson, R. E., 1939, An unusual case of intragroupagglutination, *J. Am. Med. Assoc.* **113**:126.

Levine, R. L., Fredericks, W. R., and Rapoport, S. I., 1982, Entry of bilirubin into the brain due to opening of the blood–brain barrier, *Pediatrics* **69**:255.

Lucey, J. F., Hibbard, E., Behrman, R. E., Esquivel de Gallardo, F. O., and Windle, W. F., 1964, Kernicterus in asphyxiated newborn monkeys, *Exp. Neurol.* **9**:43.

McCandless, D. W., and Abel, M. S., 1980, The effect of unconjugated bilirubin on regional cerebellar energy metabolism, *Neurobehav. Toxicol.* **2**:81.

McCluskey, S., Storey, G. N. B., Brown, G. K., 1975, Fluorometric determination of "albumin-titrable bilirubin" in the jaundiced neonate, *Clin. Chem.* **21**:1638.

McDonagh, A. F., Palma, L. A., Trull, F. R., and Lightner, D. A., 1982a, Phototherapy for neonatal jaundice configurational isomers of bilirubin, *J. Am. Chem. Soc.* **104**:6865.

McDonagh, A. F., Palma, L. A., and Lightner, D. A., 1982b, Phototherapy for neonatal jaundice. Stereospecific and regioselective photoisomerization of bilirubin bound to human serum albumin and NMR characterization of intramoleculary cyclized photoproducts, *J. Am. Chem. Soc.* **104**:6867.

Majumdar, A. P. N., Greenfield, S., and Roigaard-Petersen, H., 1973, Metabolism of plasma albumin in rats with congenital hyperbilirubinemia (Gunn), *Scand. J. Clin. Lab. Invest.* **31**:219.

Menken, M., Barrett, P. V. D., Swarm, R. L., and Berlin, N. I., 1966a, Kernicterus: Development of an experimental model using bilirubin ^{14}C, *Arch. Neurol.* **15**:68.

Menken, M., Waggoner, J. G., and Berlin, N. I., 1966b, The influence of bilirubin on oxidative phosphorylation and related reactions in brain and liver mitochondria: Effects of protein binding, *J. Neurochem.* **13**:1241.

Meriwether, L. S., Hager, H., and Scholz, W., 1955, Kernicterus: Hypoxemia; significant pathogenic factor, *AMA Arch. Neurol. Psychiatry* **73**:293.

Mollison, P. L., and Cutbush, M., 1951, A method of measuring the severity of a series of hemolytic disease of the newborn, *Blood* **6**:777.

Morphis, L., Constantopoulos, A., Matsaniotis, N., and Papaphilis, A., 1982, Bilirubin-induced modulation of cerebral protein phosphorylation in neonate rabbits *In vivo*, *Science* **218**:156.

Mugnoli, A., Manitto, P., and Monti, D., 1978, Structure of di-isopropylammonium bilirubinate, *Nature* **273**:568.

Mustafa, M. G., and King, T. E., 1970, Binding of bilirubin with lipid: A possible mechanism of its toxic reactions in mitochondria, *J. Biol. Chem.* **245**:1084.

Mustafa, M. G., Cowger, M. L., and King, T. E., 1969, Effects of bilirubin on mitochondrial reactions, *J. Biol. Chem.* **244**:6403.

Nagaoka, S., and Cowger, M. L., 1978, Interaction of bilirubin with lipids studied by fluorescence quenching method, *J. Biol. Chem.* **253**:2005.

Nelson, T., Jacobsen, J., and Wennberg, R., 1974, pH and bilirubin toxicity, *Pediatr. Res.* **8**:963.

Noir, B. A., Boveris, A., Pereira, A. M. G., and Stoppani, A. O. M., 1972, Bilirubin: A multi-site inhibitor of mitochondrial respiration, *FEBS Lett.* **27**:270.

Odell, G. B., 1959a, Studies in kernicterus. I. The protein binding of bilirubin, *J. Clin. Invest.* **38**:823.

Odell, G. B., 1959b, The dissociation of bilirubin from albumin and its clinical implications, *J. Pediatr.* **55**:268.

Odell, G. B., 1965, Influence of pH on distribution of bilirubin between albumin and mitochondria, *Proc. Soc. Exp. Biol. Med.* **120**:352.

Odell, G. B., 1966, The distribution of bilirubin between albumin and mitochondria. *J. Pediatr.* **68**:164.

Odell, G. B., 1974, Methods for measurement of the relative saturation of serum albumin with bilirubin in the management of neonatal hyperbilirubinemia, in: *Phototherapy of the newborn: An overview* (G. B. Odell, A. P. Simopoulos, and R. Schaffer, eds.), National Academy of Sciences, p. 114.

Odell, G. B., 1980, *Neonatal Hyperbilirubinemia*, Grune and Stratton, New York, p. 90.

Odell, G. B., 1982, Free bilirubin is of importance, *Pediatrics* **70**:659.

Odell, G. B., and Childs, B., 1980, Hereditary hyperbilirubinemias, *Prog. Med. Genet. New Ser.* **4**:103.

Odell, G. B., Natzschka, J. C., and Storey, G. N. B., 1967, Bilirubin nephropathy in the Gunn strain of rat, *Am. J. Physiol.* **212**:931.

Odell, G. B., Brown, R. S., and Holtzman, N. A., 1970a, Dye-sensitized photo-oxidation of albumin associated with a decreased capacity for protein binding of bilirubin, *Birth Defects* **6**:31.

Odell, G. B., Storey, G. N. B., and Rosenberg, L. A., 1970b, Studies in kernicterus. III. The saturation of serum proteins with bilirubin during neonatal life and its relationship to brain damage at five years, *J. Pediatr.* **76**:12.

Odell, G. B., Cukier, J. O., Ostrea, E. M., Jr., Maglalang, A. C., and Poland, R. L., 1977, The influence of fatty acids on the binding of bilirubin to albumin, *J. Lab. Clin. Med.* **89**:295.

Ogasawara, N., Watanabe, T., and Goto, H., 1973, Bilirubin: A potent inhibitor of NAD$^+$-linked isocitrate dehydrogenase, *Biochem. Biophys. Acta* **327**:233.

Orth, J., 1875, Veber das Vorkommen von Bilirubin-krystallen bei neugebornen Kindern, *Virchows Arch. F Pathol. Anat.* **63**:447.

Ostrea, E. M., Jr., Bassel, M., Fleury, B. A., Bartos, A., and Jesurun, A., 1983, Influence of free fatty acids and glucose infusion on serum bilirubin and bilirubin binding to albumin: Clinical implications, *J. Pediatr.* **102**:426.

Overbeek, J. Th. G., Vink, C. L. J., and Deenstra, H., 1955, The solubility of bilirubin, *Recl. Trav. Chim.* **74**:81.

Pappenheimer, J. R., Heisey, S. R., and Jordan, E. F., 1961, Active transport of diodrast and phenolsulfonphthalein from cerebrospinal fluid to blood, *Am. J. Physiol.* **200**:1.

Pappenheimer, J. R., Heisey, S. R., Jordan, E. F., and Downer, J. de C., 1962, Perfusion of the cerebral ventricular system in unanesthetized goats, Am. J. Physiol. 203:763.

Parsons, L., 1947, Clinician and Rh factor, Lancet 1:815.

Perlstein, M., 1950, Neurologic sequelae of erythroblastosis fetalis, Am. J. Dis. Child 79:605.

Petrich, C., Krieg, W., Voss, H., and Gobel, V., 1977, Effects of bilirubin on red cell metabolism, J. Clin. Chem. Clin. Biochem. 15:77.

Posner, J. B., Swanson, A. G., and Plum, F., 1965, Acid-base balance in cerebrospinal fluid, Arch. Neurol. 12:479.

Quastel, J. H., and Bickis, I. J., 1959, Metabolism of normal tissues and neoplasms in vitro, Nature 183:281.

Rapaport, S. I., 1976, Blood–Brain Barrier in Physiology and Medicine, Raven Press, New York.

Rose, A. L., and Wisniewski, H., 1979, Acute bilirubin encephalopathy induced with sulfadimethoxine in Gunn rats, J. Neuropathol. Exp. Neurol. 38:152.

Rosenthal, I. M., Zimmerman, H. J., and Hardy, N., 1956, Congenital nonhemolytic jaundice with disease of the central nervous system, Pediatrics 18:378.

Rozdilsky, B., 1966, Kittens as experimental model for study of kernicterus, Am. J. Dis. Child. 111:161.

Rozdilsky, B., and Olszewski, J., 1960, Permeability of cerebral vessels to albumin in hyperbilirubemia, Neurology 10:631.

Rozdilsky, B., and Olszewski, J., 1961, Experimental study of the toxicity of bilirubin in newborn animals, J. Neuropathol. Exp. Neurol. 20:193.

Sawasaki, Y., Yamada, N., and Nakajima, H., 1976, Developmental features of cerebellar hypoplasia and brain bilirubin levels in a mutant (Gunn) rat with hereditary hyperbilirubinemia, J. Neurochem. 27:577.

Schenker, S., McCandless D. W., and Zollman, P., 1966, Studies of cellular toxicity of unconjugated bilirubin in kernicteric brain, J. Clin. Invest. 45:1123.

Schmid, R., 1978, Bilirubin metabolism: State of the art, Gastroenterology 74:1307.

Schmid, R., Axelrod, J., Hammaker, L., and Swarm, R. L., 1958, Congenital jaundice in rats, due to a defect in glucuronide formation, J. Clin. Invest. 37:1123.

Schmorl, G., 1903, Zur Kenntnis des Ikterus neonatorum, insbesondere der dabei auf tretenden Gehirnveraenderungen, Verh. D. Detsch. Pathol. Ges. 6:109.

Schutta, H. S., and Johnson, L., 1967, Bilirubin encephalopathy in the Gunn rat. A fine structure study of the cerebellar cortex, J. Neuropathol. Exp. Neurol. 26:377.

Schutta, H. S., and Johnson, L., 1969, Clinical signs and morphologic abnormalities in Gunn rats treated with sulfadimethoxine, J. Pediatr. 75:1070.

Schutta, H. S., and Johnson, L., 1971, Fine structure observations on acute bilirubin encephalopathy in Gunn rats induced by sulfadimethoxine, Lab. Invest. 24:82.

Schutta, H. S., Johnson, L., and Neville, H. E., 1970, Mitochondrial abnormalities in bilirubin encephalopathy, J. Neuropathol. Neurol. 29:296.

Silberberg, D., Johnson, L., and Schutta, H. S., 1967, Factors influencing the toxicity of bilirubin in tissue cultures, Trans. Am. Neurol. Assoc. 92:284.

Silberberg, D. H., Johnson, L., and Ritter, L., 1970a, Factors influencing toxicity of bilirubin in cerebellum tissue culture, J. Pediatr. 77:386.

Silberberg, D. H., Johnson, L., Schutta, H., and Ritter, L., 1970b, Effects of photo degradation products of bilirubin on myelinating cerebellum cultures, J. Pediatr. 77:613.

Silverman, W. A., Andersen, D. H., Blanc, W. A., and Crozier, D. N., 1956, A difference in mortality rate and the incidence of kernicterus among premature infants allotted to two prophylactic antibacterial regimens, Pediatrics 18:614.

Stenhagen, E., and Rideal, E. K., 1939, CXCV The intraction between porphyrins and lipoid and protein monolayers, Biochem. J. 33:1591.

Strebel, L., and Odell, G. B., 1971, Bilirubin uridine diphospho-glucuronyl transferase in rat liver microsomes: Genetic variation and maturation, Pediatr. Res. 5:548.

Swatzski, A., Cheung, W. H., and Seifter, E., 1968, The effect of pH on the distribution of bilirubin in peripheral blood, cerebrospinal fluid and fat tissues, J. Pediatr. 72:700.

Thong, Y. H., and Rencis, V., 1977, Bilirubin inhibits hexose-monophosphate shunt activity of phagocytosing neutrophils, Acta Paediatr. Scand. 66:759.

Tipping, E., Ketterer, B., and Christodoulides, L., 1979, Interaction of small molecules with phospholipid bilayers, Biochem. J. 180:827.

Tran, C. D., and Beddard, G. S., 1982, Interactions between bilirubin and albumins using picosecond fluorescence and circular polarized luminescence spectroscopy, *J. Am. Chem. Soc.* **104**:6741.

Turkel, S. B., Miller, C. A., Guttenberg, M. A., Moynes, D. R., and Hodgman, J. E., 1982, A clinical pathological reappraisal of kernicterus, *Pediatrics* **69**:267.

Vaughan, V. C., III, 1946, Kernicterus in erythoblastosis fetalis, *J. Pediatr.* **29**:464.

Vaughan, V. C., III, Allen, F. H., Jr., and Diamond, L. K., 1950, Erythroblastosis fetalis. IV. Further observations on kernicterus, *Pediatrics* **6**:706.

Weil, M. L., and Menkes, J. H., 1975, Bilirubin interaction with ganglioside: Possible mechanism in kernicterus, *Pediatr. Res.* **9**:791.

Wells, R., Hammond, K., Lamola, A. A., and Blumberg, W. E., 1982, Relationships of bilirubin binding parameters, *Clin. Chem.* **28**:432.

Wiener, A. S., and Brody, M., 1946, Pathogenesis of kernicterus, *Science* **103**:570.

Williams, M., and Rodnight, R., 1977, Protein phosphorylation in nervous tissue: Possible involvement in nervous tissue function and relationship to cyclic nucleotide metabolism. *Prog. Neurobiol.* **8**:183.

Wolkoff, A. W., Chowdhury, J. R., Gartner, L. A., Rose, A. L., Biempica, L., Giblin, D. R., Fink, D., and Arias, I. M., 1979, Crigler-Najjar syndrome (Type 1) in an adult male, *Gastroenterology* **76**:840.

Youngs, J. N., and Cornatzer, W. E., 1963, Effect of oxidative phosphorylation inhibitors on synthesis of liver mitochondria lipids, *Proc. Soc. Exp. Biol. Med.* **112**:308.

Zetterstrom, R., and Ernster, L., 1956, Bilirubin an uncoupler of oxidative phosphorylation in isolated mitochondria, *Nature* **178**:1335.

Zimmerman, H. M., and Yannet, H., 1933, Kernicterus. Jaundice of the nuclear masses of the brain, *Am. J. Dis. Child.* **45**:740.

Zuelzer, W. W., and Mudgett, R. T., 1950, Kernicterus. Etiologic study based on an analysis of 55 cases, *Pediatrics* **6**:452.

11

Uremic and Dialysis Encephalopathies

PAUL E. TESCHAN and ALLEN I. ARIEFF

1. Introduction

In this chapter we first develop a definition by which the neurobehavioral syndrome, "uremic encephalopathy," and its variant presentations may be recognized as they occur clinically in integrated, whole organisms and as they may be indexed—at least to some extent—by objective and quantitative measurement techniques. These latter indices provide an essential linking "bridge of relevance" between the syndrome, itself the essential clinical reality, and the physiological and chemical data that are derived from probes into lower levels of biological organization at which the clinical, whole-organism identifiers of "uremic encephalopathy" are necessarily lost. Following this is a review of the available information concerning abnormal cerebral chemistry and metabolism in experimental animals and in human subjects with renal failure, with an attempt to validate the relevance of such observations to the defining clinical syndrome.

In addition to uremic encephalopathy, patients with end-stage renal failure are also at risk from all other causes of organic brain disease. However, two encephalopathic syndromes deserve special mention because they occur uniquely in patients treated by means of dialysis: (1) dialysis disequilibrium and (2) dialysis dementia. Again, following the clinical definitions, data derived from objective measurement are presented and the known alterations in brain composition and metabolism are reviewed.

PAUL E. TESCHAN • Division of Nephrology, Department of Medicine, Vanderbilt University School of Medicine, Nashville, Tennessee 37232.   *ALLEN I. ARIEFF* • Nephrology Section, Department of Medicine, University of California at San Francisco, San Francisco, California 94143.

2. Uremic Encephalopathy

2.1. Definition

2.1.1. Uremic Encephalopathy

Uremic encephalopathy is an acute or subacute organic brain syndrome that regularly occurs in patients with acute or chronic renal failure when levels of renal function (e.g., glomerular filtration rate) decline below 10% of normal (Bright, 1836; Addison, 1839; Schreiner, 1959; Schreiner and Maher, 1961; Tyler, 1968, 1970; Teschan, 1970a, 1983; Teschan and Ginn, 1976; Raskin and Fishman, 1976; Teschan *et al.*, 1979; Arieff, 1981). In common with other organic brain syndromes, these patients display variable disorders of consciousness, psychomotor behavior, thinking, memory, speech, perception, and emotion, not infrequently with features suggesting functional psychiatric illness. Since these features are variably represented in different patients, other verbal descriptors may sometimes be reasonably applicable in specific clinical instances, e.g., acute or chronic organic reaction, confusional state, clouding of consciousness, delirium, twilight state, coma, stupor, dementia, organic personality change, acute or chronic amnesic syndrome, and organic or toxic psychosis (Lishman, 1978).

Synonyms include "clinical uremia," the "uremic illness," or the "uremic syndrome" when these terms are used specifically to refer to the constellation of potentially or actually disabling neurobehavioral, mainly encephalopathic symptoms that are readily reversible by various modalities of dialysis treatment. In the clinical literature of nephrology, the terms *uremia* and *uremic* are variously and imprecisely used to refer to a range of abnormalities in organisms with renal failure: from modest elevations of BUN and creatinine (e.g., "a uremic serum") to any abnormality, whether symptomatic or not, detected by any means in any organ system (e.g., "uremic bone disease"; "uremic lung"; "uremic odor to the breath"; or "uremic frost" on the skin). Moreover, these terms are often used with several different implied meanings in the same communication (Schreiner and Maher, 1961; Teschan, 1983; Friedman, 1978; Bergstrom and Fürst, 1978). The concern of this text for *encephalopathy*, coupled with the early appearance and dialysis responsiveness of the albeit nonspecific symptoms of uremic encephalopathy, reinforces our choice of the more limited and precise definition. The other systemic abnormalities in patients with chronic renal failure are also separable on the grounds that they tend to appear late in the progressive clinical course rather than early, infrequently produce symptoms, are detected in cells, tissues, and organs rather than as integrated, whole-organism phenomena, and respond sluggishly or not at all to dialysis procedures (Teschan *et al.*, 1979; Teschan, 1983).

The symptoms that are reported by patients, their families, and their physicians may include: sluggishness and easy fatigue; daytime drowsiness and insomnia with a tendency toward sleep inversion; itching; inability to focus or sustain attention or to perform mental (cognitive) tasks and manipulation, or to manage ideas and abstractions; decreased flow and slurring of speech; anorexia, nausea, and (later) vomiting probably of central origin; restlessness; imprecise memory; diminished sexual interest and performance; volatile emotionality and withdrawal; feelings of coldness and hypothermia; myoclonus and "restlessness legs"; "burning feet"; asterixis; hiccoughs; paranoid thought content; disorientation and confusion with bizarre behavior; hallucinosis; muttering and mumbling; meningeal signs, amaurosis, nystagmus; vertigo and ataxia; transient pareses and aphasic episodes; torpor and preterminal coma and convulsions.

2.1.2. Certain Salient Characteristics

Certain salient characteristics of the symptoms of uremic encephalopathy are especially noteworthy:

1. They represent integrated (illness) behaviors in whole organisms.
2. As such they are generated by the nervous system and are manifested as cognitive, neuromuscular, somatosensory, and autonomic impairments.
3. Their severity and overall rates of progression vary directly with the rate at which renal function declines: uremic symptoms are generally more severe and progress more rapidly in acute than in subacute or chronic renal failure.
4. In more slowly progressive chronic renal failure, their number and severity also typically vary cyclically, in an undulating fashion with intervals of acceptable well-being, in the otherwise inexorable downhill course toward increasing disability.
5. They are readily ameliorated by dialysis procedures and suppressed by maintenance dialysis regimens.
6. They are usually relieved entirely following restoration of renal function, e.g., after promptly successful renal transplantation, especially in patients whose encounter with renal failure, dialysis, and post-transplant complications has not produced significant, residual cerebral injury.

Thus the encephalopathy of renal failure is important to recognize precisely because it is promptly and decisively treatable (i.e., within hours or days) by clinical methods that are generally available. As such, it is virtually unique among encephalopathies of man.*

2.2. Differential Diagnosis

The diagnosis of uremic encephalopathy in most patients is readily suspected by a clinical history that is positive for renal or urologic disease or injury, by such physical indicators as hypertension, anemia, pruritus, or asterixis, and by laboratory findings of azotemia or of albuminuria with or without an active urinary sediment.

---

* Other promptly reversible encephalopathies are usually acute, e.g., those accompanying severe hypercalcemia, hypoglycemia and hyperglycemia, hyposmolality and hyperosmolality, anoxia, and certain drug intoxications. Few if any slowly developing encephalopathies are so readily reversible.

However, the presenting symptoms of uremia (being nonspecific) are widely shared among many other encephalopathic states, as summarized, for example, by Lishman (1978) (Tables I and II). Thus the risk of misdiagnosis and mistreatment is significant for both the false-positive and false-negative cases (Teschan, 1983).

The problem of differential diagnosis is still more complex, however, because patients with renal failure are obviously also subject to other intercurrent illnesses and treatments that may induce other encephalopathic main effects or side effects or both. Moreover, if a drug or its metabolites with potential CNS toxicity is excreted or significantly metabolized by the kidney, the ensuing encephalopathic symptoms are *not* attributable to "uremia," but to the drug that has reached toxic levels at ordinary dose rates (Stone, 1983; Richet *et al.*, 1970). When levels of azotemia are also discovered (e.g., BUN about 100 and creatinine about 10 mg/dl or higher) that are sometimes associated with uremic encephalopathy in the absence of collateral illness, differentiation of the effects of drug versus renal failure ("uremia") may be impossible. One or more dialyses may of course restore more normal body fluid composition and also reduce

TABLE I. Causes of Acute Organic Reactions[a]

1. Degenerative	Presenile or senile dementias complicated by infection, anoxia, etc.
2. Space-occupying lesions	Cerebral tumor, subdural hematoma, cerebral abscess.
3. Trauma	"Acute post traumatic psychosis."
4. Infection	Encephalitis, meningitis, subacute meningovascular syphilis. Exanthemata, streptococcal infection, septicemia, pneumonia, influenza, typhoid, typhus, cerebral malaria, trypanosomiasis, rheumatic chorea.
5. Vascular	Acute cerebral thrombosis or embolism, episode in arteriosclerotic dementia, transient cerebral ischemic attack, subarachnoid hemorrhage, hypertensive encephalopathy, systemic lupus erythematosus.
6. Epileptic	Psychomotor seizures, petit mal status, postictal states.
7. Metabolic	Uremia, liver disorder, electrolyte disturbances, alkalosis, acidosis, hypercapnia, remote effects of carcinoma, porphyria.
8. Endocrine	Hyperthyroid crises, myxedema, Addisonian crises, hypopituitrism, hypo-, and hyperparathyroidism, diabetic precoma, hypoglycemia.
9. Toxic	Alcohol—Wernicke's encephalopathy, delirium tremens. Drugs—barbiturates (including withdrawal), bromides, salicylate intoxication, cannabis, LSD, prescribed medications (antiparkinsonian drugs, scopolamine, tricyclic and MAOI antidepressants, etc.) Others—lead, arsenic, organic mercury compounds, carbon disulphide.
10. Anoxia	Bronchopneumonia, congestive cardiac failure, cardiac dysrhythmias, silent coronary infarction, silent bleeding, carbon monoxide poisoning, postanesthetic.
11. Vitamin lack	Thiamine (Wernicke's encephalopathy), nicotinic acid (pellagra, acute nicotinic acid deficiency encephalopathy), B_{12} and folic acid deficiency.

[a] Reprinted with permission from Lishman (1978).

drug levels, so that the question remains moot while the patient recovers. Despite the possibilities that such multiple causes of encephalopathy might occur simultaneously, uremic encephalopathy may be successfully differentiated in most instances by means of the usual clinical data bases of history, physical examination, and laboratory determinations.

Meanwhile an interesting possibility derives from the fact that similar symptomatic outcomes may be produced in many etiological settings: that the brain may possess a limited response repertoire to a variety of noxious situations, identifying in fact a mechanistic "final common path" toward a relatively limited range of symptomatic expressions.

2.3. Objective Measurements of Encephalopathic Impairments

In most of the literature, uremic and other encephalopathies are characterized by verbal descriptors that usually represent authors' subjective assessments of patients' behavior. Responses to treatment are similarly noted, recording observers' comparisons of patients' present status with that recalled subjectively from some previous encounter. Especially, cognitive functions are likely to be impaired early and progressively as renal failure becomes severe. Similarly, abnormalities in the resting EEG and in evoked EEG responses were noted in such patients.

However, modern computer-based techniques facilitate acquisition and quantitative reduction of such data and permit objective comparisons between test sessions, estimates of clinical progression, and responses to treatment.

TABLE II. Causes of Chronic Organic Reactions

1. Degenerative	Senile dementia, arteriosclerotic dementia, Alzheimer's, Pick's, Huntington's, Creutzfeldt-Jakob, normal pressure hydrocephalus, multiple sclerosis, Parkinson's disease, Schilder's, Wilson's, progressive supranuclear palsy, progressive multifocal leucoencephalopathy, progressive myoclonic epilepsy.
2. Space-occupying lesions	Cerebral tumor, subdural hematoma.
3. Trauma	Posttraumatic dementia.
4. Infection	General paresis, chronic meningo-vascular syphilis, subacute and chronic encephalitis.
5. Vascular	Cerebral arteriosclerosis, "état lacunaire."
6. Epileptic	"Epileptic dementia."
7. Metabolic	Uremia, liver disorder, remote effects of carcinoma.
8. Endocrine	Myxedema, Addison's disease, hypopituitrism, hypo- and hyperparathyroidism, hypoglycemia.
9. Toxic	"Alcoholic dementia" and Korsakoff psychosis, chronic barbiturate or bromide intoxication, manganese, carbon disulphide.
10. Anoxia	Anemia, congestive cardiac failure, chronic pulmonary disease, postanesthetic, post carbon-monoxide poisoning, postcardiac arrest.
11. Vitamin lack	Lack of thiamine, nicotinic acid, B_{12}, folic acid.

Reprinted with permission from Lishman (1978).

2.3.1. Cognitive Function Measures in Uremic Encephalopathy

2.3.1a. Experimental. Learned maze behavior was studied in a spontaneously reversible model of acute renal failure in rats, using response latency (maze-running time to a food reward) as the measured endpoint (Essman, 1962, 1965). Maximal deviation of response latency from control (see Fig. 1) was noted when azotemia was maximal. Both measures returned to normal levels at about the same time. Especially interesting was the observation that the rats with renal failure were not obviously ill, had normal locomotor ability, but hesitated at the maze bifurcation in making the previously learned choice of the route to the reward. The result might have been due to anorexia, however, rather than a cognitive impairment, but the approximate congruence of azotemia and response latency established the fact that measured, overt, integrated behavior in whole organisms could be used as an outcome index of organic illness. A subsequent experiment in an operant-conditioned Rhesus monkey (see Fig. 2) confirmed and extended this assessment (Teschan, 1970a,b; Teschan *et al.*, 1970). Bilateral ureteral ligation induced levels of azotemia and

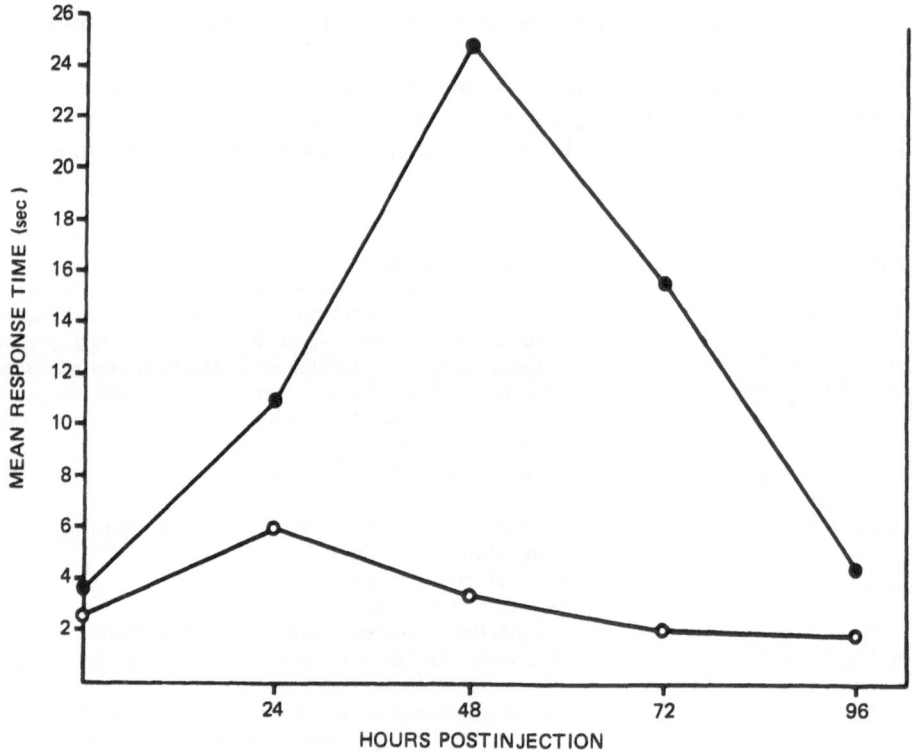

FIGURE 1. Effect of renal failure on learned maze behavior in rats. Compared with the controls (○), spontaneously reversible acute renal failure (induced by injection of a nephrotoxic material) in the experimental group (●) was associated with a trend in maze-response time that approximately paralleled the degree of renal failure as indexed by the rats' BUN concentrations. The uninjected control group maintained normal levels of performance. Reprinted by permission from Essman (1962).

other chemical changes associated with deterioration of counting behavior. The visually assessed proportion of slow EEG wave frequencies increased simultaneously. A peritoneal dialysis treatment (24 hr, automated dialysate cycler) tended to reverse both the behavioral and the EEG abnormality. In this instance, because urea was added to the dialysate to approximate its concentration in the animal's body water, little change in urea concentration occurred as a result of dialysis. The experiment suggested that the BUN concentration of 300 mg/dl did not "cause" the "toxic effect" observed in both the behavioral performance measure and the EEG. Other experimental approaches have also been explored (Teschan *et al.*, 1964; Sharp and Murphy, 1964, 1966; Murphy and Sharp, 1964).

2.3.1b. Clinical. Several measures of sustained and selective attention, speed of decision making, short-term memory, and mental manipulation of

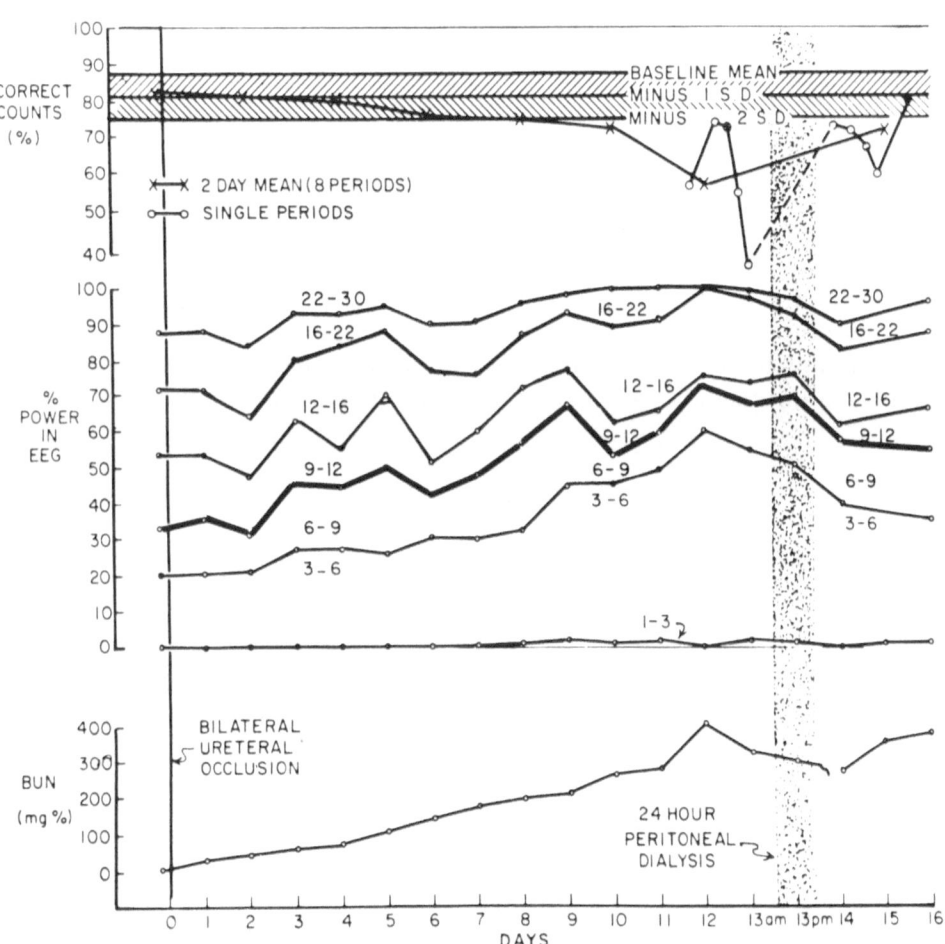

FIGURE 2. Responses of counting performance, power distribution among EEG wave frequencies, and BUN concentrations to renal failure induced by ureteral occlusion and to peritoneal dialysis in an operant-conditioned Rhesus monkey. Reprinted by permission from Teschan *et al.* (1970).

symbols have been employed in studies of patients with progressive chronic renal failure, including their response to dialysis and transplantation (Teschan *et al.*, 1979; McDaniel, 1971; Miller *et al.*, 1974; Edwards *et al.*, 1977). In Fig. 3, normalized group data from three such measures—the choice reaction time, continuous memory test, and continuous performance test—reveal abnormal mean scores in azotemic patients compared with normals, as well as a return of average scores toward normal in response to treatment by means of maintenance dialysis or successful renal transplantation (Teschan, 1981).

2.3.2. Encephalographic Indices of Uremic Encephalopathy

2.3.2a. Experimental. Figure 2 illustrates the principal, early EEG alterations of progressive renal failure, i.e., slowing of wave frequencies and the expected response to dialysis treatment. Inferentially, it should be noted that some solute concentration other than urea or some cerebral metabolic sequence not directly affected by urea concentrations in the body water (including the brain) was presumably normalized by the dialysis procedure as were the indices of encephalopathy.

2.3.2b. Clinical. Computerized EEG power spectrum analysis provided quantitative estimates of EEG slowing in azotemic, dialyzed, and "transplanted" patients, both individually and in groups, in the cited study (Teschan *et al.*, 1979) (see Fig. 3). Similarly, the latency of visually evoked potentials and the frequency loci of the EEG responses to photic driving varied congruently both with the EEG power measure and with the measures of cognitive function (Teschan, 1981). These findings were consistent with an extensive literature in which various methods of EEG analysis were utilized (Teschan *et al.*, 1979; Romano and Engel, 1944; Engel *et al.*, 1944; Engel and Romano, 1944;

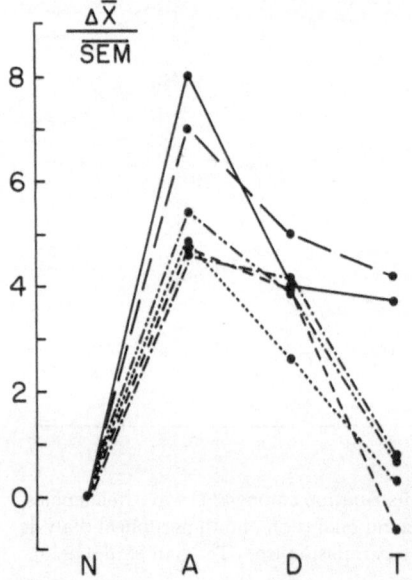

FIGURE 3. Intergroup *differences* between the means of six neurobehavioral measures in (N) normals (mean set at 0), (A) azotemic (non-dialyzed) patients with severe chronic renal failure, (D) stable hemodialyzed patients, and (T) patients following successful renal transplantation. The differences between the adjacent group means are expressed as multiples of their pooled standard errors. (—) EEG % P = % EEG power in frequency bands: (3–7 Hz)/(3–13 Hz) X 100, (———) VER = visually evoked (occipital EEG) response, latency of major negative deflection, (– – –) PD S/F = photic driving response, ratio of EEG power in subharmonic/fundamental frequency bands (–·–) CMT = continuous memory test, number of errors/110 test items, (---) CRT = choice reaction time, latency in seconds, (—–·—) CPT = continuous performance test, % errors in 300 trials. Reprinted by permission, Teschan *et al.* (1979); Teschan (1981).

Glaser, 1960, 1976; Kiley and Hines, 1963, 1975; Jacob et al., 1965; Kiley et al., 1970, 1976a,b; Klingler, 1974; Guisado et al., 1975; Bourne et al., 1975a,b; Hagstam, 1977; Spehr et al., 1977; Cooper et al., 1978; Bowling and Bourne, 1978; Hughes, 1980; Kiley, 1981; Hayman and Kooi, 1969; Lewis et al., 1978; Hamel et al., 1978).

2.3.3. Biochemical Changes in Brain

The causes of the EEG abnormalities and clinical symptoms observed in patients with either acute or chronic renal failure (ARF or CRF) are largely unknown. To investigate these abnormalities, biochemical studies have been carried out in brain of both patients and laboratory animals with renal failure. Most studies have been done in animals with ARF. Measurements have included brain intracellular pH and concentrations of Na^+, K^+, Cl^-, H_2O, Al^{3+}, Ca^{2+}, Mg^{2+}, urea, adenine nucleotides (creatine phosphate, ATP, ADP, AMP), lactate, and (Na^+,K^+)-activated adenosine triphosphatase (ATPase). In patients with acute renal failure, content of H_2O, K^+, and Mg^{2+} is normal, whereas Na^+ is modestly decreased and Al^{3+} slightly elevated. However, cerebral cortex Ca^{2+} content is almost twice the normal value (Cooper et al., 1978; Cogan et al., 1978). Similar findings have been observed in dogs with acute renal failure (Arieff and Massry, 1974). Permeability of uremic brain (in rats) to inert molecules, such as inulin and sucrose, is increased. Entry into brain of Na^+ is impaired, whereas that of K^+ is increased. The permeability of brain to weak acids, such as sulfate, penicillin, and dimethadione, is normal to low (Arieff et al., 1976; Fishman, 1970).

Alterations of cerebral metabolism that might be related to the aforementioned changes in permeability have also been studied in animals. In brain of rats with acute renal failure, van den Noort and co-workers (1968) found that creatine phosphate, ATP, and glucose were increased, but there were corresponding decreases in creatine, AMP, ADP, and lactate. Total brain adenine nucleotide content and (Na^+,K^+) ATPase activity were normal to low. The uremic brain utilized less ATP and thus failed to produce ADP, AMP, and lactate at normal rates. There was a corresponding decrease in the brain's metabolic rate, along with elevated glucose and low lactate levels (van den Noort et al., 1968). Other studies of uremic brain have shown a decrease in cerebral oxygen consumption (Arieff, 1981). Patients with chronic renal insufficiency (GFR less than 20 ml/min) have decreased brain uptake of glutamine and increased ammonia uptake. The relevance of these findings, in terms of neurotransmitters or other brain function, is unknown.

In animals with either acute or chronic renal failure, both urea concentration and osmolality are similar in brain, CSF, and plasma (Fig. 4). The solute content of brain in animals with acute renal failure is such that essentially all of the increase in brain osmolality is due to an increase of brain urea concentration (Arieff et al., 1973). However, in animals with chronic renal failure, almost half of the increase in brain osmolality is due to the presence of undetermined solute (idiogenic osmoles) (Mahoney and Arieff, 1983).

In animals who have either chronic or acute renal failure and metabolic acidosis, the intracellular pH (pH_i) of brain, and skeletal muscle as well, is normal (Fig. 5). In patients with renal failure, intracellular pH has been reported to be normal in both skeletal muscle and leukocytes, as well as in the "whole body" (Fig. 6). The pH of CSF has also been shown to be normal in both patients and laboratory animals with renal failure. Thus, despite the presence of extracellular metabolic acidemia, brain pH_i is normal in animals with either acute or chronic renal failure (Arieff *et al.*, 1976; Mahoney and Arieff, 1983).

2.3.4. Pathophysiology—Role of PTH

Although a multitude of factors contributes to uremic encephalopathy, including any of the commonly measured indicators of renal failure (BUN, creatinine, bicarbonate, pH, potassium) (Raskin and Fishman, 1976), in recent years there has been considerable discussion of the possible role of parathyroid hormone (PTH) as a uremic toxin. In particular, a substantial amount of evidence suggests that there are toxic effects of PTH on the CNS. In patients dying with acute or chronic renal failure, the calcium content in brain (cerebral cortex) is significantly elevated (Cooper *et al.*, 1978; Cogan *et al.*, 1978) (Fig. 7). Dogs with acute or chronic renal failure show increases of brain gray matter calcium and EEG changes similar to those seen in humans with renal failure (Guisado *et al.*, 1975). In dogs, both the EEG and brain calcium abnormalities can be prevented by parathyroidectomy. Conversely, they can be reproduced by administration of parathyroid hormone to normal animals, but not by elevation of the (calcium) X (phosphate) product nor by hypercalcemia induced by vitamin D intoxication. Thus PTH appears essential to produce central nervous system manifestations in the dog model of uremia.

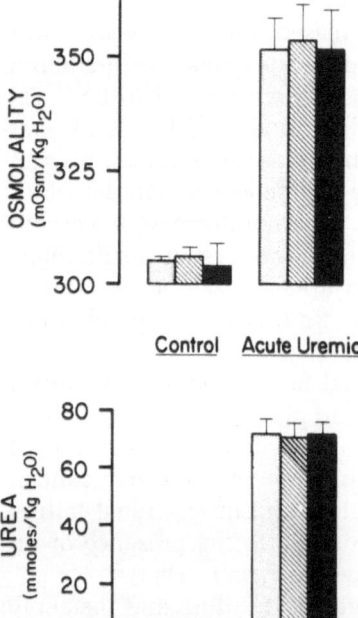

FIGURE 4. The osmolality and urea concentration of (▨) plasma, (▧) CSF, and (■) brain (cortical gray matter) are shown in both control dogs and dogs with acute uremia (serum creatinine = 11 mg/dl). The osmolality and urea concentration are similar in the three compartments in both normal and uremic dogs. Data are reproduced with permission from Arieff *et al.* (1973).

PTH is known to have CNS effects in humans even when they have normal renal function. Neuropsychiatric symptoms are frequent manifestations of primary hyperparathyroidism (Heath *et al.*, 1980) and such patients may also have EEG changes similar to those observed in patients with renal failure (Cogan *et al.*, 1978; Allen *et al.*, 1970). The common denominator appears to be elevated parathyroid hormone. In patients with acute renal failure, the EEG is abnormal within 18 hr of the onset of renal failure and is generally not affected by dialysis for periods of up to 8 weeks (Cooper *et al.*, 1978). During this interval, patients with acute renal failure have been shown to have elevated levels of PTH in plasma. In patients with either primary or secondary hyperparathyroidism, parathyroidectomy results in an improvement of both EEG and psychological testing, suggesting a direct effect of PTH on the central nervous system (Cogan *et al.*, 1978). Similarly, dialysis results in a decrement of brain (cerebral cortex) calcium toward normal in both patients and laboratory animals with renal failure, concomitant with improvement of the EEG (Arieff and Massry, 1974; Teschan *et al.*, 1979). In uremic patients, both EEG changes and psychological abnormalities are improved by parathyroidectomy or medical suppression of PTH (Arieff and Massry, 1974; Goldstein *et al.*, 1980). Thus, PTH and/or a high brain calcium content are probably responsible, at least in part, for many of the encephalopathic manifestations of renal failure.

The mechanism by which PTH might disturb central nervous system function is unclear, although the occurrence of psychosis in patients with hyperparathyroidism has been well documented (Arieff, 1981; Heath *et al.*, 1980).

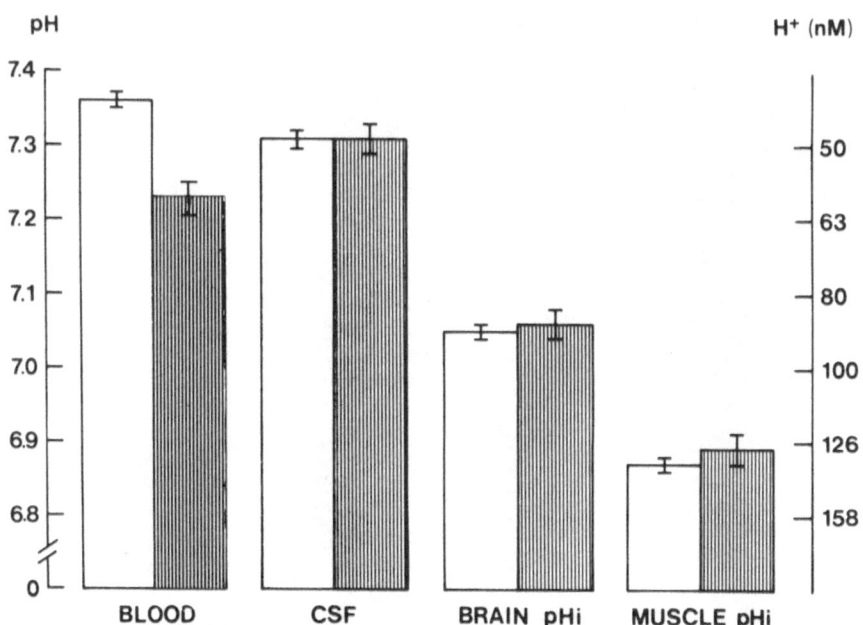

FIGURE 5. The intracellular pH (pH_i) of brain (cortical gray matter) and skeletal muscle in (□) normal dogs and those with (▥) acute renal failure (uremia, serum creatinine = 11 mg/dl). Despite a significant extracellular acidemia, pH in CSF, as well as pH_i in brain and muscle, are normal. Reproduced with permission from Arieff *et al.* (1976).

The increased calcium content in such diverse tissues as skin, cornea, blood vessels, brain, and heart in patients with hyperparathyroidism suggests that PTH may somehow facilitate the entry of Ca^{2+} into such tissues. The finding of increased calcium in the brain of both dogs and patients with chronic renal disease and secondary hyperparathyroidism is consistent with the conception that part of the central nervous system dysfunction and EEG abnormalities found in acute or chronic renal failure may be due to a PTH-mediated increase in brain calcium. Calcium is essential for the function of a large number of intracellular enzyme systems (Rasmussen and Goodman, 1977), and an increased brain calcium content could disrupt cerebral function by interfering with any of them. It is also possible that PTH itself may have a detrimental effect on the central nervous system.

In summary, of the long list of proposed uremic toxins, none have been shown to correlate better than creatinine clearance with any manifestations of uremia. A pathophysiologic role in the central nervous system dysfunction of uremia has been clearly established thus far only for parathyroid hormone. Pathologic and histologic examinations have generally shown minor, nonspecific abnormalities. However, there is a notable lack of ultrastructural investigations of the central nervous system in uremia. Neither brain edema nor intracellular acidosis occur. The functional significance of the altered brain permeability, decreased brain sodium, and presence of idiogenic osmoles is uncertain. The low turnover rate of high-energy phosphates seen in rats with

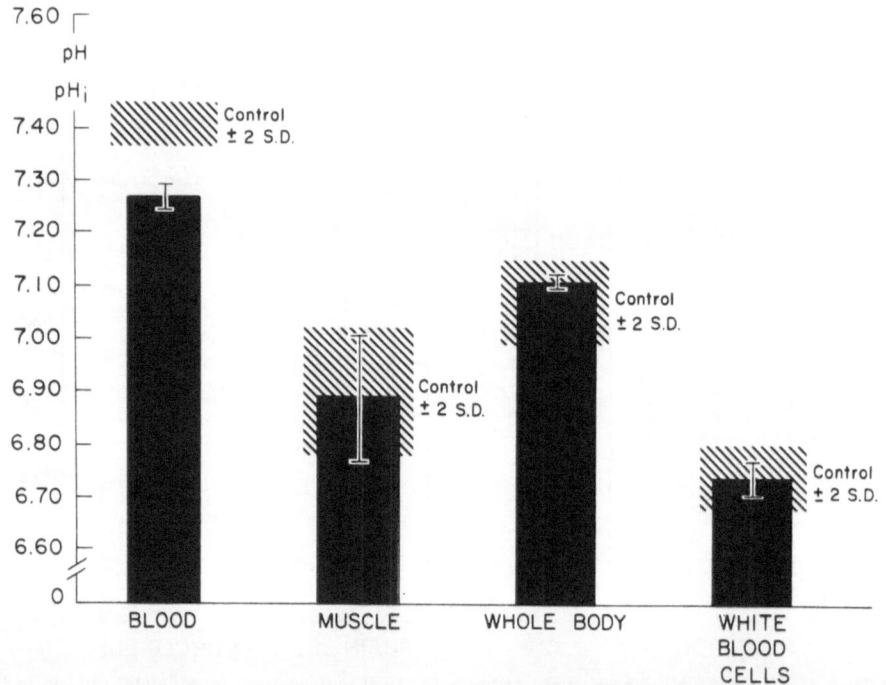

FIGURE 6. Arterial pH in patients compared with intracellular pH (pH_i) in skeletal muscle, "whole body," and white blood cells. Although there is a significant extracellular acidemia, pH_i is normal in muscle, "whole body," and white blood cells. Reproduced with permission from Arieff (1981).

acute renal failure and the decreased cerebral oxygen consumption may as easily be a result as a cause of the uremic state.

3. Dialysis-Related Encephalopathic Syndromes

3.1. Dialysis Disequilibrium Syndrome (DDS)

Although dialysis procedures and regimens relieve and suppress the symptoms of uremic encephalopathy, they also induce a transient, acute illness in which certain of the uremic symptoms appear to be simply exacerbated or in which new symptoms may appear. These occurrences are presumably attributable to the relatively rapid changes in body composition that are induced in the course of dialysis treatment sessions.

3.1.1. Clinical Description and Differential Diagnosis

The clinical manifestations of DDS have been summarized in Table I. For convenience, they have been divided into major and minor symptoms. Mild forms of DDS may be manifested by no more than restlessness and severe headache, which may occur during or soon after hemodialysis. This is commonly followed by nausea and vomiting, often accompanied by blood pressure elevation. These symptoms may be accompanied by disorientation and tremors. Seizures and cardiac arrhythmias have been reported, but are uncommon. The

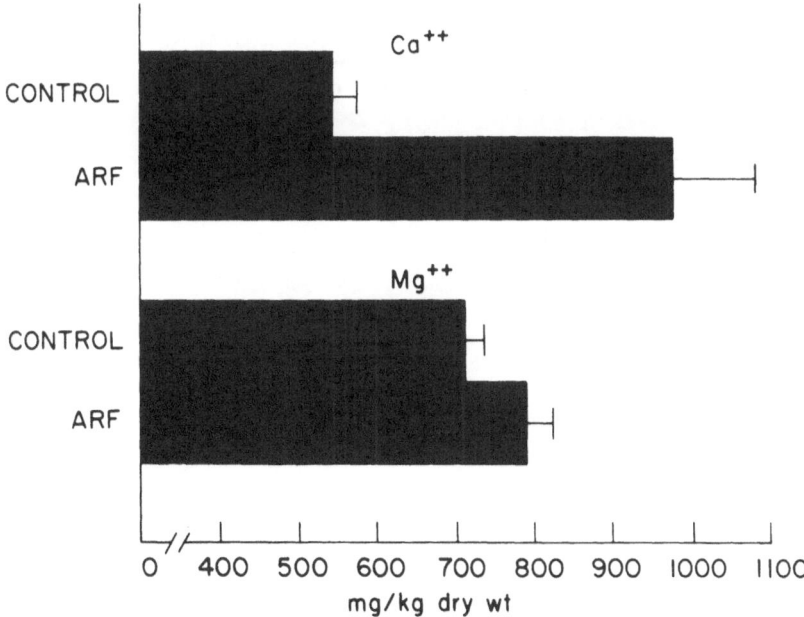

FIGURE 7. The calcium (Ca^{2+}) and magnesium (Mg^{2+}) content in brain (cortical gray matter) of control patients and those with acute renal failure (ARF). Brain Ca^{2+} is almost twice the normal value, whereas Mg^{2+} is not different from control. Figure reproduced with permission from Cooper *et al.* (1978).

seizures have usually been of the grand mal type, although focal tremors have been observed. The symptoms are usually self-limited, but recovery may take several days. In rare instances the seizures may lead to coma. It appears that modern methods of dialysis have altered the clinical picture of DDS. Most reports of seizures, coma, and death were reported prior to 1970. The symptoms of DDS as reported in the last decade (1972–1982) have generally been mild, consisting of nausea, weakness, headache, fatigue, and muscle cramps. It is also unclear whether any patient ever actually died from DDS per se or from other associated neurological complications of dialysis, such as acute stroke, subdural hematoma, subarachnoid hemorrhage, or hyponatremia.

Patients who have renal failure and are being treated with hemodialysis may manifest the symptoms of headache, nausea, emesis, muscle cramps, and seizure activity because of causes other than DDS. A differential diagnosis of such patients is shown in Table III. Recently, the diagnosis of DDS has become a "wastebasket" for a number of disorders that can occur in patients with renal failure and may affect the central nervous system. It should be stressed that the disgnosis of DDS should be one of exclusion.

3.1.2. Objective Measurements

3.1.2a. Experimental. When acute azotemic renal failure is induced in dogs, the EEG reveals disorganized alpha rhythm and increased slow-wave spike activity. These changes are accentuated with loss of alpha rhythm, more frequent spike-wave sequences, and appearance of very-high-amplitude slow waves during rapid hemodialysis (Kennedy *et al.*, 1964).

3.1.2b. Clinical. Accentuated EEG slowing and disorganization "by inspection" were reported in the early studies of the DDS (Kennedy *et al.*, 1962).

TABLE III. Manifestations of Dialysis Disequilibrium Syndrome

Minor symptoms
Headache
Nausea
Emesis
Muscle cramping
Restlessness
Fatigue
Major symptoms
Muscle tremor
Blurring of vision
Hypertension
Disorientation
Syncope
Seizures (focal or grand mal)
Cardiac arrhythmias
Laboratory findings
No consistent abnormalities
EEG abnormal, but not diagnostic
CSF pressure elevated

More recent experience using quantitative EEG analysis confirms and extends the earlier observations as illustrated in Figs. 8 and 9. Electroencephalograms were recorded predialysis on each day of the first four (daily) dialysis treatments in patients with end-stage renal disease as they began their regimens of chronic maintenance hemodialysis. The tape-recorded EEG signals were digitized, subjected to discriminant analysis, and scored in such a way that the more positive scores are more abnormal (Bowling and Bourne, 1978). Figure 8 reveals a stepwise improvement of the predialysis EEGs on successive treatment days, reflecting relief of uremic encephalopathy. However, as shown in Fig. 9, increasing transient abnormalities (disequilibrium) usually occur during all of the dialysis sessions, although with recovery and further improvement overnight (compare Fig. 8). The variance of the pre-dialysis and postdialysis differences are initially large, but decline progressively with successive treatments.

Similar studies were conducted during *chronic* maintenance hemodialysis sessions (see Fig. 10). Even so-called "stable" chronic hemodialysis patients with minimal postdialysis symptoms developed clear slowing of wave frequencies in the EEG power spectrum after 3 hr of dialysis, evidence that a degree of disequilibrium occurred. As shown, those patients who noted more

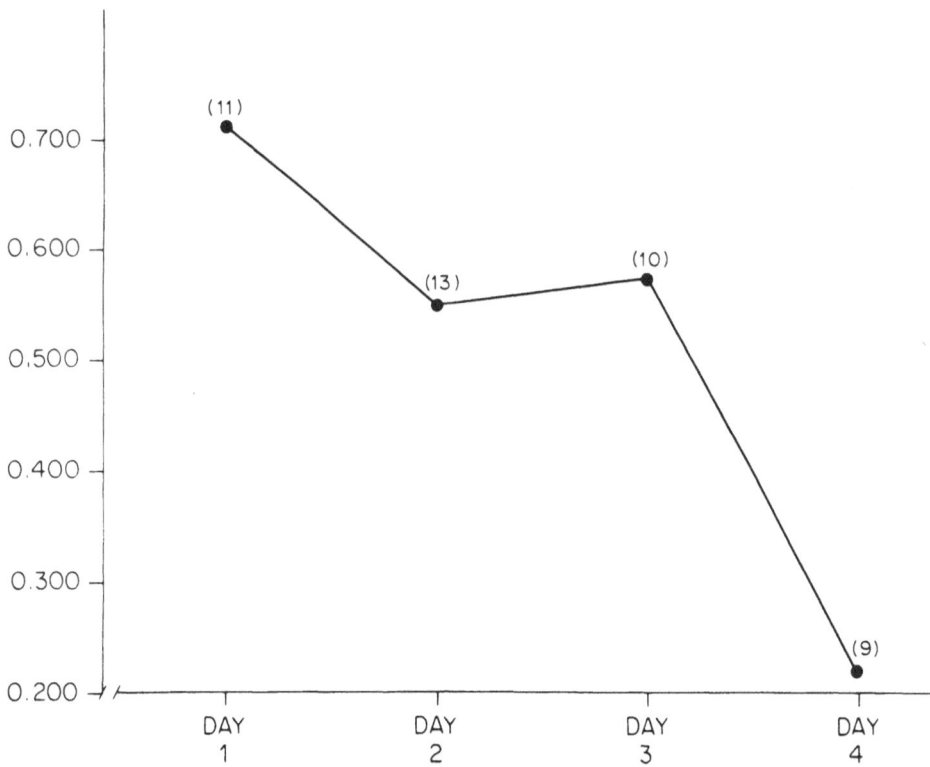

FIGURE 8. EEG discriminant scores [see Bowling and Bourne (1978)] prior to each of the first four dialyses in (n) patients beginning maintenance hemodialysis treatment. Scores among patients with severe azotemia approximate + 1.0 ± 1.0, compared with scores among normals of − 1.0 ± 1.0.

significant postdialysis symptomatology (increased weakness, "washed-out" feeling) also gave evidence of more marked EEG abnormality than did those patients without such symptoms.

3.1.3. Pathogenesis

There has been much debate about the pathogenesis of DDS. In early studies on patients treated with hemodialysis, a consistent elevation in CSF pressure was observed (Kennedy *et al.*, 1964). In a few patients who died with, but probably not because of DDS, the presence of brain edema was demonstrated (Arieff, 1981). Because of the aforementioned, it has been suggested that brain edema is a major factor underlying the pathogenesis of DDS. Early clinical investigation in patients with DDS included evaluation of pressure and composition of the CSF. In patients with DDS, findings included a persistent elevation of CSF pressure and levels of urea that were higher in CSF than in blood (Kennedy *et al.*, 1964). These findings led to the early formulation that cerebral edema was responsible for most of the manifestations of DDS.

3.1.3a. Reverse Urea Effect. The presence of brain edema in some patients with DDS, along with the observation that in patients undergoing rapid hemodialysis urea levels are higher in CSF than in blood, led to formulation of the "reverse urea" hypothesis. Simply put, the "reverse urea" effect states that steady-state concentrations of urea are similar in plasma, CSF, and brain tissue water (Fig. 4). With rapid hemodialysis, it is further assumed that clearance of urea will be more rapid from plasma than from brain. This assumption stems from the observed slower clearance of urea from CSF than from plasma (Kennedy *et al.*, 1962) in patients undergoing rapid hemodialysis. The decrement

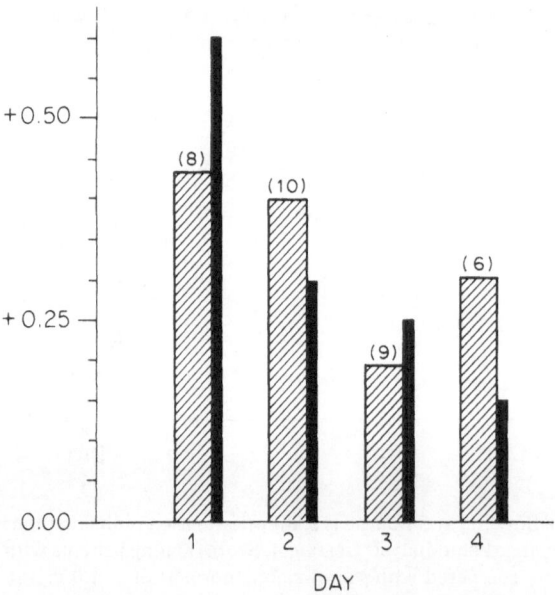

FIGURE 9. Average [▨ mean ± ■ SEM, (n) observations] changes in EEG discriminant scores [(postdialysis)-(predialysis)] during the four initial dialysis sessions in Fig. 8. The average changes toward greater abnormality in each session, but both the magnitude and the variance of the change tend to decline with successive dialyses.

of plasma urea leads to a fall in plasma osmolality. It was postulated that with a slower clearance of urea from brain, osmolality of brain tissue remained elevated above that of plasma, leading to a net movement of water from plasma into brain with resultant cerebral edema. However, the "reverse urea" hypothesis has not been supported by experimental investigations. The rate of urea clearance from brain essentially parallels that of plasma during rapid hemodialysis in uremic dogs (Arieff *et al.*, 1973).

3.1.3b. Central Nervous System Acidosis. There is an alternate explanation for the cerebral edema that does have experimental support. Along with the delayed clearance of urea from the CSF, there is evidence for a paradoxical

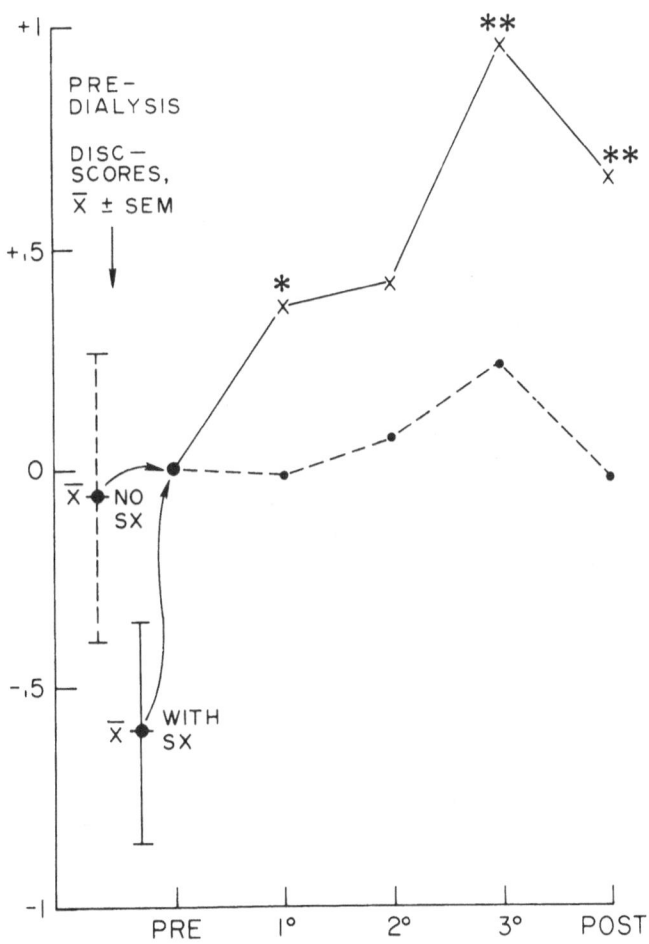

FIGURE 10. Average change in EEG discriminant scores in ten patients with end-stage chronic renal failure treated by means of maintenance hemodialysis for more than six months, five with and five without complaints of "postdialysis syndrome". Predialysis mean values, mean ± SEM are shown. (* $P < 0.05$, ** $P < 0.01$). The predialysis mean values are set at zero and the mean *differences* between the patients' scores at hourly intervals during four dialysis sessions and their respective predialysis values are shown. Note the abnormal trend (suggesting dialysis disequilibrium) in both groups, and the greater average abnormality in the patients with symptoms.

acidemia of CSF in both patients and laboratory animals with DDS. In some patients treated with rapid hemodialysis, there is a fall in pH of the CSF despite a rise in arterial pH (Arieff, 1981). A similar fall in pH of CSF occurs after rapid hemodialysis of experimental animals with acute renal failure (Arieff et al., 1976). The decrement of pH in CSF may be a major factor in the pathogenesis of DDS. Along with the decrement of pH of CSF, there is an observed decrease in the pH_i of brain. In dogs with acute renal failure, pH_i is normal in brain and skeletal muscle (Fig. 5). After rapid hemodialysis, there is no change in the pH_i of muscle, but a significant decrement is observed in the pH_i of cerebral cortical gray matter (Arieff et al., 1976). The increased H^+ ion activity in brain is accompanied by an increase of brain osmole content, which secondarily results in an increase of brain water (Arieff et al., 1973). The cerebral edema which thus occurs is probably the cause of the observed clinical manifestations of DDS. The presence of postdialysis brain swelling has recently been documented by CAT scan in patients with chronic renal failure. The source of the increased brain H^+ ion is not known.

3.2. Dialysis Dementia

3.2.1. Clinical Description

This syndrome first described by Alfrey in 1972 (Alfrey et al., 1972) consists of (1) speech disorders (slowing, stuttering, dysnomia, and dyspraxia); (2) multifocal myoclonus with or without seizures; (3) neuropsychiatric abnormalities (impaired memory and ability to concentrate, personality changes, depression, paranoid ideation, hallucinations, but with retained consciousness); and (4) typical EEGs (generalized slowing, generalized bursts of delta waves without somatic phenoma or changes in consciousness, and spike activity) in the absence of other structural, metabolic, or infectious disorders of the central nervous system (Dewberry et al., 1980). Cerebrospinal fluid dynamics are abnormal in some but not all patients (Arieff, 1981). Typically the syndrome and the EEG abnormalities are progressive during each dialysis session, in the course of the three dialysis sessions in any one week, and throughout the successive weeks of dialysis treatment. The syndrome is thus almost always progressive and fatal within a year of onset, although in a few patients the syndrome has been reversed or arrested.

3.2.2. Objective Measurements

No experimental model exists as yet. The syndrome has been documented on videotape (A. C. Alfrey, personal communication) and by means of conventional EEG recordings (Alfrey et al., 1972). The latter have not yet been quantified and are frequently contaminated with artifacts because of patients' spontaneous motor activity and increasingly and progressively limited ability to cooperate.

3.2.3. Etiology

The etiology of this syndrome remains controversial. Although an increase in brain aluminum content has been strongly implicated in some cases of dialysis dementia, the evidence is less convincing in others. Dialysis dementia most likely represents a syndrome complex that is the final common pathway for a variety of etiological agents. At this stage of our knowledge, it seems useful to subdivide dialysis dementia into three categories: (1) an epidemic form that is related to contamination of the dialysate, probably with aluminum; (2) sporadic cases in which aluminum intoxication is less likely to be a contributory factor; and (3) dementia associated with congenital or early childhood renal disease. This entity has been reported in several children who were never dialyzed or exposed to aluminum compounds. It now appears that these early childhood cases represent developmental neurologic defects resulting from exposure of the growing brain to a uremic environment (Rotundo et al., 1982).

Aluminum intoxication was first implicated in this disorder in 1976 (Alfrey et al., 1976). Aluminum content of brain gray matter was elevated 11-fold in patients with dialysis dementia and threefold in patients on chronic hemodialysis without dialysis dementia, relative to normal controls. Bone and other soft-tissue burdens of aluminum were also increased. No consistent differences were found in a wide variety of other trace elements studied. Oral phosphate-binding compounds containing aluminum Al(OH)₃ were originally suspected as the source of the elevated aluminum in brain. Significant absorption of oral aluminum can occur in patients with chronic renal failure, and may account for some of the sporadic cases occurring in undialyzed patients from areas without high dialysate aluminum (Alfrey et al., 1976; Dewberry et al., 1980). However, the weight of evidence is against oral aluminum as the major source.

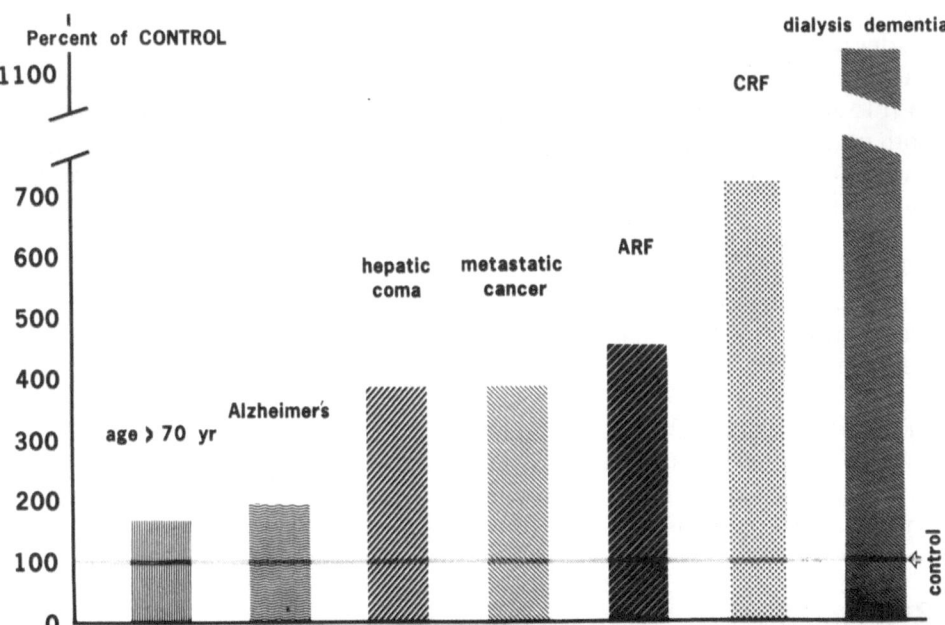

FIGURE 11. The brain (cortical gray matter) content of aluminum in patients with various medical conditions. Aluminum content is shown as a percent of control, with normal being 100% The data were compiled from a literature review [Arieff (1981)]. Brain aluminum content is above normal in all seven conditions shown, although it is highest (11 times control) in patients with dialysis dementia. Figure reproduced with permission from Arieff (1981).

3.2.4. Brain Composition

There is general agreement that brain aluminum is elevated in dialysis dementia, but it has been questioned whether brain aluminum is the cause of neurotoxicity rather than an association. Although several investigators were able to show a good separation between aluminum content of brain cortical gray matter in hemodialyzed patients with versus those without dialysis encephalopathy, other authors have noted considerable overlap (Arieff, 1981; Dewberry et al., 1980; Arieff et al., 1979). Surprisingly, brain aluminum levels are higher in patients with chronic renal failure not yet on dialysis than their dialyzed counterparts (Arieff et al., 1979). Elevated brain aluminum levels are also seen in patients with hepatic encephalopathy and metastatic cancer, although they are generally lower than in patients with dialysis encephalopathy (Fig. 11). Unfortunately, most studies have reported brain aluminum levels in normals rather than the appropriate control patients with chronic renal failure with and without hemodialysis. There is no obvious threshold level for brain aluminum that determines toxicity. In general, though, brain levels above 15 mg/kg dry weight are associated with the encephalopathy, whereas levels below 5 mg/kg are not.

4. Commentary and Prospect

The foregoing discussion and the cited references provide a survey of our knowledge at this intersection of the fields of neurology and nephrology with respect to the encephalopathy that attends severe renal failure. The thoughtful reader may well be more impressed by what is missing than by what is known, and may recognize how relatively limited have been our investigations to date in this area that is obviously epicentric to the major contemporary themes in both of the now too widely separated "parent" disciplines. Hence the methods and concepts of neurochemistry, neuropharmacology, and neuroendocrinology have yet to be applied systematically in investigations of the encephalopathy of renal failure. However, the tantalizing prospect of deriving important new information about cerebral mechanisms will hopefully stimulate such needed further investigations by collaborating neuroscientists and nephrologists who are interested in these manifestations of human disease and their response to treatment.

ACKNOWLEDGMENTS. The authors gratefully acknowledge their support for those studies included in this chapter, from colleagues, technicians, and secretaries, from contracts and grants including NIH/NIAMDD/AKCUP #NO1-AM-2-2211, and for the support of Dialysis Clinics, Inc.

References

Addison, T., 1839, On the disorders of the brain connected with diseased kidneys, Guy's Hosp. Rep. 4:1–7.
Alfrey, A. C., Mishell, J. M., Burks, J., Contiguglia, R. H., Lewin, E., and Holmes, J. H., 1972,

Syndrome of dyspraxia and multifocal seizures associated with chronic hemodialysis, *Trans. ASAIO* **18**:257–261.

Alfrey, A. C., LeGendre, G. R., Kaehney, W. D., 1976, The dialysis encephalopathy syndrome: Possible aluminum intoxication, *N. Engl. J. Med.* **294**:184–188.

Allen, E. M., Malamed, D., and Singer, F. R., 1970, Electroencephalographic abnormalities in hypercalcemia, *Neurology* **20**:15–22.

Arieff, A. I., 1981, Neurological complications of uremia, in: *The Kidney*, 2nd ed. (B. M. Brenner and F. C. Rector, Jr., eds.), W. B. Saunders, Philadelphia, pp. 2306–2343.

Arieff, A. I., and Massry, S. G., 1974, Calcium metabolism of brain in acute renal failure: Effects of uremia, hemodialysis and parathyroid hormone, *J. Clin. Invest.* **53**:387–392.

Arieff, A. I., Massry, S. G., Barrientos, A., and Kleeman, C. R., 1973, Brain water and electrolyte metabolism in uremia: Effects of slow and rapid hemodialysis, *Kidney Int.* **4**:177–187.

Arieff, A. I., Cooper, J. D., Armstrong, D., and Lazarowitz, V. C., 1979, Dementia, renal failure and brain aluminum, *Ann. Intern. Med.* **90**:741–747.

Arieff, A. I., Guisado, R., Massry, S. G., and Lazarowitz, V. C., 1976, Central nervous system pH in uremia, hemodialysis and parathyroid hormone, *J. Clin. Invest.* **58**:306–311.

Bergstrom, J., and Fürst, P., 1978, Uraemic toxins, in: *Replacement of Renal Function by Dialysis* (W. Drukker, F. M. Parsons, and J. F. Maher, eds.), Martinus Nijhoff, The Hague, pp. 334–368.

Bourne, J. R., Miezin, F. M., Ward, J. W., and Teschan, P. E., 1975a, Computer quantification of electroencephalographic data recorded from renal patients, *Comput. Biomed. Res.* **8**:461–473.

Bourne, J. R., Ward, J. W., Teschan, P. E., Musso, M., Johnston, H. B., Jr., and Ginn, H. E., 1975b, Quantitative assessment of the electroencephalogram in renal disease, *Electroencephogr. Clin. Neurophysiol.* **39**:377–388.

Bowling, P. S., and Bourne, J. R., 1978, Discriminant analysis of electroencephalograms recorded from renal patients, *IEEE Trans. Bio-Med. Eng.* **25**:12–17.

Bright, R., 1836, Cases in observations illustrative of renal disease accompanied by the secretion of albuminous urine. *Guy's Hosp. Rep.* **1**:338–379.

Cogan, M., Covey, C., Arieff, A. I., Wisniewski, A., Clark, O., Lazarowitz, V. C., and Leach, W., 1978, Central nervous system manifestations of hyperparathyroidism, *Am. J. Med.* **65**:963–970.

Cooper, J. D., Lazarowitz, V. C., and Arieff, A. I., 1978, Neurodiagnostic abnormalities in patients with acute renal failure: Evidence for neurotoxicity of parathyroid hormone, *J. Clin. Invest.* **61**:1448–1455.

Dewberry, F. L., McKinney, T. D., and Stone, W. J., 1980, Report of fourteen cases and review of the literature, *ASAIO J.* **3**:102–108.

Edwards, A. E., Kopple, J. E., Miller, J. M., Fields, L. G., and Der, D. F., 1977, Time perception and hemodialysis, *Nephron* **19**:140–145.

Engel, G. L., and Romano, J., 1944, Delirium: II. Reversibility of the EEG with experimental procedures, *Arch. Neurol. Psychiatry* **51**:378–392.

Engel, G. L., Romano, J., Ferris, E. G., Jr., Webb, J., and Stevens, C. D., 1944, A simple method of determining frequency spectrums in the electroencephalogram, *Arch. Neurol. Psychiatry* **51**:134–146.

Essman, W. B., 1962, Effects of an experimentally induced acute renal failure on a learned maze response in rats, *J. Gen. Psychol.* **67**:51–56.

Essman, W. B., 1965, Behavioral bioassay for experimentally induced uremic endotoxemia in rats, *Percept. Mot. Skills* **20**:115–120.

Fishman, R. A., 1970, Permeability changes in experimental uremic encephalopathy, *Arch. Intern. Med.* **126**:835–837.

Friedman, E. A., 1978, Oral sorbents, in: *Replacement of Renal Function by Dialysis* (W. Drukker, F. M. Parsons, and J. F. Maher, eds.), Martinus Nijhoff, The Hague, pp. 232–243.

Glaser, G. H., 1960, Metabolic encephalopathy in hepatic, renal and pulmonary disorders, *Postgrad. Med.* **27**:611–619.

Glaser, G. H. (ed.), 1976, The EEG in renal insufficiency, in: *Handbook of Electroencephalography and Clinical Neurophysiology*, Volume 15 (A. Redmond, ed.), Elsevier Scientific Publishing Company, Amsterdam, pp. 51–69.

Goldstein, D. A., Feinstein, E. I., Chui, L. A., Pattabhiraman, R., and Massry, S. G., 1980, The relationship between the abnormalities in EEG and blood levels of parathyroid hormone in dialysis patients, *J. Clin. Endocrinol. Metab.* **51**:130–134.

Guisado, R., Arieff, A. I., and Massry, S. G., 1975, Changes in the electroencephalogram in acute uremia: Effects of parathyroid hormone and brain electrolytes, *J. Clin. Invest.* **55:**738–745.

Hagstam, K. E., 1977, EEG frequency content related to chemical blood parameters in chronic uremia, *Scand. J. Urol. Nephrol.* **7**(suppl. 1):1–33.

Hamel, B., Bourne, J. R., Ward, J. W., and Teschan, P. E., 1978, Visually evoked cortical potentials in renal failure: Transient potentials, *Electroencephalogr. Clin. Neurophysiol.* **44:**606–616.

Hayman, P. R., and Kooi, K. A., 1969, Visually evoked cortical responses in renal insufficiency, *Univ. Mich. Med. Cent. J.* **35:**177–179.

Heath, H., Hodgson, S. F., and Kennedy, M. A., 1980, Primary hyperparathyroidism, *N. Engl. J. Med.* **302:**189–193.

Hughes, J. R., 1980, Correlations between EEG and chemical changes in uremia, *Electroencephalogr. Clin. Neurophysiol.* **48:**583–594.

Jacob, J. C., Gloor, P., Elwan, O. H., Dossetor, J. B., and Pateras, V. R., 1965, Electroencephalographic changes in chronic renal failure, *Neurology* **15:**419.

Kennedy, A. C., Linton, A. L., and Eaton, J. C., 1962, Urea levels in cerebrospinal fluid after hemodialysis, *Lancet* **1:**410–411.

Kennedy, A. C., Luke, R. G., and Linton, A. L., 1964, Dialysis disequilibrium syndrome, in: *Renal Failure* (S. Shaldon and G. C. Cook, eds.), Blackwell Scientific Publications, Oxford, United Kingdom, pp. 66–79.

Kiley, J. E., 1981, Residual renal and dialyzer clearance, EEG slowing, and nerve conduction velocity, *ASAIO J.* **4:**1–8.

Kiley, J. E., and Hines, O., 1963, Electroencephalographic evaluation of uremia, in: *Proceedings of the 2nd International Congress of Nephrology,* Excerpta Medica Foundation, New York, pp. 745–747.

Kiley, J. E., and Hines, O., 1975, Electroencephalographic evaluation of uremia, *Arch. Intern. Med.* **116:**67–73.

Kiley, J. E., Millora, A. B., and Woodruff, M., 1970, The quantitative EEG in uremia, in: *Proceedings, Workshop on Behavioral Bioassays in Uremia,* Washington, D.C., DHEW Publication #(NIH)72–37, U.S. Printing Office #1744-0004, pp. 41–53.

Kiley, J. E., Pratt, K. L., Gisser, D. G., and Schaffer, C. A., 1976a, Techniques of EEG frequency analysis for evaluation of uremic encephalopathy, *Clin. Nephrol.* **5:**279–285.

Kiley, J. E., Woodruff, M. W., and Pratt, K. L., 1976b, Evaluation of encephalopathy by EEG frequency analysis in chronic dialysis patients, *Clin. Nephrol.* **5:**245–250.

Klingler, M., 1974, EEG Observations in uremia (abstract), *Electroencephalogr. Clin. Neurophysiol.* **6:**519.

Lewis, E. G., Dustman, R. E., and Beck, E. C., 1978, Visual and somatosensory evoked potential characteristics of patients undergoing hemodialysis and kidney transplantation, *Electroencephalogr. Clin. Neurophysiol.* **44:**223–231.

Lishman, W. A., 1978, *Organic Psychiatry,* Blackwell Scientific Publications, London.

Mahoney, C. A., and Arieff, A. I., 1983, Central and peripheral nervous system effects of chronic renal failure, *Kidney Int.* **24:**170–177.

McDaniel, J. W., 1971, Metabolic and CNS correlates of cognitive dysfunction with renal failure, *Psychophysiology* **8:**704–713.

Miller, J. H., Kopple, J. D., Edwards, A. E., Der, D. F., Fields, L. G., and Gardner P. W., 1974, Assessment of central nervous system function in uremia, *Proc. Clin. Dial. Transplant. Forum* **4:**156–159.

Murphy, G. P., and Sharp, J. C., 1964, Timed behavior in primates during various experimental uremic states, *J. Surg. Res.* **4:**550–553.

Raskin, N. H., and Fishman, R. A., 1976, Neurologic disorders in renal failure, *N. Engl. J. Med.* **294:**143–148.

Rasmussen, H., and Goodman, D. B. P., 1977, Relationships between calcium and cyclic nucleotides in cell activation: Cellular calcium metabolism and calcium-mediated cellular processes, *Physiol. Rev.* **57:**428–441.

Richet, G., Lopez de Novales, E., and Verroust, P., 1970, Neurological episodes in chronic renal failure, *Br. Med. J.* **2:**394–395.

Romano, J., and Engel, G. L., 1944, Delirium: I. Electroencephalographic data, *Arch. Neurol. Psychiatry* **51:**356–377.

Rotundo, A., Nevins, T. E., Lipton, M., Lockman, L. A., Mauer, S. M., and Michael, A. F., 1982, Progressive encephalopathy in children with chronic renal insufficiency in infancy, *Kidney Int.* **21**:486–491.

Schreiner, G. E., 1959, Mental and personality changes in the uremic syndrome, *Med. Ann. D.C.* **28**:316–323.

Schreiner, G. E., and Maher, J. F., 1961, *Uremia, Biochemistry, Pathogenesis and Treatment,* Charles C Thomas, Springfield, Illinois.

Sharp, J. C., and Murphy, G. P., 1964, Conditioned avoidance behavior in primates during various experimental uremic states, *Nephron* **1**:172–179.

Sharp, J. C., and Murphy, G. P., 1966, A behavioral bioassay method using material from a uremic patient, *Percept. Mot. Skills* **22**:127–133.

Spehr, W., Sartorius, H., Berglund, K., Hjorth, B., Kablitz, C., Plog. V., Wiedemann, P. H., and Zapf, K., 1977, EEG and hemodialysis: A structural survey of EEG spectral analysis, Hjorth's EEG descriptors, blood variables and psychological data, *Electroencephalogr. Clin. Neurophysiol.* **43**:787–797.

Stone, W. J., 1983, Medical complications of end-stage renal disease, in: *End Stage Renal Disease, An Integrated Approach* (W. J. Stone and P. L. Rabin, eds.), Academic Press, New York, pp. 57–83.

Teschan, P. E., 1970a, On the pathogenesis of uremia, *Am. J. Med.* **48**:671–677.

Teschan, P. E., 1970b, Approaches to the study of uremic toxicity: An experimental model, in: *Proceedings of the 4th International Congress of Nephrology* Karger, Basel, pp. 242–247.

Teschan, P. E., 1981, Measurement of neurobehavioral responses to renal failure, dialysis and transplantation, in: *Psychonephrology I. Physiological Factors in Hemodialysis and Transplantation* (N. B. Levy, ed.), Plenum, New York, pp. 13–18.

Teschan, P. E., 1983, The presentation of the patient with chronic renal failure, in: *End Stage Renal Disease, An Integrated Approach* (W. J. Stone and P. L. Rabin, eds.), Academic Press, New York, pp. 31–56.

Teschan, P. E., and Ginn, H. E., 1976, The nervous system, in: *Clinical Aspects of Uremia and Dialysis* (S. G. Massry and A. L. Sellers, eds.), Charles C Thomas, Springfield, Illinois, pp. 3–33.

Teschan, P. E., Murphy, G. P., and Sharp, J. C., 1964, Investigation of behavioral performance during urine reinfusion in the male primate, *Am. J. Physiol.* **206**:510–514.

Teschan, P. E., Carter, C. B., and Taub, E., 1970, Experimental studies of toxic factors in uremic encephalopathy, *Arch. Intern. Med.* **126**:838–840.

Teschan, P. E., Ginn, H. E., Bourne, J. R., Ward, J. W., Hamel, B., Nunnally, J. C., Musso, M., and Vaughn, W. K., 1979, Quantitative indices of clinical uremia, *Kidney Int.* **15**:676–697.

Tyler, H. R., 1968, Nuerologic disorders in renal failure, *Am. J. Med.* **44**:734–748.

Tyler, H. R., 1970, Neurologic disorders seen in the uremic patient, *Arch. Intern. Med.* **126**:781–786.

van den Noort, S., Eckel, R. E., Brine, K., and Hrdlicka, J. T., 1968, Brain metabolism in uremic and adenosine-infused rats, *J. Clin. Invest.* **47**:2133–2142.

12

Epilepsy
Pathophysiology of Cerebral Dysfunction

ROBERT C. COLLINS

1. Introduction

Patients with epilepsy commonly suffer changes in mental functions. This can be caused by the seizure process itself, either focal or generalized, or related to the postictal state. Recurrent seizures, or status epilepticus, cause a profound change in mental function that often lasts several days following cessation of the motor convulsions. Side effects of anticonvulsant medicines and drug toxicity must also be considered in a differential diagnosis of encephalopathy in patients with epilepsy. Finally, a physician must often rule out other causes of mental change that are either a consequence of a seizure, such as head trauma, or are present as part of a pathophysiological disease process that includes seizures, such as the other causes of metabolic encephalopathy discussed in this book. These issues are summarized in Table I.

This chapter examines how seizures themselves change mental functions. There are three major themes. First, cerebral dysfunction in epilepsy can be understood by knowing where seizures originate and spread—the functional anatomy of epilepsy. Second, cerebral dysfunction is also an expression of the metabolic consequences of seizures. The strength and duration of discharges are important in this respect. Finally, the effect of seizures on ventilation, blood pressure, body temperature, and other systemic factors constitutes a third way seizures can lead to global brain dysfunction and encephalopathy.

2. Focal Seizures—Functional Anatomy

2.1. Seizures from Neocortex

Patients with epilepsy often have frequent interictal discharges seen as spikes or sharp waves on the EEG. These abnormalities remain localized in the

ROBERT C. COLLINS • Department of Neurology, Washington University Medical School, St. Louis, Missouri 63110.

upper layers of cortex (Collins and Caston, 1979a; Elger *et al.*, 1982) and patients are unaware of any abnormality. Once a seizure occurs, however, the process overcomes local inhibitory containment (Prince and Wilder, 1967) captures corticofugal pyramidal neurons, and spreads through orthodromic pathways to distant cortical and subcortical sites in brain. Seizures from motor cortex cause brief intense stereotyped jerking of contralateral musculature, such as face, arm, or leg. As this occurs, there is activation of ipsilateral extrapyramidal circuits (caudate, globus pallidus, subthalamic nucleus, red nucleus, and substantia nigra), thalamus (ventralis lateralis, central median, parafascicularis, nucleus reticularis), and contralateral cerebellum, facial nucleus, and spinal cord (Walker *et al.*, 1956; Kennedy *et al.*, 1975; Collins *et al.*, 1976). Seizure discharges in somatosensory or visual cortex do not cause behavioral convulsions unless motor or limbic circuits are captured into activity (Collins and Caston, 1979b). Patients become conscious of sensory abnormalities, e.g., tingling of a hand, flashing lights, and so forth, once seizures spread out from the focus. The functional anatomy responsible for this awareness is unknown.

Focal seizures also spread through corticocortical circuits to distant sites in the ipsilateral hemisphere and through transcallosal pathways to the contralateral hemisphere (Collins, 1978a). As focal seizures become stronger or longer or if a Jacksonian march occurs, then there is progressive involvement of more and more brain by the seizure process. Local cortical areas strongly activated from a distant focus can themselves fire subcortically into motor and sensory pathways and disrupt function. This results not only in increasing severity and spread of "positive symptoms" such as motor jerks, but also includes an admixture of "negative symptoms," such as loss of sensation or disruption of cognitive functions. The latter are usually not appreciated, since it is difficult to test patients having progressive focal motor convulsions.

TABLE I. Common Causes of Impaired Mental Function in Patients with Epilepsy

Focal epilepsy with recurrent seizures
  Epileptic aphasia
  Complex partial seizure—"psychomotor"
  Complex partial status epilepticus
Generalized epilepsy
  Petit mal and petit mal status—"spike wave stupor"
  Grand mal and secondarily generalized seizures
  Postictal state following generalized seizures
Anticonvulsant medicines
  Side effects
  Toxicity
Underlying etiologies
  Drug withdrawal
  Hypoxemia
  Embolic cerebrovascular disease
  Infection
Complications of seizures or status epilepticus
  Head trauma
  Pulmonary edema
  Renal failure secondary to myoglobinuria or hyperuricemia (rare)
  Intravascular coagulation (rare)

A 46-year-old right-handed man had a focal seizure disorder of his left hand associated with previous head trauma. He was admitted to the hospital following a drinking binge during which he had discontinued his anticonvulsant medication. He had recurrent focal seizures of his left upper extremity, some of which spread to involve the rest of his body over a period of 30 to 60 sec. As the seizures began, he would answer questions and deny any abnormality in response to direct sensory testing of the jerking left hand. As the clonic convulsions spread up the arm, his appreciation of pin and passive joint movement became impaired. Once the left face was involved, a disturbance in language function began. When both sides of the face were jerking he would not answer or obey simple commands. Clonic jerking became prominent on the left side with tonic posturing eventually occurring on the right. On cessation of convulsions, he was confused and disoriented for time, but knew he had experienced a seizure and could describe its course up to the point that language became disturbed and seizures spread bilaterally.

The loss of sensory function during focal motor seizures reflects at least three processes. First, there are strong reciprocal projections between motor cortex (area 4) and adjacent sensory cortex (areas 3, 1, 2) and sensory thalamus. These pathways have been well delineated by anatomic techniques (Jones, 1981a,b). Physiological studies remain inconclusive whether corticofugal projections are primarily excitatory or inhibitory with respect to normal physiological activity, but the net effect of seizures projecting into distant cortical or subcortical sites would be a profound disruption of normal afferent activity (Crowell, 1970; Schwartzkroin et al., 1974, 1975). Second, most areas in neocortex have a direct or collateral projection to nucleus reticularis, the thalamic nucleus composed of GABA inhibitory interneurons that project onto and inhibit sensory thalamus (Jones, 1977; Houser et al., 1980). Activation of this second-order thalamic circuit may explain how a focal seizure process in one restricted area of neocortex might physiologically deafferent an entire hemisphere and depress its blood flow and functional metabolism (Ueno et al., 1975; Collins, 1978a). Third, focal jerking about a single joint would send strong afferent signals through sensory pathways depressing or blocking normal sensation from adjacent joints by physiological mechanisms of feed forward and recurrent surround inhibition. Sensations from body parts adjacent to convulsive movements of muscle would not reach consciousness.

Hughlings Jackson (1870) described variable loss of consciousness whenever seizures marched from one side of the body to the contralateral side. His observations and cases similar to the presentation above (Holmes, 1927; Walshe, 1943) support experiments of Walker (1959) indicating that the anatomical basis for mental impairment during the march of focal seizures is a progressive involvement of an increasing number of local and long circuits throughout brain. This type of seizure disorder would probably also invade the core reticular activating system to some degree in the later stages so that considerations of the functional anatomy of focal seizures remain complex. For example, some patients with neocortical focal seizures have only brief localized convulsions before rapid bilateral spread and loss of consciousness. This is especially true with discharges from the frontal lobe (Goldring, 1972; Bancaud et al., 1974), probably reflecting strong bilateral descending projections from each hemisphere through basal ganglia (Kunzle, 1975; Collins, 1978b). In summary, there are two anatomical principles for understanding how focal neocortical seizures disrupt mental functions and cause encephalopathy. First, seizures can spread slowly to involve increasing areas of cortex and subcortical reciprocal sites, and second, rapid spread to subcortical extrapyramidal, thalamic, and reticular nuclei can disrupt mechanisms subserving consciousness.

One focal seizure disorder deserves special consideration. Seizure discharges in the language area of the dominant hemisphere can disrupt comprehension and speech in children and adults (Landau and Kleffner, 1957; Shoumaker et al., 1974; DePasquet et al., 1976). This can be brief and intermittent or recur and continue for several days as focal status epilepticus (Dinner et al., 1981). The clinical picture is that of a patient who appears bewildered and speaks in short telegraphic phrases, often with paraphasic errors. When an EEG reveals focal discharges in the dominant frontotemporal area, the diagnosis of epileptic aphasia can be made. During periods when seizures remain localized, clinical testing reveals that language functions alone are impaired, whereas visual-spatial tasks remain relatively normal. When seizure discharges spread bilaterally, comprehension becomes severely impaired making it impossible to test cognitive functions. In this situation, patients appear either confused or apathetic and are often mute, a situation resembling complex partial status epilepticus (see Section 2.2). The fact that even brief periods of epileptic aphasia can cause global impairment in human behavior and mental function reflects the prominent role of language in higher functions in man.

2.2. Seizures from Limbic Cortex

Focal seizures originating in entorhinal cortex, hippocampus, or amygdala do not cause overt changes in mental function unless discharges spread bilaterally. Patients with complex partial epilepsy (psychomotor epilepsy) commonly have interictal spikes in temporal or nasopharyngeal leads on their EEG. Depth electrode studies in select patients reveal that less than 20% of interictal discharges in deep limbic structures are recorded by routine surface leads (Lieb et al., 1976). Local spikes do not produce symptoms but may interfere with mechanisms subserving attention and concentration (Rausch et al., 1978). In addition, such studies reveal that most unilateral seizures in the limbic system do not disrupt behavior, but cause brief experiential auras (Gloor, 1982). Patients typically experience a short fragment of memory, such as a visual scene, or a brief emotional state, such as fear. The seeming impotence of unilateral discharges for causing more overt behavioral change is in contrast to the marked disruption of mental phenomena that occur once seizures spread bilaterally. A patient stops what he is doing, stares, loses conscious touch with environmental stimuli, and then may exhibit slow postural changes, chewing and swallowing, and purposeless fumbling automatisms. Typical episodes last less than a minute. Memory functions are suspended during this time and the episode is followed by a brief period of confusion (Delgado-Escueta et al., 1982).

Depth recordings in man, as well as experimental studies in animals, indicate two principles in understanding these mental phenomena. First, the hippocampal formation together with its connections to amygdala, septum, the extrapyramidal system, and midline periventricular thalamic nuclei play a critical role in conscious interaction with the environment. Second, if discharges disrupt only one side of the limbic system, the contralateral side is largely able to compensate. This seems especially true for maintaining normal memory functions (Wieser, 1980), observations consistent with the severe amnesia that occurs following bilateral temporal lobectomy in man (Scoville and Milner, 1957; Penfield and Milner, 1958).

Brief or recurrent transient episodes of abnormal behavior can be a manifestation of epilepsy at any age in man, but are more common in the elderly. Symptoms can be nonspecific, such as the syndrome of "ictal confusion" (Ellis and Lee, 1978), or clinical features can resemble the amnesic state of transient global amnesia, a condition that may represent an epileptic process in a limited number of patients (Gilbert, 1978). In addition, brief episodes of psychotic behavior may also reflect epileptic discharge (Wells, 1975; Lawall, 1976; Levine and Finklestein, 1982). All three of these entities probably represent bilateral seizure discharges through limbic structures, but exact circuits are unknown. The clinical diagnosis that brief encephalopathic behavior is epileptic depends on physicians recognizing the sudden onset, brief duration, and complete recovery to preictal behavior.

2.3. Recurrent Limbic Seizures—Complex Partial Status Epilepticus

Following a single psychomotor seizure, patients return to normal mental function, missing only a brief island in time. Should additional psychomotor seizures recur and a state of limbic status epilepticus develop, then the clinical picture changes dramatically. Characteristic automatisms may recur, but are often interspersed between longer periods (minutes to hours) of nonspecific encephalopathic behavior—confusion, inappropriate behavior, inattentiveness, posturing, diminished or absent speech (Delgado-Escueta et al., 1974; Markland et al., 1978; Engel et al., 1978). As there are few or no obvious tonic or clonic motor manifestations, the epileptic etiology of this encephalopathic behavior is often overlooked. In addition to occurring in patients with temporal lobe seizure disorders, the syndrome may develop during discharges from the frontal lobe (Kugoh and Hosokawa, 1977) or following repeated episodes of grand mal convulsions in patients with generalized epilepsy. The EEG commonly shows bilateral temporal high-amplitude slowing, although there are several variations. Most important, this encephalopathic syndrome responds dramatically to intravenous anticonvulsants (Delgado-Escueta and Bajorek, 1982).

Despite its relative infrequency as a clinical entity, psychomotor status is one of the most severe manifestations of epilepsy. Excessive recurrent discharges through limbic system can result in neuronal damage independent of any of the systemic changes that occur commonly with generalized status epilepticus (Ben-Air et al., 1980; Lothman and Collins, 1981; Sloviter and Diamino, 1981; Collins et al., 1983a). The cause of neuronal damage is complex, but probably related to synaptic and biophysical events within the hippocampal pyramidal cell layer (see Section 3.2). Patients recovering from psychomotor status can have deficits in mental functions and memory testing lasting for months, probably as a reflection of some permanent damage.

3. Metabolic Effects of Focal Seizures

3.1. Cerebral Energy Metabolism and Blood Flow

Seizures cause a dramatic increase in cerebral energy metabolism (for reviews, see Plum *et al.*, 1974; Siesjo, 1978; Collins *et al.*, 1983b). The stimulus occurs at the membrane surface where discharges are associated with a large flux of sodium and calcium into neurons, coupled with a potassium loss to the extracellular space (Fig. 1). These cation shifts activate ATP-dependent membrane pumps, and as these accelerate to restore ion homeostasis there is a shift in the phosphate potential with a small fall in ATP and a rise in ADP,

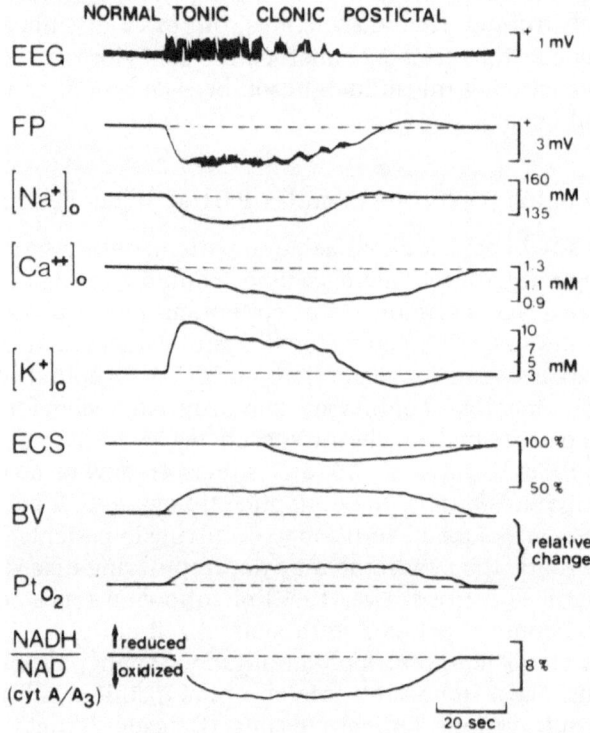

FIGURE 1. Stylized summary of the ionic and metabolic changes in cortex associated with a tonic-clonic-postictal seizure. With the onset of seizures (EEG, top trace), there is a negative shift in the field potential (FP, second trace), which probably reflects depolarization of glia by potassium (Lothman and Somjen, 1975; Greenwood *et al.*, 1981) as well as potential gradients between aligned dendrites and neuronal soma. Extracellular sodium and calcium fall during seizures (third and fourth trace) with a relatively slow correction for calcium during the postictal phase (Heinemann *et al.*, 1977; Dietzel *et al.*, 1980). Extracellular potassium rises sharply (fifth trace) with a corrective overshoot in the postictal phase (Prince *et al.*, 1973). This latter event probably reflects the electrogenic nature of the membrane pumps that contribute to the cessation of seizures and may contribute to postictal depression. Recent studies have found a decrease in extracellular volume (ECS, sixth trace; Dietzel *et al.*, 1980), perhaps reflecting passive movement of water into dendrites or cellular compartments. Concomitant with these ionic shifts, blood flow (shown as blood volume, Bv) and tissue oxygen tension (Pt_{O_2}) promptly increase and there is an oxidation of NADH and mitochrondrial coenzymes (Jobsis *et al.*, 1971; Lothman *et al.*, 1975; Kriesman *et al.*, 1981).

AMP, and phosphate. These latter substrates stimulate both glycolytic and mitochondrial enzymes and there is a prompt increase in the oxidative metabolism of glucose. This results in a rise in lactate and CO_2 and a fall in tissue pH (Caspars and Speckmann, 1972; Tenny et al., 1978). These changes, together with the increase in $[K^+]_o$, act in concert with local neurogenic factors to dilate cerebral vessels and increase cerebral blood flow.

Studies on changes in cerebral metabolism with focal seizures have emphasized the capacity of brain for maintaining a balance in energy despite the marked stimulation by seizures (Fig. 2A,B) (Collins, 1976). A key feature in this regard is that cerebral blood flow is tightly coupled to local changes in neuronal activity. With an increase in firing, there is nearly a spontaneous increase in blood flow, bringing in glucose and oxygen and removing lactate and CO_2 (Penfield, 1937; Ueno et al., 1975; Sakai et al., 1978). The amount of energy stored in brain either as glycogen or as the high-energy phosphate equivalents of glucose, ATP, and phosphocreatine can only support brain function for a few minutes when blood flow ceases. With seizures these energy stores are exhausted within 15 sec if unsufficient oxygen is delivered to brain (Fig. 2C) (Collins et al., 1970; Ferrendelli and McDougal, 1971). If blood flow homeostasis remains normal and there is no change in arterial oxygen content, then there is only a 10 to 20% drop in energy metabolites, reflecting the acceleration of metabolism in a new steady state (Fig. 2B,D) (Duffy et al., 1975; Chapman et al., 1977; Folbergrova et al., 1981).

Early observations on changes in blood flow with focal seizures suggested that brain became hypoxic. At the operating table, Penfield observed that "venous blood turned red" with focal discharges and suggested this reflected a secondary response to local tissue hypoxia caused by seizures (Penfield, 1937). Studies with oxygen-sensitive electrodes and the direct measurement of changes in mitochondrial oxidation-reduction couples have found the opposite. With single seizures there is an increase in tissue oxygen and an oxidative shift in mitochondria (Jobsis et al., 1971; Lotham et al., 1975; Kreisman et al., 1981). This begins promptly with seizures and usually outlasts discharges, reflecting prolonged metabolic requirements for restoring ion balance across membranes (Fig. 1). With recurrent seizures tissue oxygenation may fall (Dymond et al., 1976), reflecting local factors or changes in systemic oxygenation in some instances (Kriesman et al., 1983). It is remarkable that brain can maintain metabolic balance during seizures. The response is prompt and reaches metabolic rates two to four times normal, measured either as changes in glucose or oxygen metabolism or high-energy phosphates. By contrast, the metabolic response associated with strong sensory stimulation or sustained motor activity increases cerebral metabolism for glucose only 30 to 60%. As emphasized below, however, whereas the metabolic demands of single-focal seizures can be met, once recurrent seizures develop the brain's homeostatic mechanisms break down and morphological changes develop.

Recent autoradiographic studies of focal seizures in experimental animals emphasize that these metabolic changes occur not only in the focus, but throughout seizure pathways. Using the Sokoloff quantitative [14C]deoxyglucose method (Sokoloff et al., 1977), a marked increase in glucose metabolism has been found in cortical and subcortical sites (Kennedy et al., 1975; Col-

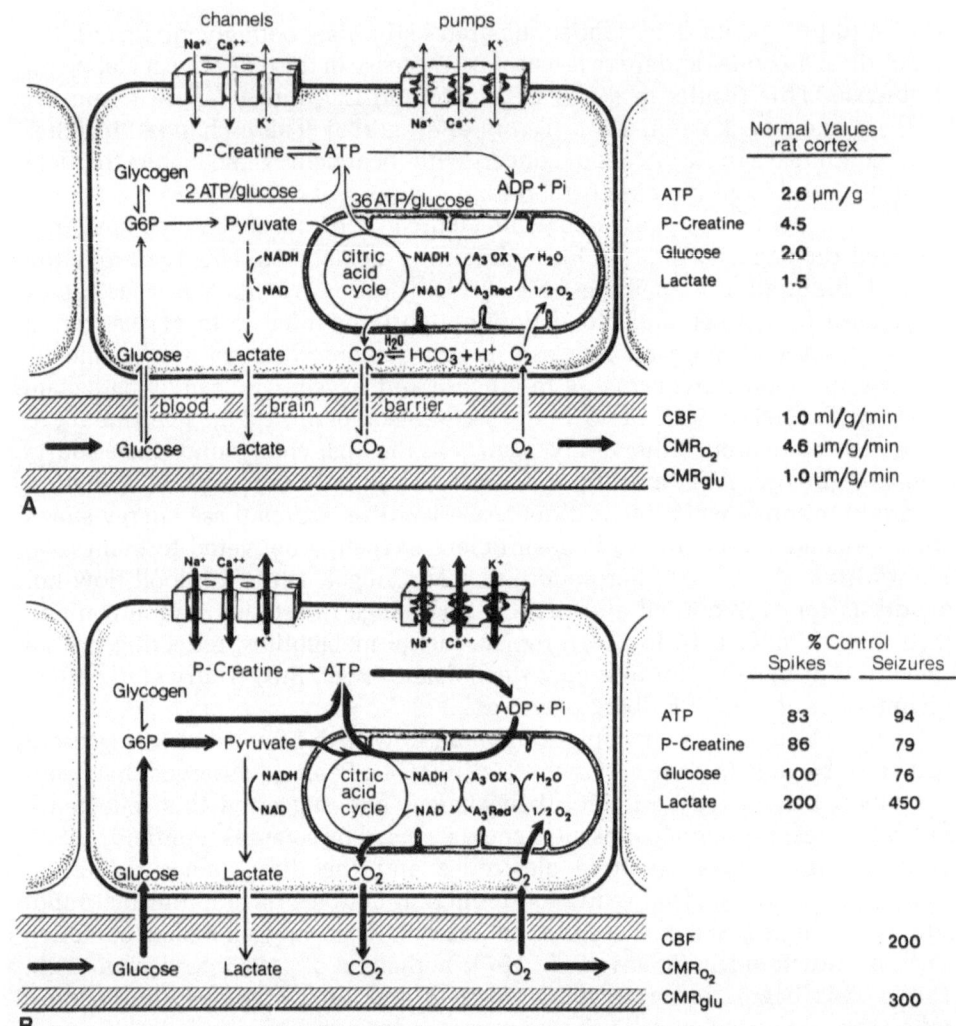

FIGURE 2. Schematic diagrams of changes in energy metabolism in brain during seizures. A: Normal state. Energy utilization in brain is devoted predominantly to ATP-dependent membrane pumps (Whittam, 1962) with the resynthesis of ATP being primarily the result of the oxidative metabolism of pyruvate in mitochondria. Since pyruvate (and lactate) do not appreciably cross the blood–brain barrier into brain, and since so little energy is stored in brain as either phosphocreatine, glycogen, or glucose, the steady-state production of ATP depends directly on a steady supply of blood glucose and oxygen. B: Focal seizures cause dramatic increase in ion fluxes across membranes in the seizure focus and its pathways. Energy-dependent pumps keep pace with these changes as a local increase in blood flow delivers more glucose and oxygen for energy synthesis. Changes in steady-state levels of substrates are mild. C: Generalized seizures that interrupt ventilation cause hypoxemia just as brain's energy demands are greatest, resulting in marked depletion of tissue substrates. Failure to pump out cations in this situation may lead to intracellular edema and swelling. D: Generalized seizures during controlled ventilation result in an increased turnover of energy substrates, but very little change in tissue levels until after status epilepticus has developed. The metabolism and blood flow values presented here, as well as the "percent of control," represent mean computed values using data from several different publications: Collins et al. (1970), Ferrendelli and McDougal (1971), Duffy et al. (1975), Collins (1976), Chapman et al. (1977), Blennow et al. (1977), Howse (1979), Horton et al. (1980), Folbergrova et al. (1981).

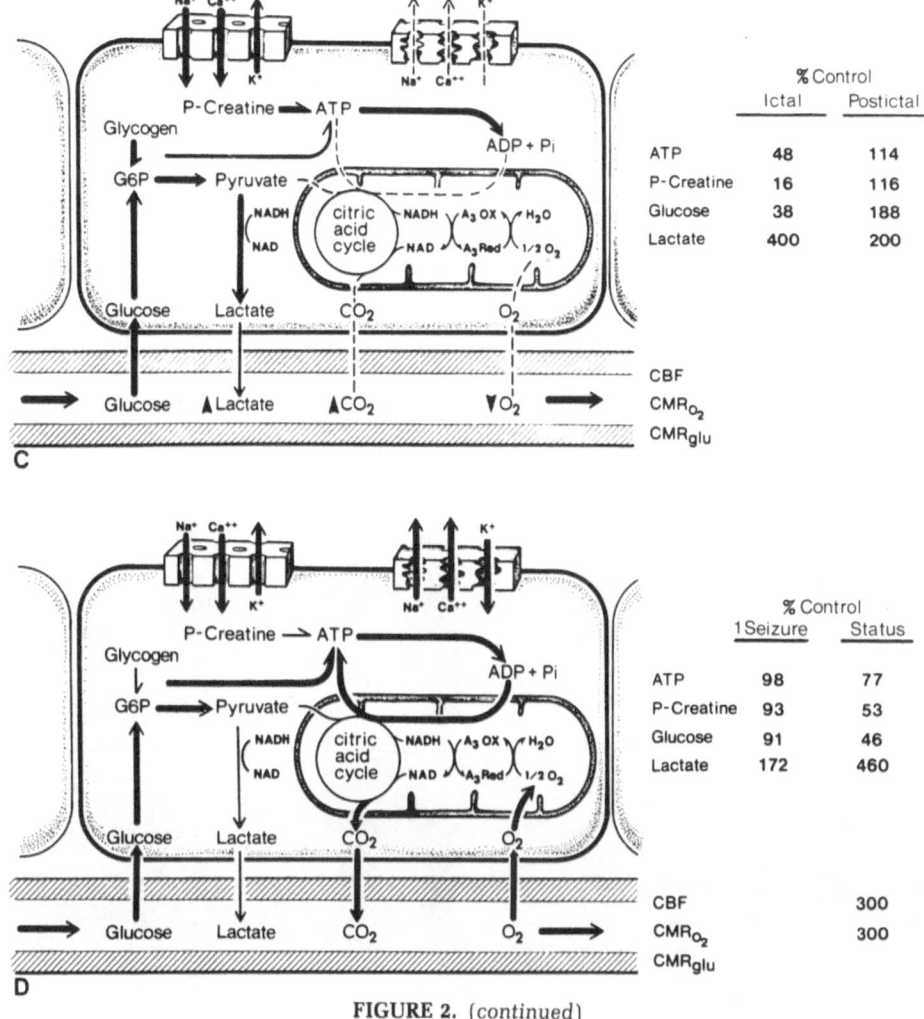

The table in panel C (% Control):

	Ictal	Postictal
ATP	48	114
P-Creatine	16	116
Glucose	38	188
Lactate	400	200

The table in panel D (% Control):

	1 Seizure	Status
ATP	98	77
P-Creatine	93	53
Glucose	91	46
Lactate	172	460
CBF		300
CMR_{O_2}		300

FIGURE 2. (continued)

lins *et al.*, 1976; Collins, 1978a,b; Collins and Caston, 1979b; Kato *et al.*, 1980). These findings have been largely confirmed in man using 18-fluorodeoxy-glucose and positron emission transaxial tomography (Kuhl *et al.*, 1980; Engel *et al.*, 1982). There is also a marked shift in amino acid metabolism in seizure pathways. With the use of [1-^{14}C]leucine as a metabolic tracer, auto-radiographic studies indicate an acceleration in amino acid oxidation and/or a concomitant inhibition in amino acid incorporation into protein (Collins and Nandi, 1982; see also Cotman *et al.*, 1971; Wasterlain, 1974). Changes in blood flow and metabolism can be as intense in sites distant from the focus as in the focus itself (Collins *et al.*, 1981) (Fig. 3). These metabolic shifts are a reflection of the intense activity of neurons driven both orthodromically and antidromically from the focus (Gutnick and Prince, 1974; Schwartzkroin *et al.*, 1975; Noebels and Prince, 1978b). Studies with ion-sensitive electrodes in distant

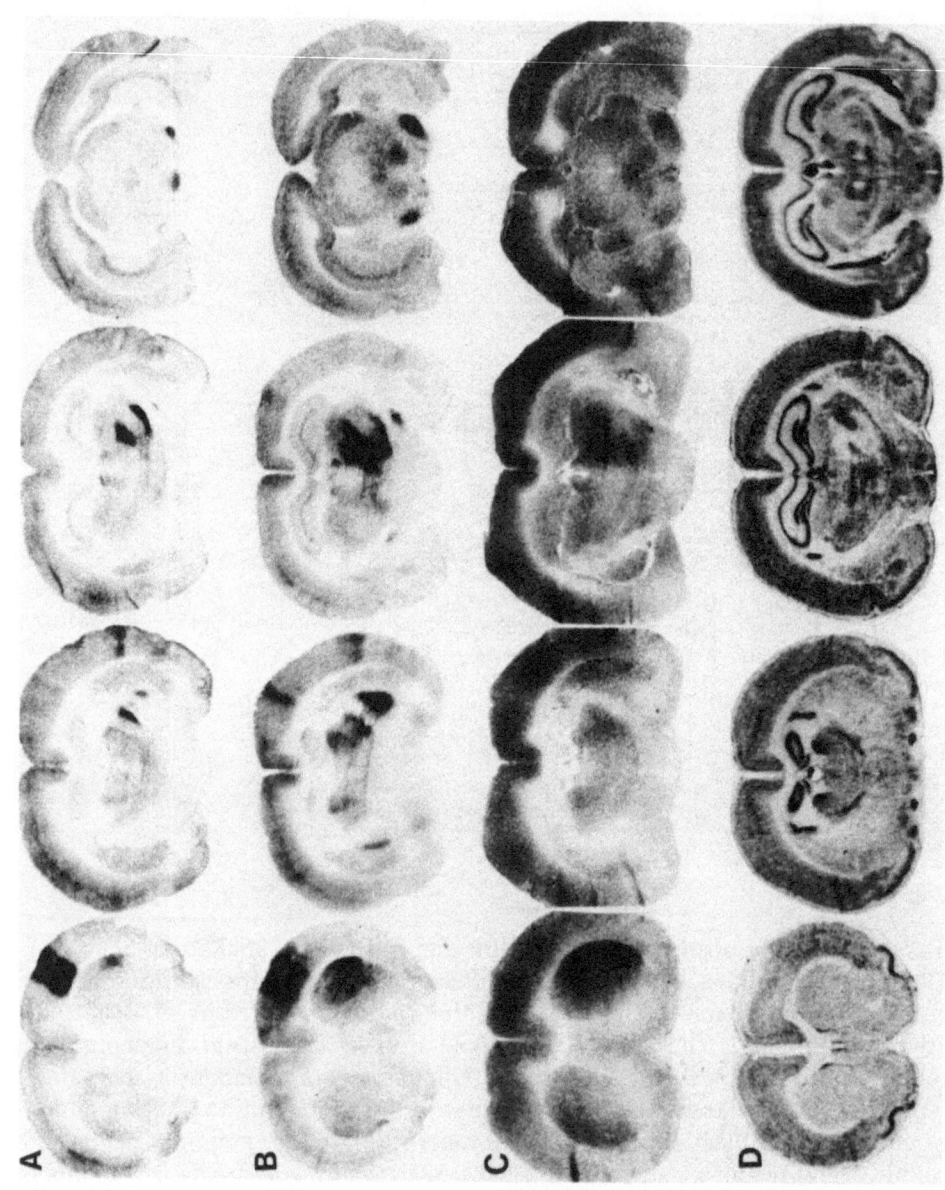

FIGURE 3. Metabolic autoradiography of focal sensorimotor seizures. In each of the four examples shown here, a seizure focus was created over the right forelimb motor cortex of rats using topical application of penicillin or bicuculline in an epidural well. Rats were awake, alert, and had contralateral forelimb jerks coincident with focal electrographic spike discharges when radioactive isotopes were given intravenously. Autoradiographic images were produced using methods developed by Sokoloff et al.(1977) (for details see Collins et al., 1976; Collins et al., 1981; Collins and Nandi, 1982]. A: [^{14}C]deoxyglucose metabolism during mild focal seizures. The panel shows increased metabolism in the cortical focus, ipsilateral caudate, globus pallidus, substantia nigra, and ipsilateral thalamic nuclei-reticularis, VL and VB. B: [^{14}C]deoxyglucose metabolism during strong focal seizures with bilateral spread. Compared with mild seizures, note the larger focus, cortical columns behind and lateral to the focus, new activity in contralateral extrapyramidal sites (particularly GP and SN) and new and larger areas of involvement in thalamus and midbrain (red nucleus, n. posterior, and superior colliculus). There is a slight depression of metabolism around the focus and throughout the ipsilateral hemisphere manifest as a decrease density in cortical layer IV. C: [^{14}C]iodoantipyrene—local cerebral blood flow changes during strong focal seizures. Local increases in blood flow are generally of the same order of magnitude as changes in deoxyglucose metabolism, but usually encompass a larger area of cortical and subcortical sites in seizure pathways. D: Autoradiography using [1-^{14}C]leucine as a metabolic precursor shows areas of pallor in seizure pathways, indicating either an acceleration of amino acid catabolism and/or an inhibition of protein synthesis. This occurs predominantly in the cortical focus and thalamic nuclei sharing a reciprocal connection with the seizure focus, specifically VB, thalamic ventrobasal complex, VL, ventralis lateralis, and nPo, nucleus posterior thalami seen here.

sites in thalamus reveal a marked shift in cations during siezure discharges (Gutnick *et al.*, 1979) indicating that the same mechanisms responsible for stimulating metabolism in the focus operate throughout seizure pathways (Fig. 1).

A consideration of these metabolic changes in the seizure focus and its pathways give some insight into neurological dysfunction that accompanies seizures. First, single-focal seizures accelerate but do not outrun cerebral energy metabolism such that dysfunction during seizures cannot be explained by energy failure. Second, energy metabolism remains accelerated for a short time following a seizure, reflecting ongoing needs to restore ion balance and membrane homeostasis. Third, during the postictal state, defined as the period when brain is refractory to subsequent stimulation, concentrations of brain energy metabolites are actually above normal (Collins *et al.*, 1970). Taken together, these results point to factors other than changes in energy metabolism as the basis for explaining the neurological dysfunction of single seizures (Passonneau, 1969).

3.2. Synaptic Function and Structure

Epileptic seizures accelerate the turnover of amino acids and neurotransmitters, although most studies indicate only small changes in steady-state tissue levels (VanGelder *et al.*, 1972; Mutani *et al.*, 1977; Bakay and Harris, 1981; Perry and Hansen, 1981). In the penicillin seizure focus, for example, there is a slight drop in the putative excitatory transmitters glutamate ($10.0 \rightarrow 8.5$ μm/g) and aspartate ($4.5 \rightarrow 3.5$ μm/g) and a slight increase in the inhibitory transmitter GABA ($1.0 \rightarrow 1.5$ μm/g) (Collins and Mehta, 1978). A methodological problem complicates these studies due to the inability to measure transmitters at the synapse, where dramatic changes in small compartments undoubtedly occur, but go undetected due to dilution of the change in the large samples needed for assay. Morphological studies, however, indicate that there would be marked changes in metabolism at the synapse.

With seizure discharges, Fischer and Langmeier (1980) as well as Nitsch and Rinne (1981) have found a rapid movement of synaptic vesicles toward and into synaptic membrane within seizure pathways. It is not clear whether membrane metabolism and energy-dependent vesicle recycling can keep pace with seizure demands. Several laboratories have found a marked increase in membrane fatty acids with seizures (Basan, 1970, Chapman *et al.*, 1980). With recurrent focal seizures, some thalamocortical axon terminals even show a pathological dilation, giving a spongiform appearance to the neuropil seen with histological stains (Collins and Olney, 1982). The cause of this is probably related to excessive accumulation of ions and water during burst discharges (Noebels and Prince, 1978a), as well as a relatively slow recycling of synaptic membrane back into vesicles. In addition, postsynaptic dendrites have also been found to undergo swelling in the seizure focus (Okada *et al.*, 1971; Scheibel *et al.*, 1974; Butler *et al.*, 1976), as well as in distant sites (Collins and Olney, 1982). This may be a pathological expression of a normal physiological process (Van Harreveld and Fifkova, 1975; Fifkova *et al.*, 1982) caused by excitotoxic effects of glutamate (Olney, 1978) or other excitatory amino acids

(Chan *et al.*, 1979). Tissue levels of glutamate are very high (10.0 mM), whereas synaptic levels are kept very low (0.002 mM) (Shank and Aprison, 1979). The release of additional amounts during seizures or the failure of reuptake and metabolism by glia could lead to excessive postsynaptic burst firing. Shifts of calcium into dendrites would first play a physiological role (Traub and Llinas, 1979), then perhaps a pathophysiological role (Heinemann and Pumain, 1980; Griffiths *et al.*, 1982) in these processes.

The marked changes in synaptic morphology at certain sites with seizures could reflect a local failure in energy metabolism within the synapse. The converse, however, seems more likely since ATP levels remain near normal if brain tissue remains oxygenated. The rapid increase in neurotransmitter and membrane metabolism together with accelerated ion and water flux could induce structural changes in synaptic membranes that open up channels or inactivate pump emzymes (Delgado-Escueta and Horan, 1980). Synaptic elements would swell in obedience to osmotic forces, but ATP levels would remain near normal if pumps could not use it. Future experiments on synaptic function and structure may give further insights as to why seizures stop, as well as why there is a prolonged dysfunction widely distributed throughout seizure pathways following prolonged intense focal discharges. These considerations are summarized in Fig. 4.

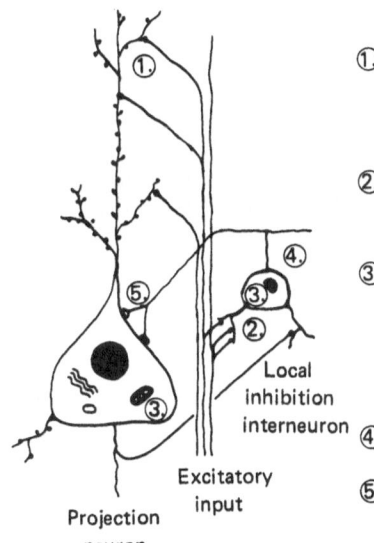

Morphological Effects of Seizures - Possible Mechanisms

1. Dendrites swell in the seizure focus and at distant sites
   a. Excito·toxic effect of glutamate neurotransmission
   b. Ion flux exceeds energy transport
   c. Direct effect of calcium
2. Axon terminals swell in the seizure focus
   a. Ion flux exceeds energy transport
   b. Orthodromic and antidromic firing
3. Swelling of endoplasmic reticulum and mitochondria
   a. Ion flux exceeds energy transport
   b. Direct effect of Ca^{2+} and/or pH

Long-Term Effects of Seizures

4. Dark cell change - preferential loss of inhibitory interneurons
5. Loss of GABAergic axosomatic synapses

Local inhibition interneuron

Excitatory input

Projection neuron

FIGURE 4. The biochemical pathogenesis of morphological changes during seizures is complex. The early swelling in dendrites and axon terminals probably reflect direct changes in membrane properties associated with channels and pumps, rather than a failure of the cell to provide ATP to the pumps. Prolonged seizures are associated with ischemic cell change—swelling of endoplasmic reticulum and mitochondria—followed by condensation of cytoplasm. Small interneurons in cortex are susceptible (Collins and Olney, 1982), and a decrease in GABAergic inhibitory interneurons and their axosomatic synapse may underlie long-term changes in excitability associated with epilepsy (Ribak *et al.*, 1979; Ribak and Riffenstein, 1982).

4. Generalized Seizures

4.1. Clinical Picture

Grand mal convulsions and status epilepticus can be etiopathic or secondary to a wide variety of systemic conditions (Table II). This latter fact simply reflects that epileptic discharges are an intrinsic property of neuronal circuits that can emerge whenever local excitatory influences override inhibition. Many causes of metabolic encephalopathy can do this, with hypoxemia, hypoglycemia, drug abuse, and uremia being the most common. During grand mal convulsions or secondarily generalized seizures, there is forced closure of the glottis with powerful contraction of thoracic muscles making ventilation impossible. This occurs just as metabolism of convulsing muscle becomes greatly increased so that oxygen is rapidly depleted from arterial blood. The brain experiences hypoxemia just at the time of its greatest energy requirements and energy stores become exhausted (Fig. 2C) (Collins *et al.*, 1970; Ferrendelli and McDougal, 1971). Blood lactic acid and CO_2 rapidly accumulate, causing a marked metabolic and respiratory acidosis. With cessation of tonic phases of generalized seizures, hyperventilation begins as a physiological attempt to restore normal oxygenation and blood pH. The increase in brain metabolism together with these systemic changes result in marked brain acidosis. Cerebrospinal fluid lactic acid remains elevated well into the postictal state (Brooks and Adams, 1976; Simpson *et al.*, 1977). Blood glucose usually rises with a generalized seizure since a stimulated release of epinephrine from the adrenal medulla causes a prompt glycogenolysis in the liver. Any rise in blood glucose depends on the quantity of liver glycogen or the immediate nutritional history of the patient. Blood glucose can fall in malnourished patients or with prolonged status epilepticus. This factor is especially important in developing brain since glucose is not only the major source of energy, but provides the carbon skeleton for lipid and protein synthesis. Long-term survival and brain

TABLE II. Etiology of Status Epilepticus in a General Hospital[a]

Etiology	Percent of total cases
Anticonvulsant irregularity	26
Alcohol withdrawal	19
Cerebrovascular disease	14
Drug abuse	9
Metabolic disorders	9
Cardiac arrest	4
Tumor	3
Trauma	3
Infection	3
Multiple causes	12
Unknown	14

[a] From Aminoff and Simon (1980).

development following repeated seizures in immature animals depend on maintaining normal blood glucose (Wasterlain and Duffy, 1976).

Despite these dramatic shifts in blood oxygen content and pH, brain metabolism rapidly returns tissue metabolites to normal once ventilation delivers sufficient oxygen. This would seem to indicate that factors other than or in addition to energy metabolism underlie the prolonged encephalopathy that not uncommonly follows a generalized convulsion. Patients are usually disoriented and confused for 5 to 15 min and often sleep for several hours. Prolonged elevations in tissue lactate, local changes in the balance of neurotransmitters, or excessive release of neurotransmitters and metabolites into the subarachnoid space may play a role.

4.2. Status Epilepticus

Recurrent generalized seizures result in progressive disruption of systemic physiological and metabolic factors that can themselves be detrimental to brain: hypoxemia, hyperthermia, hypoglycemia, and hypotension (Meldrum and Horton, 1973; Meldrum and Brierly, 1973). In this state, brain usually cannot maintain adequate energy balance and neuronal damage occurs. Whereas during sustained focal seizures, local neuronal discharges cause morphological changes at selected synaptic sites (Collins and Olney, 1982), during focal (Knopman et al., 1977) or generalized status epilepticus with hypoxemia, local and widespread "ischemic" neuronal damage can occur. There may be many metabolic causes of this type of cytopathological change (Brierly, 1976), but recent experimental studies in epilepsy suggest that factors other than ATP may be important. These include prolonged elevations in lactate and ammonia, excitotoxic effects of neurotransmitters at synapses, and the intracellular toxicity of calcium (Griffiths et al., 1982; Meldrum, 1983). Following severe generalized status epilepticus, there is laminar necrosis in neocortex, Ammon's horn of hippocampus, thalamus, and the cerebellar Purkinje layer (Corsellis and Bruton, 1983). Morbidity, mortality, and the severity of pathological damage in man are a reflection both of the intensity and duration of seizures and the degree of abnormality of systemic factors (Rowan and Scott, 1970; Celesia, 1976; Aminoff and Simon, 1980). In addition, status epilepticus can initiate secondary pathophysiological processes that can further increase morbidity, including neurogenic pulmonary edema (Darnell and Jay, 1982; Bayne and Simon, 1981), disseminated intravascular coagulation (Fischer et al., 1977), and hyperuricemia (Warren et al., 1975) or myoglobinuria (Singhal et al., 1978) with renal failure. If experimental animals are paralyzed and ventilated and measures taken so that blood pressure, body temperature, and blood glucose are controlled, then these pathological effects of status epilepticus are largely but not completely prevented (Meldrum et al., 1973; Soderfeld et al., 1981). Recovery from status epilepticus in man can take days to weeks, and permanent impairment often occurs. Psychological tests of patients with generalized epilepsy reveal cognitive deficits in proportion to the frequency of seizures (Mersky et al., 1960; Dikman and Matthews, 1977). Anticonvulsant medicines can potentially add further impairment, although this is transient and reversible (Matthews and Harley, 1975).

5. Summary

Focal seizures in neocortex can disrupt mental behavior and cause a clinical picture of encephalopathy by local interference with language functions (epileptic aphasia) or by progressive spread through widely distributed seizure pathways when convulsions become bilateral. Focal seizures that spread bilaterally in limbic systems cause profound disruption in mental behavior with loss of conscious contact with the environment. This can continue for hours in cases of limbic status epilepticus and cause profound postictal encephalopathy and amnesia. The cause of mental symptoms during these types of seizures is related to the functional anatomy of seizure spread and the derangement of ionic, metabolic, and neurotransmitter function within these sites. Basic mechanisms involved in postictal states probably relate to changes in synaptic function and structure rather than changes in energy metabolism per se since substrates are maintained near a normal level of energy balance.

During generalized seizures, and especially with convulsive status epilepticus, there are profound changes in brain energy metabolism since hypoxemia usually occurs. In this situation the brain's use of energy exceeds its supply and production and a profound tissue acidosis develops. Changes in systemic factors can accelerate deterioration in brain metabolism and a widespread ischemic type of neuronal degeneration occurs. Multiple ionic, metabolic, and neurotransmitter abnormalities interact in the pathogenesis of brain damage and the postictal encephalopathy that follows generalized convulsions.

ACKNOWLEDGMENTS. The author is pleased to acknowledge the expert technical assistance of N. Green, K. Kamer, and J. Hayes in the preparation of this manuscript. Supported by research grant from the Public Health Service, NS 14834.

References

Aminoff, M. J., and Simon, R. P., 1980, Status epilepticus—Causes, clinical features, and consequences in 98 patients, *Am. J. Med.* **69**:657–666.

Bakay, R. A. E., and Harris, A. B., 1981, Neurotransmitter, receptor and biochemical changes in monkey cortical epileptic foci, *Brain Res.* **206**:387–404.

Bancaud, J., Talairach, J., Morel, P., Bresson, M., Bonis, A., Geier, S., Hemon, H., and Buser, P., 1974, Generalized epileptic seizures elicited by electrical stimulation of the frontal lobe in man, *Electroencephalogr. Clin. Neurophysiol.* **37**:275–282.

Basan, N. G., Jr., 1970, Effects of ischemia and electroconvulsive shock on free fatty acid pool in the brain, *Biochem. Biophys. Acta* **218**:1–10.

Bayne, L. L., and Simon, R. P., 1981, Systemic and pulmonary vascular pressure during generalized seizures in sheep, *Neurology* **10**:566–569.

Ben-Ari, Y., Tremblay, E., Ottersen, O. P., and Meldrum, B. S., 1980, The role of epileptic activity in hippocampal and "remote" cerebral lesions induced by kainic acid, *Brain Res.* **191**:79–97.

Blennow, G., Nilsson, B., and Siesjo, B. K., 1977, Sustained epileptic seizures complicated by hypoxia, arterial hypotension or hyperthermia: Effects on cerebral energy state, *Acta Physiol. Scand.* **100**:126–128.

Brierly, J. B., 1976, Cerebral hypoxia, in: *Greenfield's Neuropathology* (W. Blackwood and J. A. N. Corsellis, eds.), Edward Arnold, London, pp. 43–85.

Brooks, B. R., and Adams, R. D., 1975, Cerebrospinal fluid acid-base and lactate changes after seizures in unanesthetized man. I. Idiopathic seizures, *Neurology* **25**:935–942.

Butler, A. B., Willmore, L. J., Fuller, P. M., and Blass, N. H., 1976, Focal alteration of dendrites and astrocytes in rat cerebral cortex during initiation of cobalt-induced epileptiform activity, *Exp. Neurol.* **51**:216–228.

Caspars, H., and Speckmann, E. J., 1972, Cerebral pO_2 and pCO_2 and pH: Changes during convulsive activity and their significance for the spontaneous arrest of seizures, *Epilepsia* **13**:699–729.

Celesia, G. G., 1976, Modern concepts of status epilepticus, *J. Am. Med. Assoc.* **235**:1571–1574.

Chan, P. H., Fishman, R. A., Lee, J. L., and Candelise, L., 1979, Effects of excitatory neurotransmitter amino acids on swelling of rat brain cortical slices, *J. Neurochem.* **33**:1309–1315.

Chapman, A. G., Meldrum, B. S., and Siesjo, B. K., 1977, Cerebral metabolic changes during prolonged epileptic seizures in rats, *J. Neurochem.* **28**:1025–1035.

Chapman, A. G., Ingvar, M., and Siesjo, B. K., 1980, Free fatty acids in the brain in bicuculline-induced status epilepticus, *Acta Physiol. Scand.* **110**:335–336.

Collins, R. C., 1976, Metabolic response to focal penicillin seizures in rat: Spike discharge vs. afterdischarge, *J. Neurochem.* **27**:1473–1482.

Collins, R. C., 1978a, Use of cortical circuits during focal penicillin seizures: An autoradiographic study with [^{14}C]deoxyglucose, *Brain Res.* **150**:487–501.

Collins, R. C., 1978b, Kindling of neuroanatomic pathways during recurrent focal penicillin seizures, *Brain Res.* **150**:503–517.

Collins, R. C., and Caston, T. V., 1979a, Activation of cortical circuits during interictal spikes, *Ann. Neurol.* **6**:117–125.

Collins, R. C., and Caston, T. V., 1979b, Functional anatomy of occipital lobe seizures: An experimental study in rats, *Neurology* **29**:705–716.

Collins, R. C., and Mehta, S., 1978, Effect of amino-oxyacetic acid (AOAA) on focal penicillin seizures, *Brain Res.* **157**:311–320.

Collins, R. C., and Nandi, N., 1982, Focal seizures disrupt the incorporation of amino acid into protein: An autoradiographic study using 1-14-C-leucine, *Brain Res.* **248**:102–120.

Collins, R. C., and Olney, J. W., 1982, Focal cortical seizures cause distant thalamic lesions, *Science* **218**:177–179.

Collins, R. C., Posner, J. B., and Plum, F., 1970, Cerebral energy metabolism during electroshock seizures in mice, *Am. J. Physiol.* **218**:943–950.

Collins, R. C., Kennedy, C., Sokoloff, L., and Plum, F., 1976, Metabolic anatomy of focal motor seizures, *Arch. Neurol.* **33**:536–542.

Collins, R. C., Conklin, W., Der, T., and Caston, T. V., 1981, Focal cortical seizures disrupt subcortical blood flow and metabolism, *J. Cereb. Blood Flow Metab.* **1**(suppl. 1):596–597.

Collins, R. C., Lothman, E. W., and Olney, J., 1983a, Status epilepticus in the limbic system—Biochemical and pathological changes, in: *Status Epilepticus: Mechanisms of Brain Damage and Treatment*, Volume 34, *Advances in Neurology* (A. D. Delgado-Escueta, C. G. Wasterlain, D. M. Trieman, and R. J. Porter, eds.), Raven Press, New York.

Collins, R. C., Olney, J. W., and Lothman, E. W., 1983b, Metabolic and pathological consequences of focal seizures, in: *Epilepsy* (A. A. Ward, Jr., J. K. Penry, and D. Purpura, eds.), Raven Press, New York, pp. 87–107.

Corsellis, J. A. N., and Bruton, C. J., 1983, Neuropathology of status epilepticus in humans, in: *Status Epilepticus: Mechanisms of Brain Damage and Treatment*, Volume 34, *Advances in Neurology*, (A. D. Delgado-Escueta, C. G. Wasterlain, D. M. Trieman, and R. J. Porter, eds.), Raven Press, New York, pp. 129–139.

Cotman, C. W., Banker, G., Zornetzer, S. F., and McGaugh, J. L., 1971, Electroshock effects on brain protein synthesis: Relation to brain seizures and retrograde amnesia, *Science* **173**:454–456.

Crowell, R. M., 1970, Distant effects of a focal epileptogenic process, *Brain Res.* **18**:137–154.

Darnell, J. C., and Jay, S. J., 1982, Recurrent post-ictal pulmonary edema: A case report and review of the literature, *Epilepsia* **23**:71–83.

Delgado-Escueta, A. V., and Bajorek, J. G., 1982, Status epilepticus: Mechanisms of brain damage and rational management, *Epilepsia* **23**(suppl. 1):529–541.

Delgado-Escueta, A. V., and Horan, M. P., 1980, Brain synaptosomes in epilepsy: Organization of ionic channels and the Na^+-K^+ pump, in: *Antiepileptic Drugs: Mechanisms of Action* (G. Glaser, D. K. Penry, and D. J. Woodbury, eds.), Raven Press, New York, pp. 85–126.

Delgado-Escueta, A. V., Boxley, J., Stubbs, N., Waddell, G., and Wilson, W. A., 1974, Prolonged twilight state and automatisms: A case report, *Neurology* **24**:331–339.

Delgado-Escueta, A. V., Bacsal, F. E., and Treiman, D. M., 1982, Complex partial seizures in close-circuit television and EEG: A study of 691 attacks in 79 patients, *Ann. Neurol.* **11**:292–300.

DePasquet, E. G., Gaudin, E. S., Bianchi, A., and DeMendilaharsu, S. A., 1976, Prolonged and monosymptomatic dysphasic status epilepticus, *Neurology* **26**:244–247.

Dietzel, I., Heinemann, U., Hofmeier, G., and Lux, H. D., 1980, Changes in the extracellular volume in the cerebral cortex of cats in relation to stimulus induced epileptiform afterdischarges, in: *Physiology and Pharmacology of Epileptogenic Phenomena* (M. R. Klee, H. Dieterlux, and E.-J. Speckman, eds.), Raven Press, New York, pp. 5–12.

Dikman, S., and Matthews, C. G., 1977, Effect of major motor seizure frequency upon cognitive – intellectual functions in adults, *Epilepsia* **18**:21–36.

Dinner, D. S., Lueders, H., Lederman, R., and Gretter, T. E., 1981, Aphasic status epilepticus: A case report, *Neurology* **31**:888–890.

Duffy, T. E., Howse, D. C., and Plum, F., 1975, Cerebral energy metabolism during experimental status epilepticus, *J. Neurochem.* **24**:925–934.

Dymond, A. M., and Crandall, P. H., 1976, Oxygen availability and blood flow in the temporal lobes during spontaneous epileptic seizures in man, *Brain Res.* **102**:191–196.

Elger, C. E., Speckman, E.-S., Caspers, H., and Prohaska, O., 1982, Focal interictal epileptiform discharges in the cortex of the rat: Laminar restriction and its consequences for activity descending to the cord, in: *Physiology and Pharmacology of Epileptiform Phenomena* (M. R. Klee, H. Dicter Lux, E.-J. Speckman, eds.), Raven Press, New York, pp. 13–21.

Ellis, J. M., and Lee, S. I., 1978, Acute prolonged confusion in later life as an ictal state, *Epilepsia* **19**:119–128.

Engel, J., Jr., Ludwig, B. I., and Fetell, M., 1978, Prolonged partial complex status epilepticus: EEG and behavioral observations, *Neurology* **28**:863–869.

Engel, J., Jr., Kuhl, D. E., and Phelps, M. E., 1982, Patterns of human local cerebral glucose metabolism during epileptic seizures, *Science* **218**:64–66.

Ferrendelli, J. A., and McDougal, D. B., Jr., 1971, The effect of electroshock on regional CNS energy reserves in mice, *J. Neurochem.* **18**:1197–1205.

Fifkova, E., Anderson, C. L., Young, S. J., and Van Harreveld, A., 1982, Effect of anisomycin on stimulation-induced changes in dendritic spines of the dentate granule cells, *J. Neurocytol.* **11**:183–210.

Fischer, J., and Langmeier, M., 1980, Changes in the number, size and shape of synaptic vesicles in an experimental, projected cortical epileptic focus in the rat, *Epilepsia* **21**:571–585.

Fischer, S. P., Lee, J., Zetuchne, J., and Greenberg, J., 1977, Disseminated intervascular coagulation in status epilepticus, *Thromb. Haemostasis* **38**:909–914.

Folbergrova, J., Ingvar, M., and Siesjo, B. K., 1981, Metabolic changes in cerebral cortex, hippocampus, and cerebellum during sustained bicuculline-induced seizures, *J. Neurochem.* **37**:1228–1238.

Gilbert, G. J., 1978, Transient global amnesia: Manifestations of medial temporal lobe epilepsy, *Clin. EEG* **9**:147–152.

Gloor, P., Olivier, A., Quesney, L. F., Andermann, F., and Horowitz, S., 1982, The role of the limbic system in experiential phenomena of the temporal lobe epilepsy, *Ann. Neurol.* **12**:129–144.

Goldring, S., 1972, The role of prefrontal cortex in grand mal convulsion, *Arch. Neurol.* **2**:109–120.

Greenwood, R. S., Takato, M., and Goldring, S., 1981, Potassium activity and changes in glial and neuronal membrane potentials during initiation and spread of afterdischarge in cerebral cortex of cat, *Brain Res.* **218**:279–298.

Griffiths, T., Evans, M. C., and Meldrum, B. S., 1982, Intracellular sites of early calcium accumulation in the rat hippocampus during status epilepticus, *Neurosci. Lett.* **30**:329–334.

Gutnick, M. J., and Prince, D. A., 1974, Effects of projected cortical epileptiform discharges on neuronal activities in cat VPL. I. Interictal discharge, *J. Neurophysiol.* **37**:1310–1327.

Gutnick, M. J., Heinemann, U., and Lux, H. D., 1979, Stimulus induced and seizure related changes in extracellular potassium concentration in cat thalamus (VPL), *Electroencephalogr. Clin. Neurophysiol.* **47**:329–344.

Heinemann, U., and Pumain, R., 1980, Extracellular calcium activity changes in cat sensorimotor cortex induced by iontophoretic application of amino acids, *Exp. Brain Res.* **40**:247–250.

Heinemann, U., Lux, H. D., and Gutnick, M. J., 1977, Extracellular free calcium and potassium during paroxysmal activity in the cerebral cortex of the cat, *Exp. Brain Res.* **27**:237–243.

Holmes, G., 1927, Local epilepsy, *Lancet* **1**:957–962.

Horton, R. W., Meldrum, B. S., Pedley, T. A., and Siesjo, B. K., 1980, Regional cerebral blood flow in the rat during prolonged seizure activity, *Brain Res.* **192**:399–412.

Houser, C. R., Vaughn, J. E., Barber, R. P., and Roberts, E., 1980, GABA neurons are the major cell type of the nucleus reticularis thalami, *Brain Res.* **200**:341–354.

Howse, D. C. N., 1979, Metabolic responses to status epilepticus in the rat, cat and mouse, *Can. J. Physiol. Pharmacol.* **57**:205–212.

Jackson, H., 1870, A study of convulsions, in: *Selected Writings of John Hughlings Jackson*, Volume 1 (J. Taylor, ed.), Basic Books, New York, 1958, pp. 8–36.

Jobsis, F. F., O'Connor, M., Vitale, A., and Vreman, H., 1971, Intracellular redox changes in functioning cerebral cortex. I. Metabolic effects of epileptiform activity, *J. Neurophysiol.* **34**:735–749.

Jones, E. G., 1977, Some aspects of the organization of the thalamic reticular complex, *J. Comp. Neurol.* **162**:285–308.

Jones, E. G., 1981a, Anatomy of cerebral cortex: Columnar input-output organization in: *The Organization of the Cerebral Cortex* (F. O. Schmitt, F. G. Worden, G. Adelman, and S. G. Dennis, eds.), MIT Press, Cambridge, pp. 199–236.

Jones, E. G., 1981b, Functional subdivision and synaptic organization of the mammalian thalamus, *Int. Rev. Physiol.* **25**:173–245.

Kato, M., Malamut, B. L., Caveness, W. F., Hosokawa, S., Wakisaka, S., and O'Neill, R., 1980, Local cerebral glucose utilization in newborn and pubescent monkeys during focal motor seizures, *Ann. Neurol.* **7**:204–212.

Kennedy, C., DesRosiers, M. H., Jehle, M. H., Reivich, J. W., Sharp, F., and Sokoloff, L., 1975, Mapping of functional neural pathways by autoradiographic survey of local metabolic rate with [^{14}C]deoxyglucose, *Science* **187**:850–853.

Knopman, D., Margolis, G., and Reeves, A. G., 1977, Prolonged focal epilepsy and hypoxemia as a cause of focal brain damage: A case study, *Ann. Neurol.* **1**:195–198.

Kriesman, N. R., Lamanna, J. C., Rosenthal, M., and Sick, T. J., 1981, Oxidative metabolic responses with recurrent seizures in rat cerebral cortex: Role of systemic factors, *Brain Res.* **218**:175–188.

Kriesman, N. R., Rosenthal, M., and LaManna, J. C., 1983, Cerebral oxygenation during recurrent seizures, in: *Status Epilepticus: Mechanisms of Brain Damage and Treatment*, Volume 34, *Advances in Neurology* (A. Delgado-Escueta, C. G. Wasterlain, D. M. Treiman, and R. J. Porter, eds.), Raven Press, New York, pp. 231–240.

Kugoh, T., and Hosokawa, K., 1977, Mental dullness associated with left frontal continuous focal discharge *Folia Psychiatr. Neurol. Jpn.* **31**:473–480.

Kuhl, D. E., Engel, J., Jr., Phelps, M. E., and Selin, C., 1980, Epileptic patterns of local cerebral metabolism and perfusion in humans determined by emission computed tomography of ^{18}FDG and ^{13}NH$_3$, *Ann. Neurol.* **8**:348–360.

Kunzle, H., 1975, Bilateral projections from precentral motor cortex to the putamen and other parts of the basal ganglia. An autoradiographic study in *macaca fascicularis*, *Brain Res.* **88**:195–209.

Landau, W. M., and Kleffner, F. R., 1957, Syndrome of acquired aphasia with convulsive disorder in children, *Neurology* **7**:523–528.

Lawall, J., 1976, Psychiatric presentations of seizure disorders, *Am. J. Psychiatry* **133**:321–323.

Levine, D. N., and Finklestein, S., 1982, Delayed psychosis after right temporo-parietal stroke or trauma: Relation to epilepsy, *Neurology* **32**:267–273.

Lieb, J. P., Walsh, G. O., Babb, T. L., Walter, R. D., and Crandall, P. H., 1976, A comparison of EEG seizure patterns recorded with surface and depth electrodes in patients with temporal lobe epilepsy, *Epilepsia* **17**:137–160.

Lothman, E. W., and Collins, R. C., 1981, Kainic acid induced limbic seizures: Metabolic, behavioral electroencephalographic, and neuropathological correlates, *Brain Res.* **218**:299–318.

Lothman, E. W., and Somjen, G. G., 1975, Extracellular potassium activity, intracellular and extracellular responses in the spinal cord, *J. Physiol.* **252**:115–136.

Lothman, E. W., LaManna, J., Cordingley, G., Rosenthal, M., and Somjen, G., 1975, Responses of electrical potential, potassium levels, and oxidative metabolic activity of the cerebral neocortex of cats, *Brain Res.* **88**:15–36.

Markand, O. N., Wheeler, G. L., and Pollack, S. L., 1978, Complex partial status epilepticus (psychomotor status), *Neurology* **28**:189–196.

Matthews, C. G., and Harley, P., 1975, Cognitive and motor-sensory performance in toxic and non-toxic epileptic subjects, *Neurology* **25**:184–188.

Meldrum, B. S., 1983, Metabolic factors during prolonged seizures and their relation to nerve cell death, in: *Status Epilepticus—Mechanisms of Brain Damage and Treatment*, Volume 34, *Advances in Neurology* (A. Delgado-Escueta, C. G. Wasterlain, D. M. Treema, and R. J. Porter, eds.). Raven Press, New York, pp. 261–276.

Meldrum, B. S., and Brierly, J. B., 1973, Prolonged epileptic seizures in primates—Ischemic cell change and its relation to ictal physiological events, *Arch. Neurol.* **28**:10–17.

Meldrum, B. S., and Horton, R. W., 1973, Physiology of status epilepticus in primates, *Arch. Neurol.* **28**:1–9.

Meldrum, B. S., Vogoroux, R. A., and Brierly, J. B., 1973, Systemic factors and epileptic brain damage: Prolonged seizures in paralyzed artificially ventilated baboons, *Arch. Neurol.* **29**:82–87.

Mersky, A. F., Primac, D. W., and Ajomone-Marsan, C., 1960, A comparison of the psychological test performance of patients with focal and non-focal epilepsy, *Exp. Neurol.* **2**:75–89.

Mutani, R., Durelli, L., Mazzarino, M., Valentine, C., Monaco, F., Fumero, S., and Mondurao, A., 1977, Longitudinal changes of brain amino acid content occurring before, during and after epileptic activity, *Brain Res.* **122**:513–521.

Nitsch, C., and Rinne, U., 1981, Large dense-core vessicle execytosis and membrane recycling in the mossy fiber synapses of the rabbit hippocampus during epileptiform seizures, *J. Neurocytol.* **10**:221–232.

Noebels, J. L., and Prince, D. A., 1978a, Development of focal seizures in cerebral cortex: Role of axon terminal bursting, *J. Neurophysiol.* **41**:1267–1281.

Noebels, J. L., and Prince, D. A., 1978b, Excitability changes in thalamocortical relay neurons during synchronous discharges in cat neocortex, *J. Neurophysiol.* **41**:1282–1296.

Okada, K., Ayala, G. F., and Sung, J. H., 1971, Ultrastructure of penicillin-induced epileptogenic lesions of the cerebral cortex in cats, *J. Neuropathol. Exp. Neurol.* **30**:337–353.

Olney, J. W., 1978, Neurotoxicity of excitatory amino acids, in: *Kainic Acid as a Tool in Neurobiology* (E. McGeer, J. W. Olney, and P. L. McGeer, eds.), Raven Press, New York, pp. 110–126.

Passonneau, J. V., 1969, Energy metabolites in experimental seizures, in: *Basic Mechanisms of the Epilepsies* (H. H. Jasper, A. A. Ward, and A. Pope, eds.), Little, Brown and Co., New York, pp. 98–103.

Penfield, W., 1937, The circulation of the epileptic brain. *Res. Publ. Assoc. Res. Nerv. Ment. Dis.* **18**:605–637.

Penfield, W., and Milner, B., 1958, Memory deficit produced by bilateral lesions in the hippocampal zone, *Arch. Neurol. Psychiatry* **70**:475–497.

Perry, T. L., and Hansen, S., 1981, Amino acid abnormalities in epileptogenic foci, *Neurology* **31**:872–876.

Plum, F., Howse, D. C., and Duffy, T. E., 1974, Metabolic effects of seizures, in: *Brain Dysfunction in Metabolic Disorders* (F. Plum, ed.), *Res. Publ. Assoc. Res. Nerv. Ment. Dis.* **53**:141–157, Raven Press, New York.

Prince, D. A., and Wilder, B. J., 1967, Control mechanisms in cortical epileptogenic foci—"surround inhibition," *Arch. Neurol.* **16**:194–202.

Prince, D. A., Lux, H. D., and Neher, F., 1973, Measurement of extracellular potassium activity in cat cortex, *Brain Res.* **50**:489–495.

Rausch, R., Lieb, J. B., and Crandall, P. H., 1978, Neuropsychological correlates of depth spike activity in epileptic patients, *Arch. Neurol.* **35**:699–705.

Ribak, C. E., and Riffenstein, R. J., 1982, Selective inhibitory synapse loss in chronic cortical slabs: A morphological basis for epileptic susceptibility, *J. Can. Physiol. Pharmacol.* **60**:864–870.

Ribak, C. E., Harris, A. B., Vaughn, J. E., and Robert, E., 1979, Inhibitory, GABAergic nerve terminals decrease at sites of focal epilepsy, *Science* **205**:211–240.

Rowan, A. J., and Scott, D. F., 1970, Major status epilepticus. *Acta. Neurol. Scand.* **46**:573–584.

Sakai, F., Meyer, J. F., Naritomi, H., and Hsu, M-C., 1978, Regional cerebral blood flow and EEG in patients with epilepsy, *Arch. Neurol.* **35**:648–657.

Scheibel, M. E., Crandall, P. H., and Scheibel, A. B., 1974, The hippocampal-dentate complex in temporal lobe epilepsy, Epilepsia 15:55–80.

Schwartzkroin, P. A., vanDuijn, H., and Prince, D. A., 1974, Effects of projected cortical epileptiform discharges on unit activity in the cat cuneate nucleus, Exp. Neurol. 43:106–123.

Schwartzkroin, P. A., Mutani, R., and Prince, D. A., 1975, Orthodromic and antidromic effects of a cortical epileptiform focus on ventrolateral nucleus of the cat, J. Neurophysiol. 38:795–811.

Scoville, W. B., and Milner, B., 1957, Loss of recent memory after bilateral hippocampal lesions, J. Neurol. Neurosurg. Psychiatry 20:11–21.

Shank, R. P., and Aprison, M. H., 1979, Biochemical aspects of the neurotransmitter function of glutamate, in: Glutamic Acid: Advances in Biochemistry and Physiology (L. J. Fuller, S. Garattine, M. R., Kare, W. A. Reynolds, and R. J. Wurtman, eds.), Raven Press, New York, pp. 139–150.

Shoumaker, R. D., Bennett, D. R., Bray, P. F., and Curless, R. G., 1974, Clinical and EEG manifestations of an unusual aphasic syndrome in children, Neurology 24:10–16.

Siesjo, B. K., 1978, Brain Energy Metabolism, John Wiley and Sons, New York.

Simpson, H., Habel, A. H., and George, E. L., 1977, Cerebrospinal fluid acid-base status and lactate and pyruvate concentrations after convulsions of varied duration and etiology in children, Arch. Dis. Child. 52:844–849.

Singhal, P. C., Chugh, K. S., and Gulati, D. R., 1978, Myoglobinuria and renal failure after status epilepticus, Neurology 28:200–201.

Sloviter, R. S., and Damiano, B. P., 1981, Sustained electrical stimulation of the perforant path duplicates kainate-induced electrophysiological effects and hippocampal damage in rats, Neurosci. Lett. 24:279–284.

Soderfeld, B., Kalimo, H., Olsson, Y., and Siesjo, B. K., 1981, Pathogenesis of brain lesions caused by experimental epilepsy, Acta. Neuropathol. 54:219–231.

Sokoloff, L., Reivich, M., Kennedy, C., DesRosiers, M. H., Pettigrew, K. D., Sakurada, O., and Shinohara, M., 1977, The [14-C]deoxyglucose method for the measurement of local cerebral glucose utilization: Theory, procedures and norma values in the conscious and anesthetized albino rat, J. Neurochem. 28:897–916.

Tenny, R. T., Sharbrough, F. W., Anderson, R. E., and Sundt, T. M., 1978, Correlation of intracellular redox states and pH with blood flow in primary and secondary seizure foci, Ann. Neurol. 8:564–573.

Traub, R. D., and Llinas, R., 1979, Hippocampal pyramidal cells: Significance of dendritic ion conductances for neuronal function and epileptogenesis, J. Neurophysiol. 42:476–496.

Ueno, H., Yamashita, Y., and Caveness, W. F., 1975, Regional cerebral blood flow pattern in focal epileptiform seizures in the monkey, Exp. Neurol. 47:81–96.

VanGelder, N. M., Sherwin, A. L., and Rasmussen, T., 1972, Amino acid content of epileptogenic human brain: Focal versus surround region, Brain Res. 40:385–393.

Van Harreveld, A., and Fifkova, E., 1975, Swelling of dendritic spines in the fascia dentata after stimulation of the perforant fibers as a mechanism of post-tetanic potentiation, Exp. Neurol. 49:736–749.

Walker, A. E., 1959, The state of consciousness in focal motor convulsions, Epilepsia 1:592–599.

Walker, A. E., Poggio, G. F., and Andy, O. J., 1956, Structural spread of cortically induced epileptic discharges, Neurology 6:616–626.

Walshe, F. M. R., 1943, On the mode of representation of movements in the motor cortex with special reference to "convulsions beginning unilaterally" (Jackson), Brain 66:104–139.

Warren, D. J., Leitch, A. G., and Leggett, A. J. E., 1975, Hyperuricaemic acute renal failure after epileptic seizures, Lancet 2:385–387.

Wasterlain, C. G., 1974, Inhibition of cerebral protein synthesis by epileptic seizures without motor manifestations, Neurology 24:175–180.

Wasterlain, C. G., and Duffy, T. E., 1976, Status epilepticus in immature rats—Protective effects of glucose on survival and brain development, Arch. Neurol. 33:821–827.

Wells, C. E., 1975, Transient ictal psychosis, Arch. Gen. Psychiatry 32:1201–1203.

Whittam, R., 1962, The dependence of the respiration of brain cortex on active cation transport, Biochem. J. 82:205–212.

Wieser, H. G., 1980, Temporal lobe or psychomotor status epilepticus. A case report, Electroencephalogr. Clin. Neurophysiol. 48:558–572.

IV

Metabolic Encephalopathy That May Result from Extrinsic Factors

13

Niacin-Nicotinamide Deficiency

FREDERICK C. KAUFFMAN

1. Introduction

Nicotinic acid (niacin) has been known for well over a century; however, the
biological importance of this acid and its amide (nicotinamide) did not become
apparent until the discovery of the pyridine nucleotides, NAD and NADP, and
the recognition that this vitamin serves as a precursor of the pyridine nucleotide
cofactors. The human disease caused by niacin deficiency is pellagra, a disease
that filled insane asylums all over the world before its origin was discovered.
The recognition of pellagra as an endemic disease in the United States dates
from Searcy's report in 1907 describing 88 cases of dementia in the Mount
Vernon, Alabama, Insane Asylum. In spite of our extensive knowledge about
the course of pellagra and our ability to eradicate this disease as a public health
problem, there remain great gaps in our understanding of the exact relationship
between niacin deficiency and specific pathological lesions in neural tissue.

Nutritional deficiencies among patients in mental hospitals, and in those
with senility (Gregory, 1955; Hersov, 1955; McIlwain, 1966) and behavioral
changes that are common in such patients, makes the detection of pellagra
difficult. Mental symptoms associated with pellagra often precede the der-
matitis and other effects of niacin deficiency. If the initial mental disturbances
are not remedied by administering nicotinic acid or tryptophan, from which
nicotinic acid is synthesized *in vivo*, permanent structural changes will occur
in cerebral tissue. The morphological changes observed in chronic pellagra
include central chromatolysis and degeneration, especially in the large motor
neurons of the motor cortex, brain stem, and anterior horn of the spinal cord.
These structural changes resemble those seen after axotomy (Blackwood *et al.*,
1963; Horita *et al.*, 1981) and suggest that mechanisms causing morphological
changes in neural tissue of patients with pellagra are similar, in part, to those
associated with chromatolysis induced by axonal injury. Although it is as-

FREDERICK C. KAUFFMAN • Department of Pharmacology and Experimental Therapeutics,
University of Maryland School of Medicine, Baltimore, Maryland 21201.

sumed that there is an underlying chemical defect that affects the brain in niacin deficiency, it is by no means clear what this might be. The purpose of this chapter is to review the role of niacin and nicotinamide in cerebral energy metabolism and to discuss some models that shed light on biochemical defects produced by deficiencies of niacin and nicotinamide.

2. Role of Niacin and Nicotinamide as Precursors of Cerebral Pyridine Nucleotides

Nicotinic acid derivatives in brain exist largely in the form of the active nicotinamide adenine dinucleotides, NAD and NADP. Compared with other organs, the brain is very rich in these nicotinic acid derivatives. The concentration of total niacin and derivatives of this vitamin are maintained over fairly narrow ranges in mammalian brain (Kaplan et al., 1956; Garcia-Bunuel et al., 1962; Spector, 1979; Spector and Kelly, 1979); however, the cerebral content of nicotinic acid derivatives falls considerably during deficiency (Axelrod et al., 1941; Singal et al., 1948; McIlwain, 1966). Unlike liver and kidney, brain lacks the final enzyme in the synthetic pathway for niacin, quinolate phosphoribosyltransferase (Nakamura et al., 1963); thus, niacin and niacinamide cannot be synthesized in brain from tryptophan (Ikeda et al., 1965; Krehl, 1981).

Although cerebral tissue does not synthesize niacin or nicotinamide, it has a very high capacity to form nicotinamide adenine dinucleotide (NAD) via niacinamide mononucleotide and from niacin via desamino-NMN and desamido-NAD (Ikeda et al., 1965). Cerebral tissues can also incorporate inorganic phosphate into the NAD (Heald, 1956). Administration of large doses of nicotinamide to mice will cause moderate increases in NAD in brain (Kaplan et al., 1956); however, this differs from that of liver where relatively large increases in the pyridine nucleotide cofactors occur in the presence of high concentrations of niacin or nicotinamide in plasma (Kaplan et al., 1956).

Brain contains relatively high activities of NADase or glycohydrolase (E.C. 3.2.2.5), which will exchange niacinamide in NAD and NADP moieties with unbound niacinamide (McIlwain and Rodnight, 1949; Köhler et al., 1970). This exchange process is also responsible for the formation of abnormal pyridine nucleotide analogues from derivatives of the natural vitamin such as 6-aminonicotinamide and 3-acetylpyridine (Kaplan, 1960; Köhler et al., 1970). Cerebral tissue has a high capacity to inactivate pyridine nucleotides via glycohydrolase and possibly other enzymatic processes. For example, in cerebral homogenates, loss of coenzyme occurs at rates approaching 200 μmole/g per hr, which, in the absence of resynthesis, would lead to total elimination of pyridine nucleotides in brain in 2–5 sec. This rapid loss of pyridine nucleotides does not occur if the structural integrity of neural tissue is maintained (McIlwain, 1966) suggesting either that enzymes responsible for the degradation of pyridine nucleotides and these substrates exist in separate compartments in the intact tissue or that rates of loss are balanced by rates of synthesis. The degrading enzyme that shows the properties of a deoxyribosyl transferase appears at a particular stage in development of mammals and has been extensively purified (Nemeth and Dickerman, 1960; Windmueller and Kaplan, 1962). Although the role of this enzyme in metabolism of the brain remains poorly defined, it does influence rates of glycolysis in intact cells since inhibitors, such as nicotinamide or phenosafranine, stimulate aerobic glycolysis in neuronal cells in suspension culture (Windmueller and Kaplan, 1962; McIlwain, 1966). Stimulation of aerobic glycolysis by these inhibitors may result from alterations in ratios of oxidized and reduced forms of pyridine nucleotides since the enzyme appears to be specific for the oxidized forms of NAD and NADP only. The maximal rate of degradation of pyridine nucleotides in brain occurs at about the same rate as these cofactors undergo oxidation and reduction while functioning as respiratory carriers (McIlwain, 1966). Thus, the degradation of pyridine nucleotides may be linked to their function as carriers of hydrogen ions in the intact nerve cell.

3. Relationships of Niacin and Nicotinamide to Cerebral Energy Metabolism

3.1. Glycolysis and the Pentose Phosphate Pathway

The requirement of NAD for cerebral glycolysis is indicated by the finding that glycolysis declines in cell-free suspensions of cerebral tissue that do not contain added NAD and nicotinamide (Utter, 1950). It has been known for many years that in the absence of added NAD, hexose diphosphates and triosephosphates, but not lactic acid, will accumulate (Weil-Malherbe and Bone, 1951). NAD in brain, as in other tissues, is required for the metabolism of triosephosphates via glyceraldehyde phosphate dehydrogenase (E.C.1.2.1.13) and pyruvate dehydrogenase (E.C. 1.2.4.1), as well as lactate dehydrogenase (E.C. 1.1.1.27) (Lowry and Passonneau, 1964). Activities of glyceraldehyde phosphate dehydrogenase have been assayed in brain (Laatsch, 1962) and found to have a Km for NAD of 22 μM and a requirement for orthophosphate (Lowry and Passonneau, 1964). Glyceraldehyde phosphate dehydrogenase probably contains both NAD and glutathione bound to the apoenzyme. A special property of this enzyme in brain is its sensitivity to inhibition by iodoacetate (Himwich, 1951; Samson and Dahl, 1957) which is undoubtedly related to the potent toxic effects of this substance on the central nervous system.

Lactate dehydrogenase catalyzes the reduction of pyruvate to lactate and requires NAD. Kinetic studies of the purified enzyme from brain (Winer, 1960) showed that the reduced pyridine nucleotide NADH bound to the enzyme before the formation of an enzyme–NADH–pyruvate complex. Reduction of pyruvate presumably occurs on the enzyme followed by release of lactate and finally release of NAD from the enzyme substrate complex. The affinity of this enzyme for both the pyridine nucleotide cofactor and substrate varies depending on the concentration of the other substrate or cofactor, e.g., the Km for pyruvate increases with increasing concentrations of NADH (Lowry and Passonneau, 1964). Normally, little pyruvate accumulates in brain, even anaerobically, presumably because lactate dehydrogenase has a very high activity in vivo and prevents pyruvate from accumulating more than 0.2 μmole/g of tissue, even during convulsive states in which lactate rises to very high levels

(Bain and Pollock, 1949). As in other tissues, several molecular species of lactate dehydrogenase have been identified in brain. These are identical in their physical and immunological characteristics with the enzyme from heart (Nisslbaum et al., 1964). During periods of sustained anaerobic glycolysis, lactic acid formation continues because the NAD oxidized by pyruvate is reduced by glyceraldehyde-3-phosphate (Utter, 1950). Rates of glycolysis in brain are believed to be influenced in a major way by this coupling between lactate dehydrogenase and glyceraldehyde phosphate dehydrogenase.

NADP is involved in hexose metabolism via the pentose phosphate pathway and serves as the cofactor for the two oxidative enzymes, glucose-6-phosphate dehydrogenase (E.C. 1.1.1.49) and 6-phosphogluconate dehydrogenase (E.C. 1.1.1.43). Early work demonstrated that reduction of NADP in brain homogenates required glucose-6-phosphate and magnesium (Dickens and Glock, 1951; Glock and McLean, 1954). Estimates of the activity of glucose-6-phosphate dehydrogenase made in extracts of whole brain indicate that this enzyme proceeds at a rate of about 150 μmole/g per hr (Buell et al., 1958; Roberts et al., 1958) and is more active in areas of white matter than of gray matter (Kuhlman and Lowry, 1956; McDougal et al., 1961; Luine and Kauffman, 1971). Half-maximal rates of the enzyme from rabbit brain are obtained with 1–3 μM NADP (Luine and Kauffman, 1971; Kauffman, 1972). The possibility that glucose-6-phosphate dehydrogenase activity is associated with structures having a high requirement for NADPH is suggested by the finding that this activity increases dramatically in the dorsal columns of the spinal cord during periods of active myelination (Luine and Kauffman, 1971). In contrast to glucose-6-phosphate dehydrogenase, 6-phosphogluconate dehydrogenase does not vary greatly among different regions of the central nervous system (Kauffman, 1972).

3.2. Oxidative Metabolism

The participation of the nicotinamide adenine dinucleotide cofactors in oxidative metabolism occurs at three enzymatic steps: pyruvate dehydrogenase, NAD-dependent isocitrate dehydrogenase (E.C. 1.1.1.41), and alpha-ketoglutarate dehydrogenase (E.C. 1.2.4.2). The role of these enzymes in cerebral energy metabolism are covered in other chapters in this volume and have been reviewed extensively by other authors (see, e.g., McIlwain, 1966; Lajtha et al., 1981). Cerebral tissues differ from other tissues in that pyruvate is metabolized substantially by both anaerobic and aerobic processes. Nearly 50% of pyruvate consumed may be converted to lactate. The oxidative decarboxylation of pyruvate involves thiamine pyrophosphate, lipoamide, and acetyl CoA in addition to NAD as cofactors. Energy derived from oxidation of pyruvate occurs mainly via reduction of NAD. Isocitrate dehydrogenase occurs within the inner mitochondrial matrix of mitochondria and requires NAD as the cofactor, whereas NADP$^+$-dependent isocitrate dehydrogenase (E.C. 1.1.1.42) is located primarily in the soluble portion of neural tissue (Luine and Kauffman, 1971). Although the exact role of the latter enzyme in cerebral metabolism has not been clearly defined, it has been suggested to participate in a series of reactions converting alpha-ketoglutarate to cellular acetyl CoA (Madsen et al., 1964; D'Adamo and Haft, 1965). Alpha-ketoglutarate dehydrogenase converts alpha-ketoglutarate to succinyl CoA in the presence of NAD, ATP, and acetyl CoA. Formation of a high-energy phosphate bond in this sequence of reactions derives a large part of the free energy from the reoxidation of reduced NAD.

4. Factors Influencing the Cerebral Content of Niacin and Nicotinamide

4.1. Sources of Niacin and Nicotinamide

Niacin and nicotinamide cannot be synthesized in mammalian brain because brain lacks the final enzyme in the synthetic pathway, quinolate phosphoribosyltransferase (Nakamura *et al.*, 1963). However, as with peripheral tissues such as liver and kidney, brain does have the capacity to synthesize NAD and NADP from niacinamide and niacin (Ikeda *et al.*, 1965). Brain also contains NADase or glycohydrolase which exchanges niacinamide in NAD and NADP with unbound niacinamide (McIlwain and Rodnight, 1949; Köhler *et al.*, 1970). Because brain lacks the capacity to synthesize niacin, it must receive its supply from synthesis in peripheral tissues or from dietary sources. The requirement of niacin for normal metabolism was not recognized for many years because this vitamin could be synthesized from tryptophan. At first it was thought that pellagra was due to a protein deficiency rather than the absence of a vitamin (Krehl, 1981). The discovery in 1945 that the essential amino acid tryptophan is transformed into niacin by mammalian tissues ran counter to the usual concept that essential B vitamins cannot be synthesized by higher animals. The relationship between tryptophan and niacin biosynthesis was developed in the 1940s largely from the studies of Willard Krehl, who showed that addition of 0.5% tryptophan to the diet of dogs made niacin-deficient on a high-corn diet resulted in a restoration of growth (Krehl *et al.*, 1945). Previous dietary studies had established a link between the consumption of maize as a major food stuff and the development of pellagra. Based on the extensive studies of Krehl, it was recognized that niacin deficiency induced by corn was due to the character of the protein present in corn, zein, which contained an imbalance of amino acids resulting in an increased tryptophan requirement and therefore poor growth. The concept that an amino acid imbalance could be a major factor influencing the requirement for a vitamin introduced a new nutritional concept. Work on the relationship between corn and pellagra also gave new emphasis to the concept that processing of a foodstuff may have significant effects on the development of a nutritional deficiency. For example, peasants in Mexico, who have corn as the major staple of their diet, live in extreme poverty, are exposed to strong solar radiation, lack animal foods, and exist on a diet consisting mainly of maize, do not have pellagra. This resistance to pellagra was discovered by workers in Scotland (Laguna and Carpenter, 1951), who processed corn as the Mexican peasants do in preparing tortillas, namely soaking maize in hot lime water (calcium hydroxide), and found that experimental animals maintained on this diet no longer became niacin deficient. This finding suggested that niacin or a related ingredient in corn was liberated by hydrolysis in alkali prior to preparing corn for dietary consumption. It is now known that niacin in cereals is largely present in macromolecules that are poorly digested by the enzymes of the gut (Kodicek *et al.*, 1959). Prior hydrolysis in either acid or alkali liberates niacin and tryptophan from its bound state in cereal macromolecules. A detailed account of the influence of maize processing on the induction of pellagra can be found in the review by Carpenter (1981).

The relationship between the synthesis of niacin and tryptophan metabolism was firmly established following the nutritional studies which indicated that dietary requirements for this vitamin could be met by ingestion of tryptophan. The dietary equivalent of 1 mg nicotinamide is about 60 mg of tryptophan in man (Goldsmith, 1958). The biochemical sequence leading to the synthesis of niacin from tryptophan is quite complex and involves the participation of tryptophan pyrolase, kynurenine formylase, kynurenine-3-hydroxylase, and kynureninase to yield 3-hydroxyanthranilic acid, which is the precursor for nicotinic acid (Ikeda *et al.*, 1965; Deguchi *et al.*, 1968; Jepson, 1972). 3-Hydroxyanthranilic acid is oxidized to 2-acroleyl-3-aminofumarate, which undergoes a spontaneous ring closure yielding quinolinic acid. Quinolinic acid is condensed with phosphoribosyl pyrophosphate to yield a ribonucleotide, which is then decarboxylated to yield nicotinic acid ribonucleotide. Hydrolysis of the glycosidic bond in this compound produces nicotinic acid. Although this pathway has been described mainly in liver, it is presumed to occur in other nonneuronal tissues.

Inhibitors of dopa decarboxylase, such as benserazide and carbidopa, used along with L-dopa in the treatment of Parkinson's disease, are potent inhibitors of kynurenine hydroxylase (Bender *et al.*, 1977) and consequently cause an increased reliance on dietary niacin. Biochemical evidence indicative of niacin deficiency has been detected in patients receiving these drugs (Bender *et al.*, 1979). Although dietary niacin is usually sufficient to meet the metabolic needs of most parkinsonian patients, clinicians should be alert to the possibility of pellagra in patients who receive only marginally adequate diets.

4.2. Effects of Niacin Precursors on Neuronal Structure and Function

Kynurenines are the major metabolites of tryptophan and serve as precursors of niacin and nicotinamide in nonneuronal tissue. Although it is not known whether kynurenines are formed to any extent in mammalian brain (Gal, 1974), there is evidence that such compounds have significant biological effects on the central nervous system. Injection of various kynurenines into cerebral ventricles of mice in doses ranging between 25 and 50 μg produced motor excitement and clonic convulsions (Lapin, 1978). Of the six compounds tested, quinolinic acid was the most active. Subsequent studies demonstrated that quinolinic acid was a potent excitant of neurons in rat brain possibly via interaction with receptors for excitatory dicarboxylic amino acid receptors (Stone and Perkins, 1981). In contrast to other amino-acid-related excitants, quinolinic acid excites cells in the cerebral cortex, hippocampus, and neostriatum, but not neurons in the cerebellum and spinal cord (Perkins and Stone, 1983). In view of such regional specificity and the possibility that quinolinic acid is formed in brain, interest is now being directed at exploring the role of this compound in neurotransmission and neurotoxicity.

A recent study of the neurotoxic properties of quinolinic acid by Schwarcz and his colleagues (1983) suggests that this compound may participate in certain neurodegenerative disorders. Unilateral intrastriatal injection of quinolinic acid results in a dose-dependent induction of tonic-clonic movements of the contralateral forelimb. These behavioral effects as well as the morphological changes produced by quinolinic acid are essentially the same as those noted after administration of kainic or ibotenic acids. Neuronal cell loss, in the absence of any change in glial cells, occurs around the injection site (Schwarcz et al., 1983). Most conspicuous ultrastructural changes produced by quinolinic acid were dendritic swelling and decreased numbers of synaptic complexes. Analyses of glutamic acid decarboxylase and acetylcholinesterase suggested that GABAergic neurons may be more susceptible to the toxic effects of quinolinic acid than cholinergic neurons. In contrast to quinolinic acid, injection of nicotinic acid failed to produce neuronal damage (Schwarcz et al., 1983).

4.3. Niacin and Nicotinamide Transport in the Central Nervous System

The location and mechanism for the transfer of nicotinamide and niacin into the central nervous system are beginning to be established. Nicotinamide and niacin do not readily penetrate the blood–brain barrier *in vivo* (Kaplan et al., 1956; Deguchi et al., 1968). Experiments carried out recently in rabbits demonstrate that the choroid plexus, a locus of the blood–cerebral spinal fluid barrier, contains separate mechanisms for both the uptake and release of niacin and nicotinamide. Spector and his colleagues (Spector and Kelly, 1979; Spector and Huntoon, 1981) suggest that the choroid plexus may be a major site of nicotinamide transfer to the cerebral spinal fluid. This suggestion is based on a series of experiments in which the concentrations of nicotinamide and niacin in rabbit cerebral spinal fluid and plasma were measured after injecting [^{14}C]niacin and [^{14}C]nicotinamide. Using these data, rates of entry of the labeled vitamins into the brain and the forms of labeled niacin and nicotinamide at various intervals after injection of the radioactive precursors were determined. Following 3-hr perfusion of [^{14}C]niacin or [^{14}C]nicotinamide, it was found that nicotinamide, but not niacin, readily entered the cerebral spinal fluid, choroid plexus, and brain. The uptake of labeled [^{14}C]nicotinamide into the choroid plexus and brain, but not into the cerebral spinal fluid was antagonized by the administration of unlabeled nicotinamide. When [^{14}C]niacin was injected intraventricularly, there was a rapid clearance of the material from the brain and choroid plexus. Addition of unlabeled niacin depressed the clearance of niacin from the cerebral spinal fluid. In contrast to niacin, the clearance of nicotinamide from the cerebral spinal fluid was extremely rapid and was not depressed by the injection of unlabeled nicotinamide. The results of Spector and his group suggest that total niacin and nicotinamide and the congeners of these products in brain may be regulated, at least in part, by transport mechanisms. Nicotinamide readily passes between the cerebral spinal fluid and plasma and apparently enters brain by a high-affinity accumulation system.

Under physiological conditions, much of the niacin that enters the central nervous system is converted rapidly to nicotinamide in its free or unbound forms.

The work of Spector and his colleagues presents two major findings. First, nicotinamide readily passes between the blood and cerebral spinal fluid, independent of the plasma or cerebral spinal fluid concentration. Once in the cerebral spinal fluid, however, nicotinamide enters brain cells by a concentration-dependent saturable accumulation system. Second, niacin unlike nicotinamide does not readily pass between blood and cerebral spinal fluid, but enters the brain from the cerebral spinal fluid by a concentration-dependent accumulation system. Concentrations of nicotinamide in human plasma are about 0.4 μM (Clark et al., 1975) and cisternal cerebral spinal fluid contains nicotinamide concentrations in the same range (Brunink and Wessels, 1972). Thus, rapid equilibration of nicotinamide between plasma and cerebral spinal fluid may occur via facilitated diffusion. Spector (1979) presents data consistent with this conclusion showing that even with concentrations as high as 1.8 mM nicotinamide rapidly passes from blood into the cerebral spinal fluid by a nonsaturable process. The entry of nicotinamide into the cerebral spinal fluid is much greater than that of mannitol, a neutral water-soluble molecule that is similar in size and shape to nicotinamide and achieves only 15% of the plasma concentration in the cerebral spinal fluid (Spector and Lorenzo, 1975). Nicotinamide diffuses from the cerebral spinal fluid at a much faster rate than mannitol, suggesting that facilitated diffusion of nicotinamide into and out of the cerebral spinal fluid is bidirectional. The avid uptake system for nicotinamide in brain may serve a protective role in preventing cerebral niacin deficiency. For example, nicotinamide and its congeners are maintained at normal levels in brain, but not in liver of newborn rabbits nursed by niacin- and tryptophan-deficient dams (Spector and Huntoon, 1981).

Once within the cerebral spinal fluid and the extracellular space of the brain, nicotinamide readily enters the brain in vivo and in vitro by a saturable accumulation system with affinity constants in the micromolar range. Most of the nicotinamide that enters the brain is incorporated into NAD (Deguchi et al., 1968; Spector, 1979). Results from other studies indicate that nicotinamide is not converted to niacin before incorporation into NAD (Gerber and Deroo, 1970).

5. Effects of Deficiency

Although the link between niacin deficiency and the behavioral manifestations accompanying deficiency of this vitamin have been known for many years, there is a sparsity of information concerning the precise cellular and biochemical changes underlying neurological deficits that accompany deficiencies of niacin or nicotinamide. Most of our knowledge concerning the biochemical and cellular changes due to niacin deficiency have been derived from animal models using niacin analogues, such as 6-aminonicotinamide or 3-acetylpyridine which are discussed in Section 6. Limited data obtained in animal studies of experimental niacin deficiency indicate that permanent structural changes take place in the brain if niacin deficiency is maintained for long periods of time. The morphological lesions seen with chronic niacin deficiency include central chromatolysis and degeneration, especially of large neurons in the motor cortex, brain stem, and anterior horn of the spinal cord (Blackwood et al., 1963). These lesions are seen in humans who have experienced severe niacin deficiency and have also been demonstrated in black tongue, the canine equivalent of pellagra (McIlwain, 1966).

Behavioral alterations seen in niacin deficiency derive mainly from observations of patients who have been institutionalized and include states such as depression often associated with morbid fears, dizziness, and insomnia. A prolonged deficiency brings with it apprehension, hallucinations, disorientation, and delirium. Thus, it is difficult to differentiate psychoses due to nicotinamide deficiency from other types of psychoses. Indeed, the link between niacin deficiency and behavioral alterations has led to the use of niacin and tryptophan in the treatment of both pellagra and schizophrenia. The use of such therapy in the treatment of this disorder, however, remains very controversial (Mosher, 1970; Pfeiffer, 1981). Nicotinic acid and nicotinamide have been reported to be both efficacious and harmless when used in high doses, up to 8–10 g/day, as a treatment for schizophrenia. Studies that report positive results with the use of nicotinamide in the treatment of schizophrenia (Denson, 1962; Hoffer, 1962, 1966; Osmond and Hoffer, 1962) are offset by a number of reports that dispute the efficacy of niacin in the treatment of this disease (Gallent et al., 1966; Kline et al., 1967). Although experiments with laboratory animals indicate that high doses of niacin and nicotinamide are well tolerated (Unna, 1939), the use of high quantities of these substances in the treatment of behavioral disorders is not without risk. Individuals who are particularly at risk are women of childbearing age and patients with a history of ulcer, gout, diabetes, or liver disease. Toxic effects accompanying use of high doses of nicotinamide are manifested in the skin, gastrointestinal tract, and liver. Nicotinamide appears to be twice as toxic as niacin (Unna, 1939). Duodenal ulcer and hepatic damage resulting from use of high doses of nicotinamide can be life threatening if the drug is not rapidly withdrawn (for a review of the side effects and toxicity of niacin, see Mosher, 1970).

Only a limited number of studies concerning changes in the cerebral content of pyridine nucleotides in niacin deficiency appear in the literature. Garcia-Bunuel et al. (1962) described changes in NAD^+, NADH, $NADP^+$, NADPH, and several nicotinamide-dependent enzymes in weanling rats made deficient by a low-tryptophan, niacin-free diet. With the exception of NADPH, all forms of the pyridine nucleotides were significantly reduced in brains of the niacin-deficient animals. Changes in pyridine nucleotides were much greater in livers than in brains of these animals. For example, total NAD decreased 63% in liver and 34% in brain, and total NADP decreased 38% in liver and only 15% in brain. The NAD^+/NADH ratio was significantly lowered in tissues of deficient animals, whereas $NADP^+$/NADPH ratios were slightly elevated. Activites of the NAD-dependent dehydrogenases including glutamate dehydrogenase and isocitrate dehydrogenase were not greatly altered in cerebral tissue of deficient animals; however, alpha-glycerophosphate dehydrogenase was slightly elevated in brain and depressed in liver of deficient animals. Thus, effects of niacin deficiency on pyridine nucleotide-dependent enzymes appear irregular and probably of minor consequence. This differs from flavin-containing enzymes, in which deficiencies of riboflavin result in marked depression of the activities of flavoenzymes (Burch et al., 1956).

6. Experimental Nicotinamide Deficiency

6.1. 6-Aminonicotinamide

Understanding the role of metabolites derived from niacin and nicotinamide in cerebral function has been advanced greatly through the use of analogues of the vitamin. The agent that has been used most extensively in this context is 6-aminonicotinamide, the 6-amino analogue of nicotinamide. This agent was introduced in the early 1950s as a potential antitumor drug only to be abandoned because of its marked neurotoxicity. Administration of a single injection of this compound to four different species of laboratory animals was shown to result in loss of motor control and paralysis within several hours (Johnson and McColl, 1955; Sternberg and Philips, 1958). Early studies also revealed that treatment with the antimetabolite on a chronic basis leads to degeneration of neurons in adult mammals (Sternberg and Philips, 1958) and delayed differentiation of the ependyma and choroid plexus in developing animals (Chamberlain, 1972). The cellular and biochemical sequelae produced by 6-aminonicotinamide are complex and are difficult to relate to the primary lesion(s) produced by deficiencies of niacin or nicotinamide; nevertheless, studies with this agent provide insights into the function of specific pyridine nucleotide-dependent pathways in cerebral metabolism.

6.1.1. Effects of 6-Aminonicotinamide on Cerebral Energy Metabolism

Toxicity due to 6-aminonicotinamide is believed to arise mainly from the 6-aminonicotinamide analogue of NADP rather than the antimetabolite itself. Experiments employing a single injection of ^{3}H-labeled 6-aminonicotinamide showed that the substance is taken up into various regions of the central nervous system over a four-day period and is removed from these regions at a very slow rate (Meyer-Estorf et al., 1973). Differences between the uptake and removal of 6-aminonicotinamide from the nervous system reflect the incorporation of the antimetabolite into pyridine nucleotides within nerve cells. This incorporation occurs via exchange reactions catalyzed by NAD(P)-glycohydrolase (E.C. 3.2.2.6) contained within the endoplasmic reticulum. Cytosolic NADP is converted to the 6-aminonicotinamide analogue (Willing et al., 1964; Coper et al., 1966; Herken, 1970), whereas only small amounts of NAD are converted to the corresponding analogue over the same time course (Coper et al., 1966). Nucleotides containing 6-aminonicotinamide, which are formed intracellularly, seem to be very stable and unable to penetrate the cell membrane or the blood–brain barrier. A major biochemical action of the 6-aminonicotinamide analogue of NADP (6AN-ADP) is inhibition of the second oxidative enzyme in the pentose phosphate pathway, 6-phosphogluconate dehydrogenase, which results in enormous accumulations of the substrate for this enzyme, 6-phosphogluconate (Köhler et al., 1970; Kauffman and Johnson, 1974). Early studies concerning this action of the antimetabolite are included in a review by Herken and his colleagues (1974).

The marked inhibition of 6-phosphogluconate dehydrogenase by 6-aminonicotinamide adenine dinucleotide phosphate (6AN-ADP) can be ascribed to the exceptionally high affinity of the analogue for the enzyme. The inhibitor constant, K_i, of 6AN-ADP for 6-phosphogluconate dehydrogenase in mammalian brain is about 0.1 μM, which is at least one order of magnitude lower than that for other $NADP^+$-dependent dehydrogenases (Köhler et al., 1970; Lange et al., 1970). Thus, this agent appears to be a relatively selective inhibitor of the oxidative limb of the pentose phosphate pathway (Herken et al., 1969). The selective action of 6-aminonicotinamide on 6-phosphogluconate dehydrogenase is illustrated by the nearly 200-fold increase in cerebral 6-phosphogluconate that occurs in the absence of only slight changes in substrates for other $NADP^+$-dependent dehydrogenases in brain following a single high dose (70 mg/kg) of the antimetabolite (Kauffman and Johnson, 1974). Dose-dependent accumulations of 6-phosphogluconate occur within neural tissue after exposure to 6-aminonicotinamide (Herken et al., 1969; Kauffman and Johnson, 1974; Hothersall et al., 1981). The required formation of the 6AN-ADP analogue for inhibition of 6-phosphogluconate dehydrogenase in neural tissue is reflected by a delay in the accumulation of 6-phosphogluconate following administration of 6-aminonicotinamide. For example, 6-phosphogluconate only begins to accumulate rapidly in the rat superior cervical ganglion 2 hr after administration of a single dose of the drug (Härkönen and Kauffman, 1974). Following a single injection of the antimetabolite into the subarachnoid space of the spinal cord, peak levels of 6-phosphogluconate do not occur until two days later (Deshpande et al., 1978).

The apparent selective inhibition of 6-phosphogluconate dehydrogenase by the 6AN-ADP analogue of NADP provides a useful approach to study the role of the pentose phosphate pathway in cerebral function and to examine the consequences of impairing this major NADPH-generating system in brain. In addition to inhibition of the formation of ribose-5-phosphate, required for nucleotide and nucleic acid biosynthesis, inhibition of NADPH formation at the level of 6-phosphogluconate dehydrogenase could affect a number of important biosynthetic reactions in brain including: (1) reductive steps in the synthesis of lipids and cholesterol, (2) glutamate and GABA formation, (3) synthesis of catecholamines and other hydroxyl-containing compounds, (4) detoxification of hydrogen peroxide formed in catecholamine degradation, and (5) maintenance of membrane integrity via reduced glutathione (Hothersall et al., 1981). All but the latter two functions have been shown to be compromised in neural tissue of 6-aminonicotinamide-treated animals.

Inhibition of the incorporation of ^{14}C-labeled glucose into RNA in cerebral tissue is among the first biochemical actions of 6-aminonicotinamide (Herken et al., 1969); however, the drug seems to have different effects on RNA biosynthesis in cerebral tissue from developing and adult animals. Available data suggest a preferential inhibition of RNA synthesis in glial elements in adult brain (Herken *et al.*, 1974; Sarkander *et al.*, 1978), whereas neuronal RNA biosynthesis appears to be more sensitive to inhibition by 6-aminonicotinamide in various brain regions of developing animals (Knoll-Köhler *et al.*, 1980). The effects of 6-aminonicotinamide on RNA synthesis in the cerebral cortex and cerebellum of neonatal rats is characterized by an initial depression of [^{3}H]UMP incorporation followed by a stimulation (Knoll-Köhler *et al.*, 1980). These changes in incorporation of isotope corresponded to increases and decreases in RNA initiation sites on nuclear chromatin isolated from the two areas. The decrease in RNA synthesis that occurred four days after administration of the antimetabolite also corresponded to a period of hypomotoric activity in the animals. A period of hypermotoric activity was noted 12 days after administration of the drug and this was accompanied by increased RNA synthesis. Although the relationship between RNA synthesis and motor activity is unknown, changes in cerebral RNA metabolism have been linked to variations in neuronal activity and motor activity (Edström and Grampp, 1965; Hydén and Egyhazi, 1968; Genazzani and Di Carlo, 1974).

Further studies of the metabolic effects of 6-aminonicotinamide may be useful in delineating the role of NADPH in neuronal function. NADPH is involved in both the biosynthesis and degradation of catecholamines; however, information concerning the effect of 6-aminonicotinamide on these substances as well as other putative neurotransmitters is limited. Recent studies have shown that steady-state concentrations of glutamate and GABA are reduced about 30% in brains of rats treated with 6-aminonicotinamide (Bielicki and Krieglstein, 1976). This acute action of relatively high doses of the antimetabolite may be related to reduced flux of glucose through the glutamate–GABA pathway (Hothersall *et al.*, 1981), since approximately 8% of glucose metabolized by brain occurs via this route (Balazs *et al.*, 1970). The decline in glutamate and GABA in brains of 6-aminonicotinamide-treated animals may also be related to inhibition of the NADP$^+$-dependent form of glutamate dehydrogenase, which has been reported to be sensitive to the 6AN-ADP analogue (Hothersall *et al.*, 1981). Changes in gamma-aminobutyric acid transaminase activity (E.C. 2.6.1.19) in cerebral tissue of 6-aminonicotinamide-treated animals have also been reported (Prakash and Bacquer, 1981). Activities of this enzyme were significantly depressed in the cerebrum, cerebellum, and brain stem of rats treated with 30 mg/kg 6-aminonicotinamide at 4 hr after administration of the drug (Prakash and Baquer, 1981). Since this enzyme is involved in the metabolism of GABA, an increase rather than a decrease in the neurotransmitter would be predicted. Changes in GABA noted in cerebral tissue of animals exposed to 6-aminonicotinamide may be limited to the mature nervous system. Exposure of prenatal rats to the antimetabolite did not alter the GABA content in the developing neural tube of embryonic rats (McCandless and Scott, 1981).

The effects of 6-aminonicotinamide on lipid and carbohydrate metabolism in brain are fairly well understood. The major metabolic effects of 6-aminonicotinamide can be explained by inhibition of the oxidative limb of the pentose phosphate pathway. The massive accumulation of 6-phosphogluconate that occurs leads to a secondary inhibition of glycolysis at the level of phosphohexosisomerase (Lange *et al.*, 1970; Kauffman and Johnson, 1974) and possibly phosphofructokinase (Kauffman and Johnson, 1974; Hothersall *et al.*, 1981). Inhibition of phosphohexosisomerase is known to occur with relatively low concentrations of 6-phosphogluconate which has a Ki of 5 μM for the enzyme (Kahana *et al.*, 1960). Amounts of 6-phosphogluconate in cerebral tissue of 6-aminonicotinamide-treated mice are at least 50-fold higher than the estimated inhibitor constant of this metabolite for phosphohexosisomerase (Kauffman and Johnson, 1974). Slowed glucose flux across phosphohexosisomerase is indicated by increased ratios of steady-state concentrations of glucose-6-phosphate to fructose-6-phosphate (Lange *et al.*, 1970; Kauffman and Johnson, 1974; Bielicki and Krieglstein, 1976) and by altered yields of $^{3}H_2O$ from 2-^{3}H- and 3-^{3}H-labeled glucose (Hothersall *et al.*, 1981).

6-Aminonicotinamide treatment also appears to inhibit glycolysis in neural tissue at steps below phosphohexosisomerase. Inhibition of phosphofructokinase is suggested by decreased concentrations of fructose-1,6-diphosphate and increased amounts of glucose and hexose phosphates (Kauffman and Johnson, 1974). Metabolite profiles in slices of rat cerebrum 16 hr after treatment with the antimetabolite further suggest that glycogenolysis may be inhibited at phosphoglucomutase (Hothersall *et al.*, 1981). Decreases in lactate production that occur in brains of 6-aminonicotinamide-treated animals can only be ascribed to decreased glycolytic flux and not to diffusion of lactate and pyruvate into extracellular space (Krieglstein and Stock, 1975).

A dose-dependent decrease in the utilization of ^{14}C-labeled glucose for lipid synthesis occurs in brain slices prepared from rats treated with 6-aminonicotinamide (Hothersall *et al.*, 1981). This decrease in synthesis of phospholipids from glucose involves decreases in incorporation of glucose into both the glycerol backbone and fatty acid portions of phospholipids. Incorporation of ^{14}C-labeled glycerol into neutral lipids is unaffected by 6-aminonicotinamide treatment.

Studies of metabolite changes accompanying anoxic ischemia and investigations of the fate of differentially labeled glucose indicate that inhibition of glycolysis is one of the earliest and most prominent effects of 6-aminonicotinamide intoxication (Kauffman and Johnson, 1974; Hothersall *et al.*, 1981). Utilization of glycogen and production of lactate during brief periods of anoxic ischemia are slowed significantly in brains of 6-aminonicotinamide-treated mice. Despite this apparent slowing of hexose use, the total content of high-energy phosphate was essentially the same in brains of normal and 6-aminonicotinamide-treated mice. Further, initial rates of high-energy phosphate use during the initial period of anoxic ischemia was the same in both groups. Thus, although utilization of glucose is significantly impaired in brains of 6-aminonicotinamide-treated animals. maintenance and utilization of high-energy phosphate during anoxic ischemia does not appear to be impaired.

In addition to alterations in metabolism via the pentose phosphate pathway and glycolysis, high doses of 6-aminonicotinamide have significant effects on cerebral oxidative metabolism. This is indicated by depressed formation of $^{14}CO_2$ from [2-^{14}C]pyruvate and [6-^{14}C]glucose in tissue slices prepared from cerebral cortices of 6-aminonicotinamide-treated rats (Hothersall *et al.*, 1981). Decreases in steady-state concentrations of pyruvate in brains of these animals may contribute to inactivation of pyruvate dehydrogenase (Booth and Clarke,

1968; Hothersall *et al.*, 1979). Based on yields of $^{14}CO_2$ from glucose labeled in various carbons in CO_2 and lipids of tissue slices prepared from brains of normal rats and rats treated with 6-aminonicotinamide, Hothersall *et al.* (1981) suggested the following hierarchy of responses to the antimetabolite. First, the pentose phosphate pathway and formation of lactate, glyceride-glycerol, and fatty acids from glucose are inhibited. These are followed in descending order by inhibition of the pyruvate dehydrogenase reaction, the glutamate-gamma-aminobutyrate route, and lastly by the tricarboxylic acid cycle.

6.1.2. Selective Sites of Toxicity

6-Aminonicotinamide is known to produce dose-dependent lesions in selected areas of the central nervous system of many species. The intermediate gray matter and central ventral horn regions of the spinal cord corresponding to laminae six and seven are very vulnerable, with the lumbar cord being more susceptible than the cervical. Various nuclei of the lower brain stem are also susceptible, as are midbrain and thalamic nuclei, which show degenerative changes after exposure to 6-aminonicotinamide (Wolf and Cowen, 1959; Schneider and Cervos-Navarro, 1974).

Studies of adult mammalian tissue suggest that 6-aminonicotinamide is more toxic toward glial elements than neurons (Chui and Garcia, 1979; Griffiths *et al.*, 1981). Glial changes are noted as early as 12 hr after a single intraperitoneal injection of 5–10 mg/kg of 6-aminonicotinamide; however, neuronal changes do not occur until about 36 hr later in the same animals (Schneider and Cervos-Navarro, 1974; Herken *et al.*, 1976). There appears to be an age dependence of various cellular elements to 6-aminonicotinamide toxicity. In a study of RNA metabolism in 6-aminonicotinamide-treated rats (Knoll-Köhler *et al.*, 1980), 6-aminonicotinamide was found to be more injurious to neurons than glial elements in cerebral and cerebellar nuclei of 7-day-old animals. Differences in the effect of 6-aminonicotinamide on the energy status of neural tissue is also observed in adult and embryonic tissue. Adult cerebral tissue displays decreased concentrations of ATP following administration of 6-aminonicotinamide (Kauffman and Johnson, 1974); however, this antimetabolite increases the content of ATP and phosphocreatine in neural tubes of rat embryos (McCandless and Scott, 1981).

Selective effects of 6-aminonicotinamide on central glial cells have been observed using the technique of [^{14}C]2-deoxyglucose to quantitate rates of glucose utilization in specific areas of the central nervous system (Griffiths *et al.*, 1981). Griffiths *et al.* (1981) measured local glucose utilization throughout the central nervous system at 3, 6, 12, and 24 hr after administration of 5 mg/kg of 6-aminonicotinamide. Changes in rates of local glucose utilization, as indexed by 2-deoxyglucose uptake, were compared with alterations in the histology of selected areas of the central nervous system. As early as 3 hr, local rates of glucose utilization were decreased by about 25% in the lumbar ventral horn, caudal brain stem, and cerebellum. By 12 hr, virtually all structures examined showed local rates of glucose utilization that were decreased at least 20%. Greatest reductions in rates of glucose were noted in the spinal cord, where these amounted to 40–50% of control values. By 24 hr, all major areas

examined displayed recoveries of rates of glucose utilization. The earliest histological change noted in this study (Griffiths *et al.*, 1981) was a mild swelling
of glial cells, particularly oligodendroglia, at 12 hr. More pronounced swelling
and degeneration of glial elements occurred at 24 hr. Neuronal changes were
not seen at any of these periods. Abnormalities were most marked in the ventral
gray matter of the spinal cord and in certain brain stem nuclei. The results of
Griffiths *et al.* (1981) show that there is a relationship between depression of
local rates of glucose utilization and occurrence of pathological changes. The
spinal cord, which is the most severely affected, shows the most marked decreases in local rates of glucose utilization. Unfortunately, the relative contributions of neurons and glia to glucose consumption in areas affected by the
drug are not known. In general, changes in pathology appear to follow the
degree to which 6-phosphogluconate accumulates in specific brain areas
(Kauffman and Johnson, 1974; Herken *et al.*, 1974). Although lesions produced
by 6-aminonicotinamide have a selective distribution in the central nervous
system, there does not appear to be a specific localization of 6-aminonicotin-
amide uptake in these areas. Uptake of 6-aminonicotinamide appears to be
relatively uniform throughout the central nervous system (Meyer-Estorf *et al.*,
1973).

In view of the higher susceptibility of glial cells to the toxicity of 6-ami-
nonicotinamide, it is tempting to suggest that the pentose phosphate pathway
makes a greater contribution to glial metabolism. In line with this possibility,
Keller *et al.* (1976) found that cultured rat C6-astrocytoma cells are severely
affected by exposure to 6-aminonicotinamide; that is, rates of glucose consumption and lactate production were severely reduced, while 6-phospho-
gluconate and gluconate are markedly elevated in these cells.

Neural toxicity of 6-aminonicotinamide is not limited to the central nervous system. Administration of the antimetabolite to developing animals
(Brzoska and Adhami, 1975; Frieda and Bischhausen, 1978) caused a selective
swelling of the inner Schwann cell cytoplasmic process, followed by disruption
of this process, resulting in a large fluid-filled periaxonal space. The periaxonal
edematous space differs from that seen in isoniazid neuropathy, where the
inner Schwann cell cytoplasmic process persists and remains adherent to the
axolemma (Jacobs *et al.*, 1979). Thus, edema produced with 6-aminonicotin-
amide appears to be extracellular and in some instances penetrates for a short
distance into the myelin. The peripheral visual system appears to be a very
sensitive target to the toxicity of 6-aminonicotinamide (Grant, 1980). 6-Ami-
nonicotinamide has been shown to severely affect the ultrastructure of the optic
nerve, probably by splitting the first inner lamellae leaving the axon unchanged
(Meyer-König, 1973). Studies of the rat superior cervical ganglion indicate that
the antimetabolite also acts on mammalian autonomic neural tissue (Härkönen
and Kauffman, 1974).

6.1.3. Actions on the Developing Central Nervous System

The sensitivity of the mammalian central nervous system to 6-aminoni-
cotinamide is complicated by age-dependent changes. Following injection of
a single 10 mg/kg dose of 6-aminonicotinamide, the anterior horn cells of 20-

to 25-month-old rats increase more in size and recover slower from chromatolytic changes than those observed in 3-month-old rats (Horita *et al.*, 1981). In the acute stage of 6-aminonicotinamide intoxication, reactive and degeneration changes in glial and mesenchymal elements were more conspicuous in 3-month-old rats than in the older animals. However, the lesions that appeared tended to disappear by day 14 and were accompanied by a prominent proliferation of hypertrophic astrocytes. Older rats showed less intensity of the initial response and a slower progression of the recovery, as well as the late inflammatory response. In the acute stage, reactive and degenerative changes of glial cells and proliferation of phagocytes were clearly more prominent in 3-month-old rats. Persistent neuronal abnormalities are induced by exposure of prenatal animals to 6-aminonicotinamide (Chamberlain and Nelson, 1963). Despite the severe abnormalities induced in developing embryos by this antimetabolite, there seem to be limited effects on energy metabolism in the rat embryo neural tube (McCandless and Scott, 1981). Based on their finding that ATP and phosphocreatine increased in embryonic neural tubes, McCandless and Scott suggested that energy-utilizing reactions may be compromised in developing embryos exposed to 6-aminonicotinamide.

6.1.4. Functional Changes Induced by 6-Aminonicotinamide

Recordings of extracellular muscle action potentials indicate that the spastic paresis induced by 6-aminonicotinamide is accompanied by activation of electromyographs, which continue throughout the period of spastic paresis produced by the drug (Herken *et al.*, 1976). Activation of the frequency of the electromyographs corresponds temporally with the destruction of interneurons in the spinal cord by the antimetabolite. The chemical destruction of interneurons with 6-aminonicotinamide and the accompanying rigidity and spasticity of the hindlimbs have been suggested to provide a useful model for testing antispastic drugs. Para-chlorophenylalanine, GABA, and chlorpromazine were found to produce temporary reduction of the excitation processes on the electromyograph of animals exposed to 6-aminonicotinamide (Herken *et al.*, 1976).

A detailed electrophysiological, biochemical, histological study of animals treated with 6-aminonicotinamide administered via the subarachnoid space indicated that many changes that occur in muscle resemble those observed after surgical transection of the spinal cord (Deshpande *et al.*, 1978). Administration of 50 μg of 6-aminonicotinamide into the subarachnoid space at L2 and L4 of female adult rats produced a rigid paralysis in both hindlimbs at 24–36 hr. This model is very useful because the general systemic effects of 6-aminonicotinamide are limited by the local application of the drug to the spinal cord. In extensor muscles, the membrane potential decreased significantly at about 48 hr after injection; this partial depolarization took somewhat longer in soleus muscles. Spontaneous transmitter release as indicated by the frequency and amplitude of miniature endplate potentials was significantly affected beginning at day 7 after injection. At the onset of paralysis, 6-phosphogluconate concentrations were increased more than 1000-fold in the lumbar region of the spinal cord and returned to control levels 14 days after injection. Beginning at about three days, an increase in small phagocytic cells was noted in both the cervical and lumbar regions of the cord. These proliferating cells were highly localized to spinal gray matter where maximal numbers were detected at eight days. Few viable motor neurons were present in the lumbar region of the cord at this time. There was a close correlation between accumulation of small inflammatory cells and β-glucuronidase activity in the spinal cord. Cellular changes noted in the spinal cord after 6-aminonicotinamide were maintained for as long as 540 days after exposure to the drug. At this time, most of the electrophysiological properties of the extensor muscle were restored to normal, suggesting that collateral sprouts from residual motor neurons may have reinnervated the denervated muscle fibers. This model of neuronal injury may be extremely useful because it provides a means of approaching several important neuropathological questions. For example, the highly localized infiltration of small inflammatory cells in the anterior horn region provides a model to begin to explore mechanisms associated with inflammation in the central nervous system. The model is also interesting from the standpoint of exploring such diverse neural phenomena as neurotrophic effects, rigidity, collateral sprouting, and cellular responses to neuronal injury.

Application of 6-aminonicotinamide to the spinal cord via subarachnoid injections also caused a complete inhibition of axoplasmic flow in motor fibers, but not sensory fibers of the sciatic nerve (Boegman and Albuquerque, 1980). This inhibition occurred within 2 hr after administration of the drug and was not reversible. The action of 6-aminonicotinamide differed from that of the depolarizing neurotoxin, batrachotoxin, which reversibly inhibited axoplasmic flow in both sensory and motor fibers after administration into the subarachnoid space. Although the difference in the action of the two drugs was ascribed to different patterns of distribution after subarachnoid injection, the possibility remains that motor and sensory neurons differ in sensitivity to 6-aminonicotinamide. Mechanisms responsible for the marked inhibition of axoplasmic flow after 6-aminonicotinamide treatment are not known; however, blockade of protein biosynthesis does not appear to be involved since incorporation of [^{3}H]leucine into protein was not inhibited by the antimetabolite (Boegman and Albuquerque, 1980).

6.2. 3-Acetylpyridine

In view of the extensive studies made with 6-aminonicotinamide, it is surprising that other analogues of nicotinamide or niacin have not been used more to study mechanisms of neurotoxicity and possibly the role of niacin in neuronal function. 3-Acetylpyridine is a substance that was introduced at about the same time as 6-aminonicotinamide; however, this substance has not been studied as widely as 6-aminonicotinamide. Early reports indicated that 3-acetylpyridine was injurious to the hypothalamus (Hicks, 1955) and to the hippocampus (Coggeshall and MacLean, 1958). Damage to the hypothalamus could be prevented by coadministration of nicotinamide with 3-acetylpyridine (Hicks, 1955). An analysis of 3-acetylpyridine nucleotides in different brain regions of rats treated with 3-acetylpyridine indicated that highest accumulations of the analogue occurred in the hippocampus (Willing *et al.*, 1964). About 45% of the NADP in this structure was converted to the 3-acetylpyridine derivative, whereas only a small portion of NAD was converted to the analogue. The drug has also been reported to produce hypothalamic lesions and abnormal gait when given to rats (Woolley, 1952; Kaplan, 1960).

In addition to producing hypothalamic lesions, administration of 3-acetylpyridine causes degeneration of the inferior olives, olivocerebellar fibers, lower cranial nerve nuclei, and areas of the pons and nigra (Desclin and Escubi, 1974). Such structural changes and the long-lasting disturbances in gait resemble alterations seen in human olivopontocerebellar atrophies (Plaitakis *et al.*, 1980). In contrast to most areas of the brain that show an increase in local rates of glucose utilization, 3-acetylpyridine treatment causes a decrease in glucose utilization in the inferior olivary nuclei (Gibson *et al.*, 1983). Based on similarities between animals treated with 3-acetylpyridine and the human degenerative disorder, Plaitakis *et al.* (1980) examined a number of $NADP^+(H)$-dependent enzymes in cultured fibroblasts from patients afflicted with olivopontocerebellar atrophy and found a selective and marked reduction in glutamic dehydrogenase. It is not clear whether similar biochemical mechanisms underlie neuronal deficits in animals exposed to 3-acetylpyridine and the human disorder. Nevertheless, possible relationships between altered glutamate metabolism and neuronal dysfunction in the two conditions should be considered.

Two substances closely related to nicotinamide, 3-aminopyridine, and 4-aminopyridine have been used to block voltage-dependent potassium conductances in excitable cells (Llinàs *et al.*, 1976; Adams *et al.*, 1980; Hermann and Gorman, 1981). Although these substances are related structurally to niacin and nicotinamide, there is little information concerning their metabolic actions and it is not certain how specific they are as inhibitors of potassium conductances.

7. Summary and Conclusions

The involvement of niacin and nicotinamide in cerebral function is undoubtedly related to their role as precursors of the pyridine nucleotide cofactors. Deficiencies of the vitamin or treatment with analogues, such as 6-aminonicotinamide, result in profound functional and structural changes in the central and peripheral nervous systems. We know a great deal more about the biochemical changes produced by treatment of laboratory animals with analogues of niacin and nicotinamide than we do about changes that take place in niacin and nicotinamide deficient states. Because of this, it is difficult to ascribe changes that occur after treatment with the antimetabolites to alterations that are related directly to changes in niacin and nicotinamide content in brain. Thus, there is a need to focus further research on the biochemical and physiological sequelae produced by niacin and nicotinamide deficiency. Although it has been known for years that many patients with pellagra have a major psychosis, we do not have any understanding of the underlying biochemical defect affecting the brain.

Research employing 6-aminonicotinamide as a nicotinamide analogue has contributed substantially to understanding the role of $NADP^+$-dependent enzymes in cerebral metabolism. Results of such studies suggest an important role for the oxidative enzymes of the pentose phosphate pathway in both neurons and glia. The greater sensitivity of glial elements in the mature nervous system to the antimetabolite suggests that glia cells may be more dependent on metabolism via the pentose phosphate pathway than neurons. Changes in the dependence of various neural elements on the oxidative pentose phosphate pathway occur with development and this appears to be reflected in alterations in the selective toxicity of 6-aminonicotinamide. Studies using this antimetabolite have provided insight into the role of the pentose phosphate pathway in supplying reducing equivalents for various biosynthetic pathways such as fatty acid biosynthesis. It is important to establish the relationship, if any, between inhibiting these pathways and the structural alterations that occur in specific brain regions after exposure to 6-aminonicotinamide. To date, a number of functional changes have been described in neural tissue of animals treated with 6-aminonicotinamide; however, considerably more work needs to be done to establish links between the biochemical actions of the antimetabolite and physiological function. Comparative studies employing other nicotinamide analogs, such as 3-acetylpyridine, which may have a different specificity than 6-aminonicotinamide, could be valuable in delineating mechanisms related to alterations in structure and function produced by niacin analogs.

ACKNOWLEDGMENT. The author's studies cited above and this work were supported in part by NIH Grants NS-08157 and HD-16506. Mrs. Roxanne Evans is thanked for her excellent assistance in assembling the manuscript.

References

Adams, D. A., Smith, S. J., and Thompson, S. H., 1980, Ionic currents in molluscan soma, *Annu. Rev. Neurosci.* **3**:141–167.

Axelrod, A. E., Spies, T. D., and Elvehjam, C. A., 1941, The effect of a nicotinic acid deficiency upon the coenzyme I content of the human, *J. Biol. Chem.* **138**:667–676.

Bain, J. A., and Pollock, G. H., 1949, Normal and seizure levels of lactate, pyruvate and acid soluble phosphates in the cerebellum and cerebrum, *Proc. Soc. Exp. Biol. Med. N.Y.* **71**:495–497.

Balazs, R., Machiyama, Y., Hammond, B. J., Julian, T., and Richter, D., 1970, The operation of the gamma-aminobutyrate bypath of the tricarboxylic acid cycle in brain tissue *in vitro*, *Biochem. J.* **116**:445–467.

Bender, D. A., Smith, W. R. D., and Humm, R. P., 1977, Effects of benserazide on tryptophan metabolism in the mouse, *Biochem. Pharmacol.* **26**:1619–1623.

Bender, D. A., Earl, C. J., and Lees, A. J., 1979, Niacin depletion in Parkinsonian patients treated with L-dopa, benserazide and carbidopa, *Clin. Sci.* **56**:89–93.

Bielicki, L., and Krieglstein, J., 1976, Inhibition of glucose phosphorylation in rat brain by thiopental, *Naunyn-Schmied. Arch. Pharmacol.* **293**:25–29.

Blackwood, W., McMenemey, W. H., Meyer, A., Norman, R. M., and Russell, D. S., 1963, *Greenfield's Neuropathology*, Arnold, London.

Boegman, R. J., and Albuquerque, E. X., 1980, Axonal transport in rats rendered paraplegic following a single subarachnoid injection of either batrachotoxin or 6-amino-nicotinamide into the spinal cord, *J. Neurobiol.* **11**:283–290.

Booth, R. F. G., and Clarke, J. B., 1978, The control of pyruvate dehydrogenase in isolated brain mitochondria, *J. Neurochem.* **30**:1003–1008.

Brunink, H., and Wessels, E. J., 1972, The determination of nicotinic acid by fluorometric densitometry, *Analyst* **97**:258–259.

Brzoska, H.-R., and Adhami, H., 1975, Electron microscopic study of the effect of 6-AN on the sciatic nerve in newborn rats, *Acta Neuropathol.* **33**:59–66.

Buell, M. V., Lowry, O. H., Roberts, N. R., Chang, M-L. W., and Kapphahn, J. I., 1958, The quantitative histochemistry of the brain. V. Enzymes of glucose metabolism, *J. Biol. Chem.* **232**:979–993.

Burch, H. B., Lowry, O. H., Padilla, A. M., and Combs, A. M., 1956, Effects of riboflavin deficiency and realimentation on flavin enzymes of tissues, *J. Biol. Chem.* **233**:29–45.

Carpenter, K. I., 1981, Effects of different methods of processing maize on its pellagragenic activity, *Fed. Proc.* **40**:1531–1535.

Chamberlain, J. G., 1972, 6-Aminonicotinamide (6-AN)-induced abnormalities of the developing ependyma and choroid plexus as seen with the scanning electron microscope, *Teratology* **6**:281–286.

Chamberlain, J. G., and Nelson, M. M., 1963, Multiple congenital abnormalities in the rat resulting from acute maternal niacin deficiency during pregnancy, *Proc. Soc. Exp. Biol. Med.* **112**:836–840.

Chui, E., and Garcia, H. J., 1979, Pathogenesis of 6-aminonicotinamide Neurotoxicity: New structural analysis, in: *Progress in Neuropathology*, Volume 4 (H. M. Zimmerman, ed.), Raven Press, New York, pp. 341–359.

Clark, B. R., Halpern, R. M., and Smith, R. A., 1975, A fluorimetric method for quantitation in the picomole range of N^1-methylnicotinamide and nicotinamide in serum, *Anal. Biochem.* **68**:54–61.

Coggeshall, R. E., and MacLean, P. D., 1958, Hippocampal lesions following administration of 3-acetylpyridine, *Proc. Soc. Exp. Biol. Med.* **98**:687–689.

Coper, H., Hadass, H., and Lison, H., 1966, Untersuchungen zum Mechanismus zentralnervöser Funktionsstörungen durch 6-Aminonicotinamid, *Naunyn-Schmied. Arch. Pharmakol. Exp. Pathol.* **255**:96–106.

D'Adamo, A. F., Jr., and Haft, D. E., 1965, An alternate pathway of alpha-ketoglutarate catabolism in the isolated, perfused rat liver. I. Studies with DL-glutamate-2- and -5-^{14}C, *J. Biol. Chem.* **240**:613–617.

Deguchi, T., Ichiyama, A., Nishizuka, Y., and Hayaishi, O., 1968, Studies on the biosynthesis of nicotinamide adenine dinucleotide in the brain, *Biochim. Biophys. Acta* **158**:382–393.

Denson, R., 1962, Nicotinamide in the treatment of schizophrenia, *Dis. Nerv. Syst.* **23**:162–172.

Desclin, J. C., and Escubi, J., 1974, Effects of 3-acetylpyrine on the central nervous system of the rat, as demonstrated by silver methods, *Brain Res.* **77**:349–364.

Deshpande, S. S., Albuquerque, E. X., Kauffman, F. C., and Guth, L., 1978, Physiological, biochemical and histological changes in skeletal muscle, neuromuscular junction and spinal cord of rats rendered paraplegic by subarachnoidal administration of 6-aminonicotinamide, *Brain Res.* **140**:89–109.

Dickens, F., and Glock, G. E., 1951, Direct oxidation of glucose-6-phosphate, 6-phosphogluconate and pentose-5-phosphates by enzymes of animal origin, *Biochem. J.* **50**:81–95.

Edström, J.-E., and Grampp, W., 1965, Nervous activity and metabolism of ribonucleic acids in the crustacean stretch receptor neuron, *J. Neurochem.* **12**:735–741.

Frieda, R. L., and Bischhausen, R., 1978, How do axons control myelin formation? The model of 6-aminonicotinamide neuropathy, *J. Neurol. Sci.* **35**:341–353.

Gal, E. M., 1974, Cerebral tryptophan-2,3-dioxygenase (pyrrolase) and its induction in rat brain, *J. Neurochem.* **22**:861–863.

Gallent, M., Bishop, M., and Steele, G., 1966, DPN (NAD oxidized form): A preliminary evaluation in chronic schizophrenic patients, *Ann. Ther. Residency* **8**:542.

Garcia-Bunuel, L., McDougal, D. B., Jr., Burch, H. B., Jones, E. M., and Touhill, E., 1962, Oxidized and reduced pyridine nucleotide levels and enzyme activities in brain and liver of niacin deficient rats, *J. Neurochem.* **9**:589–594.

Genazzani, E., and Di Carlo, R., 1974, Interference of neurologically active drugs with metabolism of RNA in brain, in: *Central Nervous System. Studies on Metabolic Regulation and Function* (E. Genazzani and H. Herken, eds.), Springer-Verlag, Berlin, Heidelberg, New York, pp. 217–222.

Gerber, G. B., and Deroo, J., 1970, Metabolism of labelled nicotinamide coenzyme in different organs of mice and rats. *Proc. Soc. Exp. Biol. Med.* **134**:689–693.

Gibson, G. E., Glantz, S., Duffy, T. E., and Blass, J. P., 1983, Regional brain glucose utilization and behavior during niacin deficiency, *Trans. Am. Soc. Neurochem.* **14**:121.

Glock, G. E., and McLean, P., 1954, Levels of enzymes of the direct oxidative pathway of carbohydrate metabolism in mammalian tissues and tumours, *Biochem. J.* **56**:171–175.

Goldsmith, G. A., 1958, Niacin-tryptophan relationships in man and niacin requirement, *Am. J. Clin. Nutr.* **6**:479–486.

Grant, W. M., 1980, The peripheral visual system as a target, in: *Experimental and Clinical Neurotoxicology* (P. S. Spencer and H. H. Schaumburg, eds.), Williams and Wilkins, Baltimore, pp. 77–91.

Gregory, I., 1955, The role of nicotinic acid (niacin) in mental health and disease, *J. Ment. Sci.* **101**:85–109.

Griffiths, I. R., Kelly, P. A. T., and Grome, J. J., 1981, Glucose utilization in the central nervous system in the acute gliopathy due to 6-aminonicotinamide, *Lab. Invest.* **44**:547–552.

Härkönen, M. A., and Kauffman, F. C., 1974, Metabolic alterations in the axotomized superior cervical ganglion of the rat. II. The pentose phosphate pathway, *Brain Res.* **65**:141–157.

Heald, P. J., 1956, Effects of electrical pulses on the distribution of radioactive phosphate in cerebral tissues, *Biochem. J.* **63**:242–249.

Herken, H., 1970, Antimetabolic action of 6-aminonicotinamide on the pentose phosphate pathway in the brain, in: *A Symposium on Mechanisms of Toxicity* (W. N. Aldridge, ed.), MacMillan and Co., London, pp. 189–203.

Herken, H., Lange, K., and Kolbe, H., 1969, Brain disorders induced by pharmacological blockade of the pentose phosphate pathway, *Biochem. Biophys. Res. Commun.* **36**:93–100.

Herken, H., Lange, K., Kolbe, H., and Keller, K., 1974, Antimetabolic action on the pentose phosphate pathway in the central nervous system induced by 6-aminonicotinamide, in: *Central Nervous System. Studies on Metabolic Regulation and Function* (E. Genazzani and H. Herkin, eds.), Springer-Verlag, Berlin, Heidelberg, New York, pp. 41–54.

Herken, H., Meyer-Estorf, G., Halbhübner, K., and Loos, D., 1976, Spastic paresis after 6-aminonicotinamide: Metabolic disorders in the spinal cord and electromyographically recorded changes in the hind limbs of rats, *Naunyn-Schmied. Arch. Pharmacol.* **293**:245–255.

Hermann, A., and Gorman, A. L. F., 1981, Effects of 4-aminopyridine on potassium currents in a molluscan neuron, *J. Gen. Physiol.* **78**:63–86.

Hersov, L. A., 1955, A case of childhood pellagra with psychosis, *J. Ment. Sci.* **101**:878–883.

Hicks, S. P., 1955, Pathological effects of antimetabolites. I. Acute lesions in the hypothalamus, peripheral ganglia, and adrenal medulla caused by 3-acetylpyridine and prevented by nicotinamide, *Am. J. Pathol.* **31**:189–199.

Himwich, H. E., 1951, *Brain Metabolism and Cerebral Disorders*, Williams and Wilkins, Baltimore.

Hoffer, A., 1962, *Niacin Therapy in Psychiatry*, Charles C. Thomas, Springfield, Illinois.

Hoffer, A., 1966, The effect of nicotinic acid on the frequency and duration of rehospitalization of schizophrenic patients, a controlled comparison study, *Int. J. Neuropsychiatry* **2**:234–240.

Horita, N., Ishii, T., and Izumiyama, Y., 1981, Ultrastructure of 6-aminonicotinamide (6-AN)-induced lesions in the central nervous system of rats. III. Alterations of the spinal gray matter lesion with aging, *Acta Neuropathol.* **53**:227–235.

Hothersall, J. S., Baquer, N. Z., Greenbaum, A. L., and McLean, P., 1979, Alternative pathways of glucose utilization in brain. Changes in the pattern of glucose utilization in brain during development and the effect of phenazine methosulfate on the integration of metabolic routes, *Arch. Biochem. Biophys.* **198**:478–492.

Hothersall, J. S., Zubairu, S., McLean, P., and Greenbaum, A. L., 1981, Alternative pathways of glucose utilization in brain; Changes in the pattern of glucose utilization in brain resulting from treatment of rats with 6-aminonicotinamide, *J. Neurochem.* **37**:1484–1496.

Hydén, H., and Egyhazi, E., 1968, The effect of tranylcypromine on synthesis of macromolecules and enzyme activities in neurons and glia, *Neurology* **18**:732–736.

Ikeda, M., Tsuji, H., Nakamura, S., Ichiyama, A., Nishizuka, Y., and Hayaishi, O., 1965, Studies on the biosynthesis of nicotinamide adenine dinucleotide. II. A role of picolinic carboxylase in the biosynthesis of nicotinamide adenine dinucleotide from tryptophan in mammals, *J. Biol. Chem.* **240**:1395–1401.

Jacobs, J. M., Miller, R. H., Whittle, A., and Cavanagh, J. B., 1979, Studies on the early changes in acute isoniazid neuropathy in the rat, *Acta Neuropathol.* **47**:85–92.

Jepson, J. B., 1972, Hartnup disease, in: *The Metabolic Basis of Inherited Disease*, 3rd ed. (J. B. Stanbury, J. B. Wyngaarden, and D. S. Fredrickson, eds.), McGraw Hill, New York, pp. 1486–1503.

Johnson, W. J., and McColl, J. D., 1955, 6-Aminonicotinamide, a potent nicotinamide antagonist, *Science* **122**:834.

Kahana, S. E., Lowry, O. H., Schulz, D. W., Passonneau, J. V., and Crawford, E. J., 1960, The kinetics of phosphoglucoisomerase, *J. Biol. Chem.* **235**:2178–2184.

Kaplan, N. O., 1960, in: *Neurochemistry of Nucleotides and Amino Acids* (R. O. Brady and D. B. Tower, eds.), Wiley, New York, p. 70.

Kaplan, N. O., Goldin, A., Humphreys, S. R., Ciotti, M. M., and Stolzenbach, F. E., 1956, Pyridine nucleotide synthesis in the mouse, *J. Biol. Chem.* **219**:287–298.

Kauffman, F. C., 1972, The quantitative histochemistry of enzymes of the pentose phosphate pathway in the central nervous system of the rat, *J. Neurochem.* **19**:1–9.

Kauffman, F. C., and Johnson, E. C., 1974, Cerebral energy reserves and glycolysis in neural tissue of 6-aminonicotinamide-treated mice, *J. Neurobiol.* **5**:379–392.

Keller, K., Kolbe, H., Herken, H., and Lange, K., 1976, Glycolysis and glycogen metabolism after inhibition of hexose monophosphate pathway in C_6-glial cells, *Naunyn-Schmied. Arch. Pharmacol.* **294**:213–215.

Kline, N. S., Barclay, G. L., Cole, J. O., Esser, A. H., Lehmann, H., and Wittenborn, J. R., 1967, Diphosphopyridine nucleotide (DPN) in the treatment of schizophrenia, *J. Am. Med. Assoc.* **200**:881–882.

Knoll-Köhler, E., Wojnorowicz, F., and Sarkander, H.-J., 1980, Correlated changes in neuronal cerebral rat brain RNA synthesis and hypo- and hypermotoric disorders induced by 6-aminonicotinamide (6-AN), *Exp. Brain Res.* **38**:173–179.

Kodicek, E., Braude, R., Kon, S. K., and Mitchell, K. G., 1959, The availability to pigs of nicotinic acid in tortilla baked from maize treated with lime-water, *Br. J. Nutr.* **13**:363–384.

Köhler, E., Barrach, H-J., and Neubert, D., 1970, Inhibition of NADP dependent oxidoreductases by the 6-aminonicotinamide analogue of NADP, *Febs Lett.* **6**:225–228.

Krehl, W. A., 1981, Discovery of the effect of tryptophan on niacin deficiency, *Fed. Proc.* **40**:1527–1530.

Krehl, W. A., Teply, L. J., and Elvehjem, C. A., 1945, Corn as an etiological factor in the production of nicotinic acid deficiency in the rat, *Science* **101**:283.

Krieglstein, J., and Stock, R., 1975, Decreased glycolytic flux rate in the isolated perfused rat brain after pretreatment with 6-aminonicotinamide, *Naunyn-Schmied. Arch. Pharmacol.* **290**:323–327.

Kuhlman, R. E., and Lowry, O. H., 1956, Quantitative histochemical changes during the development of the rat cerebral cortex, *J. Neurochem.* **1**:173–180.

Laatsch, R. H., 1962, Glycerol phosphate dehydrogenase activity of developing rat central nervous system, *J. Neurochem.* **9**:487–492.

Laguna, J., and Carpenter, K. J., 1951, Raw versus processed corn in niacin-deficient diets, *J. Nutr.* **45**:21–28.

Lajtha, A. L., Maker, H. S., and Clarke, D. D., 1981, Metabolism and transport of carbohydrates and amino acids, in: *Basic Neurology* (G. J. Siegel, R. W. Albers, B. W. Agranoff, and R. Katzman, eds.), Little, Brown, and Co., Boston, pp. 329–354.

Lange, K., Kolbe, H., Keller, K., and Herken, H., 1970, Der Kohlenhydratstoffwechsel des Gehirns nach Blockade des Pentose-Phosphat-Weges durch 6-Aminonicotinsäureamid, *Hoppe-Seyler's Z. Physiol. Chem.* **351**:1241–1252.

Lapin, I. P., 1978, Stimulant and convulsive effects of kynurenines injected into brain ventricules in mice, *J. Neural Transm.* **42**:37–43.

Llinàs, R., Walton, K., and Bohr, V., 1976, Synaptic transmission in squid giant synapse after potassium conductance blockage with external 3- and 4-aminopyridine, *Biophys. J.* **16**:83–86.

Lowry, O. H., and Passonneau, J. V., 1964, The relationships between substrates and enzymes of glycolysis in brain, *J. Biol. Chem.* **239**:31–42.

Luine, V. N., and Kauffman, F. C., 1971, Triphosphopyridine nucleotide-dependent enzymes in the developing spinal cord of the rabbit, *J. Neurochem.* **18**:1113–1124.

Madsen, J., Abraham, S., and Chaikoff, I. L., 1964, The conversion of glutamate carbon to fatty acid carbon via citrate. I. The influence of glucose in lactating rat mammary gland slices, *J. Biol. Chem.* **239**:1305–1309.

McCandless, D. W., and Scott, W. J., 1981, The effect of 6-aminonicotinamide on energy metabolism in rat embryo neural tube, *Teratology* **23**:391–395.

McDougal, D. B., Jr., Schultz, D. W., Passonneau, J. V., Clark, J. R., Reynolds, M. A., and Lowry, O. H., 1961, Quantitative studies of white matter. I. Enzymes involved in glucose-6-phosphate metabolism, *J. Gen. Physiol.* **44**:487–498.

McIlwain, H., 1966, *Biochemistry and the Central Nervous System*, 3rd ed., J. and A. Churchill, London, pp. 176–181.

McIlwain, H., and Rodnight, R., 1949, Breakdown of cozymase by a system from nervous tissue, *Biochem. J.* **44**:470–477.

Meyer-Estorf, G., Schulze, P. E., and Herken, H., 1973, Distribution of ^{3}H-labelled 6-aminonicotinamide and accumulation of 6-phosphogluconate in the spinal cord, *Naunyn-Schmied. Arch. Pharmacol.* **276**:235–241.

Meyer-König, E., 1973, Ultrastruktur der Glia- und Axonschädigung durch 6-Aminonicotinamid (6-AN) am Sehnerv der Ratte, *Acta Neuropathol.* **26**:115–126.

Mosher, L. R., 1970, Nicotinic acid side effects and toxicity: A review, *Am. J. Psychiatry* **126**:1290–1296.

Nakamura, S., Ikeda, M., Tsuji, H., Nishizuka, Y., and Hayaishi, O., 1963, Quinolinate transphosphoribosylase: A mechanism of niacin ribonucleotide formation from quinolinic acid, *Biochem. biophys. Res. Commun.* **13**:285–290.

Nemeth, A. M., and Dickerman, H., 1960, Pyridine nucleotides and diphosphopyridine nucleotidase in developing mammalian tissues, *J. Biol. Chem.* **235**:1761–1764.

Nisslbaum, J. S., Packer, D. E., and Bodansky, O., 1964, Comparison of the actions of human brain, liver, and heart lactic dehydrogenase variants on nucleotide analogues and on substrate analogues in the absence and in the presence of oxalate and oxamate, *J. Biol. Chem.* **239**:2830–2834.

Osmond, H., and Hoffer, A., 1962, Massive niacin treatment of schizophrenia: Review of a nine year study, *Lancet* **1**:316–319.

Perkins, M. N., and Stone, T. W., 1983, Quinolinic acid: Regional variations in neuronal sensitivity, *Brain Res.* **259**:172–176.

Pfeiffer, C. C., 1981, Extranutrients and mental illness, *Biol. Psychiatry* **16**:797–799.

Plaitakis, A., Nicklas, W. J., and Desnick, R. J., 1980, Glutamate dehydrogenase deficiency in three patients with spinocerebellar syndrome, *Ann. Neurol.* **7**:297–303.

Prakash, M. R., and Baquer, N. Z., 1981, Inhibition of gamma-aminobutyric acid transaminase with 6-aminonicotinamide in regions of the rat brain, *Biochem. Pharmacol.* **30**:663–664.

Samson, F. E., Jr., and Dahl, N. A., 1957, Cerebral energy requirement of neonatal rats, *Am. J. Physiol.* **188**:277–280.

Sarkander, H.-I., Knoll-Köhler, E., and Cervos-Navarro, J., 1978, Repression of glial RNA transcription during the development of 6-aminonicotinamide (6-AN)-induced acute gliopathy, *J. Pharmacol. Exp. Ther.* **205**:503–514.

Schneider, H., and Cervos-Navarro, J., 1974, Acute gliopathy in spinal cord and brain stem induced by 6-aminonicotinamide, *Acta Neuropathol.* **27**:11–23.

Schwarcz, R., Whetsell, W. O., Jr., and Mangano, R. M., 1983, Quinolinic acid: An endogenous metabolite that produces axon-sparing lesions in rat brain, *Science* **219**:316–318.

Searcy, G. H., 1907, An epidemic of acute pellagra, *J. Am. Med. Assoc.* **49**:37.

Singal, S. A., Sydenstricker, V. P., and Littlejohn, J. M., 1948, The nicotinic acid content of tissues of rats on corn rations, *J. Biol. Chem.* **176**:1069–1073.

Spector, R., 1979, Niacin and niacinamide transport in the central nervous system. *In vivo* studies, *J. Neurochem.* **33**:895–904.

Spector, R., and Huntoon, S., 1981, No effect of maternal niacin deficiency on niacin metabolism in newborn brain, *Neurochem. Res.* **6**:475–483.

Spector, R., and Kelly, P., 1979, Niacin and niacinamide accumulation by rabbit brain slices and choroid plexus *in vitro*, *J. Neurochem.* **33**:291–298.

Spector, R., and Lorenzo, A. V., 1975, Myo-inosital transport in the central nervous system, *Am. J. Physiol.* **228**:1510–1518.

Sternberg, S. S., and Philips, F. S., 1958, 6-Aminonicotinamide and acute degenerative changes in the central nervous system, *Science* **127**:644–646.

Stone, T. W., and Perkins, M. N., 1981, Quinolinic acid: A potent endogenous excitant at amino acid receptors in central nervous system, *Eur. J. Pharmacol.* **72**:411–412.

Unna, K., 1939, Studies on the toxicity and pharmacology of nicotinic acid, *J. Pharmacol. Exp. Ther.* **65:**95–103.

Utter, M. F., 1950, Mechanism of inhibition of anaerobic glycolysis of brain by sodium ions, *J. Biol. Chem.* **185:**499–517.

Weil-Malherbe, H., and Bone, A. D., 1951, Studies on hexokinase. I. The hexokinase activity of rat brain extracts, *Biochem. J.* **49:**339–347.

Willing, F., Neuhoff, V., and Herken, H., 1964, Der Austausch von 3-Acetylpyridin gegen Nicotin-säureamid in den Pyridinnucleotiden verschiedener Hirnregionen, *Naunyn-Schmied. Arch. Pharmacol.* **247:**254–266.

Windmueller, H. G., and Kaplan, N. O., 1962, Solubilization and purification of diphosphopyridine nucleotidase from pig brain, *Biochim. Biophys. Acta* **56:**388–391.

Winer, A. D., 1960, Fluorescent studies of ox-brain lactic and malic dehydrogenase, *Biochem. J.* **76:**5p–6p.

Wolf, A., and Cowen, D., 1959, Pathological changes in the central nervous system produced by 6-aminonicotinamide, *Bull. N.Y. Acad. Med.* **35:**814–817.

Woolley, D. W., 1952, *A Study of Antimetabolites*, Chapman and Hall, London.

Thiamine Deficiency and Cerebral Energy Metabolism

DAVID W. McCANDLESS

1. Introduction

The present chapter will examine the effects of experimentally produced thiamine deficiency on various aspects of cerebral energy metabolism. Thiamine deficiency is a common finding in chronic alcoholics, and since Leigh's disease patients may have an alteration in thiamine metabolism, consideration of cerebral energetics in these disorders may elucidate underlying mechanisms. This chapter will therefore address the normal role of thiamine both as a coenzyme as well as its possible role in nerve conduction. Following a consideration of the effects of thiamine deficiency on enzyme systems and metabolism will be a discussion of two major clinical entities thought to be associated with thiamine deficiency.

The original classic work on thiamine deficiency was done by Sir Rudolf Peters working on thiamine-deficient pigeons (Kinnersley and Peters, 1929). Previous investigators had attributed the opisthotonic posturing of symptomatic pigeons to muscular abnormality and had been looking in vain for attending biochemical alterations. Peters realized that the symptoms of thiamine-deficient pigeons could have been central in origin. He was soon able to demonstrate that brain brei from thiamine-deficient pigeons had a lower oxygen uptake than controls. He further was able to reverse this effect in vitro by the addition of thiamine to the brei. These results were soon repeated in rats (Peters, 1936). From these studies from Peters' laboratory came the concept that thiamine deficiency produces a biochemical lesion in oxidative metabolism that precedes the pathological lesion and is reversible.

Soon, from the work of Lohmann and Schuster (1937), the active form of thiamine, thiamine pyrophosphate, was discovered. Over the next few years,

DAVID W. McCANDLESS • Department of Neurobiology and Anatomy, University of Texas Medical School at Houston, Houston, Texas 77025.

subsequent studies showed that thiamine pyrophosphate served as a coenzyme for the decarboxylation of pyruvic acid, and that the blockage of this reaction in thiamine-deficient brain resulted in decreased oxygen uptake. Not until 1953 was thiamine pyrophosphate shown to serve as a coenzyme in a transketolation reaction in the hexosemonophosphate shunt (Horecker and Symrniotis, 1953). Soon thereafter, rats suffering thiamine deficiency were shown to have a severe decrease in transketolase activity (Brin et al., 1958), the rate-limiting step in this shunt (Kauffman, 1972). Later, Dreyfus showed that the largest decrease in thiamine-dependent enzyme activity occurred in the brain region with the most severe structural changes, the lateral pontine tegmentum (Dreyfus, 1965). Dreyfus also measured pyruvate decarboxylase activity in various cerebral regions (Dreyfus and Hauser, 1965). He found that whereas transketolase activity was highest in myelinated areas, pyruvate decarboxylase activity was highest in gray matter. In thiamine-deficient rats, there was a decrease in pyruvate decarboxylase only in the brain stem, and the degree of depression was not as great as the degree of depression of transketolase.

The early studies served to set a base for many subsequent experiments on the mechanism of thiamine deficiency. Before considering the results of experimental thiamine deficiency, current concepts regarding the role of thiamine in the normal brain will be reviewed.

2. Role of Thiamine in Metabolism

Extensive studies have established a variety of roles for the phosphorylated form of thiamine, thiamine pyrophosphate, in metabolism. Many of these appear to occur in bacteria, and the role of these reactions in brain remain unclear (Krampitz, 1969). This discussion, therefore, will be limited to a discussion of the role of thiamine pyrophosphate in the decarboxylation of alpha-keto acids and in transketolation. In addition, evidence will be presented for a role for thiamine in nerve conduction.

2.1. Alpha-Keto Acid Dehydrogenase Decarboxylation

In 1937, it was shown that the pyrophosphate ester of thiamine served as coenzyme for the decarboxylation of pyruvate (Lohmann and Schuster, 1937). Subsequent studies have shown that thiamine pyrophosphate also serves as coenzyme in the decarboxylation of alpha-ketoglutarate. These two enzymes are quite different from the standpoint of molecular weight, structure, and especially in the fact that the thiamine pyrophosphate moiety is tightly bound in the case of alpha-ketoglutarate dehydrogenase (Gubler, 1961) The thiamine pyrophosphate is highly dissociable from pyruvate dehydrogenase, and therefore the overall activity is totally dependent on the pyrophosphate (Linn et al., 1969). The activity of the dehydrogenase complex is inhibited by reaction products, such as acetyl-CoA and NADH, and stimulated by CoA and NAD$^+$ (Linn et al., 1969). Because of the key role of pyruvate and its subsequent decarboxylation in energy metabolism, this reaction has been extensively studied in thiamine deficiency. Since there is tight binding of the coenzyme to the apoen-

zyme, alpha-ketoglutarate dehydrogenase, one might expect protection from
the deleterious effects of thiamine deficiency. But this was not the case. The
activity of alpha-ketoglutarate dehydrogenase was reduced to half in both the
brain and the liver of pyrithiamine-treated animals (Holowach et al., 1968). In
pyrithiamine-treated mice, alpha-ketoglutarate increased to very high levels
(Holowach et al., 1968; Collins et al., 1970; Seltzer and McDougal, 1974). Similar findings were seen in thiamine-deprived mice (J. H. Holowach and R. E.
Hauhart, unpublished data). In fact, in these animals the increase in brain
alpha-ketoglutarate levels was higher than the increase in pyruvate, ninefold
versus threefold.

337

*THIAMINE
DEFICIENCY AND
CEREBRAL
ENERGY
METABOLISM*

2.2. Transketolation

The enzyme transketolase participates in two separate reactions in the
hexosemonophosphate shunt:

1. Xylulose-5-PO4 + Ribose-5-PO4 $\rightleftharpoons$
   Sedoheptose-7-PO4 + Glyceraldehyde-3-PO4
2. Xylulose-5-PO4 + Erythrose-4-PO4 $\rightleftharpoons$
   Fructose-6-PO4 + Glyceraldehyde-3-PO4

The first of these two reactions serves as the base for estimation of transketolase activity (Dreyfus and Moniz, 1962). The main purpose for the hexosemonophosphate shunt appears to be the generation of five carbon moieties
for nucleotide synthesis and for the formation of reducing power in the form
of NADPH (Dreyfus, 1976). The actual flux of molecules through this shunt in
adult brain is the subject of some controversy, but most investigators agree that
it represents about 5–8% of flow. The most widely used technique to measure
flow through the shunt is to compare $^{14}CO_2$ evolution from [1-^{14}C] glucose and
[6-^{14}C]glucose (Landau and Katz, 1964). A high yield of $^{14}CO_2$ from [1-
^{14}C]glucose relative to that from [6-^{14}C]glucose is taken to indicate a high flow
through the shunt, whereas an equal yield indicates little or no flow through
the shunt. In any case, these methods are subject to error, and are only an
approximation (Katz et al., 1966).

Flow through the shunt is higher than adult levels in young myelinating
animals (Novello and McLean, 1968). This increase in activity is thought to
represent the increased need for reducing power for fatty acids for myelination.

2.3. Nerve Conduction

Whereas the role of thiamine pyrophosphate as a coenzyme in metabolism
is well established, the role of thiamine triphosphate is still unknown. The
early observation by Minz (1938) and subsequent experiments by von Muralt
(1939) provided evidence that on electrical stimulation, thiamine was released
into the media from nerve preparation. These data were later confirmed for
spinal cord (Cooper et al., 1963). The exact meaning of these observations in
terms of conduction are, however, unclear.

Other types of studies implicate thiamine as participating in neural function in a role other than as coenzyme. It has been shown, for example (Fox and Duppel, 1975), that the decrease in sodium and potassium currents in the node of Ranvier with time could be prevented with addition of thiamine or its esters. These investigators concluded that thiamine or its esters were active compounds in nerve membrane and acted perhaps by controlling the number of ionic channels by stabilizing the density of negative surface changes. There are other studies that suggest that thiamine may play a role in neuronal function other than its role as coenzyme, and the reader is referred to an excellent recent review of this subject (Cooper and Pincus, 1979).

3. Thiamine Deficiency

The following section will deal with experimentally induced thiamine deficiency. There are generally two animal models of thiamine deficiency in wide use. In the first model, animals are placed on a thiamine-deficient diet, whereas in the second model, in addition to the deficient diet, a thiamine antimetabolite is also administered. In mice and rats, the first model is more chronic; about 4–6 weeks are required for the development of overt neurological symptoms. In the antimetabolite model, pyrithiamine (or in some cases oxythiamine) is coadministered, and this seems to shorten the time of onset of symptoms to around ten days.

Both animal models are excellent in that they are fully reversible, and have many features in common with human thiamine-related disorders. Features common to both human and animal thiamine deficiencies include:

1. Length of illness—both the human disorder and the dietary-only model in animals are chronic disorders relative to the life span of the individual.
2. Clinical symptoms in the most widely seen human thiamine deficiency associated disorder, Wernicke's disease, and experimental thiamine deficiency include ataxia, opisthotonos, and stupor.
3. Anatomical lesions in both human and animal deficiencies are localized to the thalamus, mammillary bodies, floor of the 4th ventricle, nucleus of the 8th cranial nerve, and cerebellum. Microscopically, these lesions include vacuolization, degeneration of glial and neural elements, and small areas of hemorrhagic necrosis (see Fig. 1A,B).
4. In both animals and man, the early symptoms are rapidly reversible with thiamine administration. In laboratory animals this means that the biological significance of observed changes in neurochemical parameters can be evaluated by noting if these changes revert toward normal as the symptoms are reversed with thiamine.

Thus, although extrapolation of results from animal experimentation to man is difficult, many features of the rat/mouse model of thiamine deficiency correlate closely with those seen in Wernicke's encephalopathy. Therefore, study of the intracerebral mechanisms of murine thiamine deficiency may have relevance to the human disorder.

339

THIAMINE
DEFICIENCY AND
CEREBRAL
ENERGY
METABOLISM

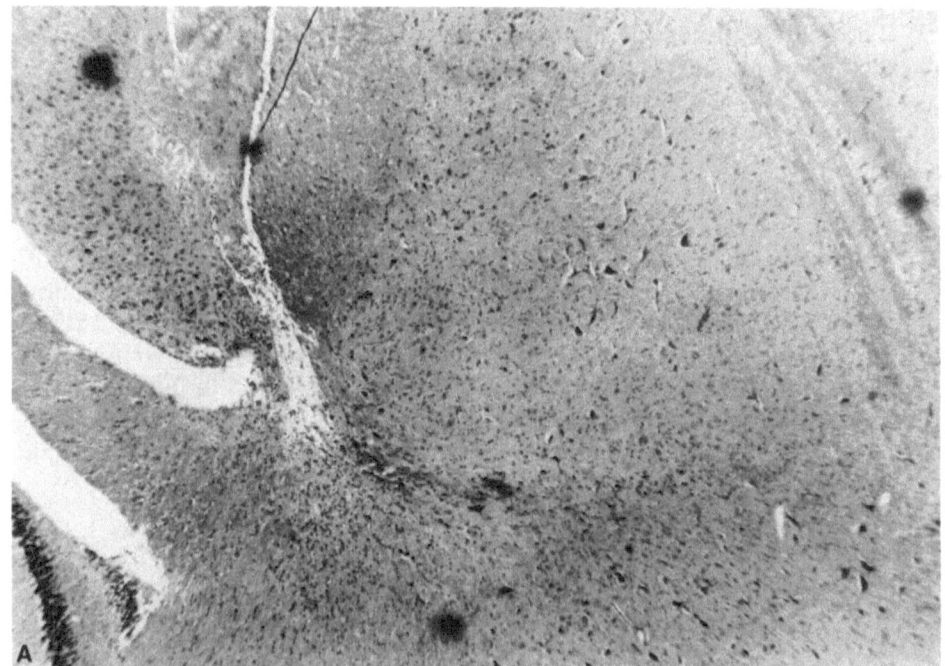

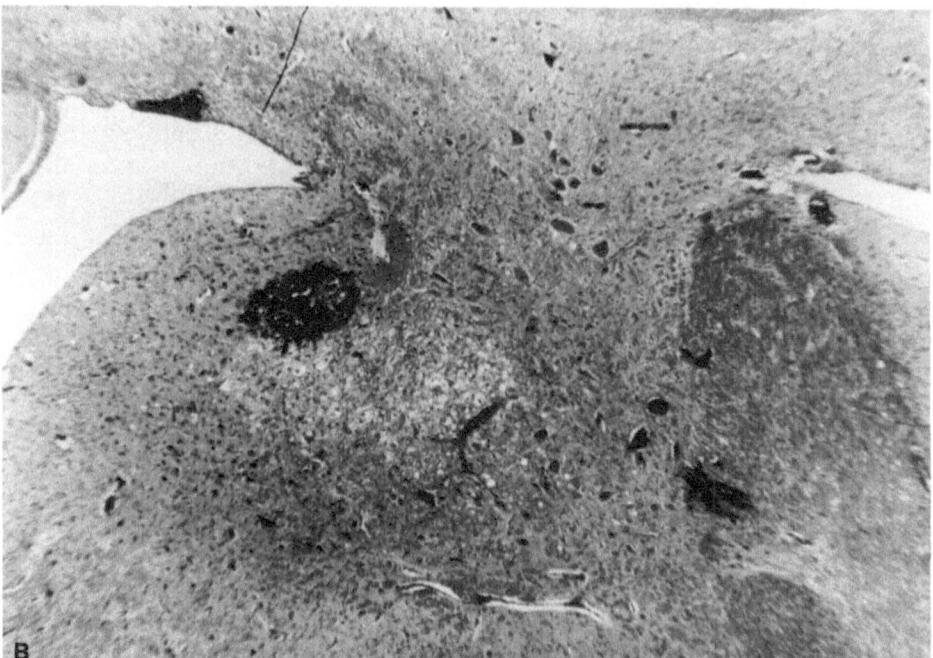

FIGURE 1. A: Light microscopic section through the lateral vestibular nucleus showing appearance of pair-fed control rat. (H and E stain, 35×.) B: Light microscopic section through the lateral vestibular nucleus showing multiple areas of hemorrhagic necrosis in severe dietary induced thiamine deficiency. (H and E stain, 35×.)

3.1. Mechanisms—Dietary versus Antagonist

The standard experimental procedure for producing thiamine deficiency in laboratory animals is to place them on a commercial diet replete with all essential components except thiamine. Since animals soon suffer a decreased appetite and therefore consume less food, pair-fed controls are an important additional group.

Many investigators have utilized thiamine antagonists in the production of thiamine deficiency. The major antagonists used are oxythiamine and pyrithiamine; since oxythiamine does not enter brain, pyrithiamine has been utilized almost exclusively in the study of the neurological aspects of thiamine deficiency. The rate of appearance of symptoms is dependent in some measure on the ratio of pyrithiamine to thiamine. Pyrithiamine is thought to have a deleterious effect via a blockage of the formation of thiamine pyrophosphate (Steyn-Parve, 1967). The majority of symptoms are similar in animals made deficient by dietary means only or by pyrithiamine administration. The major difference in the model seems to be the time necessary to produce these symptoms.

3.2. Generalized Effects—Body Weight, Neurological Symptoms, and Pathological Changes

The effect of dietary-induced thiamine deficiency on the growth pattern of rats is depicted in Fig. 2. These animals weighed about 70 g initially and gained weight at a normal rate for about the first two weeks after which their gain diminished and they soon were losing weight. It is important to note that even with careful pair feeding, the weight of pair-fed controls does not fall as dramatically as that of the deficient animals.

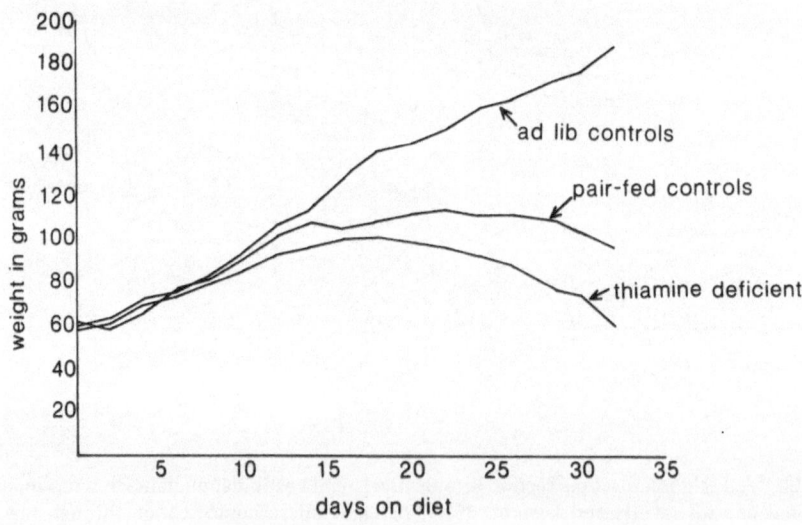

FIGURE 2. Effect of dietary-induced thiamine deficiency on the growth pattern of rats.

Symptoms other than weight loss begin to appear in about 2–2½ weeks. These consist of decreased activity, hair loss, and a hunched posture. Soon, deficient animals exhibit a wobbly gait. By 4–4½ weeks, animals exhibit severe symptoms that include opisthotonic posturing and almost complete loss of appetite. Most striking in these animals is an uncontrolled series of rolling over movements that follow being suspended by the tail. When animals have reached this severe stage, they always die within 24–48 hr unless treated with thiamine. Interestingly the symptoms observed in rats and nearly identical symptoms described in monkeys (Mesulum et al., 1977) are very similar to those observed in chronic alcoholics.

With pyrithiamine the symptoms are similar, but the time course is much shorter. Thus, symptoms such as weight and hair loss may begin as soon as 4–6 days after treatment and severe neurological symptoms usually appear in 9–12 days. As in the case of dietary thiamine deficiency, death occurs unless the animals are "reversed" with thiamine treatment. Reversal, usually accomplished by the intraperitoneal injection of 100–500 µg of thiamine HCl, results in an observable inprovement in the neurological status of the animals within 24 hr. Twelve hours after injection, animals may be neurologically intact. This "reversible" feature of the disorder has been utilized to examine the biological importance of changes in enzymes and metabolites that are altered in the symptomatic stage.

Structural changes accompany thiamine deficiency, and these have been demonstrated by both light and electron microscopy. In thiamine-deficient rats, for example, it has been found that the early phases of recurrent thiamine deficiency involve glial cell swelling and vacuolization, and this process involved both foot processes and perikarya (1967). In another study (Robertson et al., 1968), the early lesions of dietary–induced thiamine deficiency consisted of swelling of perivascular glial processes. Subsequently there was swelling of cell bodies as well as glial processes. Neurons appeared to be unaffected in these early lesions. The appearance of hemorrhagic lesions in areas such as mammillary bodies and pontine tegmentum have been noted by several investigators (Prickett, 1934; Watanabe et al., 1981).

3.3. Effects on Enzymatic Systems

There have been numerous studies showing a decrease in the activity of thiamine-dependent enzymes in the brains of thiamine-deficient animals. These studies have focused almost exclusively on transketolase and on the pyruvate and alpha-ketoglutarate dehydrogenase activities.

Table I represents the effect of thiamine deficiency induced by various means on the levels of cerebral pyruvate decarboxylase and transketolase. In most instances, in in vivo studies, the decrease in transketolase was greater than that of pyruvate decarboxylase. The fact that the transketolase decrease is greater in the dietary-induced form of thiamine deficiency has led many investigators to suggest that the myelin changes seen microscopically in this form of thiamine deficiency may be causally related to an alteration in the pentose phosphate pathway. This line of thinking persisted until later studies on both metabolites of the pentose phosphate pathway and flux through the pathway showed rather conclusively the importance of this system in thiamine deficiency. These data are discussed below.

In spite of decreased transketolase (about 60%) in the brain stem, the activities of G-6-PDH and 6-PGDH as measured by the method of Glock and McLean (1953) were unaltered in thiamine-deficient as compared with pair-fed controls (McCandless et al., 1976). These enzyme data are especially significant in light of studies which indicate that these first two oxidative steps in the hexose monophosphate (HMP) shunt are the regulatory ones (Eggleston and Krebs, 1975; Krebs and Eggleston, 1973). Transketolase may be thought of as rate limiting due to it having less activity than the other enzymes of the HMP shunt. Regulation of the flow of molecules through transketolase activity is greatly reduced in the absence of change in G-6-PDH and 6-PGDH. This emphasizes the large enzymatic reserve capacity of the shunt and is in general agreement with earlier observations on the "excess" enzyme capacity of this shunt.

It is also of some interest to note that thiamine deficiency may be associated with magnesium deficiency (Friend et al., 1973) in the liver. The significance of this is that the concomitant presence of Mg_2 deficiency acts to decrease the response to thiamine treatment of thiamine-deficient rats (Zieve et al., 1968a). Specifically, transketolase activity in the liver does not recover as rapidly with thiamine treatment in rats that are both thiamine and Mg_2 deficient (Zieve et al., 1968b).

Although the quantitative decrease in pyruvate decarboxylase is less than that of transketolase, the fact that on symptom reversal with thiamine administration the decarboxylase returns more nearly to normal suggested it may be more directly related to symptom production (McCandless and Schenker, 1968). Coupled with this is the concept that since the symptoms can be reversed in 3–4 hr, the accompanying biochemical changes would have to be at least as rapid. This argued in favor of an energy-associated change (pyruvate de-

TABLE I. **Thiamine Deficiency and Cerebral Thiamine Dependent Enzymes**

Pyruvate decarboxylase		Transketolase		
% Decrease	% Decrease with realimentation	% Decrease	% Decrease with realimentation	Author
75	59	50	47	Holowach et al. (1968)[a]
29	15	62	54	McCandless et al. (1976)[b]
37	30	40	36	Pincus et al. (1971)[c]
98	—	44	—	Schwartz and McCandless (1976)[d]
25	—	56	—	Dreyfus and Hauser (1965)[e]
20	—	—	—	Gubler and Johnson (1967)[f]

[a] Whole brain, pyrithiamine-induced thiamine deficiency.
[b] Brain stem, dietary-induced thiamine deficiency.
[c] Pons, dietary-induced thiamine deficiency.
[d] Neuroblastoma, tissue culture.
[e] Pons, dietary-induced thiamine deficiency.
[f] Whole brain, dietary-induced thiamine deficiency.

carboxylase) as compared with a myelin change in which the turnover times are measured in days or weeks rather in just a few hours.

3.4. Effects on Energy Metabolites and Flux

Before considering the effect of thiamine deficiency on energy metabolites, it might be of interest to consider the manner in which thiamine is depleted from brain. Figure 3 depicts the effect of dietary-induced thiamine deficiency on whole-brain thiamine levels. Control values were derived from rats fed control diet in the amount that the thiamine-deficient animals consumed the previous day (pair-fed controls). Note that there is a gradual depletion of thiamine throughout the duration of the deficiency. Symptoms do not appear until whole-brain thiamine levels are less than 20% of the controls. Similar results have been demonstrated by other investigators.

Since the measurement of enzymes is only an indication of the capacity of tissue to catalyze a particular reaction, assessment of metabolite levels, metabolite turnover, and flux through appropriate pathways is a more meaningful indication of the effect of thiamine deficiency. To this end, we have completed a series of studies in which we examined these parameters in thiamine-deficient animals (McCandless *et al.*, 1976).

As regards transketolase and the pentose phosphate pathway, the methods of Katz and co-workers (Katz and Wood, 1960; Landau and Katz, 1964; Katz *et al.*, 1966) can be used to estimate flux through the pentose phosphate cycle.

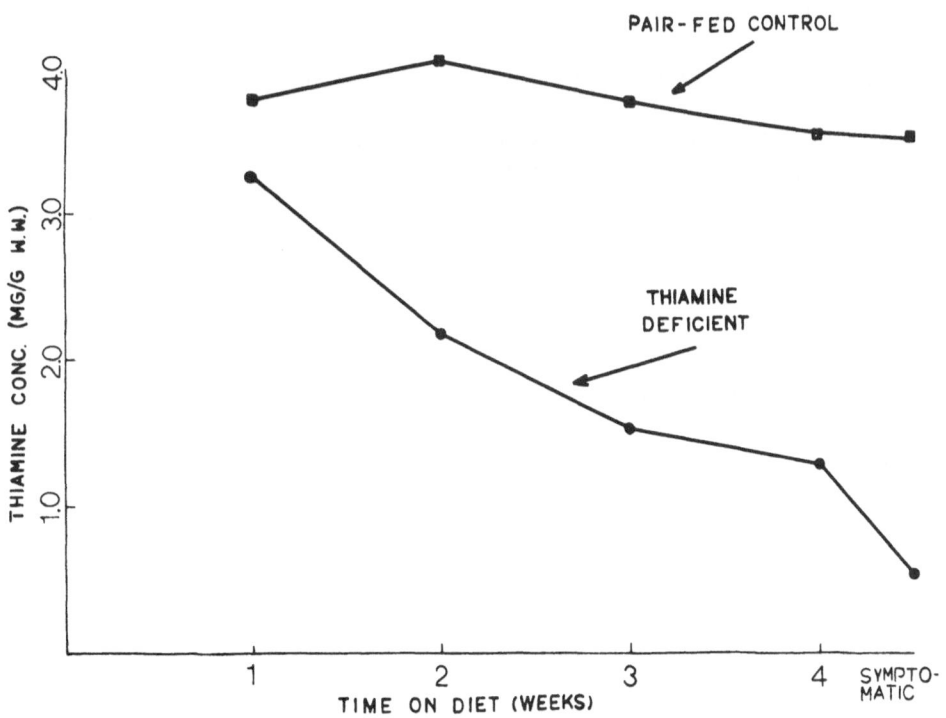

FIGURE 3. Effect of dietary-induced thiamine deficiency on whole brain thiamine content.

In this method, brain is incubated with glucose labeled in either the C1 or C6 portion with ^{14}C. C1/C6 ratios are calculated by dividing the specific yield of $^{14}CO_2$ from [1-^{14}C]glucose by the specific yield of $^{14}CO_2$ from [6-^{14}C]glucose. Results from these studies in thiamine-deficient rats showed no decrease in the C1/C6 ratio in thiamine-deficient as compared with pair-fed controls (McCandless *et al.*, 1976). In fact, the thiamine-deficient animal's ratio was slightly higher than the control. When the net levels of the pentose phosphate pathway metabolites ribose-5-phosphate, ribulose-5-phosphate, and xylulose-5-phosphate were measured in thiamine-deficient brain, they were not significantly different from those found in pair-fed controls (D. W. McCandless and C. E. Cassidy, unpublished results). These data, coupled with the enzyme data, do not support the hypothesis that in thiamine-deficient brain the decrease in transketolase activity leads to a decreased pentose phosphate pathway activity.

In the case of the pyruvate dehydrogenase activity, the correlation of the thiamine-induce reversal of symptoms with reversal of enzyme activity lends credence to the importance of this step in cerebral dysfunction. However, net ATP and phosphocreatine levels are not changed as compared with controls (McCandless and Schenker, 1968). Furthermore, when microsections weighing as little as 100 ng were analyzed from the lateral vestibular nucleus, again, as in the case of whole brain stem, there were no significant differences between controls and thiamine deficient animals (McCandless, 1982). Finally, turnover of the adenosine nucleotides ATP, ADP, and AMP is not affected in thiamine-deficient brain stem (McCandless and Cassidy, 1976). It is possible that some very small brain region is affected in terms of ATP metabolism and that inclusion of contiguous nonaffected tissue has masked the change, but this seems unlikely. It would seem, therefore, that if the decreased pyruvate decarboxylase is significant in symptom production, it does not work through altered ATP metabolism.

In a recent study, the 2-deoxyglucose method has been used to determine the local glucose utilization in the brains of thiamine-deficient rats (Hakim and Pappius, 1983). These investigators found a gradual decrease in glucose utilization in 40 discrete brain regions as thiamine deficiency progressed. At a time when neurological symptoms began, there was an increase in glucose utilization in certain regions followed by focal areas of depressed glucose utilization. These workers propose that the selective rise in glucose utilization is somehow linked to the subsequent clinical symptoms. Previous studies from the same laboratory (Hakim and Pappius, 1981) has shown no significant change in regional cerebral blood flow after six weeks of dietary-induced thiamine deficiency.

3.5. Acetylcholine

Several groups of investigators have examined the effects of thiamine deficiency on acetylcholine. Results from these studies have been equivocal. When net levels of acetylcholine have been examined, one group of investigators have found no effect (Speeg *et al.*, 1970; Hosein *et al.*, 1966; Stern and Igic, 1970; Reynolds and Blass, 1975), whereas other workers have found a decrease in the neurotransmitter (Cheney *et al.*, 1969; Heinrich *et al.*, 1973; Vorhees *et al.*, 1978; Lissak *et al.*, 1943). One problem has been considerable variation in sacrifice methods and assay technique. However, even when these variables are optimal, the results vary between laboratories and this area remains one of some controversy. There is generally more agreement concerning the effects of thiamine deficiency on acetylcholine production. It has been shown in vitro, that acetylcholine formation is decreased in pigeon brain (Mann and Quastel, 1940) and in rat brain (Bhagat and Lockett, 1962). Additional studies have shown a decrease in acetylcholine utilization (Voorhees et al., 1977) and coupled with decreased synthesis could explain normal net levels.

345

THIAMINE
DEFICIENCY AND
CEREBRAL
ENERGY
METABOLISM

It is important that there appears to be a rather close correlation between energy metabolism and acetylcholine metabolism. Even though only about 1% of the flow of molecules from pyruvate is directed to acetylcholine synthesis, it is quite sensitive to variations (Blass et al., 1981; Gibson et al., 1981). In support of the concept that altered acetylcholine metabolism may be of significance in thiamine deficiency is the finding that the acetylcholinesterase inhibitor physostigmine acts to prolong the onset of the severe neurological symptoms in thiamine-deficient rats (Cheney et al., 1969). For a further discussion of acetylcholine metabolism and thiamine deficiency, the reader is referred to a recent excellent paper by Gibson et al. (1982).

3.6. Thiamine Triphosphate

In vitro studies have shown that the role of thiamine triphosphate (TTP), quite apart from the role of thiamine pyrophosphate as a coenzyme, may be in the maintenance of neuronal membrane potentials. This is based on the observation that with electrical stimulation, TTP is released from intact nerve preparations and that TTP is apparently localized to nerve membranes as opposed to axoplasm (Itokawa and Cooper, 1970). Thiamine triphosphatase has also been shown to be associated with brain membrane fractions and further supports a role for thiamine in nerve conduction (Barchi and Braun, 1972). How these observations relate to in vivo nerve function is, however, unclear. Arguing against this hypothesis is the observation that in murine thiamine deficiency, thiamine pyrophosphate decreases in the areas of brain most severely affected (brain stem), whereas TTP remained unchanged (Pincus and Grove, 1970). Although this study measured net levels and not turnover rates of TTP, these results do not support a biologically significant role for TTP in symptomatic thiamine-deficient rats. In a more recent in vivo study, the effects of electrical stimulation and potassium depolarization on thiamine ester uptake, synthesis, and efflux were examined (Berman and Fishman, 1975). It was found that these experimental manipulations produced no significant changes in TTP metabolism These results do not support a nonmetabolic role for TTP in nerve conduction. It should be emphasized, however, that this is an area of active investigation, and further studies may find a more important role for thiamine in nerve condition.

4. Wernicke's Encephalopathy

Thiamine deficiency in man contributes to the Wernicke–Korsakoff syndrome seen in many alcoholics (Phillips et al., 1952; Victor and Adams, 1961), and in rats produces a neurological disorder characterized by loss of righting response, ataxia, opisthotonos, and death (Church, 1935; Dreyfus, 1961). In each instance, many of the early symptoms are fully and rapidly reversible, and from this derives the now classic concept that thiamine deficiency in its early stage is a metabolic, i.e., biochemical lesion.

Wernicke's encephalopathy is characterized clinically by ataxia, ocular palsies, nystagmus, and mental changes that may include confusion, apathy, and stupor. With thiamine treatment, many of these symptoms are reversed. However, most patients then pass into a stage of confabulation and amnesia, which is called Korsakoff's psychosis. The amnesia involves failure to remember facts formed from the time of illness back many months or years (retrograde amnesia) and an inability to learn new information. Although some improvement with therapy may occur, the majority of these patients have permanently impaired mentation. Although the symptoms of Korsakoff's psychosis are not so readily reversible with thiamine as those of Wernicke's encephalopathy, this does not indicate different underlying cerebral mechanisms. In fact, most investigators agree that the symptoms of Korsakoff's psychosis, which invariably follow Wernicke's encephalopathy, represent permanent structural changes induced by chronic alcoholism and thiamine deficiency (Victor et al., 1971).

5. Subacute Necrotizing Encephalomyelopathy

5.1. General Features

Subacute necrotizing encephalomyelopathy (Leigh's disease) was first discribed in 1951 by Denis Leigh (Leigh, 1951). Since then, well over 100 cases have been described and perhaps as many as 100 more diagnosed (at postmortem) and not recorded. Leigh's disease is an autosomal recessive disease characterized clinically by an insidious onset at 1 to 2 years of age, a slow progressive failure to thrive, and severe mental retardation. Leigh's disease carries a 100% mortality rate with the majority of cases dying before 5 years of age. More acute and more chronic cases have been reported; clinically there is a wide variability in symptoms and in the course of the disease. The diagnosis during life is difficult and presently is based largely on the presence in the urine of Leigh's disease patients of a glycoprotein that inhibits the enzyme ATP-TPP-phosphoryl transferase (Cooper et al., 1969). This inhibitor has been found in the urine of all Leigh's disease cases and is also present in carriers (parents) of Leigh's disease. Unfortuantely, the assay for inhibitor is difficult; only a few laboratories in the country currently assay this inhibitor.

The precise etiology of Leigh's disease is unknown. The only unifying feature of Leigh's disease is the presence in urine of the inhibitor. The enzyme that is blocked by the inhibitor catalyzes the production of TTP. When TTP was analyzed in the brains of children who had died of Leigh's disease, it was found to be absent (Pincus et al., 1971). However, the role of TTP in mammalian brain is unknown, and so the relation between the inhibitor (or blocked TTP production) and the symptoms of Leigh's disease is unclear.

The location of the cerebral lesion of Leigh's disease (diencephalon and brain stem) and the histological appearance of the lesion (focal areas of hemorrhagic necrosis with relative sparing of neurons) so closely resemble the lesion of Wernicke's disease as to suggest to many workers that altered thiamine metabolism may be a feature of Leigh's disease, as has been shown in Wernicke's disease (Victor and Adams, 1961). In addition, there has been described an absence of hepatic pyruvate carboxylase in some, but not all, patients with Leigh's disease (Hommes *et al.*, 1968). Pyruvate carboxylase is a key enzyme in gluconeogenesis, and exactly how an absence in the liver or brain of this enzyme could produce cerebral lesions is also vague.

In short, there are scant data relating to Leigh's disease. The clinical features are variable and the biochemical test for the urinary inhibitor is difficult, making the diagnosis one frequently made at postmortem. The etiology of the disease is unknown; no *proven* treatment exists and the mortality rate is 100% for untreated cases.

5.2. Biochemical Features

Biochemical changes in Leigh's disease imply that a defect in thiamine-related metabolism may be responsible for the clinical and pathological alteration. In 1968, an absence of hepatic pyruvate carboxylase activity was described in a patient with Leigh's disease (Hommes *et al.*, 1968). Prior to 1968, the only significant and consistent changes seen in Leigh's disease were elevated blood lactate and pyruvate levels. A decrease in pyruvate metabolism was suspected due to increased lactate, pyruvate, and alanine values and low fasting glucose levels.

Other enzymes associated with carbohydrate metabolism analyzed and found normal in the liver of Leigh's disease patients at autopsy include transketolase, phosphoenol pyruvate carboxykinase, and pyruvate decarboxylase (Pincus, 1972; Grover *et al.*, 1972). However, liver values at autopsy may not reflect regional cerebra levels *in vivo*. Cerebral as well as hepatic pyruvate carboxylase have been shown to be absent in the brain in a Leigh's disease patient at autopsy (Grover *et al.*, 1972). The precise way in which a decrease in pyruvate carboxylase (a gluconeogenic enzyme) could result in the neurological symptoms and demise of the patient is totally unclear. The biological significance of elevated lactate, pyruvate, and alanine in symptom production has been recently challenged. It has been shown that the degree of increase in lactate and pyruvate can be produced by mild hyperventilation. Since hyperventilation is a feature of Leigh's disease, elevated lactate and pyruvate may be a result only of the hyperventilation (Montpetit *et al.*, 1971).

In 1969, there was described, in the urine of a patient with Leigh's disease, a glycoprotein that inhibited the enzyme ATP-TPP-phosphoryl transferase (Cooper *et al.*, 1969). This inhibitor has subsequently been shown to be present in the blood and CSF of Leigh's disease patients and has been found in all Leigh's disease patients. In fact, the diagnosis of Leigh's disease before death is made almost exclusively on the presence in the urine of this inhibitory substance (false positives are rare). Further, it has been shown that the parents of Leigh's disease children also have the inhibitor.

The enzyme ATP-TPP-phosphoryl transferase catalyzes a reaction in which a third phosphate group is added to TTP. Thiamine triphosphate has been shown to be absent in the brains of several patients who have died of Leigh's disease, whereas content of other thiamine esters in the brain are normal (Pincus *et al.*, 1971). However, a precise *in vivo* role of TTP in mammalian brain has never been discovered. There is some evidence to support the view that TTP may play a role in membrane conduction.

ACKNOWLEDGMENT. The author wishes to express his gratitude to Ms. Diana Parker for her expert secretarial assistance.

References

Barachi, R., and Braun, P., 1972, A membrane associated thiamine triphosphate from rat brain, *J. Biol. Chem.* **247:**7668–7675.

Berman, K., and Fishman, R. A., 1975, Thiamine phosphate metabolism and possible coenzyme independent functions of thiamine in brain, *J. Neurochem.* **24:**457–465.

Bhagat, B., and Lockett, M. F., 1962, The synthesis of acetylcholine by dried powders from the brains of normal rats and of thiamine deficient rats, *J. Pharm. Pharmacol.* **14:**37–40.

Blass, J. R., Gibson, G. E., Duffy, T. E., and Plum, F., 1981, Cholinergic dysfunction: A common denominator in metabolic encephalopathies, in: *Cholinergic Mechanisms: Phylogenetic Aspects, Central and Peripheral Synapses, and Clinical Significance* (G. Pepeu and H. Ladinsky, eds.), Plenum Press, New York, pp. 921–928.

Brin, M., Schohert, S. S., and Davidson, C. S., 1958, The effect of thiamine deficiency on the glucose oxidative pathway of rat erythrocytes, *J. Biol. Chem.* **230:**319–326.

Cheney, D. L., Gubler, C. J., and Jaussi, A. W., 1969, Production of acetylcholine in rat brain following deprivation and treatment with thiamine antagonists, *J. Neurochem.* **16:**1283–1291.

Church, C. F., 1935, Functional studies of the nervous system in experimental beri-beri, *Am. J. Physiol.* **111:**660–667.

Collins, R. C., 1967, Glial cell changes in the brain stem of the thiamine deficient rat, *Am. J. Pathol.* **50:**791–814.

Collins, R. C., Kirkpatrick, J. B., and McDougal, D. B., Jr., 1970, Some regional pathologic and metabolic consequences in mouse brain of pyrithiamine-induced thiamine deficiency, *J. Neuropathol. Exp. Neurol.* **29:**57–69.

Cooper, J. R., and Pincus, J. H., 1979, The role of thiamine in nervous tissue, *Neurochem. Res.* **4:**223–239.

Cooper, J. R., Roth, R. H., and Kini, M. M., 1963, Biochemical and physiological function of thiamine in nervous tissue, *Nature* **199:**609–610.

Cooper, J. R., Itokawa, Y., and Pincus, J., 1969, Thiamine triphosphate deficiency in subacute necrotizing encephalomyelopathy, *Science* **164:**74–76.

Dreyfus, P. M., 1961, The quantitative histochemical distribution of thiamine in deficient rat brain, *J. Neurochem.* **8:**139–145.

Dreyfus, P. M., 1965, The regional distribution of transketolase in the normal and the thiamine deficient nervous system, *J. Neuropathol. Exp. Neurol.* **24:**119.

Dreyfus, P. M., 1967, Transketolase activity in the nervous system, in: *Thiamine Deficiency* (G. E. W. Wolstenholme, ed.), Little Brown and Co., Boston, pp. 103–111.

Dreyfus, P. M., 1976, Thiamine deficiency encephalopathy: Thoughts on its pathogenesis, in: *Thiamine* (C. J. Gubler, M. Fujiwara, and P. M. Dreyfus, eds.), John Wiley and Sons, New York, pp. 229–244.

Dreyfus, P. M., and Hauser, G., 1965, The effect of thiamine deficiency on the pyruvate decarboxylase system of the central nervous system, *Biochim. Biophys. Acta* **104:**78–84.

Dreyfus, P. M., and Moniz, R., 1962, The quantitative histochemical estimation of transketolase in the nervous system of the rat, *Biochim. Biophys. Acta* **65:**181–189.

Eggleston, L. V., and Krebs, H. A., 1974, Regulation of the pentose phosphate cycle, *Biochem. J.* **138:**425–435.

349

*THIAMINE
DEFICIENCY AND
CEREBRAL
ENERGY
METABOLISM*

Fox, J. M., and Duppel, W., 1975, The action of thiamine and its di- and triphosphates on the slow exponential decline of the ionic currents in the node of Ranvier, *Brain Res.* **89**:287–302.

Friend, B. A., Williams, R. V., Mehlman, M. A., and Tobin, R. B., 1973, Effect of thiamine deficiency on liver and kidney metal ion concentrations in rat, *Nutr. Rep. Int.* **8**:33–88.

Gibson, G. E., Ksiezak, H. J., and Duffy, T. E., 1981, Acetylcholine synthesis and glucose oxidation with varying oxygen levels in vivo and in vitro, in: *Cholinergic Mechanisms: Phylogenetic Aspects, Central and Peripheral Synapses, and Clinical Significance* (G. Pepeu and H. Ladinsky, eds.), Plenum Press, New York, pp. 443–450.

Gibson, G. E., Barclay, L., and Blass, J., 1982, The role of the cholinergic system in thiamine deficiency, *Ann. N.Y. Acad. Sci.* **378**:382–403.

Glock, G. E., and McLean, P., 1953, Further studies on the properties and assay of glucose-6-phosphate dehydrogenase and 6-phosphogluconate dehydrogenase of rat liver, *Biochem. J.* **55**:400–408.

Grover, W., Averback, V., and Patel, M., 1972, Biochemical studies and therapy in subacute necrotizing encephalomyelopathy, *J. Pediatr.* **81**:39–56.

Gubler, C. J., 1961, Studies on the physiological functions of thiamine, *J. Biol. Chem.* **236**:3112–3120.

Gubler, C. J., and Johnson, L. R., 1967, Enzyme studies in thiamine deficiency, in: *Thiamine Deficiency* (G. E. W. Wolstenholme, ed.), Little Brown and Co., Boston, pp. 54–66.

Hakim, A. M., and Pappius, H. M., 1981, The effect of thiamine deficiency on local cerebral glucose utilization, *Ann. Neurol.* **9**:334–339.

Hakim, A. M., and Pappius, H. M., 1983, Sequence of metabolic, clinical, and histologic events in experimental thiamine deficiency, *Ann. Neurol.* **13**:365–375.

Heinrich, C. P., Stadler, H., and Weiser, H., 1973, The effect of thiamine deficiency on the acetylcoenzyme-A and acetylcholine levels in the rat brain, *J. Neurochem.* **21**:1273–1281.

Hommes, F., Polman, H., and Reerink, J., 1968, Leigh's encephalomyelopathy: An inborn error of gluconeogenesis, *Arch. Dis. Child.* **43**:423–431.

Holowach, J., Kauffman, F., Ikossi, M. G., Thomas, C., and McDougal, D. B., Jr., 1968, The effects of a thiamine antagonist, pyrithiamine, on levels of selected metabolic intermediates and on activities of thiamine-dependent enzymes in brain and liver, *J. Neurochem.* **15**:621–631.

Horecker, B. L., and Smyrniotis, P. Z., 1953, The coenzyme function of thiamine pyrophosphate in pentose metabolism, *J. Am. Chem. Soc.* **75**:1009–1010.

Hosein, E. A., Chabrol, J. G., and Freedman, G., 1966, The effect of thiamine deficiency in rats and pigeons on the content of materials with acetylcholine-like activity in brain, heart, and skeletal muscle, *Rev. Can. Biol.* **25**:129–134.

Itokawa, Y., and Cooper, J., 1970, Ion movements and thiamine, *Biochem. Biophys. Acta* **196**:274–278.

Katz, J., and Wood, H. G., 1960, The use of glucose-C^{14} for the evaluation of the pathways of glucose metabolism, *J. Biol. Chem.* **235**:2165–2177.

Katz, J., Landau, B., and Bartsch, G., 1966, The pentose cycle triose phosphate isomerization, and lipogenesis in rat tissue, *J. Biol. Chem.* **241**:727–740.

Kauffman, F. C., 1972, The quantitative histochemistry of enzymes of the pentose phosphate pathway in the central nervous system of the rat, *J. Neurochem.* **19**:1–9.

Kinnersley, H., and Peters, R., 1929, Observations on carbohydrate metabolism in birds, *Biochem. J.* **23**:1126–1136.

Krampitz, L. O., 1969, Catalytic functions of thiamine diphosphate, *Ann. Rev. Biochem.* **38**:213–240.

Krebs, H. A., and Eggleston, L. V., 1973, The regulation of the pentose phosphate cycle in rat liver, in: *Advances in Enzyme Regulation*, Volume 12 (G. Weber, ed.), Pergamon Press, Elmsford, New York, pp. 421–434.

Landau, B., and Katz, J., 1964, A quantitative estimation of the pathways of glucose metabolism in rat adipose tissue in vitro, *J. Biol. Chem.* **239**:697–704.

Leigh, D., 1951, Subacute necrotizing encephalomyelopathy in an infant, *J. Neurol. Neurosurg. Psychiatry.* **14**:216–219.

Linn, T. C., Pettit, F. H., and Reed, L. J., 1969, Alpha keto acid dehydrogenase complex, *Proc. Natl. Acad. Sci. U.S.A.* **62**:234–241.

Lissak, K., Kovacs, T., and Nagy, E. K., 1943, Acetylcholin-und cholinesterasegehalt von organen B1-avitaminotischer und normaler ratten, *Pflug. Arch. Eur. J. Physiol.* **247:**124–131.

Lohmann, K., and Schuster, P., 1937, Untersuchungen uber die cocarboxylase, *Biochem. Ztschr.* **294:**188–214.

Mann, P. J. G., and Quastel, J. H., 1940, Vitamin B$_1$ and acetylcholine formation in isolated brain, *Nature* **145:**856–857.

McCandless, D. W., 1982, Energy metabolism in the lateral vestibular nucleus in pyrithiamine deficiency, *Ann. N.Y. Acad. Sci.* **378:**355–364.

McCandless, D. W., and Cassidy, C. E., 1976, Adenine nucleotides in thiamine deficient rat brain, *Res. Commun. Chem. Pathol. Pharmacol.* **14:**579–582.

McCandless, D. W., and Schenker, S., 1968, Encephalopathy of thiamine deficiency: Studies of intracerebral mechanisms, *J. Clin. Invest.* **47:**2268–2280.

McCandless, D. W., Curley, A. D., and Cassidy, C. E., 1976, Thiamine deficiency and the pentose phosphate cycle in rats: Intracerebral mechanisms, *J. Nutr.* **106:**1144–1151.

Mesulam, M., Van Hoesen, G. W., and Butters, N., 1977, Clinical manifestations of chronic thiamine deficiency in the Rhesus monkey, *Neurology* **27:**239–245.

Minz, B., 1938, Sur la liberation de la vitamin par leé tronc isole du nerf pneumogastrique soumis a la excitation électrique, *C.R. Seances Soc. Biol.* **127:**1251–1253.

Montpetit, V., Andermann, F., Carpenter, S., Fawcett, J. C., Zborowska, J., and Gibberson, S., 1971, Subacute necrotizing encephalomyelopathy: A review and study of two families, *Brain* **94:**1–6.

Novello, F., and McLean, P., 1968, The pentose phosphate pathway of glucose metabolism: Measurement of the nonoxidative reactions of the cycle, *Biochem. J.* **107:**775–791.

Peters, R. A., 1936, The biochemical lesion in vitamin B$_1$ deficiency, *Lancet* **1:**1161–1165.

Phillips, G. B., Victor, M., Adams, R. D., and Davidson, C. S., 1952, A study of the nutritional defect in Wernicke's syndrome, *J. Clin. Invest.* **31:**359–370.

Pincus, J. H., 1972, Subacute necrotizing encephalomyelopathy (Leigh's disease): A consideration of clinical features and etiology, *Dev. Med. Child Neurol.* **14:**87–101.

Pincus, J. H., and Grove, I., 1970, Distribution of thiamine phosphate esters in normal and thiamine deficient brain, *Exp. Neurol.* **28:**477–483.

Pincus, J. H., Solitaire, G., and Itokawa, Y., 1971, Thiamine, thiamine triphosphate, and nervous system lesions in subacute necrotizing encephalomyelopathy, *Neurology* **21:**444–457.

Prickett, C. O., 1934, The effect of deficiency of vitamin B$_1$ upon the central and peripheral nervous systems of the rat, *Am. J. Physiol.* **107:**459–470.

Reynolds, S., and Blass, J. P., 1975, Normal levels of acetyl-coenzyme A and acetylcholine in the brains of thiamine deficient rats, *J. Neurochem.* **24:**185–186.

Robertson, D. M., Wasan, S. M., and Skinner, D. B., 1968, Ultrastructural features of early brain stem lesions of thiamine deficient rats, *Am. J. Pathol.* **52:**1081–1097.

Schwartz, J. P., and McCandless, D. W., 1976, Glycolytic metabolism in cultured cells of the nervous system, *Mol. Cell. Biochem.* **13:**49–53.

Seltzer, J. L., and McDougal, D. B., Jr., 1974, Temporal changes of regional cocarboxylase levels in thiamine-depleted mouse brain, *Am. J. Physiol.* **227:**714–718.

Speeg, K. V., Chen, D., McCandless, D. W., and Schenker, S., 1970, Cerebral acetylcholine in thiamine deficiency, *Proc. Soc. Exp. Biol. Med.* **134:**1005–1009.

Stern, P., and Igic, R., 1970, The control of material with acetylcholine-like activity in the brains of animals following thiamine deprivation and treatment with pyrithiamine, in: *Cholinergic Mechanisms* (F. Heilbronn and A. Winters, eds.), Forvarets Forskningsanstalt, Stockholm, pp. 419–427.

Steyn-Parve, E., 1967, The mode of action of some thiamine analogs with antivitamin activity, in: *Thiamine Deficiency* (G. E. W. Wolstenholme, ed.), Little Brown and Co., Boston, pp. 26–41.

Victor, M., and Adams, R. D., 1961, On the etiology of the alcoholic neurologic diseases with special reference to the role of nutrition, *Am. J. Clin. Nutr.* **9:**379–389.

Victor, M., Adams, R. D., and Collins, G., 1971, *The Wernicke-Korsakoff Syndrome*, F. A. Davis, Philadelphia, pp. 71–131.

von Muralt, A., 1939, Gibt es Aktionssubstanzen der Nervenerregung, *Naturwissenschaften* **27:**265–270.

Vorhees, C. V., Schmidt, D. E., Barrett, R. J., and Schenker, S., 1977, Effects of thiamine deficiency on acetylcholine levels and utilization *in vivo* in rat brain, *J. Nutr.* **107**:1902–1908.

Vorhees, C. V., Schmidt, D. E., and Barrett, R. J., 1978, Effects of pyrithiamine and oxythiamine on acetylcholine levels and utilization in rat brain, *Brain Res. Bull.* **3**:493–496.

Watanabe, I., Tomita, T., Hung, K. S., and Iwasaki, Y., 1981, Edematous necrosis in thiamine deficient encephalopathy of the mouse, *J. Neuropathol. Exp. Neurol.* **40**:454–471.

Zieve, L., Doizaki, W. M., and Stenroos, L. E., 1968a, Effect of magnesium deficiency on growth response to thiamine of thiamine deficient rats, *J. Lab. Clin. Med.* **72**:261–267.

Zieve, L., Doizaki, W. M., and Stenroos, L. E., 1968b, Effect of magnesium deficiency on blood and liver transketolase activity and on the recovery of enzyme activity in thiamine deficient rats receiving thiamine, *J. Lab. Clin. Med.* **72**:268–277.

351

THIAMINE
DEFICIENCY AND
CEREBRAL
ENERGY
METABOLISM

Thiamine Deficiency

Cerebral Amino Acid Levels and Neurologic Dysfunction

JEAN HOLOWACH THURSTON,
RICHARD E. HAUHART, JOHN A. DIRGO, and
DAVID B. McDOUGAL, JR.

1. Introduction

Ferrari (1957) was the first to report a decrease of glutamate levels in the whole brain of thiamine-deficient rats. Glutamate levels were also reduced in the brains of pyrithiamine-treated mice (Holowach *et al.*, 1968). Subsequently, decreases in threonine, γ-aminobutyrate (GABA), aspartate, and serine and increases in glycine were also seen in the brains of thiamine-deprived animals and/or of those treated with pyrithiamine (Gaitonde and Nixey, 1974; Gubler *et al.*, 1974; Gaitonde and Fayein, 1975).

There is increasing evidence that glutamic and aspartic acids function as excitatory synaptic transmitters in the central nervous system and that GABA and glycine act as inhibitory transmitters (see Curtis and Johnston, 1970, for a review). Taurine has also been shown to have an inhibitory action on certain neurons (Curtis and Watkins, 1960; Haäs and Hösli, 1973). In view of the changes in the amino acid content of whole brain of thiamine-deficient animals, especially the decreased levels of the excitatory amino acids glutamate and aspartate and the increased levels of the inhibitory amino acid, glycine, it seemed worthwhile to determine the effects of thiamine deficiency on the levels of these amino acids in cerebrum, cerebellum, and brain stem of weanling mice. Similar studies have been performed in rats (Butterworth, 1982; Plaitakis *et al.*, 1982). Amino acid concentrations in brain stem would be of particular interest. Although the location of pathologic lesions of thiamine deficiency is somewhat variable from

JEAN HOLOWACH THURSTON, RICHARD E. HAUHART, and JOHN A. DIRGO • Edward Mallinckrodt Department of Pediatrics, Division of Neurology, St. Louis Children's Hospital, St. Louis, Missouri 63178.   *DAVID B. McDOUGAL, JR.* • Department of Pharmacology, The Beaumont-May Institute of Neurology, Washington University School of Medicine, St. Louis, Missouri 63110.

species to species, the vestibular nucleus in the medulla (brain stem) seems to be affected in all of the species studied so far (Collins *et al.*, 1970). Collins *et al.* (1970) correlated regional anatomical and biochemical studies on non-amino acid metabolites in the brains of mice made thiamine deficient using pyrithiamine. In comparison to whole brain, metabolite aberrations were greater in areas susceptible to microscopic lesions and smaller in areas in which no lesions were found. These changes were found to occur quite abruptly late in the course of the development of the deficiency, and appeared to depend on relatively small differences in thiamine pyrophosphate levels (Seltzer and McDougal, 1974). In the work that forms the basis of the present report, mice made thiamine deficient by diet alone were used in addition to our standard model of pyrithiamine-assisted thiamine deficiency. Because of the difference in the rate of progress of the development of the deficiency in the two groups of animals, it was necessary to use the development of neurological signs as the indication for the time of tissue sampling. This ensured correspondence between the two groups and reproducibility within them. No statistically significant differences were found between the two groups and the pooled results are reported.

2. Methods

Thiamine deficiency was produced in weanling Swiss-Webster mice by means of a thiamine-free diet (Nutritional-Biochemical Corporation) with or without injection of pyrithiamine. Weight controls (not pair-fed controls) were fed the thiamine-deficient diet plus thiamine (20 μg per gram of diet). Other controls received thiamine plus the diet without dietary restrictions. To induce thiamine deficiency by the use of pyrithiamine, weanling mice received a basal diet consisting of 68% sucrose, 18% casein, 10% vegetable oil, and 4% USP salt mixture #2, supplemented with all vitamins except thiamine. Experimental animals received daily intraperitoneal injections of 50 μg of pyrithiamine plus 1 μg thiamine. Controls received intraperitoneal injections of 64 μg of thiamine daily. The mice were killed when signs of neurologic abnormality appeared: after 23 days in the animals on the thiamine-deficient diet and in seven days in those treated with pyrithiamine.

3. Distribution of Amino Acids in Normal Brain

In normal weanling mice, the regional distribution of glutamate was fairly uniform and similar to that seen in adult rats (Duffy *et al.*, 1972; Cutler and Dudzinski, 1974) (Table I). Values were highest in cerebrum, 14% less in cerebellum, and 26% less in brain stem. Aspartate values in cerebrum and cerebellum were almost identical (Table I). In brain stem, aspartate levels were 20% higher [a considerably smaller difference than that found in adult rats (Cutler and Dudzinski, 1974)]. The higher aspartate concentration in brain stem and the lower levels of glutamate in the same area are of interest. As in adult rats (Shaw and Heine, 1965; Shank and Aprison, 1970), glycine values were low in cerebrum, only 38% higher in cerebellum, but 300% higher in brain stem. GABA levels in

TABLE I. Regional Levels of Brain Amino Acids in Normal Weanling Mice[a]

Amino acid	Cerebrum	P value[b]	Cerebellum	P value[c]	Brain stem	P value[d]
Aspartate	3.22 ± 0.10 (14)	NS	3.26 ± 0.16 (14)	0.006	3.89 ± 0.13 (13)	< 0.001
Glutamate	12.49 ± 0.34 (14)	< 0.001	10.80 ± 0.25 (14)	0.002	9.27 ± 0.38 (13)	< 0.001
Glycine	0.89 ± 0.03 (10)	0.008	1.23 ± 0.11 (10)	< 0.001	3.59 ± 0.23 (13)	< 0.001
GABA	2.33 ± 0.09 (12)	< 0.001	1.70 ± 0.07 (6)	0.004	2.22 ± 0.12 (6)	NS
Taurine	10.87 ± 0.51 (5)	NS	10.52 ± 0.76 (5)	< 0.001	5.85 ± 0.26 (5)	< 0.001
Glutamine	7.10 ± 0.34 (10)	NS	7.62 ± 0.26 (10)	0.003	6.18 ± 0.34 (10)	NS

[a] After decapitation the head was allowed to fall directly into liquid nitrogen with rapid stirring. Cerebellum, cerebrum and brain stem (from intercollicular sulcus to obex) were dissected in a cryostat at −35 °C. Tissue samples (20 to 25 mg) were powdered and weighed in a cold room at −22 °C and perchloric acid extracts were prepared (Lowry and Passonneau, 1972). Aspartate and glutamate were measured by the methods of Lowry and Passonneau (1972); glycine by the method of Berger et al. (1975); and GABA by a minor modification of the method of Hirsch and Robins (1962). Taurine was measured by the fluorescamine method of Orr et al. (1976). The method of glutamine assay was that of Young and Lowry (1966). Amino acid levels in the different groups of controls were not significantly different, therefore results were pooled. Values in mmole/kg; mean ± SE. The number of animals is given in parentheses. NS = not significant (p > 0.05).

[b] Cerebrum versus cerebellum.

[c] Cerebellum versus brain stem.

[d] Brain stem versus cerebrum.

cerebrum and brain stem were similar. In cerebellum, values were 25% higher. Similarly, in normal adult rats, brain GABA levels in these three regions did not appear to differ greatly (Cutler and Dudzinski, 1974; Shaw and Heine, 1965; Shank and Aprison, 1970). The levels of taurine in the cerebrum and cerebellum of normal weanling mice are extraordinary; values were almost as high as those for glutamate. As others have noted (Cutler and Dudzinski, 1974; Shaw and Heine, 1965; Shank and Aprison, 1970), taurine levels in normal brain stem were only half as high.

4. Effects of Thiamine Deficiency

Compared with controls, thiamine deficiency produced highly significant reductions of aspartate and glutamate concentrations in all of the three regions of brain examined (Fig. 1). Decreases in aspartate levels were greater than those in glutamate and reductions of both were somewhat greater in brain stem than in cerebrum or cerebellum. In contrast, glycine levels in brain stem were somewhat elevated (22%), whereas cerebral and cerebellar levels of glycine were not affected by thiamine deficiency (Fig. 1). In all three regions, levels of GABA and taurine were not significantly different from the control values.

The amino acid levels shown in Table I were determined by specific enzymic procedures (see legend, Table I). Levels of GABA were also determined in the same samples in the laboratory of Dr. Ralph Bradshaw, Department of Biochemistry, by the use of an automatic amino acid analyzer. Again, no regional differences in GABA levels were seen in thiamine deficient mice when compared with controls.

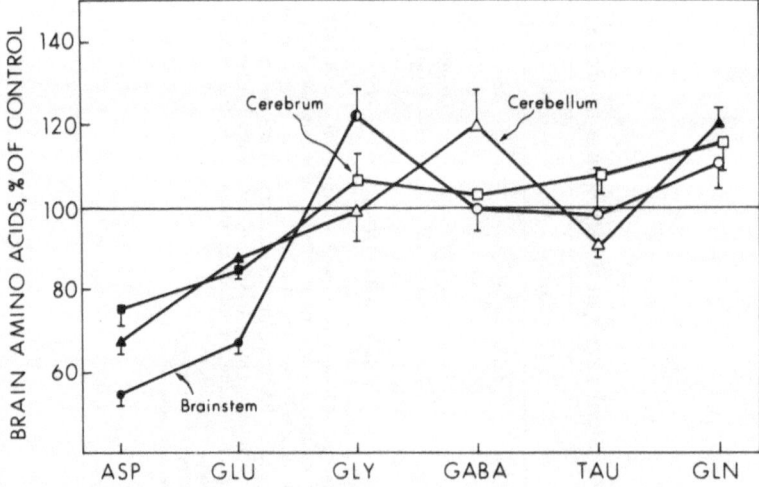

FIGURE 1. Amino acid levels in three regions of brains of the pyrithiamine-treated, thiamine-deficient mice. The values are expressed as percentage of control levels (Table I). Each point represents the mean value of 5 to 25 mice. The length of the vertical line represents ± SE. Values marked with a filled symbol are significantly changed from the control value at $P < 0.001$. Glycine in brain stem was significantly changed at $P = 0.015$. Values marked with an open symbol were not significantly changed ($P > 0.05$).

5. Discussion

There were some notable differences between the changes we found in weanling mice and those reported by Hamel *et al.* (1979) in thiamine-deficient adult rats. In the weanling mice glycine levels increased significantly in brain stem and were unchanged in cerebellum, whereas in the adult animals glycine levels decreased in both medulla oblongata and cerebellum. In the adult animals glutamine levels were reduced in cerebellum; in the weanlings the glutamine content of cerebellum was significantly increased. Furthermore, although taurine levels were reduced in the cerebellum of the thiamine-deficient adult rat, no significant changes were seen in the taurine content of any of the regions examined in the young mice.

Abnormalities of the amino acid content in the thiamine-deficient weanling mice followed the same trend as that seen for pyruvate, α-ketoglutarate, and 6-P-gluconate concentrations: changes were greatest in those areas of brain shown to be most consistently and most severely vulnerable to pathologic lesions by thiamine deficiency (Collins *et al.*, 1970). An exception to this rule was glutamine; values increased 20% in cerebellum, but brain stem levels were unchanged.

The reason for the decreases in glutamate and aspartate in whole brain or in regions of the brains of thiamine-deficient mice is not clear. In view of the greatly elevated pyruvate and α-ketoglutarate levels resulting from a decrease in the activities of the corresponding dehydrogenases (Holowach *et al.*, 1968), one might expect to see increases in brain alanine and glutamate concentrations. In other studies, alanine in whole brain was unchanged by thiamine deficiency (Gaitonde and Fayein, 1975; J. H. Thurston *et al.*, unpublished experiments) and glutamate was, in fact, reduced, as many have shown.

It is tempting to attach etiologic significance to the 20% to 50% decreases in aspartate and glutamate levels (excitatory neurotransmitters) and the increase in glycine (an inhibitory neurotransmitter) in the area of the brain most vulnerable to thiamine deficiency—the brain stem. However, in whole brain, Gaitonde and Fayein (1975) found that marked changes in these amino acids were not associated with neurotoxicity unless GABA and serine levels were significantly reduced as well. Serine levels were not measured in our study, but in whole brain (unpublished data) and in the cerebrum, cerebellum, and brain stem (present study), GABA levels were unchanged despite profound neurologic disturbance (Fig. 1). Gubler *et al.* (1974) concluded that the moderate decrease in the brain GABA level was not responsible for the neurologic disturbance since the change occurred in both thiamine-deprived and pyrithiamine-treated rats, but only the latter exhibited the classical neurologic signs. Hamel *et al.* (1979) found a 35% decrease in the cerebellar GABA content of adult thiamine-deficient rats, but no changes were seen in medulla or midbrain.

Despite these inconsistencies, it seems probable that the changes in the two potential excitatory neurotransmitter amino acids and one, or perhaps two, inhibitory ones are involved in producing the neurologic picture in thiamine-deficient animals. It may be that studies of metabolic compartmentation of these amino acids in thiamine-deficient animals would help to resolve the problems involved in their participation in the pathogenesis of the neurologic changes observed. However, there is as yet no way to measure the fraction of an amino acid involved in its function as transmitter separately from that which is involved in energy or protein metabolism, although some notable progress has been made, in the case of glutamate, for example. Since glycinergic synapses appear to be rare in cerebrum (see Snyder, 1975, for review), the 1 mmole/kg found there must be enough to carry out the necessary functions of cell metabolism. The excess over 1 mmole/kg found in other areas, notably brain stem and spinal cord (Aprison et al., 1969), may be available for use in synaptic transmission in these regions. This amount, 2.5 mmole/kg in brain stem, more in the case of spinal cord (Aprison et al., 1969), seems very high for a neurotransmitter. Furthermore, it is certainly possible that regional differences in such multifunctional compounds as amino acids may occur in response to tissue demands other than those of synaptic transmission. However that may be, proper assignment of the changes in amino acid concentrations occurring in thiamine deficiency, the elevation of brain stem glycine, and the decreases in brain stem aspartate and glutamate to a synaptic or other role will be necessary before a definitive interpretation of the meaning of these changes to the function of the organism can be made.

It should perhaps be mentioned that the enormous changes in the state of oxidation of the pyridine nucleotides and/or the intracellular pH implied by the unbalanced increases in both pyruvate and α-ketoglutarate found in thiamine-deficient brain (Holowach et al., 1968) must have far-reaching metabolic reverberations. To date, the changes have not been directly assessed, nor have the reverberations been systematically explored. It has seemed more economical to pursue the changes in transmitters, cyclic nucleotides, and the like than to describe the metabolic perturbations one by one as they extend "outward" from the thiamine-deprived enzymes. Whether this will in fact prove the quickest way to shed light on the unsolved mechanism of neurologic dysfunction in thiamine deficiency cannot be said.

References

Aprison, M. H., Shank, R. P., Davidoff, R. A., and Werman, R., 1969, The distribution of glycine, a neurotransmitter suspect, in the central nervous system of several vertebrate species, *Life Sci.* **7**:583–590.

Berger, S. J., Carter, J. A., and Lowry, O. H., 1975, An enzymatic method for glycine, *Anal. Biochem.* **65**:232–240.

Butterworth, R. F., 1982, Regional amino acid neurotransmitter distribution in thiamine deficiency, *Ann. N.Y. Acad. Sci.* **378**:464–465.

Collins, R. C., Kirkpatrick, J. B., and McDougal, D. B., Jr., 1970, Some regional pathologic and metabolic consequences in mouse brain of pyrithiamine-induced thiamine deficiency, *J. Neuropathol. Exp. Neurol.* **29**:57–69.

Curtis, D. R., and Watkins, J. C., 1960, The excitation and depression of spinal neurons by structurally related amino acids, *J. Neurochem.* **6**:117–141.

Curtis, R. C., and Johnston, G. A. R., 1970, Amino acid neurotransmitters, in: *Handbook of Neurochemistry*, Volume 4 (Lajtha, A., ed.), Plenum Press, New York, pp. 115–134.

Cutler, R. W. P., and Dudzinski, D. S., 1974, Regional changes in amino acid content in developing rat brain, *J. Neurochem.* **23**:1005–1009.

Duffy, T. E., Nelson, S. R., and Lowry, O. H., 1972, Cerebral carbohydrate metabolism during acute hypoxia and recovery, *J. Neurochem.* **19**:959–977.

Ferrari, V., 1957, The metabolic changes in thiamine deficiency as reflected on the individual free amino acids in tissues. I. Observations on brain, *Acta Vitaminol. Enzymol.* **11**:53–56.

Gaitonde, M. K., and Fayein, N. A., 1975, Decreased metabolism *in vivo* of glucose into amino acids of the brain of thiamine-deficient rats after treatment with pyrithiamine, *J. Neurochem.* **24**:1215–1223.

Gaitonde, M. K., and Nixey, R. W. K., 1974, The effect of deficiency of thiamine on the metabolism of [U-^{14}C]glucose and [U-^{14}C]ribose and the levels of amino acids in rat brain, *J. Neurochem.* **22**:53–61.

Gubler, C. J., Adams, B. L., Hammond, B., Yuan, E. C., Guo, S. M., and Bennion, M., 1974, Effect of thiamine deprivation and thiamine antagonists on the level of γ-aminobutyric acid and on 2-oxoglutarate metabolism in rat brain, *J. Neurochem.* **22**:831–836.

Haäs, H. L., and Hösli, L., 1973, The depression of brain stem neurones by taurine and its interaction with strychnine and bicuculline, *Brain Res.* **52**:399–402.

Hamel, E., Butterworth, R. F., and Barbeau, A., 1979, Effect of thiamine deficiency on levels of putative amino acid transmitters in affected regions of the rat brain, *J. Neurochem.* **33**:575–577.

Hirsch, H. E., and Robins, E., 1962, Distribution of γ-aminobutyric acid in the layers of the cerebral and cerebellar cortex. Implications for its physiological role, *J. Neurochem.* **9**:63–70.

Holowach, J., Kauffman, F., Ikossi, M. G., Thomas, C., and McDougal, D. B., Jr., 1968, The effects of a thiamine antagonist, pyrithiamine, on levels of selected metabolic intermediates and on activities of thiamine-dependent enzymes in brain and liver, *J. Neurochem.* **15**:621–631.

Lowry, O. H., and Passonneau, J. V., 1972, *A Flexible System of Enzymatic Analysis*, Academic Press, New York, pp. 121–128, 146–218.

Orr, H. T., Cohen, A. I., and Lowry, O. H., 1976, The distribution of taurine in the vertebrate retina, *J. Neurochem.* **26**:609–611.

Plaitakis, A., Hwang, E. C., Van Woert, M. H., Szilagyi, P. I. A., and Berl, S., 1982, Effect of thiamine deficiency on brain neurotransmitter systems, *Ann. N.Y. Acad. Sci.* **378**:367–380.

Seltzer, J. L., and McDougal, D. B., Jr., 1974, Temporal changes of regional cocarboxylase levels in thiamine-depleted mouse brain, *Am. J. Physiol.* **227**:714–718.

Shank, R. P., and Aprison, M. H., 1970, The metabolism *in vivo* of glycine and serine in eight areas of the rat central nervous system, *J. Neurochem.* **17**:1461–1475.

Shaw, R. K., and Heine, J. D., 1965, Ninhydrin positive substances present in different areas of normal rat brain, *J. Neurochem.* **12**:151–155.

Snyder, S. H., 1975, The glycine synaptic receptor in the mammalian central nervous system, *Br. J. Pharmacol.* **53**:473–484.

Young, R. L., and Lowry, O. H., 1966, Quantitative methods for measuring the histochemical distribution of alanine, glutamate and glutamine in brain, *J. Neurochem.* **13**:785–793.

16

Alcohol-Induced Encephalopathy

RODERICK K. ROBERTS, ANASTACIO M. HOYUMPA, JR.,
GEORGE I. HENDERSON, and STEVEN SCHENKER

1. Introduction

Ethyl alcohol through its effects on cerebral function has been a source of recreation to mankind for centuries and this has changed little over recent years. Despite a major effort, an understanding of the mechanism(s) by which ethanol exerts its central nervous system effects eludes medical research. This probably relates more to the complexity and inaccessibility of the human brain than to the nature of ethanol and its actions. The heterogeneity of the brain in terms of both macroanatomy and microanatomy confounds efforts at understanding the effects of agents such as ethanol. Are the changes observed in a given anatomical location following ethanol administration a direct consequence of ethanol effect, a change in electrical input to the region as a consequence of remote effects of ethanol, an effect of metabolites of ethanol, or a consequence of some more generalized effect such as altered blood flow, oxygen supply, or temperature change (Kalant, 1975)? Are any effects seen specific for ethanol or common to various neurodepressants? Are the changes seen causal or consequences of some other more basic effect? Questions such as these render research in this area both challenging and frustrating.

The present review will deal initially with the clinical aspects of ethanol on the brain, separating acute intoxication from the sequelae of chronic ethanol ingestion. An attempt will then be made to relate these clinical observations to experimental data to give some understanding of the pathogenesis of ethanol-related syndromes. Although the clinical discussion implies a constellation of signs and symptoms that are characteristic of ethanol intoxication, there is a similarity between responses to ethanol and those to other psychoactive drugs.

RODERICK K. ROBERTS, ANASTACIO M. HOYUMPA, JR., GEORGE I. HENDERSON, and STEVEN SCHENKER • Departments of Medicine and Pharmacology, The University of Texas Health Science Center at San Antonio; and Research Service, Audie L. Murphy Memorial Veterans' Hospital, San Antonio, Texas 78284.

In fact, there is evidence that users may not be able to distinguish ethanol effects from those of other agents when appropriately blinded to the drug administered (Jones and Stone, 1970). In addition, although this introduction has implied that ethanol itself induces the clinical syndrome, the role of metabolites of ethanol will also be discussed. Fortunately, ethanol is a simple molecule with a well-described metabolic fate (Lieber, 1982), which simplifies this aspect of the review. Finally, the management of the various neurological disorders associated with ethanol ingestion will be discussed.

2. Clinical Effects of Ethanol on the Brain

2.1. Acute Effects

The acute effects of ethanol on the brain are well known to most readers. These effects may be conceived as an initial "excitement" stage and later a stage of slowed mentation that may progress to coma. Although there is a rough correlation between the acute effect and the level of blood ethanol, factors such as prior exposure, genetic and racial characteristics, rate of increase of blood ethanol levels, and psychological factors modify response (Vessell, 1972; Kalant et al., 1971; Wolff, 1972). Concomitant administration of other drugs, acting pharmacokinetically [at the site(s) of absorption, distribution, and/or degradation of ethanol] and pharmacodynamically at the cerebral level may increase or decrease the effects of ethanol. The latter considerations are important both with therapeutic drug usage and in the setting of intentional or rarely accidental drug overdose, where concomitant ethanol ingestion is frequently a complicating factor.

Ethanol early in the course of intoxication induces feelings of exhilaration, but judgment, concentration, and insight wane. As a consequence of this, confidence rises, expressed as loudness and talkativeness, and emotional control diminishes. Perhaps fortunately, short-term memory is also often impaired. As blood alcohol levels rise further, central nervous system depression and eventually stupor and then coma may develop. The physical signs consist initially of facial flushing and a rise in pulse rate. Spinal reflexes are heightened, but motor performance diminishes. Pupils may be dilated. Progressively, nystagmus, often diplopia, and dysarthria develop and gait is impaired. Most of these manifestations are present at blood alcohol levels of 150–200 mg/100 ml. Further increases in blood alcohol lead to diminution and then loss of spinal reflexes, reduced perfusion with hypotension and hypothermia, and pupillary constriction, progressing in the terminal stages of ethanol overdose to fixed dilated pupils. Many of the above effects are common to a number of anesthetic agents acting presumably via the reticular activating system and exerting effects on the hypothalamus.

2.2. Chronic Effects

With chronic ingestion of ethanol in substantial quantities, several related clinical syndromes may develop. The expression of these will depend on the length of exposure and average intake of ethanol, as well as individual susceptibility. Tolerance is observed when repeated administration of ethanol leads to a decreasing effect for a given dose (Fraser, 1957) and is a well-documented phenomenon that is lost after several months of abstinence (Kalant et al., 1971; Isbell et al., 1955; Mirsky et al., 1941; Newman, 1941). The mechanisms may be both pharmocokinetic and pharmacodynamic (Kalant et al., 1971). A detailed discussion of this complex phenomenon is beyond the scope of the present review, but several comments are appropriate. Although ethanol tolerance has a pharmacokinetic component, the pharmacodynamic (central) component is probably more important in terms of the final cerebral response to a given total dose of ethanol. Pharmacodynamically, tolerance seems to have early and late components that can be distinguished in animal studies (Victor and Adams, 1953). The development of the central component of tolerance will be further discussed when mechanisms of ethanol action are considered. A very dramatic consequence of long-term ethanol abuse is delirium tremens, a manifestation of physical dependence and withdrawal (Victor and Adams, 1953; Behnke, 1976). On the other hand, the most devastating, often lasting, consequence is organic brain disease.

The classic triad of delirium tremens is delirium, tremulousness, and hallucinations, but the degree of each varies according to the stage of the syndrome and the individual patient. The fully developed syndrome does not always occur and the milder manifestions are best termed the ethanol withdrawal syndrome. The clinical expression is best appreciated by a description of the progression of the withdrawal syndrome. Approximately 8 hr after the cessation or the reduction of ethanol intake, mild tremulousness, anorexia, nausea, and anxiety develop. The tremor is coarse and although it may fluctuate markedly, it is generally worse on purposeful movement. There may be some disorientation on careful interview, but the sufferer appears inappropriately alert. A mild tachycardia is usually present. Over the ensuing 12 to 48 hr but occasionally over several (up to 8) days, hyperactivity and tremor may become exaggerated; the patient has insomnia and nightmares and delusions or hallucinations may develop. Grand mal convulsions may occur and are a distinguishing feature of worsening withdrawal. If the withdrawal syndrome progresses, it may be expressed in its most severe form, delirium tremens. This occurs between three and five days after cessation of ethanol intake, but has occurred as late as 12 days. Confusion, delusions, agitation, and tremor are severe, and vivid hallucinations occur. There is autonomic hyperactivity evidenced by tachycardia, dilated pupils, diaphoresis, and fever. There are no focal neurological signs and respiratory alkalosis and hypomagnesemia are common. Cerebrospinal fluid lactate may be increased (Behnke, 1976; Brooks and Adams, 1975). The syndrome of delirium tremens characteristically runs a course of four days or less and at least in recent years has carried a mortality of under 10% (Thompson et al., 1975). Mortality, however, is higher when there is a high fever, convulsions, severe gastrointestinal disturbance, liver disease, pneumonia (aspiration), or pancreatitis. Old age and chronic disease also worsen prognosis.

The organic brain disease consequent on chronic ethanol ingestion has no distinguishing features. There is progressive memory disturbance, loss of intellectual capacity, and altered behavior and emotional responses (Freund, 1973). There may be tremor, diminished pupillary response to light and dysarthria. Whereas the relationship between ethanol intake and the ethanol withdrawal syndromes is clear, the same is not true of the organic brain syndrome.

Other factors may influence brain function in the chronic ethanol abuser including poor nutrition, repeated head injury, advancing age, and other disease processes such as atherosclerosis. Careful observation of selected groups of alcoholics, however, suggests that ethanol alone may lead to irreversible brain damage (Freund, 1973). Additionally, the cerebral atrophy seen in alcoholics on computed tomography is at least partially reversible with abstinence (Carlen et al., 1978). Similarly, a prolonged cerebrospinal fluid acidosis that weakly correlates with intellectual impairment is present in recently abstinent alcoholics (Carlen et al., 1980). There are also studies in rodents that suggest an independent effect of chronic ethanol on brain function (Freund, 1970; Freund and Walker, 1971; Walker and Freund, 1971), although there is debate as to the significance of the observed changes.

Although the clinical syndromes associated with ethanol ingestion are well described, the mechanisms underlying these various syndromes are poorly understood. In the following sections, current knowledge of the pathogenesis of both the acute and chronic effects of ethanol on the brain will be reviewed.

3. Mechanism of Action of Ethanol

3.1. General Concepts

A unifying hypothesis as to the effects of ethanol on the brain must reconcile the different responses to acute and chronic administration, the development of tolerance, and the response to ethanol withdrawal. When interpreting data from experimental studies, particularly of the acute effects of ethanol, some appreciation of the pharmacokinetics of ethanol is important. Ethanol is a small, highly water-soluble molecule (partition coefficient 0.1, lipid: water), which normally is completely absorbed. Absorption rate depends on the rate of gastric emptying, presence of foodstuff in the stomach, and congeners in the alcoholic beverage consumed. First-pass elimination is significant when ethanol concentrations in portal blood are low and this may relate to the dose-dependence of ethanol metabolism. Once absorbed, ethanol distributes in total body water. Equilibration between blood and a given organ will be a function of the water content of the organ and its blood flow. The relatively high blood flow to the brain and the facility with which ethanol crosses the blood–brain barrier may result in brain ethanol concentrations exceeding those in peripheral venous blood during the rapid postabsorption rise in ethanol blood concentration. Ethanol metabolism is a complex process involving a cytosolic alcohol dehydrogenase, a microsomal ethanol-oxidizing system, and possibly a catalase (Lieber, 1982). Virtually all metabolism occurs in the liver and specifically there is little evidence for a quantitatively significant ethanol oxidation in the brain (Raskin and Sokoloff, 1968), although such oxidation may be important microregionally. It is unlikely that greater understanding of ethanol metabolism will shed further light on the nature of its actions on the brain.

When analyzing the potential mechanisms by which ethanol may modify brain function, the following considerations are important. Is ethanol itself or its metabolites, or both, responsible for clinical syndromes? Does ethanol have an effect generally on brain function, for example, through membrane changes or an effect on intermediary metabolism, or is the major influence exerted in a localized area or areas? What is the specific nature of ethanol-induced changes at a cellular level?

3.2. Role of Ethanol versus Metabolites

There is circumstantial evidence that acetaldehyde per se does not mediate the acute effects of ethanol on the brain. The behavioral responses to ethanol intoxication follow blood ethanol levels closely (Goldberg, 1943). As indicated above, there is also little evidence for significant acetaldehyde production by alcohol dehydrogenase locally in the brain (Raskin and Sokolof, 1968), and a search for indirect evidence of this by measurement of the NADH to NAD ratio has yielded conflicting results with one study (Rawat et al., 1973) reporting an increase and two other studies (Veloso et al., 1977) using more sophisticated techniques observing no change. Modern methods indicate also that blood acetaldehyde levels after ethanol ingestion are very low. Furthermore, if the conversion of ethanol to acetaldehyde is blocked by pyrazole, the neurological response to ethanol remains (Goldstein and Pal, 1971). The effects of pyrazole itself on the brain are quite separate from, but additive to, those of ethanol (Goldberg et al., 1972). Another line of evidence pointing to the importance of ethanol over acetaldehyde as the mediator of neurological effect is the effect of other alcohols. When alcohol was administered orally to produce maximum intoxication, 2.5–3.0 g of tertiary butanol/kg body weight produced a similar degree of impairment as 6–9 g of ethanol/kg body weight daily. When tertiary butanol or ethanol were administered to rats over a three-week period in an animal model of withdrawal, the withdrawal syndromes following both alcohols were similar. As tertiary butanol is not oxidized to acetaldehyde, these data suggest that acetaldehyde plays no major role in acute ethanol intoxication or withdrawal. Interestingly, tertiary butanol produced a more pronounced diuresis and weight loss than ethanol (Wallgren, 1973). Further evidence against the role of acetaldehyde in acute intoxication is provided by the disulfiram reaction. This reaction was first observed by serendipity (Hald et al., 1948) and is characterized by severe gastrointestinal symptoms, flushing, tachycardia, and intense anxiety. Later, marked hypotension may occur. These effects, except for the hypotension which may be caused by the disulfiram, have been ascribed to the sympathomimetic effects of acetaldehyde (Eade, 1959) that is present in high concentration during concomitant administration of ethanol and disulfiram (Asmussen et al., 1948). Whether or not acetaldehyde is responsible for all these responses, the clinical syndrome after acetaldehyde administration is quite different from that produced by acute ethanol intoxication. Because of the superficial similarities of the effect described after acetaldehyde administration to the ethanol withdrawal syndrome, a role for acetaldehyde in this syndrome has been postulated (Ortiz et al., 1974). However, as stated earlier, the evidence that tertiary butanol, which is metabolized to formaldehyde, produces a similar withdrawal syndrome to that of ethanol argues against this (Wallgren et al., 1974).

It is also possible that acetaldehyde may exert an indirect effect on brain function. Aldehydes are known to generate nonenzymatic condensation products with catecholamines to form alkaloids. For example, dopamine in vitro has been shown to condense with its own aldehyde to form tetrahydropapaveroline (Davis and Walsh, 1970) and with acetaldehyde to form salsolinol (Cohen and Collins, 1970), both being examples of tetrahydroisoquinoline alkaloids (TIQs). At least two hypotheses attempt to relate TIQs to the effect of ethanol administration and withdrawal. First, TIQs may act as false neurotransmitters competing centrally and peripherally for catecholamine receptors, resulting in an adaptive (compensatory) increase in catecholamine synthesis. A sudden decrease in TIQ concentration might then lead to overstimulation of the catecholamine receptor, producing some of the manifestations of withdrawal. There is some supportive evidence for this concept based on measurement of brain norepinephrine turnover in ethanol-dependent rats (Hunt and Majchrowicz, 1974a) and in the exaggeration of the withdrawal reaction in ethanol-dependent mice following TIQ infusion into CSF (Blum et al., 1976a).

A second hypothesis suggests that the more complex TIQs (such as tetrahydropapaveroline), which are precursors of the opium alkaloids may per se, because of their similarity to morphine, induce dependence and withdrawal (Davis and Walsh, 1970; Santi et al., 1967; Turner et al., 1974). Although this is an attractive hypothesis, it has major problems. It is possible to identify salsolinol or other TIQs after perfusion of bovine adrenal with ethanol, acetaldehyde, and dopamine (Cohen and Collins, 1970; Cohen, 1971, 1973) or by blocking catechol-o-methyl transferase with pyrogallol in ethanol-dependent rats (Collins and Bigdel; 1975), as well as in the brain of rats (Turner et al., 1974) and the urine of patients receiving L-dopa (Sandler et al., 1973). However, ethanol administration per se has not resulted in detectable levels of salsolinol in brain despite fairly sensitive assay methodology (limit of 8 ng/g) (O'Neill and Rahwan, 1977). To be sure, even this approach may not be sensitive enough when one considers the data of Melchior et al. (1978), wherein very small amounts of acetaldehyde or TIQs injected into the lateral ventricle of rats induced a marked ethanol preference and a withdrawal state when ethanol administration ceased. There are two other pieces of indirect evidence against an important role for TIQs in ethanol intoxication, dependence, and withdrawal. The data for tertiary butanol that is not metabolized to acetaldehyde has already been cited (Wallgren, 1973). In addition, the opiate antagonist naloxone does not produce the same withdrawal reaction in ethanol-dependent mice as it does in morphine-dependent mice (Goldstein and Judson, 1971).

Thus, the role of acetaldehyde and its condensation products relative to that of ethanol itself remains uncertain. There is little evidence for its role in acute intoxication and uncertain and controversial evidence for a role in tolerance, dependence, and withdrawal.

3.3. Specific Effects of Ethanol

3.3.1. Cerebral Blood Flow and Energy Metabolism

Cerebral blood flow, the blood–brain barrier, and cerebral energy metabolism are closely interrelated in terms of the supply of oxygen, as well as transport to and utilization by neurons of essential nutrients. Administration of large doses of ethanol to the rat reduces cerebral blood flow and oxygen and glucose utilization (Hemmingsen and Barry, 1979; Nielsen et al., 1975). It has been more difficult to demonstrate similar changes with smaller doses of ethanol in man (Battey *et al.*, 1953), although minor redistributions in blood flow patterns have been shown in some studies (Newlin *et al.*, 1982; Shimojyo *et al.*, 1967; Berglund and Risberg, 1981). Ethanol has also been shown to induce spasm in rat cerebral arterioles *in vivo* and in isolated canine basilar and middle cerebral arteries (Altura *et al.*, 1983). Although it is possible that such acute effects of ethanol play a role in the pathogenesis of cerebrovascular accidents in man (particularly where the cerebral circulation is already compromised), it seems unlikely that such effects explain the entire spectrum of ethanol-induced brain malfunction. Reduction in cerebral blood flow, both total and regional, may affect the delivery of oxygen and essential nutrients (predominantly glucose) to the brain, although the relationship between regional blood flow and oxygen and glucose utilization is not as close as one might expect (Lassen *et al.*, 1978; Reivich *et al.*, 1975). The evolution of newer techniques for studying brain metabolic function such as ^{31}P nuclear magnetic resonance may shed light on regional differences in response to ethanol (Prichard *et al.*, 1968).

The data as to the effects of ethanol withdrawal on cerebral blood flow and metabolism are contradictory, perhaps a function of the timing of studies in relation to withdrawal (Kruger *et al.*, 1980). Whatever the explanation, ethanol withdrawal has been shown to both reduce cerebral metabolic rate (Eisenberg, 1968; Berglund and Risberg, 1977, 1981) and to increase cerebral metabolism (Hemmingson *et al.*, 1979, 1980; Campbell *et al.*, 1982).

Long-term ethanol intake may lead to irreversible brain damage in which memory disturbance is a prominent feature. This observation focused attention on the effects of ethanol on cerebral protein metabolism (Noble and Tewari, 1977). Ethanol administered acutely or chronically depresses the incorporation of precursor amino acids into cerebral protein (Tewari and Noble, 1971; Jarlstedt, 1972; Jarlstedt and Hamberger, 1972), an effect that is accentuated during withdrawal (Tewari *et al.*, 1977). Ethanol may induce these changes at several points in the protein synthetic sequence. These changes include impaired transfer RNA synthesis (Tewari and Noble, 1971; Fleming *et al.*, 1975) and impaired polyribosomal function. The latter is consequent on abnormal synthesis and availability of messenger RNA, as well as a defect in polypeptide synthesis in the presence of normal amounts of messenger RNA (Noble and Tewari, 1977). Incorporation of [5-^{3}H]orotic acid into transfer, ribosomal, and polyribosomal RNA is decreased, nuclear release of RNA impaired, and polyadenylation at the periribosomal level reduced (Tewari and Noble, 1971).

In summary, the effect of ethanol on brain metabolism, during both intoxication and withdrawal, is complex. More research is required using ethanol concentrations closer to intoxicating levels seen in man, focusing on critical brain regions rather than the whole organ, and relating metabolism to blood flow. Whether the changes observed are a direct consequence of the effects of ethanol on the brain or are a nonspecific response to ethanol-induced depression of cerebral function remains to be resolved.

3.3.2. Membrane Effects

The initial observations that the potency of anesthesic agents could be related to their lipid solubility were made in the early part of this century (Overton, 1901; Meyer, 1937). The effectiveness of the alcohols generally as anesthetic agents increases with increasing thermodynamic activity coefficient relative to ethanol (Wallgren, 1973), this coefficient being related to molecular conformation and size. Membrane penetration by alcohols as measured by effectiveness as nerve blockers or as stabilizers of red blood cell membranes is related to molecular size as well (Seeman, 1972). These observations suggest that ethanol (and other alcohols and anesthetic agents) could exert their central effects by a general interaction with cell membranes to increase fluidity. This physical change in membrane structure may lead to disruption of membrane-associated proteins responsible for transmission of impulses with consequent impairment of function. When sophisticated techniques, such as nuclear magnetic resonance, capable of determining changes in membrane structure (Metcalfe et al., 1968) were applied, it became clear that alcohols indeed could alter membrane fluidity. Electron paramagnetic resonance confirms that biological membranes are largely fluid (McConnell et al., 1972) and that ethanol increases the degree of fluidity of a variety of membranes including those of mammalian cells (Chin and Goldstein, 1977; Curran and Seeman, 1978). The fact that relatively high concentrations of ethanol are necessary to achieve a biological effect such as coma also suggests an effect on membranes generally rather than interaction with a specific receptor. Another technique for assessing membrane fluidity, fluorescence polarization, also suggests an effect of acute ethanol to increase membrane fluidity (Johnson et al., 1979).

It will clearly be difficult to design in vivo studies to validate the membrane fludity hypothesis. However, there are several lines of evidence based on in vitro studies that are seemingly confirmatory. Indirect evidence supporting an effect of ethanol and anesthetic agents on membrane structures comes from the use of pressure to reverse anesthesia. Increased pressure (up to 200 atmospheres) will reverse anesthesia in experimental animals. For example, treatment with hyperbaric oxygen reverses ethanol narcosis in mice (Alkana and Malcolm, 1980). It is known from in vitro studies that high pressure forces model membranes back into order (Trudel et al., 1973). It is worth pointing out, however, that the fluidity changes seen are small and the techniques for measuring these are relatively new so that while the membrane hypothesis is attractive it remains far from proven. In fact, not all in vitro studies support the membrane hypothesis. The magnitude of membrane changes induced by intoxicating doses of ethanol may be similar to those caused by normal daily body temperature changes that clearly do not alter the level of consciousness (Johnson et al., 1980). In addition, when the effects of ethanol, other anesthetic agents, and increased hydrostatic pressure on action potential and synaptic transmission in rat cervical ganglion were studied, results were conflicting. Ethanol reduced the action potential height, an effect that was antagonized by increased pressure; however, both ethanol and pressure independently decreased synaptic transmission with effects that were additive (Kendig et al., 1975).

The membrane hypothesis of alcohol action has been extended further in an attempt to explain the effects of ethanol administered chronically, in particular on the development of tolerance and the withdrawal reaction. As with other hypotheses developed to explain chronic ethanol effects, to be discussed below, that relating to the membrane effect of ethanol is a variant of the concept of homeostatic or adaptive response (Kalant et al., 1971). This concept argues that when alterations in membrane fluidity occur, homeostatic mechanisms are brought into play in an attempt to correct these changes. Microorganisms have been extensively studied in relation to membrane responses to environmental perturbations. *Escherichia coli* incorporates more unsaturated fatty acid into its membrane at low than at high temperatures (Marr and Ingraham, 1962). That this phenomenon apparently allows the membrane to function normally at the ambient temperature may be inferred from viscosity measurements (Sinensky, 1974). This is apparently accomplished by changing substrate specificity of an acyl transferase with change in temperature. There are examples of changes in membrane fatty acid composition in fish (Roots, 1968) and fat composition of aquatic mammals (Henriques and Hansen, 1901), which are also temperature dependent. Similar adaptive responses occur to ethanol. *Escherichia coli* responds to ethanol by adaptive changes in the saturated/unsaturated fatty acid ratio in a direction that would increase rigidity (Ingram, 1976; Buttke and Ingram, 1978). *Escherichia coli* membrane may not be an appropriate model for the membrane effects of ethanol in man, however, since it lacks cholesterol and has a lower proportion of unsaturated fatty acid than mammalian membrane. In addition, alcohol concentrations used in these studies were higher than those in most mammalian experiments. However, when three applications of ethanol were made to synapses of the abdominal ganglion of *Aplysia californica*, the change in posttetanic potentiation was diminished when compared with acute ethanol and this effect persisted for 11 hr after the last ethanol treatment (Traynor et al., 1976). Similar adaptive changes were seen in the phrenic nerve terminals of rats after long-term ethanol treatment (Curran and Seeman, 1977). When rats were rendered ethanol tolerant, erythrocyte and brain synaptosomal membranes were more resistant to ethanol-induced fluidization, suggesting adaptation (Chin and Goldstein, 1977). There is evidence that the mechanisms for such adaptation involve changes in membrane structure. When Swiss mice are treated chronically with ethanol, there is an increase in the ratio of saturated to unsaturated fatty acid in the membrane which thus becomes more rigid and more resistant to the fluidization effects of acute doses of ethanol (Littleton et al., 1979). If the temperature at which phase transition of the membrane occurs (i.e., between fluid and gel phases) is compared in sucrose control and ethanol-tolerant rat brain microsomes, preparations from ethanol-tolerant animals are more resistant to the fluidizing effect of ethanol than in controls (Rangaraj and Kalant, 1982). In contrast, chronic ethanol treatment caused an even greater decrease in phase transition temperature than controls *in vitro*, implying a more fluid membrane in these animals (Levental and Tabakoff, 1980; Rangaraj and Kalant, 1982). In these experiments, phase transition temperature was measured by change in activity of sodium potassium ATP transferases. One potential explanation for this discrepancy may be that changes in membrane fluidity at the site of fluorescence or electron-spin resonance probes are different from those in the immediate vicinity of these enzymes. Changes in reactive groups of membranes have also been seen in response to ethanol. Brain microsomes from rats treated chronically (6–8 weeks) with ethanol contained more reactant sulfhydryl groups than those from rats treated acutely, although there was no change in the protein composition or total number of sulfhydryl groups. This suggested a conformational change of some kind (Gruber *et al.*, 1977). Lastly, cultured hamster astroblasts released increased amounts of sialic acid after prolonged ethanol exposure, suggesting some modification of surface glycoprotein (Noble *et al.*, 1976). Similarly, two weeks of ethanol treatment is reported to cause a 70% increase in exposed sialic acid in rat brain membranes (Ross *et al.*, 1977), and in a model of ethanol tolerance and dependence in mice, an increase in surface glycoprotein with and without sialic acid can be demonstrated (Goldstein *et al.*, 1983).

The evidence taken in its entirety suggests that ethanol acutely may induce membrane changes in the direction of increased fluidity. Adaptive responses may then occur with chronic ethanol to increase membrane rigidity, perhaps by an increase in the saturated fatty acid components or to reduce the effect of any given concentration of ethanol by changes in surface components or reactive groups. This may be a possible explanation for tolerance. Once ethanol is withdrawn, the membrane shifts to a more excitable state because of these adaptive changes. This may be the counterpart, at a cellular level, of the withdrawal syndromes. Clearly, further correlative work will be needed to verify this interpretation.

3.3.3. Effects on Electrolyte Transport

3.3.3a. NA$^+$ and K$^+$ Transport and (NA$^+$,K$^+$)ATPase. The physiological function of neurons is intimately related to their electrical properties, in particular to the transport of cations by the neuronal membrane. In turn, cation transport by the cell membrane is closely related to energy metabolism and membrane structure so that changes in these have the potential for inducing changes in ion flux. When the brain is exposed acutely to ethanol, three main effects are seen: (1) there is a change in NA$^+$ and K$^+$ flux across the neuronal membrane, (2) (Na$^+$,K$^+$)-stimulated ATPase activity is depressed, and (3) calcium is released from neuronal membranes.

Several studies have demonstrated Na$^+$ and K$^+$ movement across membranes through discrete channels during excitation, with the Na$^+$ channel regulated from outside of the cell (Hille, 1970; Rojas and Armstrong, 1971). Ethanol in concentrations of 500 mg/100 ml (Wallgren *et al.*, 1974) and 1 and 2 g/100 ml (Israel *et al.*, 1975) depressed the entry of Na$^+$ into the membrane interior with little K$^+$ loss during electrical stimulation. Similarly, ethanol affects ion flux during action potentials in squid axons (Armstrong and Binstock, 1964; Moore *et al.*, 1964), lobster axons (Houck, 1969), isolated *Aplysia* ganglion (Bergmann *et al.*, 1974), mouse cerebellar explants (Calvet *et al.*, 1974), the isolated sympathetic ganglion (Montoya *et al.*, 1977), rat brain cortical slices (Seir *et al.*, 1977; Sunahara and Kalant, 1980), and cultured spinal cord neurons (Gruol, 1980). The levels of ethanol in these studies are much higher than those seen in alcohol intoxication in man, but they do suggest an effect on the outer surface of the membrane so as to inhibit Na$^+$ flux and thus the generation of the action potential. However, when dissociated dorsal root ganglion neurons from embryonic rats were exposed to concentrations of ethanol varying from 50 to 300 mg/100 ml (levels seen in intoxication in man), there was no effect seen on resting membrane potential or action potentials. At the lower ethanol concentrations, the major effect was during repolarization. (Na^+,K^+)-stimulated ATPase functions predominantly at the stage of repolarization and active ion transport, and there are a number of studies demonstrating an effect of ethanol on this enzyme. For example, ethanol administered acutely inhibits microsomal and synaptosomal (Na^+,K^+)-activated ATPase in animal (Israel et al., 1965) and human brain (Sun and Samorajski, 1975). A similar effect on other tissues has been demonstrated, e.g., in rat gut (Hoyumpa et al., 1977). Ethanol might act by a membrane effect to induce conformational changes in the enzyme (Grisham and Barnett, 1972; Kimelberg and Papahadjopoulas, 1974; Kalant et al., 1978). These data and more recent studies (Rangaraj and Kalant, 1982) indicate that changes in fluidity alone may not entirely account for kinetic changes in the enzyme and that some change in protein structure may be involved as well. The changes that occur in cation transport and in (Na^+,K^+)-activated ATPase in chronic ethanol ingestion differ from those during acute ethanol ingestion. For example, whereas acute ethanol consumption depresses (Na^+,K^+)ATPase activity in brain, as discussed above when ethanol is withdrawn after chronic ingestion, an increase in activity of this enzyme occurs in some studies (Israel et al., 1970; Wallgren, 1975; Knox et al., 1972; Rangaraj and Kalant, 1978), although others have not confirmed these findings (Akera et al., 1973; Goldstein and Israel, 1972). It may be that changes in (Na^+,K^+)ATPase, if they do occur, are secondary to stress-related catecholamine release in ethanol withdrawal rather than any adaptive response (Rangaraj and Kalant, 1978). There is a need for further study of this enzyme as well as of Na^+, K^+, and Ca^{2+} fluxes in ethanol tolerance and withdrawal, in resting and stimulated states, and in regional brain tissue.

3.3.3b. Effect of Ca^{2+} Flux. Calcium ions have an important function in membrane structure and conductance and regulate neurotransmitter release (Haycock and Meligeni, 1977). The effect of ethanol and several other drugs on Ca^{2+} distribution and flux have been studied and the results are, in part conflicting. Ethanol and morphine, but not pentobarbital, cause a depletion of regional brain Ca^{2+} and the effect of ethanol is dose related. Salsolinol has a similar action. This effect of morphine and salsolinol is blocked by naloxone (Ross et al., 1974). Although acute morphine has been shown to decrease synaptosomal Ca^{2+}, chronic morphine administration causes an increase (Yamamoto et al., 1978). Similarly, chronic administration of ethanol increases membrane-bound Ca^{2+} while decreasing Ca^{2+} binding to synaptosomes (Ross, 1977; Michaelis and Myers, 1979). The observations suggest an adaptive response to chronic ethanol administration. Not all studies support these observations. For example, other investigators have failed to confirm a decrease in whole brain or regional calcium after acute ethanol administration (Ferko and Bobyock, 1980; Hood and Harris, 1979). More recently, the effects of ethanol on synaptosomal Ca^{2+} influx has been investigated in an attempt to relate changes in Ca^{2+} flux to neurotransmitter alterations. In two reports, ethanol is reported to reduce synaptosomal Ca^{2+} uptake at alcohol concentrations down to 50 mM (Ross, 1980; Harris and Hood, 1980, Ross and Garrett, 1983). In contrast, a third study demonstrated enhanced synaptosomal Ca^{2+} uptake after longer ethanol exposure. Further investigation will be required to resolve these differences and relate these changes to alterations in flux of other ions, especially K^+, and membrane changes, as well as neurotransmitter status (Friedman *et al.*, 1980).

3.3.4. Neurotransmitter Alterations

Ethanol could alter brain function by an effect on neurotransmitters. One of the difficulties in reviewing this subject is the profusion of research publications in the area of functional neurochemistry generally and particularly in relation to the effects of ethanol (Noble and Tewari, 1977). Contradictions in results observed at times may be related to differing doses and duration of ethanol use, methods of brain preparation, choice of whole brain or regional areas, species differences, and whether the studies in question measure simply single concentrations of transmitter or turnover rates. Changes in status of an individual neurotransmitter could be secondary to alterations in another critical transmitter, changes in electrolyte status, or of membranes. Where data are available, each neurotransmitter will be dealt with individually.

3.3.4a. Acetylcholine. Perhaps the best understood of the neurotransmitters influenced by ethanol is acetylcholine. Interestingly, the effect of ethanol on the acetylcholine synapse contrasts peripherally and centrally. There is evidence that peripherally ethanol stimulates release of acetylcholine from cholinergic terminals at the neuromuscular junction (Gage, 1965) and that this effect is subject to the development of tolerance (Curran and Seeman, 1978). Centrally, the effect of ethanol in the acetylcholine terminal is quite different. *In vivo*, ethanol raises brain acetylcholine and decreases its utilization in rats (Parker *et al.*, 1977). Similarly, in cat brain acetylcholine release *in vivo* is diminished by ethanol with the fairly rapid development of tolerance in this system occurring over 30 min (Sinclair and Lo, 1978). *In vitro* ethanol, in usually intoxicating concentrations, inhibits the efflux of acetylcholine from electrically stimulated rat cortical slices (Israel *et al.*, 1975). It is generally agreed that these effects of ethanol are secondary since they are shared by a number of drugs including central nervous system depressants. In addition, such changes in acetylcholine turnover are not accompanied by predictable behavioral changes (Graham and Erickson, 1974).

3.3.4b. The Biogenic Amines. The biogenic amines, including the catecholamines, serotonin, and dopamine, as well as GABA, act centrally as neurotransmitters. The effect of ethanol on these various transmitters differs and they will be discussed separately.

3.3.4c. Catecholamines. There is very little change in norepinephrine or dopamine content of whole brain or brain regions (where these have been studied) after acute administration of ethanol. The early report of a decrease in levels (Gursey and Olsen, 1960) was not substantiated by later work (Hunt and Majchrowicz, 1974a; Rawat, 1974; Pohorecky, 1974; Carlsson *et al.*, 1973), although striatal dopamine levels rise transiently after ethanol administration

(Wajda *et al.*, 1977). Using o-methyl-p-tyrosine to inhibit tyrosine hydroxylase, or by measuring metabolites, catecholamine turnover (assumed to parallel neuronal activity) can be calculated. If catecholamine turnover is measured after ethanol given in a single large dose, there is no early effect (within 15 min) on dopamine turnover, but there is an acceleration of norepinephrine turnover (Corrodi *et al.*, 1966). At a later time, both dopamine and norepinephrine turnover are depressed (Hunt and Majchrowicz, 1974a). There may also be regional differences in the acute effect of ethanol on catecholamine turnover in that reduction in dopamine levels is greatest in the caudate nucleus and substantia nigra (Pohorecky and Brick, 1977; Bacapoulos *et al.*, 1978). Chronic ethanol administration has been reported to induce varying changes in the catecholamines, depending apparently on duration of ethanol use. Three to five days of ethanol treatment produced no change in dopamine or norepinephrine levels, but increased dopamine turnover in the withdrawal period (Hunt and Majchrowicz, 1974a). In other studies, ten days of ethanol treatment produced either no change (Griffiths *et al.*, 1974; Littleton *et al.*, 1974; Ahtec and Svartström-Fraser, 1975) or a marked increase in norepinephrine and dopamine content (Ortiz *et al.*, 1974). Ethanol administration for periods of several weeks produces either no change or a depression in whole brain content (Rawat, 1974), although such treatment is reported to increase striatal levels (Chopde *et al.*, 1977); twelve months of ethanol treatment is reported to increase regional dopamine and norepinephrine content (Post and Sun, 1973). When catecholamine turnover is measured during chronic ethanol administration, it is generally found to be increased (Hunt and Majchrowicz, 1976; Pohorecky, 1976; Pohorecky *et al.*, 1976; Karoum *et al.*, 1976). The withdrawal period after chronic ethanol administration is associated with inconsistent changes in catecholamines. As well as the changes described above, ethanol withdrawal is reported to cause no initial change in norepinephrine content after 10 hr of cessation of intake, but a rise after 20 and 30 hr (Pohorecky *et al.*, 1978). Dopamine turnover may be accelerated in the prodromal period of withdrawal, but falls during the fully developed withdrawal syndrome (Hunt and Majchrowicz, 1974a; Karoum *et al.*, 1976). When cerebral norepinephrine content was lowered by 50% using 6-hydroxy-dopamine the degree of withdrawal was unchanged (Tabakoff and Ritzmann, 1977), whereas potentiation of withdrawal has been seen with norepinephrine depletion (Goldstein, 1973). Finally, it has been suggested that increased α-adrenergic receptor binding in rat brain contributes to the hyperadrenergic state in ethanol withdrawal (Banerjee *et al.*, 1978).

Overall, it is difficult to ascribe a primary role to the catecholamines in mediating the events in ethanol intoxication or withdrawal. Generally, increased adrenergic activity is found in withdrawal, but it is unclear if this is a secondary phenomenon or is causally related.

3.3.4d. Serotonin. Ethanol does not alter serotonin content or turnover in brain in any consistent way. Most studies fail to show any change in brain content of serotonin after chronic (Rawat, 1974; Hunt and Majchrowicz, 1974b; Kuriyama *et al.*, 1971; Tabakoff and Boggan, 1974; Frankel *et al.*, 1974) or acute (Rawat, 1974; Hunt and Majchrowicz, 1974b; Tabakoff and Boggan, 1974) ethanol administration. When serotonin turnover is measured, results are conflicting, there being either a decrease (Hunt and Majchrowicz, 1974b) or increase (Palaic *et al.*, 1971) or no change (Kuriyama *et al.*, 1971). When serotonin content in brain is depleted, the ethanol withdrawal reaction in rats is diminished, but this may simply be a nonspecific depressant effect (Wood, 1980). Interestingly, tryptophan hydroxylase, the rate-limiting enzyme in serotonin synthesis, is normally unsaturated so that serotonin levels are determined by free tryptophan content in blood perfusing brain. Recent studies in man and laboratory animals indicate an absolute decrease in tryptophan and a decrease relative to amino acids competing with it for transport into the brain in chronic alcohol ingestion (Branchey *et al.*, 1981).

Finally, there has been continuing interest in the role of serotonin in the development of tolerance. Depletion of brain serotonin levels tends to reduce ethanol consumption by rats, whereas elevated levels of serotonin are found regionally in strains of ethanol-preferring rats (Myers, 1978). However, the data in this area as a whole are conflicting, making interpretation difficult. For example, the administration of p-chlorophenylalanine, which depletes serotonin stores impairs the development of tolerance in rats chronically fed ethanol (Frankel *et al.*, 1974). However, methysergide, a serotonin antagonist, increases ethanol sleep times (Blum *et al.*, 1974), although this drug does potentiate withdrawal convulsions (Blum *et al.*, 1976b). The role of serotonin in the development of ethanol tolerance in experimental animals and man remains, therefore, unclear.

3.3.4e. GABA. Gamma-aminobutyric acid acts electrophysiologically to hyperpolarize neurons (Curtis and Johnson, 1974) and to depolarize primary afferent terminals (Levy, 1977). It is generally conceived as predominantly an inhibitory neurotransmitter. Its central actions appear closely linked with those of the benzodiazepines, barbiturates, and possibly ethanol to the GABA–benzodiazepine–chloride–ionophore receptor complex. There is also a large body of evidence linking GABA with seizure phenomena (Wood, 1975). These observations have suggested a possible role for GABA in mediating some of the effects of ethanol. Research in this area has produced results that are at times conflicting and for a complete review of this data the reader is referred to Hunt (1983). In summary, there is evidence that ethanol may exert its actions via an effect on the GABAergic neurons, although exactly how this occurs is unclear. It is possible that by a membrane effect ethanol modifies the GABA–benzodiazepine–chloride–ionophore receptor complex. Such a hypothesis would relate the effect of the benzodiazepines and barbiturates, both of which have actions at this membrane site, to those of ethanol and perhaps explain the cross-tolerance observed. There is evidence that barbiturate withdrawal and ethanol withdrawal both reduce GABA levels, that ethanol and benzodiazepines act similarly at a molecular level, and that each of these drugs have similar presynaptic actions (Hunt, 1983). Further research, however, is needed before the complex nature of these interactions can be understood.

3.3.4f. Endorphins and Encephalins. These opiatelike peptides are thought to influence dopaminergic activity in the corpus striatum (Pert, 1978).

Morphine increases levels of dopaminergic transmitters in the striatum (Urwyler and Tabakoff, 1981), but this effect is attenuated in ethanol-withdrawn animals (Tabakoff et al., 1981). A related and intriguing observation by one group is that naloxone blocks ethanol-induced intoxication (Jeffcoate et al., 1979). Although these are early observations, they point to the need for further investigation of the role of endogenous opiates in mediating the effects of ethanol.

3.3.4g. Cyclic Nucleotides. There seems little alteration in cyclic AMP concentration regionally with either the acute or chronic administration of ethanol (Redos et al., 1976), although some regional differences have been reported by other laboratories (Volicer and Hunter, 1977). Although a single dose of ethanol does not effect adenylate cyclase activity (Kuriyama and Israel, 1973), one study on mouse brain demonstrates reduced activity of this enzyme after chronic ethanol administration (Kuriyama, 1977).

In contrast to cyclic AMP, there is a clear effect of ethanol on cyclic GMP. Ethanol in single doses produces a dose-related reduction in cyclic GMP levels, especially in cerebellum (Volicer and Hunter, 1977). There may be partial development of tolerance with rebound of cyclic GMP levels on withdrawal (Volicer and Hunter, 1977), although this effect was not seen by all investigators (Hunt et al., 1977).

3.3.5. Conclusions

The significance of these observations regarding the cyclic nucleotides is unclear. Further research is also needed in the area of neurotransmitters generally and their interactions with ethanol and other centrally active drugs before an understanding of their role in the observed clinical response emerges.

4. Treatment of Ethanol-Related Syndromes

4.1. Treatment of Acute Intoxication

The management of acute ethanol intoxication will vary widely according to the particular circumstance and will depend on the severity of central nervous system symptoms (particularly respiratory depression) and the presence of major accompanying medical illnesses, the associated ingestion of other drugs (especially sedatives), and the patient's home environment.

When intoxication is moderate, it may only be necessary to protect the patient and others from him or herself, and in general no drug therapy is necessary. When intoxication is severe and there is evidence of coma, respiratory depression, or hypotension, in-patient management is necessary. The severely intoxicated patient is at risk of head injury with short- and long-term sequelae, as well as of other complications of long-term alcohol abuse such as cirrhosis with hepatic encephalopathy and the Wernicke-Korsakoff syndrome. Various biochemical abnormalities may contribute to the altered mental state. Malnourished individuals or those who have been fasting are prone to hypoglycemia. Hyponatremia is not an uncommon biochemical abnormality in the intoxicated chronic alcoholic and this may lead to depression of cerebral function.

Management will depend on identifying and correcting biochemical abnormalities, providing adequate nursing care when the patient is comatose, especially to prevent pulmonary aspiration, and providing respiratory support where appropriate. Rarely, some attempt may be made to increase alcohol elimination. Hemodialysis or peritoneal dialysis have been used particularly where alcohol levels are very high, where there is the presumed or proved presence of other dialyzable toxins, and where there is associated severe acidosis. Attempts at accelerating ethanol elimination by increasing its hepatic oxidation have generally not been effective. Glucocorticoids, insulin, thyroxin, epinephrine, and fructose have been used. Only fructose in large doses enhanced ethanol elimination and then only to a small extent. This and the undesirable effects of hyperuricemia, lactic acidosis, and gastrointestinal upset following fructose limit its use. An alternative approach is to attempt to modify the pharmacodynamic effect of ethanol so as to reduce the effect of a given blood level of ethanol on the brain. Although several substances have been tested in experimental animals and in man and some have yielded encouraging results, including physostigmine, aminophylline, and L-dopa (Alkana *et al.*, 1977), a functionally important inhibition of the cerebral effects of high blood alcohol has not as yet been demonstrated in man.

4.2. Management of Ethanol Withdrawal

The clinical supportive care has been described earlier. The patient must be carefully assessed for the possible ingestion of other sedatives that may lead to withdrawal syndromes, for trauma (especially to the head), and for associated diseases, particularly cardiovascular, respiratory, hepatic, or pancreatic. Laboratory assessment will include a search for electrolyte disturbance, hypoglycemia, and evidence of liver disease. Examination of the cerebral spinal fluid is often necessary as are sputum and urine cultures, depending on clinical status. Acid base balance must be assessed and a search for other potential toxins initiated. The causes of death in patients with delirium tremens include cardiovascular collapse, hyperpyrexia, infection, especially of the central nervous system, arrythmias, and occasionally fat emboli when trauma has occurred (Thompson, 1978; Tavel *et al.*, 1961).

Supportive care of the patient with severe delirium tremens includes management of the risk factors identified above and correction of any specific biochemical abnormalities detected. Fluid replacement in the patient with delirium tremens should be based on the degree of dehydration and must be individualized, but will usually amount to 4 to 6 liters during the first 24-hr period. The choice of fluid will be dictated by the patient's electrolyte status with due recognition of electrolyte disturbances that may complicate the presentation. Hypokalemia is common and may be worsened by administration of intravenous glucose; replacement according to biochemical measurements is necessary and may be of the order required in diabetic ketoacidosis. Severe magnesium deficiency should be treated by intravenous magnesium sulphate

2 g every 6 to 8 hr during the initial 24 hr. Care should be taken in the patient with renal insufficiency as toxicity may develop rapidly, with evidence of coma and hypotension. A practical point is that this salt should not be mixed with bicarbonates. Phosphate depletion is common in chronic alcoholics (Knochel, 1977) and requires correction if severe (< 1 mg/100 ml). It may be treated orally with a 3:8 (w/w) mixture of $NaH_2PO_4.Na_2HPO_4$ 15 to 30 ml three times each day orally, or if necessary, as KH_2PO_4/K_2HPO_4 [(2.74:1 w/w) = 2 nmole of phosphorus/ml)] intravenously. Hypoglycemia may require correction by intravenous glucose. This should be given in conjunction with thiamine hydrochloride, 100 mg/day intravenously, since in the absence of thiamine a latent deficiency syndrome may be expressed in the presence of glucose repletion. Such patients usually have multiple vitamin deficiencies and a multivitamin preparation is usually administered prophylactically either orally or intravenously.

4.3. Drug Therapy

Until the exact pathophysiology of ethanol interaction with the central nervous system is known, it is not possible to rationally prescribe for delirium tremens, i.e., to specifically reverse the consequences of long-term alcohol usage. Ethanol is theoretically attractive as treatment and anecdotally effective, but has a low margin of safety and its use is undesirable in the overall management of alcoholism. Alcohol use in such patients likewise may interfere with long-term attempts to treat alcoholism. It has, however, been studied in the treatment of alcohol withdrawal both orally (Smith, 1953; Golbert et al., 1967) and intravenously in man (Lereboullet et al., 1960).

The relationship between the delirium tremens and cessation or reduction of chronic ingestion of ethanol was suggested in the early studies of Victor and Adams (1953), a hypothesis that was subsequently tested by others (Isbell, 1955; Mendelson and LaDou, 1964) and in large part confirmed. However, the incidence of fully developed delirium tremens was low in these studies, an observation corresponding to the experience of most centers. Since the latter part of the last century when chloral hydrate and paraldehyde were first used in patients suffering the ethanol withdrawal syndrome, almost 150 agents have been used, most being of the ethanol/barbiturate group. However, it was several years before it became apparent that an essential element in management was replacement of ethanol by another central nervous system depressant. The principle of drug therapy has remained that of first replacing ethanol by a sedative drug and then gradually reducing the dosage of the latter. Although this approach has proved effective for controlling the early and less severe manifestations of ethanol withdrawal, delirium tremens itself is not as readily controlled (Hemmingsen et al., 1979). There are a plethora of clinical trials on the drug therapy of alcohol withdrawal but few stand up to critical analysis (Moskowitz et al., 1983). Psychoactive drugs used in the treatment of ethanol withdrawal syndromes fall into three groups: (1) sedative/hypnotics mostly of the alcohol/barbiturate type (including benzodiazepines, paraldehyde, barbiturates, ethanol, and clormethiazole), as well as benactyzine and hydroxyzine; (2) the phenothiazines and related compounds so effective in managing major psychoses of other etiologies; and (3) anticonvulsants, the best example of which is phenytoin.

4.3.1. The Benzodiazepines

Since their introduction in the early 1960s, the benzodiazepines have become the most frequently prescribed drugs in the ethanol withdrawal syndrome (Favazza and Martin, 1974). The benzodiazepines, particularly chlordiazepoxide and diazepam have been demonstrated to be superior to placebo and to phenothiazine in controlled studies. In a large study comparing chlordiazepoxide, chlorpromazine, hydroxyzine, thiamine, and placebo (Kaim *et al.*, 1969), chlordiazepoxide administration was shown to be clearly superior to all other treatment in preventing the onset of delirium tremens and convulsions. Chlorpromazine was associated with an unacceptably high seizure rate as will be discussed below. On the other hand, it has not been possible to clearly demonstrate superiority of chlordiazepoxide (or other sedatives of the benzodiazepine type) over ethanol, barbiturates, or a combination of paraldehyde and chloral hydrate (Golbert *et al.*, 1967). In fact, the latter combination proved marginally superior, although no clear difference between chlordiazepoxide and paraldehyde alone could be demonstrated in a subsequent study (Kaim and Klett, 1972). Similarly, no clear difference emerged between diazepam and paraldehyde in terms of the development of delirium tremens (Thompson *et al.*, 1975), although two deaths occurred in patients receiving paraldehyde rectally. There are no strong arguments for the use of one benzodiazepine over another, although the onset of action of diazepam may be more rapid than of chlodiazepoxide even when both are given intravenously (Brown *et al.*, 1972).

Further comment about the benzodiazepines in two areas is appropriate to the present discussion. Liver disease is frequently a concomitant problem in the chronic alcoholic and is known to affect the disposition of many drugs including the benzodiazepines. Diazepam (Klotz *et al.*, 1975) and chlordiazepoxide (Roberts *et al.*, 1978) elimination is impaired in liver disease, whereas that of lorazepam (Kraus *et al.*, 1978) and oxazepam (Shull *et al.*, 1976) is largely unaffected. The latter drugs may on the surface seem preferable not only because their disposition is unaltered in the presence of liver disease, but also because their metabolites, unlike those of chlordiazposide and diazepam, are pharmacodynamically inert. However, oxazepam is not available for intravenous use and the dose of lorazepam is difficult to titrate against desired response. Finally, as long as the benzodiazepine chosen is used with discretion and the dose titrated to produce the desired effect of modest sedation, problems should be averted. Careful observations of the patient is essential and the drugs should not be administered on a *PRN* basis.

The second consideration is the pharmacokinetic interaction of the benzodiazepine with ethanol. Alcohol causes a moderate impairment of the elimination of diazepam (Sellers *et al.*, 1980), chlordiazepoxide (Desmond *et al.*, 1980), and lorazepam (Hoyumpa *et al.*, 1981). Conversely, chronic ethanol ingestion may nonspecifically enhance mixed functional oxidation. Again, discretion and careful monitoring of the patient should limit any unwanted effects.

4.3.2. Paraldehyde

Paraldehyde was for many years the mainstay of treatment of ethanol withdrawal states. This compound is a trimer of acetaldehyde, metabolized to CO_2 through acetaldehyde as intermediate (Zaleska and Gessner, 1983). It is a clear liquid with a characteristic pungent odor that is partially soluble in water, its solubility being greatest (12.8) v/v at 12 °C and decreasing above and below this point, being about 7.8% at 37 °C. The adverse reactions related to paraldehyde administration may relate to the fact that it was at times injected pure or as a 10% solution, to its relative instability in the presence of air and electromagnetic radiation (it oxidizes to acetic acid), and to an erratic and slow rectal and intramuscular absorption. Most studies of paraldehyde in alcohol withdrawal syndromes have found it to be as effective (particularly in combination with chloral hydrate) as the benzodiazepines. No study has evaluated intravenous paraldehyde, however, perhaps because of the toxic potential and little has been published on this drug in ethanol withdrawal since the late 1960s. Although questions remain, it may be that an effective drug has been maligned because of poor understanding of its pharmacy and pharmacokinetics. At present, it cannot be recommended as first-line treatment of ethanol withdrawal.

4.3.3. The Barbiturates

The barbiturates are considered by some to be the drug treatment of choice in the management of dependence on other drugs of the alcohol-barbiturate type (Smith and Wesson, 1970) and have been shown to be as or more effective than benzodiazepines in the management of ethanol withdrawal (Nielsen, 1965; Kramp and Rafaelsen, 1978). Their use is limited by the concern that in the alcoholic they may lead to barbiturate addiction. Whether these concerns are real or not is unclear, but it is unlikely that these drugs will replace benzodiazepines in the near future.

4.3.4. Chlormethiazole

Chlormethiazole, which has chemical similarities to thiamine and pharmacodynamic similarities to the ethanol-barbiturate group of drugs, is widely used in Europe for the treatment of ethanol withdrawal syndromes. In one study, chlormethiazole was as effective and perhaps more effective than chlordiazepoxide in the management of ethanol withdrawal (McGrath, 1975), and other evidence clearly shows its value in managing ethanol withdrawal. It does have the advantage of a relatively short duration of action, but the disadvantage of being an effective suicidal agent (Horder, 1977). Furthermore, because of its high hepatic clearance, bioavailability of orally administered drug may increase dramatically in the presence of significant liver disease, especially with extensive portal-systemic shunting. Even in the absence of liver disease, there will be a large increment between appropriate oral dose and intravenous dosage. Importantly, acute ethanol has been shown not to impair chlormethiazole elimination in man (Burg *et al.*, 1983).

4.3.5. Major Tranquilizers of the Phenothiazine Type

The major tranquilizers were first used for alcohol withdrawal in the 1950s and initially achieved widespread popularity (Laties *et al.*, 1958). However, it became apparent with time that these drugs may lower the seizure threshold and that they were associated with a higher mortality in delirium tremens than the benzodiazepines, paraldehyde, or, in fact, ethanol itself (Golbert *et al.*, 1967; Thomas and Friedman, 1964). Several other subsequent studies confirmed these findings, and it is clear that phenothiazines are absolutely contraindicated in the patient at risk for severe withdrawal syndromes. They should probably also be avoided in milder syndromes as it remains difficult to predict which patient will progress to the more severe manifestations of withdrawal.

4.3.6. Anticonvulsants

The use of anticonvulsants in the patient with ethanol withdrawal is probably unnecessary in milder withdrawal states unless there is a past history of withdrawal seizure. However, in severe withdrawal states, seizures are not uncommon and worsen the prognosis. The evidence that pharmacotherapy can avert such seizure activity is controversial. It is probable that in those without a prior history of seizure activity, benzodiazepines alone are sufficient. Phenytoin is not effective in controlling barbiturate withdrawal seizures in animals (Okamoto *et al.*, 1977; Essig and Carter, 1964; Gessner, 1974) or in man (Essig, 1966). There are two reports regarding its value in controlling ethanol withdrawal seizures in man. One study (Rothstein, 1973) excluded patients with seizure activity in the two weeks prior to entry and found that chlordiazepoxide and phenytoin administration was associated with no seizure activity, whereas historically 20% had previously experienced withdrawal seizures. When phenytoin and chlordiazepoxide were administered to patients with recurrent admissions for ethanol withdrawal usually associated with seizures, 2 of 70 experienced seizure, significantly lower than the incidence of 11 of 66 controls (Sampliner and Iber, 1974). Phenytoin levels in this study were below the therapeutic range usually accepted for the treatment of other forms of seizures. Primidone is reported to be more effective than phenytoin in controlling ethanol withdrawal seizures (Smith, 1976). There is also evidence that dipropylacetate (sodium valproate) may be effective, but the studies are not adequately controlled (BonLiglio *et al.*, 1977).

One recommended regimen for the drug therapy of ethanol withdrawal reactions consists of diazepam, 20 mg every 6 hr reducing over four days by 20 mg/day or chlordiazepoxide, 100 mg every 6 hr reducing by 100 mg/day over four days. Phenytoin should be administered orally as a loading dose of 1 g (reduced if the patient is under 50 kg) or infused intravenously at a dose of 10 mg/kg body weight with maintenance of 300–400 mg/day where there is a history of seizure (Sellers and Kalant, 1982). An alternative, simpler regimen has been proposed more recently (Sellers *et al.*, 1983).

5. Conclusion

It is evident from this brief review that the clinical features of alcohol intoxication and withdrawal are well defined and their treatment fairly well established. The organic brain syndrome that may follow chronic excessive ethanol consumption is much less clear-cut and its treatment, other than cessation of alcohol intake and repletion of any vitamin deficiencies, is far from satisfactory. The demonstration of reversal of cerebral abnormalities on CAT scanning in some chronic alcoholics with alcohol withdrawal is of major interest both as to pathophysiology of the disorder and hope for future therapeutic intervention.

The mechanism(s) of the effects of ethanol on the brain both acutely and chronically clearly are still incompletely understood. Perhaps the most exciting hypothesis relates to the membrane-altering effects of ethanol and its consequences on ion and neurotransmitter homeostasis. Even this area, however, requires more study as to cause–effect relationships. The difficulties in this field relate to lack of complete understanding of normal brain function in terms of its neurochemistry and physiology and to the heterogeneity and complexity of the brain. However, alcohol should be one of the easier neurodepressants to examine since its structure and metabolism are simple and excellent animal models of acute intoxication and withdrawal from ethanol are available.

Finally, it should be emphasized that a major concept, not discussed in this review, pertains to the predisposition of some individuals to alcoholism. This genetic proclivity has now been reproduced in experimental animals, verified in some human studies, and may, in the future, be testable by the use of evoked potential measurement and possibly other markers for developing alcoholism. The translation of this genetic predisposition mechanistically into alcoholism is likely to occupy our efforts and challenge our resources for years to come, and ultimately may be a most important key to the understanding of alcoholism.

ACKNOWLEDGMENTS. Supported by funds from Veterans Administration (Research Service Audie Murphy Memorial Veterans' Hospital) and NIH Grant (NIAAA 7R01AA05814-01).

References

Ahtee, L., and Svartström-Fraser, M., 1975, Effect of ethanol dependence and withdrawal on the catecholamines in rat brain and heart, *Acta Pharmacol. Toxicol.* **36**:289–298.

Akera, T., Rech, R. H., Marquis, W. J., Tobin, T., and Brody, T. M., 1973, Lack of relationship between brain (Na$^+$ + K$^+$)-activated adenosine triphosphatase and the development of tolerance to ethanol in rats, *J. Pharmacol. Exp. Ther.* **185**:594–601.

Alkana, R. L., and Malcolm, R. D., 1980, Antagonism of ethanol narcosis in mice by hyperbaric pressures of 4–8 atmospheres, *Alcohol. Clin. Exp. Res.* **4**:350–353.

Alkana, R. L., Parker, E. S., Cohen, H. B., Birch, H., and Noble, E. P., 1977, Reversal of ethanol intoxication in humans: An assessment of the efficacy of L-DOPA aminophylline and ephedrine, *Psychopharmacology* **55**:203–212.

Altura, B. M., Altura, B. T., and Gebrewo, A., 1983, Alcohol-induced spasms of cerebral blood vessels: Relation to cerebrovascular accidents and sudden death, *Science* **220**:331–333.

Armstrong, C. M., and Binstock, L., 1964, The effects of several alcohols on the properties of the squid giant axon, *J. Gen. Physiol.* **48**:265–277.

Asmussen, E., Hald, J., and Larsen, V., 1948, The pharmacological effects of acetaldehyde on the human organism, *Acta Pharmacol.* **4**:311–320.

Bacapoulos, N. G., Bhatnager, R. K., and Van Orden, III, L. S., 1978, The effects of subhypnotic doses of ethanol on regional catecholamine turnover, *J. Pharmacol. Exp. Ther.* **204**:1–10.

Banerjee, S. P., Sharma, V. K., and Khanna, J. M., 1978, Alterations in β-adrenergic receptor binding during ethanol withdrawal, *Nature* (*London*) **276**:407–409.

Battey, L. L., Heyman, A., and Patterson, J. L., Jr., 1953, Effects of ethyl alcohol on cerebral blood flow and metabolism, *J. Am. Med. Assoc.* **152**:6–10.

Beard, J. D., and Sargent, W. Q., 1979, Water and electrolyte metabolism following ethanol intake and during acute withdrawal from ethanol, in: *Biochemistry and Pharmacology of Ethanol*, Volume 2 (E. Majchrowicz and E. P. Noble, eds.), Plenum Press, New York, pp. 3–16.

Behnke, R. H., 1976, Recognition and management of the alcohol withdrawal syndrome, *Hosp. Pract.* **11**:79–84.

Berglund, M., and Risberg, J., 1981, Regional cerebral blood flow during alcohol withdrawal, *Arch. Gen. Psychiatry.* **38**:351–355.

Bergmann, M. C., Klee, M. R., and Faber, D. S., 1974, Different sensitivities to ethanol of three early transient voltage clam currents of *Aplysia* neurons, *Pflüg. Arch. Eur. J. Physiol.* **348**:139–153.

Blum, K., Wallace, J. E., Calhoun, W., Tabor, R. G., and Eubanks, J. D., 1974, Ethanol narcosis in mice: Serotonergic involvement, *Experientia* **30**:1053–1054.

Blum, K., Eubanks, J. D., Wallace, J. E., Schwertner, H., and Morgan, W. N., 1976a, Possible role of tetrahydroisoquinoline alkaloids in post-alcohol intoxication states, *Ann. N.Y. Acad. Sci.* **273**:234–246.

Blum, K., Wallace, J. E., Schwertner, H. A., and Eubanks, J. D., 1976b, Enhancement of ethanol-induced withdrawal convulsions by blockade of 5-hydroxytryptamine receptors, *J. Pharm. Pharmacol.* **28**:832–835.

BonLiglio, G., Falli, S., and Pacini, A., 1977, Proposta di introduzione nella practica ospedaliera di una nuova sostanza: il dipropilacetato di sodio nella prevenzione e terapia del delirium tremens alcoholico, *Minerva Med.* **68**:4233–4245.

Branchey, L., Shaw, S., and Lieber, C. S., 1981, Chronic ethanol feeding results in a decrease of brain tryptophan and serotonin, *Alcohol. Clin. Exp. Res.* **5**:144.

Brooks, B. R., and Adams, R. D., 1975, Cerebrospinal fluid acid-base and lactate changes after seizures in unanesthetized man. II. Alcohol withdrawal seizures, *Neurology* **25**:943–948.

Brown, J. H., Moggey, D. E., and Shane, F. H., 1972, Delirium tremens: A comparison of intravenous treatment with diazepam and chlordiazepoxide, *Scott. Med. J.* **17**:9–12.

Burg, R. W., Desmond, P. V., Mashford, M. L., Westwood, B., Shaw, P., and Breen, K. J., 1983, The effect of ethanol administration on the disposition and elimination of chlormethiazole, *Eur. J. Clin. Pharmacol.* **24**:383–385.

Buttke, T. M., and Ingram, L. O., 1978, Mechanism of ethanol-induced changes in lipid composition of *Escherichia coli*: Inhibition of saturated fatty acid synthesis *in vivo*, *Biochemistry* **17**:637–644.

Calvet, M. C., Drian, M. J., and Privat, A., 1974, Spontaneous electrical patterns in cultured Purkinje cells grown with an antimitotic agent, *Brain Res.* **79**:285–290.

Campbell, G. A., Eckardt, M. J., Majchrowicz, E., Marietta, C. A., and Weight, F. F., 1982, Ethanol-withdrawal syndrome associated with both general and localized increases in glucose uptake in rat brain. *Brain Res.* **237**:517–522.

Carlen, P. L., Wortzman, G., Holgate, R. C., Wilkinson, D. A., and Rankin, J. G., 1978, Reversible cerebral atrophy in recently abstinent chronic alcoholics measured by computed tomography scans, *Science* **200**:1076–1078.

Carlen, P. L., Kapur, B., Huszar, L. A., Lee, M. A., Moddel, G., Singh, R., and Wilkinson, D. A., 1980, Prolonged cerebrospinal fluid acidosis in recently abstinent chronic alcoholics, *Neurology* **30**:956–962.

Carlsson, A., Magnusson, T., Svensson, T. H., and Wardell, B., 1973, Effects of ethanol on the metabolism of brain catecholamines, *Psychopharmacologia* **30**:27–36.

Chin, J. H., and Goldstein, D. B., 1977, Effects of low concentrations of ethanol on the fluidity of spin-labelled erythrocyte and brain membranes, *Mol. Pharmacol.* **13**:435–441.

Chopde, C. T., Brahmankar, D. M., and Shripad, V. N., 1977, Neurochemical aspects of ethanol dependence and withdrawal reactions in mice, *J. Pharmacol. Exp. Ther.* **200**:314–319.

Cohen, G., 1971, Tetrahydroisoquinoline alkaloids in the adrenal medulla after perfusion with "blood concentrations" of ^{14}C acetaldehyde, *Biochem. Pharmacol.* **20**:1757–1761.

Cohen, G., 1973, Tetrahydroisoquinoline alkaloids uptake storage and secretion by the adrenal medulla and by adrenergic nerves, *Ann. N.Y. Acad. Sci.* **215**:116–119.

Cohen, G., and Collins, M., 1970, Alkaloids from catecholainines in adrenal tissue: Possible role in alcoholism, *Science* **167**:1749–1751.

Collins, M. A., and Bigdeli, M. G., 1975, Biosynthesis of tetrahydroisoquinoline alkaloids in brain and other tissues of ethanol-intoxicated rats, *Adv. Exp. Med. Biol.* **59**:79–91.

Corrodi, H., Fuxe, K., and Hökfelt, T., 1966, The effect of ethanol on the activity of central catecholamine neurons, *J. Pharm. Pharmacol.* **18**:821–823.

Curran, M., and Seeman, P., 1977, Alcohol tolerance in a cholinergic nerve terminal: Relation to the membrane expansion-fluidation theory of ethanol action, *Science* **197**:910–911.

Curran, M. J., and Seeman, P., 1978, Mechanisms of ethanol tolerance in a cholinergic nerve terminal, presented at a symposium on "Theories of Tolerance and Dependence on Ethanol: Mechanistic Approaches," July 7–11, Chexbres, Switzerland.

Curtis, D. R., and Johnson, G. A., 1974, Amino Acid transmitters in the mammalian central nervous system, *Ergebn. Physiol.* **69**:97–188.

Davis, V. E., and Walsh, M. J., 1970, Alcohol, amines and alkaloids: A possible biochemical basis for alcohol addiction, *Science* **167**:1005–1007.

Desmond, P. V., Patwardhan, R. V., Schenker, S., and Hoyumpa, A. M., 1980, Short-term ethanol administration impairs the elimination of chlordiazepoxide (Librium) in man, *Eur. J. Clin. Pharmacol.* **18**:275–278.

Eade, N. R., 1959, Mechanism of the sympathomimetic action of aldehydes, *J. Pharmacol. Exp. Ther.* **127**:29–34.

Eisenberg, S., 1968, Cerebral blood flow and metabolism in patients with delirium tremens. *Clin. Res.* **16**:71.

Essig, C. F., 1966, Non-narcotic addiction newer sedative drugs that can cause states of intoxication and dependence of barbiturate type, *J. Am. Med. Assoc.* **196**:714–717.

Essig, C. F., and Carter, W. W., 1962, Failure of diphenylhydantoin in preventing barbiturate withdrawal convulsions in the dog, *Neurology* **12**:481–484.

Favazza, A. R., and Martin, P., 1974, Chemotherapy of delirium tremens: A survey of physician's preferences, *Am. J. Psychiatry,* **131**:1031–1033.

Ferco, A. P., and Bobyock, E., 1980, A study on regional brain calcium concentrations following acute and prolonged administration of ethanol in rats, *Toxicol. Appl. Pharmacol.* **55**:179–187.

Fleming, E. W., Tewari, S., and Noble, E. P., 1975, Effect of chronic ethanol ingestion on brain amino-acyl-tRNA synthetases and tRNA, *J. Neurochem.* **24**:553–560.

Frankel, D., Khanna, J. M., Kalant, H., and LeBlanc, A. E., 1974, Effect of acute and chronic ethanol administration on serotonin turnover in rat brain, *Psychopharmacologia* **37**:91–100.

Fraser, H. E., 1957, Tolerance to and physical dependence on opiates, barbiturates, and alcohol, *Annu. Rev. Med.* **8**:427–440.

Freund, G., 1970, Impairment of shock avoidance learning after long-term alcohol ingestion in mice, *Science* **168**:1599–1601.

Freund, G., 1973, Chronic central nervous system toxicity of alcohol, *Annu. Rev. Pharmacol.* **13**:217–227.

Freund, G., and Walker, D. W., 1971, Impairment of avoidance learning by prolonged ethanol consumption in mice, *J. Pharmacol. Exp. Ther.* **179**:284–292.

Friedman, M. B., Erickson, C. K., and Leslie, S. W., 1980, Effects of acute and chronic ethanol administration on whole mouse brain synaptosomal calcium influx, *Biochem. Pharmacol.* **29**:1903–1908.

Gage, P. W., 1965, The effects of methyl, ethyl and n-propyl alcohol on neuromuscular transmissions in the rat, *J. Pharmacol. Exp. Ther.* **150**:236–243.

Gessner, P. K., 1974, Failure of diphenylhydantoin to prevent alcohol withdrawal convulsions in mice, *Eur. J. Pharmacol.* **27**:120–129.

Golbert, T. M., Sanz, C. J., Rose, H. D., and Leitshul, T. H., 1967, Comparative evaluation of treatments of alcohol withdrawal syndromes, *J. Am. Med. Assoc.* **201**:99–102.

Goldberg, L., 1943, Quantitative studies on alcohol tolerance in man, *Acta Physiol. Scand.* **5**(Suppl.):1–128.

Goldberg, L., Hollstedt, C., Neri, A., and Rydberg, U., 1972, Synergistic action of pyrazole on ethanol incoordination: Differential metabolic and central nervous system effects, *J. Pharm. Pharmacol.* **24**:593–601.

Goldstein, A., and Judson, B. A., 1971, Alcohol dependence and opiate dependence: Lack of a relationship in mice, *Science* **172**:290–292.

Goldstein, D. B., 1973, Alcohol withdrawal reactions in mice: Effects of drugs that modify neurotransmission, *J. Pharmacol. Exp. Ther.* **186**:1–9.

Goldstein, D. B., and Israel, Y., 1972, Effects of ethanol on mouse-brain (Na + K)-activated adenosine triphosphatase, *Life Sci.* **11**(II):957–963.

Goldstein, D. B., and Pal, N., 1971, Alcohol dependence produced in mice by inhalation of ethanol: Grading the withdrawal reactions, *Science* **172**:288–290.

Goldstein, D. B., Hungund, B. C., and Lyon, R. C., 1983, Increased surface glycoconjugates of synaptic membranes in mice during chronic ethanol treatment, *Br. J. Pharmacol.* **78**:008–010.

Graham, D. T., and Erickson, C. K., 1974, Alteration of ethanol-induced C.N.S. depression: Ineffectiveness of drugs that modify cholinergic transmission, *Psychopharmacologia* **34**:173–180.

Griffiths, P. J., Littleton, J. M., and Ortiz, A., 1974, Changes in monoamine concentrations in mouse brain associated with ethanol dependence and withdrawal, *Br. J. Pharmacol.* **50**:489–498.

Grisham, C. M., and Barnett, R. E., 1972, The interrelationship of membrane and protein structure in the functions of the (Na$^+$ + K$^+$)-activated ATP-ase, *Biochim. Biophys. Acta* **266**:613–624.

Gruol, D. L., 1980, Ethanol alters synaptic activity in cultured spinal cord neurons, *Brain Res.* **243**:25–33.

Gruber, B., Dinovo, E. C., Noble, E. P., and Tewari, S., 1977, Ethanol-induced conformational change in rat brain microsomal membranes, *Biochem. Pharmacol.* **26**:2181–2185.

Gursey, D., and Olsen, R. E., 1960, Depression of serotonin and norepinephrine levels in brain stem of rabbit by ethanol, *Proc. Soc. Exp. Biol. Med.* **104**:280–281.

Hald, J., Jacobsen, E., and Larsen, V., 1948, The sensitizing effect of tetraethylthiuram-disulphide (Antabuse) to ethyl alcohol, *Acta Pharmacol.* **4**:285–296.

Harris, R. A., and Hood, W. F., 1980, Inhibition of synaptosomal Ca^{2+} uptake by ethanol, *J. Pharmacol. Exp. Ther.* **213**:562–568.

Haycock, J. W., and Meligeni, J. A., 1977, Neurotransmitter accumulation and calcium dependent release from different regions of rat brain, *Life Sci.* **21**:1837–1844.

Hemmingsen, R., and Barry, D. I., 1979, Adaptive changes in cerebral blood flow and oxygen consumption during ethanol intoxication in the rat, *Acta Physiol. Scand.* **106**:249–255.

Hemmingsen, R., Kramp, P., and Rafaelsen, O. J., 1979, Delirium tremens and related clinical states. Aetiology, pathophysiology and treatment, *Acta Psychiatr. Scand.* **59**:337–369.

Hemmingsen, R., Barry, D. I., and Chapman, A. G., 1980, Cerebral blood flow and cerebral metabolic state during abstinence following ethanol intoxication in the rat, *Acta Neurol. Scand.* (suppl.) **78**:157–166.

Henriques, V., and Hansen, C., 1901, Vergleichonae untersuchungen uber die chemische zusammensetzung ans thierischen fette, *Skand. Arch. Physiol.* **11**:151–165.

Hille, B., 1970, Ionic channels in nerve membranes, *Prog. Biophys. Mol. Biol.* **21**:1–32.

Hood, W. F., and Harris, R. A., 1979, Effect of pentobarbital, ethanol and morphine on subcellular localization of calcium and magnesium in brain, *Biochem. Pharmacol.* **28**:3075–3080.

Horder, J. M., 1977, Fatal chlormethiazole poisoning in chronic alcoholics, *Br. Med. J.* **2**:614.

Houck, D. J., 1969, Effect of alcohols on potentials of lobster axons, *Am. J. Physiol.* **216**:364–367.

Hoyumpa, A. M., Nichols, S. G., Wilson, E. A., and Schenker, S., 1977, Effect of ethanol on intestinal (Na,K) ATP-ase and intestinal thiamine transport in rats, *J. Lab. Clin. Med.* **90**:1086–1095.

Hoyumpa, A. M., Patwardhan, R., Maples, M., Desmond, P., Johnson, R., Sinclair, A., and Schenker, S., 1981, Effect of short-term ethanol administration on lorazepam clearance, *Hepatology* **1**:47–53.

Hunt, W. A., 1983, The effect of ethanol on GABAergic transmission, *Neurosci. Behav. Rev.* **7**:87–195.

Hunt, W. A., and Majchrowicz, E., 1974a, Alterations in the turnover of brain norepinephrine and dopamine in alcohol-dependent rats, *J. Neurochem.* **23**:549–552.

Hunt, W. A., and Majchrowicz, E., 1974b, Turnover rates and steady-state levels of brain serotonin in alcohol-dependent rats, *Brain Res.* **72**:181–184.

Hunt, W. A., Redos, J. D., Dalton, T. K., and Catravas, G. N., 1977, Alterations in brain cyclic quanosine 3′: 5′ monophosphate levels after acute and chronic treatment with ethanol, *J. Pharmacol. Exp. Ther.* **201**:103–109.

Ingram, L. O., 1976, Adaptation of membrane lipids to alcoholics, *J. Bacteriol.* **125**:670–678.

Isbell, H., Fraser, H. F., Wikler, A., Belleville, R. E., and Eisenman, A. J., 1955, An experimental study of the etiology of "Rum Fits" and delirium tremens, *Q. J. Stud. Alcohol* **16**:1–33.

Israel, Y., Kalant, H., and Laufer, I., 1965, Effects of ethanol on Na, K, Mg-stimulated microsomal ATPase activity, Biochem. Pharmacol. 14:1803–1814.

Israel, Y., Kalant, H., LeBlanc, E., Bernstein, J. C., and Salazar, I., 1970, Changes in cation transport and (Na + K)-activated adenosine triphosphatase produced by chronic administration of ethanol, J. Pharmacol. Exp. Ther. 174:330–336.

Israel, Y., Carmichael, F. J., and MacDonald, J. A., 1975, Effects of ethanol on electrolyte metabolism and neurotransmitter release in the C.N.S., Adv. Exp. Med. Biol. 59:55–64.

Jarlstedt, J., 1972, Experimental alcoholism in rats: Protein synthesis in subcellular fractions from cerebellum, cerebral cortex and liver after long-term treatment, J. Neurochem. 19:603–608.

Jarlstedt, J., and Hamberger, A., 1972, Experimental alcoholism in rats. Effect of acute ethanol intoxication on the in vitro incorporation of [^{3}H]leucine into neuronal and glial proteins, J. Neurochem. 19:2299–2306.

Jeffcoate, W. J., Cullen, M. H., Herbert, M., Hastings, A. G., and Walder, C. P., 1979, Prevention of effects of alcohol intoxication by naloxone, Lancet 2:1157–1159.

Johnson, D. A., Lee, N. M., Cooke, R., and Loh, H. H., 1979, Ethanol-induced fluidization of brain lipid bilayers: Required presence of cholesterol in membranes for the expression of tolerance, Mol. Pharmacol. 15:739–746.

Johnson, D. A., Lee, N. M., Cooke, R., and Loh, H., 1980, Adaptation to ethanol-induced fluidization of brain lipid bilayers: Cross tolerance and reversibility, Mol. Pharmacol. 17:52–55.

Jones, R. T., and Stone, G. C., 1970, Psychological studies of marijuana and alcohol in man, Psychopharmacologia 18:108–117.

Kaim, S. C., and Klett, C. J., 1972, Treatment of delirium tremens, Q. J. Stud. Alcohol 33:1065–1072.

Kaim, S. C., Klett, C. J., and Rothfeld, B., 1969, Treatment of the acute alcohol withdrawal: A comparison of four drugs, Am. J. Psychiatry 125:1640–1646.

Kalant, H., 1975, Direct effect of ethanol on the nervous system, Fed. Proc. 34:1930–1941.

Kalant, H., Leblanc, A. E., and Gibbins, R. J., 1971, Tolerance to, and dependence on, some non-opiate psychtropic drugs, Pharmacol. Rev. 23:135–191.

Kalant, H., Woo, N., and Endrenyi, L., 1978, Effect of ethanol on the kinetics of rat brain (Na$^+$ + K$^+$) ATPase and K$^+$-dependent phosphatase with different alkali ions, Biochem. Pharmacol. 27:1353–1358.

Karoum, F., Wyatt, R. J., and Majchrowicz, E., 1976, Brain concentrations of biogenic amino metabolites in acutely treated and ethanol-dependent rats, Br. J. Pharmacol. 56:403–411.

Kendig, J. J., Trudell, J. R., and Cohen, E. N., 1975, Effects of pressure and anaesthetics on conduction and synaptic transmission, J. Pharmacol. Exp. Ther. 195:216–224.

Kimelberg, H. K., and Papahadjopoulos, D., 1974, Effects of phospholipid acyl chain fluidity, phase transitions, and cholesterol on (Na$^+$ + K$^+$)-stimulated adenosine triphosphatase, J. Biol. Chem. 249:1071–1080.

Klotz, V., Avant, G. R., Hoyumpa, A., Schenker, S., and Wilkinson, G. R., 1975, The effect of age and liver disease on the disposition and elimination of diazepam in adult man, J. Clin. Invest. 55:347–359.

Knochel, J. P., 1977, The pathophysiology and clinical characteristics of severe hypophosphatemia, Arch. Intern. Med. 137:203–220.

Knox, W. H., Perrin, R. G., and Sen, A. K., 1972, Effect of chronic administration of ethanol on (Na + K)-activated ATPase in six areas of cat brain, J. Neurochem. 19:2881–2884.

Kramp, P., and Rafaelsen, O. J., 1978, Delirium tremens: A double-blind comparison of diazepam and barbital treatment, Acta Psychiatr. Scand. 58:174–190.

Kraus, J. W., Desmond, P. V., Marshall, J. P., Johnson, R. A., Schenker, S., and Wilkinson, G. R., 1978, The effect of aging and liver disease on the elimination of lorazepam in man, Clin. Pharmacol. Ther. 24:411–419.

Kruger, G., Haubitz, I., Weinhardt, F., and Hoyer, S., 1980, Brain oxidative metabolism and blood flow in alcoholic syndromes, Subs. Alcohol Actions Misuse 1:295–307.

Kuriyama, K., 1977, Ethanol induced changes in activities of adenylate cyclase, quanylate cyclase and cyclic adenosine 3', 5' monophosphate dependent protein kinase in the brain and the liver, Drug Alcohol Depend. 2:335–348.

Kuriyama, K., and Israel, M. A., 1973, Effect of ethanol administration on cyclic 3', 5'-adenosine monophosphate metabolism in brain, Biochem. Pharmacol. 22:2919–2922.

Kuriyama, K., Rauscher, G. E., and Sze, P. Y., 1971, Effect of acute and chronic administration of ethanol on the 5-hydroxytryptamine turnover and tryptophan hydroxylase activity of the mouse brain, *Brain Res.* **26**:450–454.

Lassen, N. A., Ingvar, D. H., and Skinhoj, E., 1978, Brain functions and blood flow, *Sci. Am.* **239**:62–71.

Laties, V. C., Lasagna, L., Gross, G. M., Hitchman, I. L., and Flores, J., 1958, A controlled trial of chlorpromazine and promazine in the management of delirium tremens, *Q. J. Stud. Alcohol* **19**:238–243.

Lereboullet, J., Benda, P., and Poisson, M., 1960, Le meprobamate injectable. Traitement efficace des delires alcoholigues aigus, *Presse Med.* **68**:473–475.

Levental, M., and Tabakoff, B., 1980, Sodium-potassium activated adenosine triphosphatase activity as a measure of neuronal membrane characteristics in ethanol-tolerant mice, *J. Pharmacol. Exp. Ther.* **212**:315–319.

Levy, R. A., 1977, The role of GABA in primary afferent depolarization, *Prog. Neurobiol.* **9**:211–267.

Lieber, C. S., 1982, Medical disorders of alcoholism; pathogenesis and treatment, in: *Major Problems in Internal Medicine*, Volume XXII, (L. H. Smith, Jr., ed.), W. B. Saunders, Philadelphia, pp. 1–42.

Littleton, J. M., Griffiths, P. J., and Ortiz, A., 1974, The induction of ethanol dependence and the ethanol withdrawal syndrome: The effects of pyrazole, *J. Pharm. Pharmacol.* **26**:81–91.

Littleton, J. M., John, G. R., and Grieve, S. J., 1979, Alterations in phospholipid composition in ethanol tolerance and dependence, *Alcohol. Clin. Exp. Res.* **3**:50–56.

Majchrowicz, E., 1973, Induction of physical dependence on alcohol and the associated metabolic and behavioral changes in rats, *Pharmacologist* **15**:159.

Marr, A. G., and Ingraham, J. L., 1962, Effects of temperature on the composition of fatty acids in *Escherichia coli, J. Bacteriol.* **84**:1260–1267.

McComb, J. A., and Goldstein, D. B., 1979, Quantitative comparison of physical dependence on tertiary butanol and ethanol in mice: Correlation with lipid solubility, *J. Pharmacol. Exp. Ther.* **208**:113–117.

McConnell, H. M., Wright, K. L., and McFarlane, B. G., 1972, The fraction of the lipid in a biological membrane that is in a fluid state: A spin label assay, *Biochem. Biophys. Res. Commun.* **47**:273–281.

McGrath, S. D., 1975, A controlled trial of chlormethiazole and chlordiazepoxide in the treatment of the acute withdrawal phase of alcoholism, *Br. J. Addict.* **70**(suppl.):81–90.

Melchior, C. L., Mueller, A., and Deitrich, R. A., 1978, Half-life of tetrahydropapaveroline and salsolinol following injection into the cerebral ventricle of rats, *Fed. Proc.* **37**:420.

Mendelson, J. H., and LaDou, J., 1964, Experimentally induced chronic intoxication and withdrawal in alcoholic subjects, *Q. J. Stud. Alcohol* **25**:1–8.

Metcalfe, J. C., Seeman, P., and Burgen, A. S. V., 1968, The proton relaxation of benzyl alcohol in erythrocyte membranes, *Mol. Pharmacol.* **4**:87–95.

Meyer, K. H., 1937, Contributions to the theory of narcosis, *Trans. Faraday Soc.* **33**:1062–1068.

Michaelis, E. K., and Myers, S. L., 1979, Calcium binding to brain synaptosomes. Effects of chronic ethanol intake, *Biochem. Pharmacol.* **28**:2081–2087.

Mirsky, I. A., Piker, P., Rosenbaum, M., and Lederer H., 1941, "Adaptation" of the central nervous system to varying concentrations of alcohol in the blood, *Q. J. Stud. Alcohol.* **2**:35–45.

Montoya, G. A. Riker, W. K., and Russell, N. J., 1977, Stimulus-bound repetitive synaptic firing caused by ethanol in sympathetic ganglion, *J. Pharmacol. Exp. Ther.* **200**:320–327.

Moore, J. W., Ulbricht, W., and Takata, M., 1964, Effect of ethanol on the sodium and potassium conductances of the squid axon membrane, *J. Gen. Physiol.* **48**:279–295.

Moskowitz, G., Chalmers, T. C., Sacks, H. S., Fagerstrom, R. M., and Smith, H., 1983, Deficiencies of clinical trials of alcohol withdrawal. *Alcoholism: Clin. Exp. Res.* **7**:42–46.

Myers, R. D., 1978, Psychopharmacology of alcohol, *Annu. Rev. Pharmacol. Toxicol.* **18**:125–144.

Myers, R. D., and Melchior, C. L., 1977, Alcohol drinking: Abnormal intake caused by tetrahydropapaveroline in brain, *Science* **196**:554–556.

Newlin, D. B., Golden, C. J., Quaife, M., and Graber, B., 1982, Effect of alcohol ingestion on regional cerebral blood flow, *Int. J. Neurosci.* **17**:145–150.

Newman, H. W., 1941, Acquired tolerance to ethyl alcohol, *Q.J. Stud. Alcohol* **2**:453–463.

Nielsen, J., 1965, Delirium tremens in Copenhagen, *Acta Psychiatr. Scand.* (suppl.) **187**:5–92.

Nielsen, R. H., Hawkins, R. A., and Veech, R. L., 1975, The effects of acute ethanol intoxication on cerebral energy metabolism, *Adv. Exp. Med. Biol.* **59**:93–109.

Noble, E. P., and Tewari, S., 1977, Metabolic aspects of alcoholism in the brain, in: *Metabolic Aspects of Alcoholism* (C. S. Lieber, ed.), University Park Press, Baltimore, pp. 149–187.

Noble, E. P., Synapin, P. J., Vingran, R., and Rosenberg, A., 1976, Neuraminidase-releasable surface sialic acid of cultured astroblasts exposed to ethanol, *J. Neurochem.* **27**:217–221.

Okamoto, M., Rosenberg, H. C., and Boisse, N. R., 1977, Evaluation of anti-convulsants in barbiturate withdrawal, *J. Pharmacol. Exp. Ther.* **202**:479–489.

O'Neill, P. J., and Rahwan, R. G., 1977, Absence of formatin of brain salsolinol in ethanol-dependent mice, *J. Pharmacol. Exp. Ther.* **200**:306–313.

Ortiz, A., Griffiths, P. J., and Littleton, J. M., 1974, A comparison of the effects of chronic administration of ethanol and acetaldehyde to mice: Evidence for a role of acetaldehyde in alcohol dependence, *J. Pharm. Pharmacol.* **26**:249–260.

Overton, E., 1901, Studien uber die narcose: zugleichein bietrag zur allgemeinen pharmacologie, Jena, Verlag von Gustaf Fisher.

Palaic, D. J., Desaty, J., Albert, J. M., and Panisset, J. C., 1971, Effects of ethanol on metabolism and subcellular distribution of serotonin in rat brain, *Brain Res.* **25**:381–386.

Parker, T. H., Roberts, R. K., Vorhees, C. V., Schmidt, D. E., and Schenker, S., 1977, The effects of acute and subacute ammonia intoxication on regional brain acetylcholine levels in rats, *Biochem. Med.* **18**:235–244.

Pert, A., 1978, The effect of opiates on nigrostriatal dopaminergic activity, in: *Characteristics and Functions of Opiates* (J. M. vanRee and L. Terenius, eds.), Elsevier North-Holland Biomedical, Amsterdam, pp. 389–401.

Pohorecky, L. A., 1974, Effects of ethanol on central and peripheral noradrenergic neurons, *J. Pharmacol. Exp. Ther.* **189**:380–391.

Pohorecky, L. A., and Brick, J., 1977, Activity of neurons in the locus coeruleus of the rat: Inhibition by ethanol, *Brain Res.* **131**:174–179.

Pohorecky, L. A., Jaffe, L. S., and Berkeley, H. A., 1974, Ethanol withdrawal in the rat: Involvement of noradrenergic neurons, *Life Sci.* **15**:427–437.

Pohorecky, L. A., Newman, B., Sun, J., and Bailey, W. H., 1978, Acute and chronic ethanol ingestion and serotonin metabolism in rat brain, *J. Pharmacol. Exp. Ther.* **204**:424–432.

Post, M. E., and Sun, A. Y., 1973, The effect of chronic ethanol administration on the levels of catecholamines in different regions of the rat brain, *Res. Commun. Chem. Pathol. Pharmacol.* **6**:887–894.

Prichard, J. W., Alger, J. R., Behar, K. L., Petroff, O. A. C., and Shulman, R. G., 1983, Cerebral metabolic studies *in vivo* by ^{31}P NMR, *Proc. Natl. Acad. Sci. U.S.A.* **80**:2748–2751.

Rangaraj, N., and Kalant, H., 1978, Effects of ethanol withdrawal, stress and amphetamine on rat brain (Na$^+$ + K$^+$)-ATPase, *Biochem. Pharmacol.* **27**:1139–1144.

Rangaraj, N., and Kalant, H., 1982, Effect of chronic ethanol treatment on temperature dependence and norepinephrine sensitization of rat brain (Na$^+$ + K$^+$)-adenosine triphosphatase, *J. Pharmacol. Exp. Ther.* **223**:536–539.

Raskin, N. H., and Sokoloff, L., 1968, Brain alcohol dehydrogenase, *Science* **162**:131–132.

Rawat, A. K., 1974, Brain levels and turnover rates of presumptive neurotransmitters as influenced by administration and withdrawal of ethanol in mice, *J. Neurochem.* **22**:915–922.

Rawat, A. K., Kuriyama, K., and Mose, J., 1973, Metabolic consequences of ethanol oxidation in brains from mice chronically fed ethanol, *J. Neurochem.* **20**:23–33.

Redos, J. D., Hunt, W. A., and Catravas, G. N., 1976, Lack of alteration in regional brain adenosine - 3′, 5′-cyclic monophosphate levels after acute and chronic ethanol administration, *Life Sci.* **18**:989–992.

Reivich, M., Sokoloff, L., Kennedy, C., and Des Rosier, M., 1975, An autoradiographic method for the measurement of local glucose metabolism in the brain, in: *Brain Work: The Coupling of Function, Metabolism and Blood Flow in the Brain* (D. H. Ingvar, and N. A. Lassen, eds.), Munksgaard, Copenhagen, pp. 377–384.

Roberts, R. K., Branch, R. A., Wilkinson, G. R., and Schenker, S., 1978, The effect of age and chronic liver disease on the disposition and elimination of chlordiazepoxide (Librium), *Gastroenterology* **75**:429–485.

Rojas, E., and Armstrong, C., 1971, Sodium conduction activation without inactivation in pronase-perfused axons, *Nature (London)* **229**:177–178.

Roots, B. I., 1968, Phospholipids of goldfish (Carassius Auratus L.) brain: The influence of environmental temperature, *Comp. Biochem. Physiol.* **25**:457–466.

Ross, D. H., 1977, Adaptive changes in Ca^{2+}-membrane interactions following chronic ethanol exposure, *Adv. Exp. Med. Biol.* **85A**:459–471.

Ross, D. H., 1980, Molecular aspects of calcium-membrane interactions: A model for cellular adaptation to ethanol, in: *Behavioral Pharmacology of Ethanol Tolerance and Dependence* (H. Rigter and J. C. Grabbe, eds.), Elsevier-North Holland, New York, pp. 227–240.

Ross, D. H., and Garrett, K. M., 1983, Acute pharmacological actions of ethanol on the central nervous system, in: *The Pathogenesis of Alcoholism* (B. Kissin and H. Begleiter, eds.), Plenum Press, New York, pp. 57–75.

Ross, D. H., Medina, M. A., and Cardenas, H. L., 1974, Morphine and ethanol: Selective depletion of regional brain Ca^{2+}, *Science* **186**:63–65.

Ross, D. H., Kibler, B. C., and Cardenas, H. L., 1977, Modification of glycoprotein residues as Ca^{2+} receptor sites after chronic ethanol exposure, *Drug Alcohol Depend.* **2**:305–315.

Rothstein, E., 1973, Prevention of alcohol withdrawal seizures: The roles of diphenylhydantoin and chlordiazepoxide, *Am. J. Psychiatry.* **130**:1381–1382.

Sampliner, R., and Iber, F. L., 1974, Diphenylhydantoin control of alcohol withdrawal seizures, *JAMA* **230**:1430–1432.

Sandler, M., Carter, S. B., Hunter, K. R., and Stern, G. M., 1973, Tetrahydroisoquinoline alkaloids: In vivo metabolites of L-dopa in man, *Nature (London)* **24**:439–443.

Santi, R., Ferrari, M., Toth, C. E., Contessa, A. R., Fassina, G., Bruni, A., and Luciani, S., 1967, Pharmacological properties of tetrahydropapaveroline, *J. Pharm. Pharmacol.* **19**:45–51.

Seeman, P., 1972, The membrane actions of anaesthetics and tranquilizers, *Pharmacol. Rev.* **24**:583–655.

Seil, F. J., Leiman, A. L., Herman, M. M., and Fish, R. A., 1977, Direct effects of ethanol on central nervous system cultures: An electrophysiological and morphological study, *Exp. Neurol.* **55**:390–404.

Sellers, E. M., and Kalant, H., 1982, Alcohol withdrawal and delirium tremens, in: *Encyclopaedic Handbook on Alcoholism* (E. M. Pattison, and E. Mansell, eds.), New York, Gardner Press.

Sellers, E. M., Naranjo, C. A., Giles, H. G., Frecker, R. C. Khouw, V., Beeching, M., and Fan, T., 1980, Mechanisms of ethanol-diazepam pharmacokinetic interaction, *Clin. Pharmacol. Ther.* **27**:286.

Sellers, E. M., Naranjo, C. A., Harrison, M., Devenyi, P., Roach, C., and Sykora, K., 1983, Diazepam loading: Simplified treatment of alcohol withdrawal. *Clin. Pharmacol. Ther.* **34**:822–826.

Shimojyo, S., Scheinberg, P., and Reinmuth, O., 1967, Cerebral blood flow and metabolism in the Wernicke-Korsakoff syndrome, *J. Clin. Invest.* **46**:849–854.

Shull, H. J., Jr., Wilkinson, G. R., Johnson, R., and Schenker, S., 1976, Normal disposition of oxazepam in acute viral hepatitis and cirrhosis, *Ann. Intern. Med.* **84**:420–425.

Sinclair, J. G., and Lo, G. F., 1978, Acute tolerance to ethanol on the release of acetylcholine from cat cerebral cortex, *Can. J. Physiol. Pharmacol.* **56**:668–670.

Sinensky, M., 1974, Homeoviscous adaptation—A homeostatic process that regulates the viscosity of membrane lipids in *Escherichia coli*, *Proc. Natl. Acad. Sci. U.S.A.* **71**:522–525.

Smith, D. E., and Wesson, D. R., 1970, A new method for treatment of barbiturate dependence, *J. Am. Med. Assoc.* **213**:294–295.

Smith, J. A., 1953, Methods of treatment of delirium tremens, *J. Am. Med. Assoc.* **152**:384–387.

Smith, R. Y., 1976, Relative effectiveness of primidone (Nysoline) and diphenylhydantoin (Dilantin) in the management of sedative withdrawal seizures, *Ann. N.Y. Acad. Sci.* **273**:378–380.

Syn, A. Y., and Samorajski, T., 1975, The effects of age and alcohol on $(Na^+ + K^+)$-ATPase activity of whole homogenate and synaptosomes prepared from mouse and human brain, *J. Neurochem.* **24**:161–164.

Sunahara, G. I., and Kalant, H., 1980, Effect of ethanol on potassium-stimulated and electrically stimulated acetylcholine release in vitro from rat cortical slices, *Can. J. Physiol. Pharmacol.* **58**:706–711.

Tabakoff, B., and Ritzmann, R. F., 1977, The effects of 6-hydroxy dopamine on tolerance to and dependence on ethanol, *J. Pharmacol. Exp. Ther.* **203**:319–331.

Tabakoff, B., Urwyler, S., and Hoffman, P. L., 1981, Ethanol alters kinetic characteristics and function of striatal morphine receptors, *J. Neurochem.* **37**:518–521.

Tavel, M. E., Davidson, W., and Batterton, T. D., 1961, A critical analysis of mortality associated with delirium tremens. A review of 39 fatalities in a 9-year period, *Am. J. Med. Sci.* **242**:18–29.

Tewari, S., and Noble, E. P., 1971, Ethanol and brain protein synthesis, *Brain Res.* **26**:469–474.

Tewari, S., Fleming, E. W., and Noble, E. P., 1975, Alterations in brain RNA metabolism following chronic ethanol ingestion, *J. Neurochem.* **24**:561–569.

Tewari, S., Goldstein, M. A., and Noble, E. P., 1977, Alterations in cell free brain protein synthesis following ethanol withdrawal in physically dependent rats, *Brain Res.* **126**:509–518.

Thomas, D. W., and Friedman, D., 1964, Treatment of the alcohol withdrawal syndrome, *J. Am. Med. Assoc.* **188**:316–318.

Thompson, W. L., 1978, Management of alcohol withdrawal syndromes, *Arch. Intern. Med.* **138**:278–283.

Thompson, W. L., Johnson, A. D., Maddrey, W. L., and Osler Medical Housestaff, 1975, Diazepam and paraldehyde for the treatment of severe delirium tremens, *Ann. Intern. Med.* **82**:175–180.

Traynor, M. E., Woodson, P. B. J., Schlapfer, W. T., and Barondes, S. H., 1976, Sustained tolerance to a specific effect of ethanol on posttetanic potentiation in *Aplysia*, *Science* **193**:510–511.

Trudell, J. R., Hubbell, W. L., and Cohen, E. N., 1973, Pressure reversal of inhalation anaesthetic-induced disorder in spin-labelled phospholipid vesicles, *Biochem. Biophys. Acta* **291**:328–334.

Turner, A. J., Baker, K. M., Algeri, S., Frigerio, A., and Garattini, S., 1974, Tetrahydropapaveroline: Formation *in vivo* and *in vitro* in rat brain, *Life Sci.* **14**:2247–2257.

Urwyler, S., and Tabakoff, B., 1981, Stimulation of dopamine synthesis and release by morphine and D-ala^2-D leu^5-encephalin in the mouse striatum, *Life Sci.* **28**:2277–2286.

Veloso, D., Passonneau, J. V., and Veech, R. L., 1977, The effects of intoxicating doses of ethanol upon intermediary metabolism in rat brain, *J. Neurochem.* **19**:2679–2686.

Vessell, E. S., 1972, Ethanol metabolism; regulation by genetic factors in normal volunteers under a controlled environment and the effect of chronic ethanol administration, *Ann. N.Y. Acad. Sci.* **197**:79–88.

Victor, M., and Adams, R. D., 1953, The effect of alcohol on the nervous system, *Assoc. Res. Nerv. Ment. Dis. Res. Publ.* **32**:526–573.

Volicer, L., and Hurter, B. P., 1977, Effects of acute and chronic ethanol administration and withdrawal on adenosine 3′, 5′-monophosphate and quanosine 3′, 5′-monophasphate levels in rat brain, *J. Pharmacol. Exp. Ther.* **200**:298–305.

Wajda, I. J., Maniqualt, I., and Hudick, J. P., 1977, Dopamine levels in the striatum and the effect of alcohol and reserpine, *Biochem. Pharmacol.* **26**:653–655.

Walker, D. W., and Freund, G., 1971, Impairment of shuttle box avoidance learning following prolonged alcohol consumption in rats, *Physiol. Behav.* **7**:773–778.

Wallgren, H., 1973, Neurochemical aspects of tolerance to and dependence on ethanol, *Adv. Exp. Med. Biol.* **35**:15–31.

Wallgren, H., Nikander, P., Boguslawsky, P. von, and Linkola, J., 1974, Effects of ethanol, tertiary butanol and chlomethiazole on net movements of sodium and potassium in electrically stimulated cerebral tissue, *Acta Physiol. Scand.* **91**:83–93.

Wolff, P. H., 1972, Ethnic differences in alcohol sensitivity, *Science* **175**:449–450.

Wood, J. D., 1975, The role of α-aminobutyric acid in the mechanism of seizures, *Prog. Neurobiol.* **5**:77–95.

Wood, J. M., 1980, Effect of depletion of brain 5-hydroxytryptamine by 5,7-dihydroxytryptamine on ethanol tolerance and dependence in the rat, *Psychopharmacologia* **67**:67–72.

Yamamoto, H., Harris, R. A., Loh, H. H., and Way, E. L., 1978, Effects of acute and chronic morphine treatments on calcium localization and binding in brain, *J. Pharmacol. Exp. Ther.* **205**:255–264.

Zaleska, M. M., and Gessner, P. K., 1983, Metabolism of [^{14}C]paraldehyde in mice *in vivo*, generation and trapping of acetaldehyde, *J. Pharmacol. Exp. Ther.* **224**:614–619.

Heavy Metal Toxicity and Energy Metabolism in the Developing Brain
Lead as the Model

JOHN J. O'NEILL and DAVID HOLTZMAN

LEAD ENCEPHALOPATHY

1. Introduction

1.1. The Clinical Problems and Animal Models

It has long been recognized that lead poisoning produces devastating neurologic damage to the CNS in children (encephalopathy) that is characterized by cerebral edema, convulsion, and coma (Aub *et al.*, 1925; Byers and Lord, 1943). Poisoning in adults results largely from occupational exposure and produces clinical or subclinical signs of peripheral neuropathy usually without CNS involvement. The reasons(s) for this maturational change in brain sensitivity to lead toxicity are uncertain, but perhaps can be related to the development of the mammalian nervous system (Reiter, 1982). In the protracted period of CNS development, there are specific time-related processes termed "growth spurts" by Davison and Dobbing (1968), who related brain growth to differing time scales depending on the species, i.e., days in rats and mice, months for human development. Since such estimates fail to take into account individual growth rates for neurons and glia, Rodier (1979) described the prenatal and postnatal time course of neuron proliferation. His studies showed that although different neuronal systems develop at different rates within a single species, and the exact timing may vary, the sequence of development is similar for all

JOHN J. O'NEILL • Department of Pharmacology, Temple University School of Medicine, Philadelphia, Pennsylvania 19140.   *DAVID HOLTZMAN* • Department of Psychiatry and Neurology and Department of Pediatrics, Tulane University School of Medicine, New Orleans, Louisiana 70112.

species studied. In addition to such neuroanatomic correlates of growth, periods of neurochemical transition termed "critical periods" have been described by Himwich (1951). Toxic damage following exposure to noxious chemicals has been related in specific cases (e.g., methyl mercury) to these stages of nervous system development (Suzuki, 1980).

In recent years, there has been a growing awareness of the toxic dangers to the developing nervous system from prolonged exposure to low lead levels (e.g., Lin-Fu, 1980). The encephalopathic effects of low-level lead exposure may be expressed as deficits in IQ, learning problems, and poor classroom behavior (de la Burde and Choate, 1975; Beattie et al., 1975; Landrigan et al., 1975). The difficulty encountered in establishing a causal relationship between exposure and toxicity may be due to the lack of analytical data relating extent of exposure to specific periods of CNS development. Some attempts have been made by measuring lead content in hair samples (Pihl and Parkes, 1977) and in deciduous teeth (Needleman et al., 1979). Needleman's group described a comprehensive study of Boston elementary school children and reported deficits in psychologic testing and poor classroom behavior that correlated with cumulative lead content in dentine of the child's first teeth. The data are persuasive that an association exists between lead exposure and psychosocial behavior, but such studies may suffer from the inherent weakness of not being able to relate the age and duration of lead exposure to these deficits.

In rat, as in man, acute lead encephalopathy occurs predominantly at young ages and rarely in adults (Goldstein et al., 1974; Clasen et al., 1974). Based on the pioneering work of Pentschew and Garro (1966), many studies using the rat as a model have reported toxic effects of high-dose lead feedings on the developing nervous system. The pathologic changes are similar in the rat pup and young children including hemorrhage, edema, neuronal necrosis, and capillary hypertrophy most marked in the cerebellum, but present also in basal ganglia, neocortex, and hippocampus at younger ages (Clasen et al., 1974; Press, 1977a,b). Largely through the efforts of Michaelson (1973), Patel et al. (1974), Krigman and Hogan (1974), Maker et al. (1975), and Holtzman and Hsu (1976), much has been established concerning the neurochemical effects of lead on brain growth and metabolism.

Neuropathologic changes also have been reported to occur following exposure to low lead doses in drinking water. The brains of rat pups, exposed to lead through their dam's consumption of lead in drinking water (10 mg/ml

TABLE I. Number of Spines per Unit Length of Dendrites of Hippocampal Pyramidal Cells after Neonatal Pb Treatment of Rats[a]

Age of animals (days)	Control		Pb treated		Difference in percent
	Number of cells counted	Spine density[b]	Number of cells counted	Spine density[b]	
20	71	29.31 ± 0.88	70	18.07 ± 0.75[c]	38.35
56	73	32.32 ± 1.45	85	19.80 ± 0.89[c]	38.74

[a] From Kiraly and Jones (1982).
[b] Mean ± SE number of spines per 100 μm of dendrite.
[c] $P < 0.001$, compared with value of the corresponding control, Student's two-tailed t test.

393

*HEAVY METAL
TOXICITY AND
ENERGY
METABOLISM IN
THE DEVELOPING
BRAIN*

lead acetate) from postnatal day 2 onward, were examined histologically after fixation and staining by Golgi-Cox impregnation (Fig. 1) at 20 days of age (Kiraly and Jones, 1982). A significant decrease in pyramidal CA-1 neuronal dendritic spine density is noted after lead treatment (Fig. 1 and Table I). In similar studies, dams were exposed to 2 mg/ml lead acetate in drinking water at parturition and the brains of 15-day-old pups were examined by light and electron microscopy (Campbell *et al.*, 1982). The dose of lead ingested by the pups was estimated by the method of Bornschein *et al.* (1977) to be approximately 0.2 μg lead/g body weight per day from day 1 to days 15–20. Unlike high lead ingestion, low-level lead exposure allows normal somatic growth, but results in impairment of neuropil development and synaptogenesis (Table II). Vascular damage and other signs of cytotoxicity are not seen.

Several mechanisms of cellular lead toxicity have been proposed. Campbell *et al.* (1982) speculate that lead exposure may cause developmental impairment by direct and indirect actions. Because of the unique distribution of zinc and other transition metals within the hippocampus, lead, if present, could displace these metals from essential sites. Postnatal zinc deficiency has been demonstrated to impair growth of the hippocampus in rat (Buell *et al.*, 1977). Campbell *et al.* (1982) have interpreted their results as indicating that low-level lead exposure during development preferentially affects the later-appearing structures rather than affecting structures already mature. Several groups also have suggested that lead–calcium interactions may be important in the pathogenesis of lead encephalopathy (Goldstein. 1977; Holtzman *et al.*, 1980b; Pounds *et al.*, 1982a,b). The relationship between the pathogenesis of lead

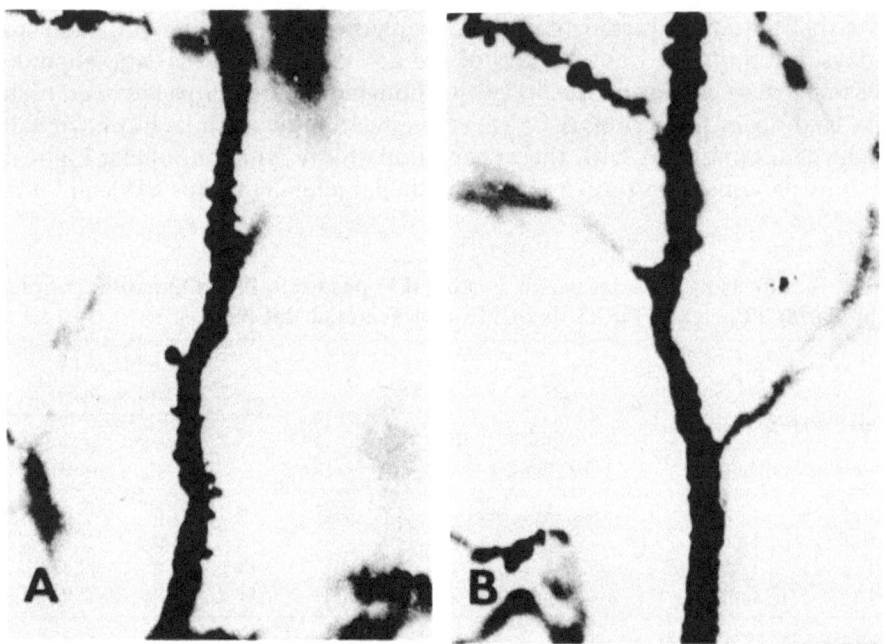

FIGURE 1. Apical dendritic segment of hippocampal pyramidal neurons in (A) control and (B) Pb-treated rats at 20 days of age. Golgi-Cox impregnation, × 1180. From Binah *et al.* (1978).

encephalopathy and effects of lead on cellular aerobic energy metabolism in the immature brain that was originally proposed by Pentschew and Garro (1966) has recently received comprehensive treatment (Holtzman *et al.*, 1982). It is the purpose of the present chapter to emphasize the effect of lead and other heavy metals (e.g., mercury) on energy metabolism and neutotransmitter metabolism in the developing nervous system.

1.2. Specific or Enhanced Susceptibility of the Developing Brain

The rat pup is sensitive to the overt encephalopathic effects if high-dose lead feedings are begun before 18 or 20 days of age (Krigman *et al.*, 1977; Holtzman *et al.*, 1982). In pups begun on lead feedings early in the first week of life, the pathologic changes are similar in cerebrum and cerebellum (Press, 1977b). In a recent study, pups were given lead acetate feedings by means of an esophageal feeding tube once daily beginning at various ages (Holtzman *et al.*, 1982). The encephalopathy resulting from these feedings, as evidenced by cerebellar swelling and discoloration and by hindleg paresis, occurs at lower daily lead doses in pups receiving lead beginning at 14 days compared with 16 or 18 days of age (Table III). By light microscopy, the cerebrum is normal while the cerebellum shows edema, focal hemorrhages, necrosis of small neurons, proliferation of astrocytes, and occasional hypertrophy of capillary endothelial cells. With very high daily lead doses, the incidence of encephalopathy approaches 100% at each sensitive age. The encephalopathic lead doses (i.e., ED_{50}) at each sensitive age result in failure of weight gain and anemia.

In contrast to the younger animals, pups begun on lead feedings at 20 or 24 days of age show no signs of gross cerebellar encephalopathy. At daily lead doses that are much higher than the encephalopathic dose for pups fed from 18 days, the pups fed from 20 days of age do suffer weight loss and anemia to the same extent as the encephalopathic younger animals. Pups fed even higher daily lead doses from 24 days of age also develop this anemia, but have better weight gain compared with the encephalopathic younger animals. Light microscopy demonstrates only a patchy cerebellar edema in pups fed lead for two

TABLE II. Effects of Prenatal versus Postnatal Exposure to Pb on Synaptic Counts in the Parietal Cortex of the 15-day-Old Cross-Fostered Rat Pup[a]

Pup exposure period		Synaptic density $\times 10^8/mm^3 \pm$ SEM	$(N)^b$	Pup blood Pb (μg/dl $\pm$ SEM)	
Prenatal	Postnatal			Birth	15 days postpartum
Control	Control	4.15 ± 0.36	(11)^d	6.7 ± 1.0	4.6 ± 0.9
Control	Lead	3.97 ± 0.57	(5)	6.6 ± 1.0	24.8 ± 1.3
Lead	Lead	2.92 ± 0.42^c	(14)^d	88.0 ± 5.7	30.3 ± 2.9
Lead	Control	2.78 ± 0.51^c	(5)	69.2 ± 8.0	6.9 ± 0.9

[a] From McCauley *et al.* (1982).
[b] N is the number of litters examined, one pup per litter.
[c] $P < 0.05$ when compared with group CT/CT.
[d] Includes data from two studies.

395

HEAVY METAL
TOXICITY AND
ENERGY
METABOLISM IN
THE DEVELOPING
BRAIN

weeks beginning at 20 days of age and no pathologic changes in cerebellums of pups fed lead from 24 days (Holtzman *et al.*, 1982).

1.3. Cellular Energy Metabolism as the Potential Site of Lead Action

When the action of heavy metals on cellular energy metabolism is studied, it is well to reflect on earlier reports concerned with metabolic aspects in the developing nervous system. The late Harold E. Himwich (1951) stressed the importance of an awareness of developmental changes occurring in the CNS of differing species. Structural development of the brain is thought to consist of a series of scheduled events known to contain "critical" periods when both anatomical and metabolic changes occur. Metabolic adaptations in brain, as in other organs, are undoubtedly essential for the organism's survival.

Energy processes may be thought to consist of those needed to support syntheses and those required for neural function. Although growth in cell size and differentiation precedes onset of myelination and function, many of these processes are occurring in temporal association depending on the brain region and species studied. This species distinction is best illustrated when considering differences between "precocial" animals in which brain maturation is

TABLE III. Effects of Lead Acetate Feedings for 14 Days Beginning at Various Ages in the Rat Pup[a]

Age of start of feedings (days)	Daily lead dose (μg Pb/g body) weight	Deaths (%)	Encephalopathy[b]	# of animals
14	0[c]	12.5	0	(16)
	400[d]	62.5	62.5	(24)
	800	75	62.5	(8)
	1200	100	100	(8)
16	0	25	0	(16)
	400	12.5	0	(8)
	800	38	63	(24)
	1600	100	75	(8)
18	0	0	0	(8)
	400	0	0	(8)
	800	25	25	(24)
	2400	87	75	(8)
20	0	0	0	(8)
	1600	0	0	(24)
	2000	6	0	(24)
	2400	12.5	0	(8)
24	0	0	0	(16)
	2400	0	0	(8)
	3200	0	0	(16)

[a] Data taken from Holtzman *et al.* (1982).
[b] Animals were considered encephalopathic if they showed at least two of the following characteristics: hindleg paresis, brownish discoloration of the cerebellum, or cerebellar swelling. The results shown are the number of autopsy-verified encephalopathic animals as a percent of the number of animals fed the respective lead doses (shown in parentheses).
[c] All control animals received Na acetate (1200 μg/g body weight) in about 0.25 ml H_2O.
[d] The encephalopathic doses, defined as the lowest tested daily dose of lead acetate which produced the cerebellar encephalopathy in at least 50% of the pups in the respective age group, are indicated.

nearly complete at birth, e.g., horse, sheep, other herd animals, and the guinea pig and "nonprecocial" animals of which man and rat are prime examples. In rat, cell division is complete in the brain by day 21 postnatally (Winick and Noble, 1965). In guinea pig, there is little cell division in brain after birth (Mandel et al., 1965). In human brain, cell division continues until the end of the first year of life. According to Dobbing and Sands (1970), two peaks of DNA synthesis occur in human brain. The first at 26 weeks of gestation may signal neuronal division and a second near birth, which may indicate the rate of glial cell division.

The energy metabolism in brain undergoes time-dependent changes starting at the earliest fetal period measurable and continuing throughout life. It is likely that these changes reflect changes in the physiologic uses and dependence of the brain on aerobic energy and metabolism. It is important in discussing heavy metal toxicity that such developmental changes be appreciated in the interpretation of the pathologic effects of altered energy metabolism. The differences in energy metabolism between mature brain and the developing nervous system in the rat are summarized in the succeeding paragraphs.

The activities of oxidative systems are lower in brain tissue of newborn as compared with adult rats (Tyler and Van Harreveld, 1942; Greengard and McIlwain, 1955). These activities begin to increase shortly after birth and reach maximal levels more than 20 days later in the case of the rat. It has been reported that with brain development there is an increase in the number of mitochondria per gram whole brain, the rate of increase being greatest between days 10 and 21 postpartum (Samson et al., 1960). Those estimates suffered by being derived from brain homogenate data following ultracentrifugal separation into mitochondrial-rich fractions. At least a portion of the apparent increase in mitochondrial numbers is thought to be due to the contamination by myelin fragments containing mitochondria. This age period, 10–20 days, also contains the greatest increase in myelination in rat. The recognized heterogeneity and variable stability of at least two mitochondrial populations (i.e., nerve ending and cell body mitochondria) creates a further problem in interpreting such data.

A more precise method for the estimation of mitochondrial number in the various subcellular fractions as applied to newborn and adult rats was described by Gregson and Williams (1969). This method visualizes mitochondria at the electron microscopic level. The respiratory enzyme content per mitochondrion of adult brain was found by direct measurement to be 3.5 times greater than for the neonate. The succinic dehydrogenase activity per liver mitochondrion shows no such developmental change from neonate to adult. In contrast to brain, in liver there is an increase in mean mitochondrial population per gram tissue over this interval. In brain, the average mitochondrial number per cell shows an approximate doubling per cell (776 to 1480/cell) based on DNA but not per tissue weight (36.09 to 36.62/g, wet wt.). All estimates assume that the recovered mitochondria are representative of the total population. The major difference in Samson's results compared with those of Gregson and Williams is the limit resolution imposed by light versus electron microscopy. The results indicate that although the numbers of mitochondria per gram brain remain relatively the same, there is a tenfold increase in wet weight of brain from neonate to adulthood and the absolute number of mitochondria in brain must increase therefore. Gregson and Williams' study presents a sense of the "average" differences in mitochondrial number and enzyme content existing between neonatal and mature brain.

The existence of at least two subpopulations of brain mitochondria was emphasized by van den Berg and collaborators (1969, 1975) and by Holtzman et al. (1979). Distribution of mitochondrial enzymes between synaptosomes and neuronal cell bodies in 15-day-old and adult rat brain was studied in some detail by Greengard and McIlwain (1955). The importance of this study, in which six mitochondrial enzyme activities were compared, is that in immature and in adult brain the enzyme composition of the synaptosomal and perikaryal mitochondria differ from each other and undergo separate changes with age. In both instances, but especially in synaptosomal mitochondria, increase in enzyme activity per gram protein was greater than that of total protein. This is intepreted as an increase in the relative specific activity of respiratory enzymes.

Brain mitochondrial enzyme activity also varies with age according to the brain region selected. Although most studies have employed cerebral cortex as the source of mitochondria, Ryder (1980) compared mitochondrial enzyme activities from frontal and medial cortex and striatum from 3-month and 1-year-old rats. Glutamate dehydrogenase, which was present to the same extent at 3 months in all areas, declined to approximately half that level at 1 year of age. In contrast, succinic dehydrogenase, which was highest in striatum at 3 months, increased in all areas by 1 year. Highest activity was demonstrated by NAD-malate dehydrogenase, being approximately 100-fold higher than glutamic dehydrogenase; the latter shows a decline at 1 year to approximately 66% of the 3-month level. Although exhibiting bimodal distribution between mitochondria and cytosol in brain, a larger proportion of activity resides in mitochondria, especially in nerve-ending fractions (Salganicoff and Koeppe, 1968). Gregson and Williams (1969), in comparing the cytochrome content in newborn mitochondria with those from adults, found an increase in adult brain. Chipelinsky and de Lores Arnaiz (1970) extended these findings to include measurements of ubiquinone (CoQ) and found the greatest changes in cytochromes and CoQ to occur between day 10 and day 20 of postnatal development, a period that would include myelination and glial as well as neuronal maturation.

In vivo studies of lead effects on energy metabolism have emphasized the developing nervous system. Patel et al. (1974) studied lead effects on [^{14}C]glucose metabolism in CNS of rat pups 7 to 14 days postnatal. They found that the incorporation of ^{14}C into amino acids increases markedly over this period and beyond in control animals. In sharp contrast, lead treatment significantly depresses incorporation of label into amino acids. Although during the nursing and weaning period the developing brain utilizes ketone bodies to varying degrees, the possibility of a shift in pathway from glucose to ketone bodies was ruled out by their finding that incorporation of label from substrates other than glucose also is decreased. The decline in glutamate and aspartate levels suggested to these authors an effect on tricarboxylic acid cycle oxidation. This may be a direct effect on energy metabolism or due to delayed developmental demands for amino acid precursors in protein synthesis.

The importance of lead effects on cellular energetics and resultant pathogenesis of lead-induced encephalopathy was emphasized in the studies of Holtzman and co-workers. Attention was drawn to the fact that the effects on mitochondrial respiration show relative lead concentration and substrate specificities similar to those seen with *in vitro* exposure of isolated brain mitochondria to lead (Holtzman *et al.*, 1978a,b,1980a). In an earlier study nursing dams were fed 4% lead carbonate from 14th postnatal day and lead was maintained in the pup diet to 28th postnatal day (Holtzman and Hsu, 1976). This late initiation of lead exposure contrasts with many studies in which lead was started at birth. By the latter time, the rudimentary blood–brain barrier, consisting of capillary foot processes, is in place. After 14 days of lead feedings, half the litter develop the encephalopathy. Cerebellar mitochondria from these pups show a marked increase in ADP-independent state 4 respiration with the NAD-linked substrates, glutamate and malate, early in the course of feedings. In later stages of feeding, the ADP-dependent increase in respiration that is normally seen (state 3) is inhibited (Fig. 2). There is a parallel loss of efficiency of oxidative phosphorylation.

Thus, in feeding studies with 14-day-old rat pups, effects of lead on NAD-

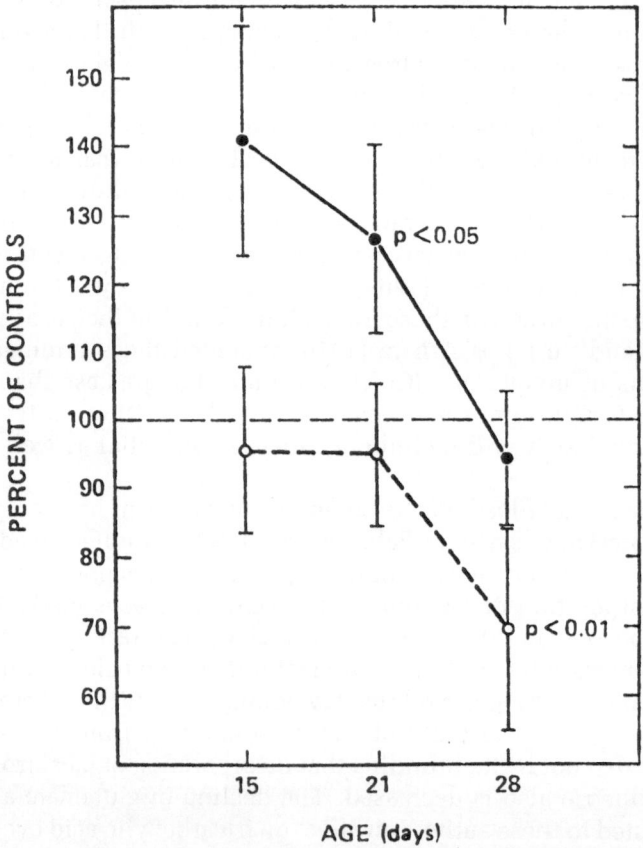

FIGURE 2. Influence of lead on respiratory control ration (RCR) [(○) state 3, (●) state 4] in brain mitochondria. From Holtzman *et al.* (1978b).

linked substrate-supported respiration, e.g., glutamate and malate, include an early increase in ADP-independent respiration (state 4) followed later in the course of feedings by inhibition of state 3 respiration. In cerebellar mitochondria, succinate-supported respiration is refractory to the presence of lead at these times (Table IV). In mitochondria isolated from cerebral cortex of the same animals, respiratory rates with either NAD-linked substrates or succinate are not affected. Mitochondria isolated from cerebral cortex and cerebellum of adult rats on the same lead regimen fail to show any effects with any substrates tested (Holtzman and Hsu, 1976). Thus, respiratory changes are found only in mitochondria from regions in which pathologic changes are found in animals fed lead from 14 days onward. This proposal is supported by the finding that, in pups fed lead for two weeks from birth, mitochondria from cerebellum and cerebral cortex show decreased respiratory control ratios due to inhibition of state 3 respiration with NAD-linked substrates (Holtzman et al., 1978a). Rat pups fed lead from birth show more extensive cerebral damage than when lead feeding is initiated later (Press, 1977a). Although lead has been shown to affect heme synthesis under some circumstances, none of these changes in respiratory rates can be ascribed to decreases in electron transport-related cytochromes (Holtzman et al., 1981). Differences in mitochondrial sensitivity to lead are not due to some qualitative variation in mitochondria from cerebellum as opposed to those from cerebral cortex. This is inferred from the observation that in vitro lead effects are the same in mitochondria isolated from cerebral or cerebellar cortex from immature (14 day on) or adult animals (Holtzman et al., 1978b).

In vitro experiments demonstrate that with either succinate or NAD-linked substrates, lead produces a concentration-dependent increase in respiration

399

HEAVY METAL
TOXICITY AND
ENERGY
METABOLISM IN
THE DEVELOPING
BRAIN

TABLE IV. Respiratory Control Ratios (Mean ± SEM) with Glutamate plus Malate (G + M) or with Succinate as Substrate in Cerebral and Cerebellar Mitochondria Isolated from Rats Fed PbCO$_3$ from 14 Days of Age (Pups) or from 60 Days of Age (Adults)[a]

		Respiratory control ratios[b]			
		Cerebellum		Cerebrum	
Duration of lead feedings (days)		G + M (%)	succinate (%)	G + M (%)	succinate (%)
Pups	2	65.8 ± 7.9 (11)[c] p 0.005[d]	95.5 ± 5.6 (4)	99.9 ± 7.7 (13)	100.6 ± 9.1 (4)
	7	72.4 ± 8.3 (20) p 0.005	105.5 ± 5.4 (9)	91.1 ± 5.0 (20)	103.8 ± 10.4 (10)
	14	77.6 ± 8.7 (16) p 0.025	98.7 ± 5.5 (9)	93.9 ± 6.6 (21)	111.8 ± 9.4 (14)
Adults	14	84.7 ± 15.6 (10)	102.6 ± 3.9 (10)	111.2 ± 8.2 (10)	111.8 ± 5.3 (10)
	28	107.0 ± 11.3 (9)	109.1 ± 5.2 (8)	113.5 ± 7.3 (8)	111.1 ± 5.6 (8)

[a] Table is taken from Holtzman et al. (1980a).
[b] Respiratory control ratio is defined as the ratio of the respiratory rate in presence of ADP to the rate after consumption of the added ADP. Experimental RCRs are expressed as a percentage of RCRs measured in mitochondria isolated from age-matched control animals.
[c] Number of experiments given in parenthesis.
[d] P values (<0.05) are given for differences between results in mitochondria from experimental and matched control animals.

(Holtzman et al., 1978b). The stimulated respiration requires inorganic phosphate (Pi). With the NAD-linked substrates, but not with succinate, lead exposure then inhibits both the increase in respiratory rate and ADP-dependent (state 3) respiration. In other studies, the order in which lead and Pi are added appeared to play a significant role. In heart mitochondria, lead exerts a greater degree of inhibition if it is added prior to Pi (Parr and Harris, 1976). In experiments on isolated brain mitochondria from young adult rats, maintenance of high-energy phosphates is affected by the presence of Pb^{2+} in which a similar ordered sequence also was employed (Table V) (Gmerek et al., 1981). In kidney, as in brain mitochondria, low Pb^{2+} levels caused an increase in state 4 respiration, but higher levels also significantly reduced state 3 respiration (van Rossum and Kapoor, 1984). These authors concluded that lead acts to stimulate state 4 by uncoupling oxidative phosphorylation, an effect which is oligomycin insensitive.

The cellular effects of lead may directly or indirectly (i.e., by inhibiting aerobic energy metabolism) alter cellular calcium homeostasis. Lead inhibits calcium uptake by isolated mitochondria (Goldstein, 1977). Ruthenium red, a mucopolysaccharide stain that inhibits energy-dependent mitochondrial calcium transport (Moore, 1971; Ash and Bygrave, 1977), inhibits both the lead-induced respiratory stimulation and the delayed respiratory inhibition in isolated brain mitochondria (Holtzman et al., 1980b). These effects of lead on mitochondrial calcium transport or binding may be important in the altered cellular calcium uptake and release demonstrated in isolated brain capillaries (Goldstein et al., 1977) and in primary cultured hepatocytes (Pounds et al., 1982a,b).

1.4. Effects on Neurotransmitters and Calcium Homeostasis

Subclinical lead toxicity or the sequelae of lead encephalopathy in children is manifested by evidence of behavior and learning disorders (e.g., Lin-Fu, 1980). From studies in lead-treated animals, it has been proposed that changes in central catecholamine function may be responsible (Sauerhoff and Michaelson, 1973; Silbergeld and Goldberg, 1975). In a series of papers, Trabucchi and co-workers (Lucchi et al., 1981) have studied the effects of chronic lead treatment on the dopaminergic system. They found in earlier studies that dopamine turnover was affected, but that dopamine receptor response related to adenylate cyclase activity was not modified (Govoni et al., 1979). Recently, they (Lucchi et al., 1981) have presented evidence that there are drugs, known to be selective for one type of dopamine receptor, that antagonize the hyperactivity in lead-exposed rats. It is known that when there is diminished dopamine output presynaptically, there are postsynaptic changes in the dopamine population of receptors, described as being of the subtypes D_1 and D_2. There is evidence showing that the D_1 subtype is associated with adenylate cyclase activity, in contrast to the D_2 subtype. Administration of the dopamine receptor-binding drug tiapride, according to Lucchi et al., (1981), has proved palliative in children suffering various forms of hyperkinetic behavior. This drug apparently does not affect adenylate cyclase-related dopamine activity. Similarly, D_2 binding is now well characterized by a benzamide homolog, $(-)$sulpiride. Lucchi

401

HEAVY METAL
TOXICITY AND
ENERGY
METABOLISM IN
THE DEVELOPING
BRAIN

et al. found that this drug is more potent in producing sedation in lead-treated animals than in controls. This contrasts with the D_1-receptor-binding drug haloperidol, which produces sedation at equal doses in control and lead-treated animals. Neurochemical evidence based on $(-)[^3H]$sulpiride binding studies (a D_2-ligand-binding drug) provides a basis for the suggestion that lead causes preferential effects on D_2 receptors, effects that vary with the brain area examined. For example, D_2-receptor activity is increased in striatum, but decreased in the nucleus accumbens.

There is evidence that acetylcholine (ACh) release in some regions of the CNS is regulated by presynaptic D_2 receptors. Several studies on lead actions have reported inhibition of ACL release in peripheral neural tissue (Kostial and Vouk, 1957; Silbergeld, 1973; Silbergeld and Goldberg, 1974) and in the CNS (Hrdina *et al.*, 1976; Carroll *et al.*, 1977; Silbergeld and Adler, 1978). Shih and Hanin (1978) demonstrated *in vivo*, that Pb^{2+} treatment decreases brain turnover rate without affecting ACh content. Chronic exposure to a 1% lead acetate solution in drinking water of rats results in significant changes centrally in the enzymes choline acetyltransferase and acetylcholinesterase and in ACh content, according to Modak *et al.* (1975). In a recent study in adult rats, chronic (20 day) exposure to lead in drinking water produced effects on brain glucose metabolism and Ach synthesis (Sterling *et al.*, 1982). Following exposure, rats were sacrificed and cerebral cortex slices were incubated with 10 mM glucose labeled with tracer amounts of $[6-^3H]$glucose and either $[6-^{14}C]$glucose or $[3-^{14}C]d,l-\beta$-hydroxybutyrate. Incorporation of mixed labeled glucose into lactate, citrate, and ACh was considerably decreased, although no changes were noted in the $^3H/^{14}C$ ratio (Table VI). A change in the ratio would occur if some step in ACh synthesis had been affected. Similar effects of lead were found when ^{14}C-labeled d,l-β-hydroxybutyrate was substituted for $[^{14}C]$glucose (Table VI). It would appear from an analysis of these and other data (Table VI) that Pb^{2+} exerts a generalized effect on energy metabolism and not on a specific step in the synthetic pathway from glucose to ACh. It has been proposed that the mechanism of Pb^{2+}-induced changes involves competitive replacement of Ca^{2+} at presynaptic sites, affecting ACh release or the availability of choline (Silbergeld and Goldberg, 1975; Carroll *et al.*, 1977; Silbergeld, 1977; Silbergeld and Adler, 1978). However, Pb^{2+} also binds to rat brain mitochondria, significantly decreasing Krebs cycle activity and energy metabolism (Bull *et al.*, 1975; Holtzman *et al.*, 1978b; Gmerek *et al.*, 1981). This effect on mitochondrial metabolism may limit availability of acetylcoenzyme A (AcCoA) essential to ACh production.

The information regarding the effects of chronic lead exposure on ACh receptor pharmacology is sparse. *In vitro* studies by Aronstam *et al.* (1978) indicate that in the presence of lead, there are changes in muscarinic receptor binding activity. These effects were measured by radioligand binding with ^3H-labeled quinuclidine benzillate $[(^3H)-QNB]$, an atropinelike compound. Binding appears to be increased by low doses of lead and impaired at higher levels suggesting some form of allosteric change.

When pregnant rat dams, in the second day of gestation, are placed on lead-containing drinking water (600 ppm) and spinal cords dissected free from rat fetuses from day 15 of gestation onward, lead can be shown to affect the

TABLE V. Influence of Low Levels of Pb^{2+} on Phosphorylation[a]

[Pb^{2+}] (μg/mg protein)	[Mitochondrial protein] (mg/ml)	ATP	ADP	AMP	Total nucleotides (μmoles)	Number of samples
Controls [NaAc]	6 ± 0.6	87 ± 2.3	3 ± 1.2	10 ± 1.5	3.9 ± 0.49	7
0.4 – 0.5	8 ± 0.4	60 ± 6.0[b]	22 ± 3.7[b]	15 ± 0.7	5.0 ± 0.51	6
1.0 – 2.5	5 ± 0.8	30 ± 2.4[b]	57 ± 6.0[b]	14 ± 1.1	5.7 ± 0.25	7

[a] From Gmerek et al. (1981). Relative amounts of adenine nucleotides calculated as the percent of total per sample, given as the mean ± SEM; [P$_1$] = 4.5 mm.
[b] $P < 0.005$, calculated by Student's t test.

TABLE VI. Conversion of ^{3}H from [6-^{3}H]Glucose and ^{14}C from [3-^{14}C]β-Hydroxybutyrate to Metabolic Intermediates and Acetylcholine[a]

Product	Control (n = 7): Radioactivity recovered in product			Pb^{2+}-treated (n = 6): Radioactivity recovered in product (% change from control)		
	^{3}H	^{14}C	^{3}H/^{14}C	^{3}H	^{14}C	^{3}H/^{14}C
Lactate	21.4 ± 0.80	0.33 ± 0.01	64.8 ± 3.4	14.8 ± 0.3[b] (−30.9%)	0.25 ± 0.02 (−24.3%)	59.2 ± 3.6 (−8.7%)
Citrate (total)	0.079 ± 0.006	0.693 ± 0.045	0.11 ± 0.01	0.045 = 0.002[b] (−43.1%)	0.303 ± 0.033[b] (−56.3%)	0.15 ± 0.01 (+36.3%)
Citrate (4.5)	0.054 ± 0.002	0.178 ± 0.010	0.30 ± 0.01	0.026 ± 0.002[b] (−51.9%)	0.108 ± 0.012[b] (−39.4%)	0.24 ± 0.02 (−20.0%)
Malate	0.061 ± 0.002	0.411 ± 0.045	0.15 ± 0.02	0.042 ± 0.003[b] (−31.2%)	0.272 ± 0.028 (−33.9%)	0.15 ± 0.01 (−0—)
Acetylcholine	0.017 ± 0.002	0.041 ± 0.005	0.41 ± 0.02	0.010 ± 0.001[b] (−41.2%)	0.027 ± 0.003[a] (−34.2%)	0.37 ± 0.02 (−9.2%)

[a] From Sterling et al. (1982).
[b] $P < 0.005$.
[c] $P < 0.001$, compared with controls.

cholinergic system (J.J. O'Neill et al., unpublished data). Employing [^{3}H]-QNB to measure specific muscarinic binding and [^{125}I]alpha bungarotoxin as an index of nicotinic receptor binding, there is a 1–2 day delay in changes in specific binding for both types of receptors up to postnatal day 20. A similar delay in appearance of increases in the enzyme choline acetyltransferase parallels these changes. These effects are not explained by a generalized effect on protein synthesis since all results are normalized for protein content. Attempts to correlate cholinergic changes in spinal cord development with synaptogenesis were inconclusive.

McCauley et al. (1979, 1982) (Table II), utilizing an electron microscopy technique that employs ethanolic phosphotungstic acid (Aghajanian and Bloom, 1967), reported that low lead (200 ppm) in drinking water of female rats during gestation and lactation causes a delay in maturation of the cerebral cortex in offspring. In thin sagittal sections prepared from 15-day postnatal rat brain, there was a statistically significant decrease in the number and qualitative appearance of stained synaptic figures when compared with control brains. The initial studies chose postnatal day 15 because as Bull et al. (1979) (Table VII) had demonstrated earlier, delays in the postnatal increase in the various cerebral cytochromes was greatest at 15-day's time.

In a subsequent study reported by these same authors (McCauley et al., 1982), attempts were made to correlate changes in synaptogenesis with blood Pb during gestation and in the postnatal period to 21 days. Time points for electron microscopy were 11,15, and 21 days of age, respectively. According to the authors, day 11 was chosen as the time used by Aghajanian and Bloom (1967) before initiation of synaptogenesis in layer 1 of the parietal cortex in rat. Similarly, day 21 marked the end of postpartum synaptogenesis in this region. The tissues sliced for subsequent electron microscopy consisted of cortical layers 1, 2, and 3. Lead exposure in utero caused a marked decrease in synaptic densities ($2.92 \pm 0.42 \times 10^8$/mm^3) when compared with tissue from control brain (4.15 ± 0.36) or with tissue from pups exposed to lead postnatally ($3.97 \pm 0.57 \times 10^8$/mm^3). Brains from pups exposed to lead in utero but crossfostered with nonlead dams postnatally also had depressed synaptic numbers ($2.78 \pm 0.51 \times 10^8$/mm^3). The effects only correlated well with blood lead levels at birth since at day 15, pups on lead throughout the prenatal and postnatal periods had blood Pb levels similar to animals only exposed to lead post-

403

HEAVY METAL
TOXICITY AND
ENERGY
METABOLISM IN
THE DEVELOPING
BRAIN

TABLE VII. The Effect of Prenatal and Postnatal Treatment with Pb on the Normal Increase in Cerebral Cytochrome $c + c_1$ Concentrations between 10 to 15 Days of Age in the Rat Pup and Blood Pb Concentrations at Weaning[a]

	Number	Blood Pb, 21 days of age	OD, 550 nm, 10–15 days
Control	25	8.1 ± 2.0[b]	100.0 ± 8.5[c]
5 mg Pb/1	12	11.7 ± 2.3	89.9 ± 10.0
30 mg Pb/1	10	21.3 ± 1.9	82.4 ± 10.9
200 mg Pb/1	20	35.7 ± 3.4	70.8 ± 8.1

[a] From Bull et al. (1979).
[b] μg Pb/100 g $\pm$ SEM.
[c] Percent of control change $\pm$ SEM.

partum. Blood lead from rat pups exposed only *in utero* had 15-day levels similar to control animals, i.e., 6.9 ± 0.9 versus 4.6 ± 0.9 µg/dl ± SEM. The authors conclude that lead effects occur from *in utero* exposure, but the effects on synaptic delay (Fig. 3) are reversible since there were no significant differences between treated and control brains observed either at day 11 or day 21 of age. The results in brains obtained from control animals compare favorably with the earlier reports of Aghajanian and Bloom (1967) and with Krigman and Hogan (1974) using more standard techniques of transmission electron microscopy. In the latter study, there were decreased numbers of synapses and synapses per cortical neuron in 30-day postpartum rats. Since much higher lead exposure was used (4% lead carbonate) for the full 25 days of nursing, these differences appear to be dose related. It is concluded that prenatal exposure to lead causes subsequent postnatal synaptic delay, possibly due to the abnormally elevated blood Pb levels (Fig. 4) during premigration mitosis of neuroblasts that are later found in cortical layers 2 and 3.

The role of calcium ions in transmitter function is well known (Rahamimoff, 1976). The effects of heavy metals on neurotransmission are less well established. Silbergeld and Adler (1978) in describing possible subcellular mechanisms of lead neurotoxicity suggest that interference with Ca^{2+} ions

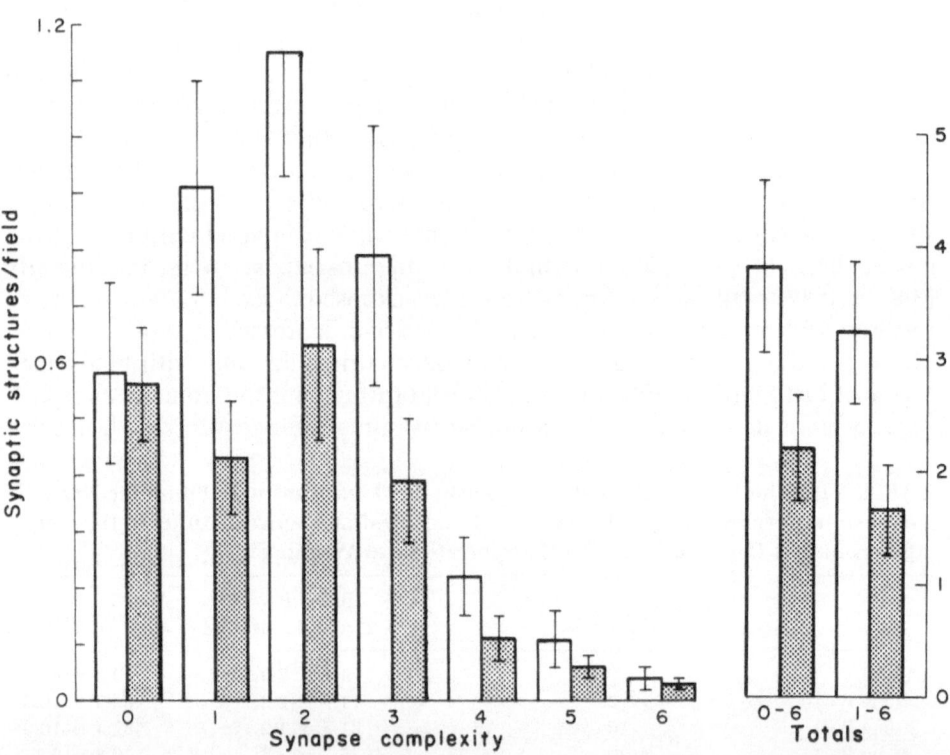

FIGURE 3. Pb-induced depression of synaptic figures in the 15-day-old rat cerebral cortex. Level of complexity was judged by the number of presynaptic dense projections. Results of 7 (□) control and 7 (▨) Pb litters ± SEM. From McCauley *et al.* (1982).

405

*HEAVY METAL
TOXICITY AND
ENERGY
METABOLISM IN
THE DEVELOPING
BRAIN*

results in qualitatively different effects depending on the neurotransmitter. Lead blocks ^{45}Ca binding to peripheral cholinergic ganglia, but increases ^{45}Ca binding to synaptosomes from caudate nucleus. In the CNS, Pb^{2+} ions appear to increase intracellular Na$^+$ ion content that relates to the triggering of Na$^+$–Ca^{2+} exchange reaction at the level of brain mitochondria (Lowe *et al.*, 1976). In addition, Pb^{2+} has been shown to inhibit the Na$^+$-induced release of ^{45}Ca from isolated mitochondria from brain. This, Silbergeld and Adler (1978) have suggested, would result in increased transmembrane flux of exogenous Ca^{2+} and increased exocytotic events at the synapse. Norepinephrine and dopamine release which are calcium-dependent exocytotic events are reported to be increased in lead-exposed animals (Golter and Michaelson, 1975; Silbergeld and Goldberg, 1975; Silbergeld, 1977) and *in vitro*. In the CNS, inhibition of ACh release may be a direct result of Pb^{2+}–Ca^{2+} antagonism or an indirect result of dopamine presynaptic inhibition of ACh release.

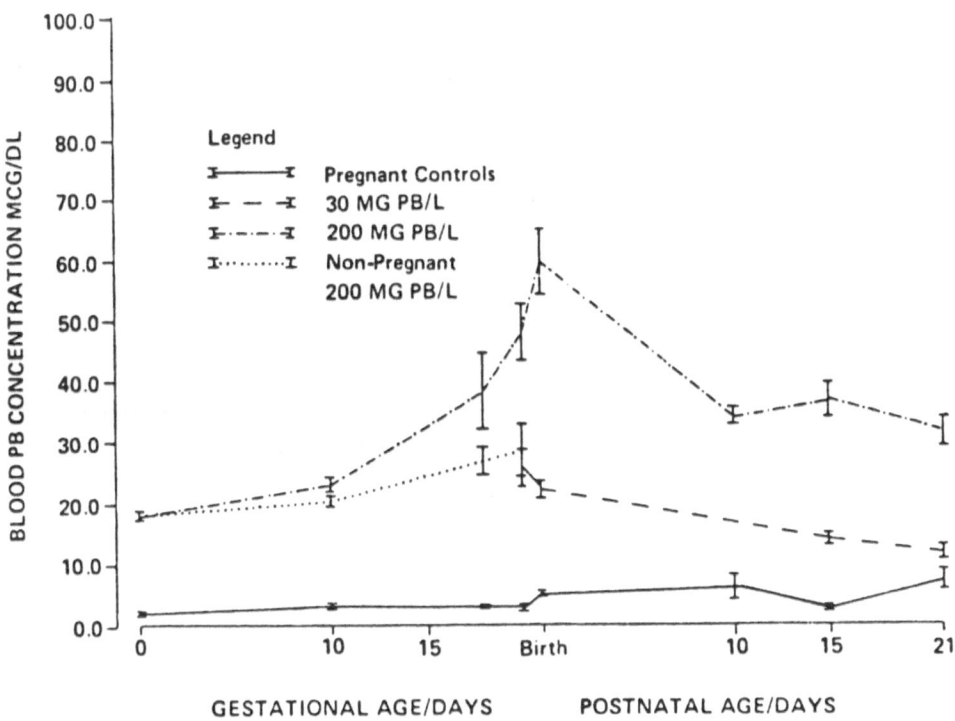

FIGURE 4. Temporal changes of blood Pb concentrations of nonpregnant, pregnant, and lactating female rats exposed to 0, (– – –) 30, or (–·–·) 200 mg Pb/liter drinking water. Pb treatment began 14 days prior to breeding of animals and continued through the 21st postnatal day. Note that the dotted line (····) indicating blood lead levels in nonpregnant rat treated with 200 mg Pb/liter intersects a dashed line 1 day before birth. This dashed line indicated blood lead levels for dam treated with 30 mg Pb/liter drinking water. Data for nonpregnant control dams are not represented for clarity's sake, since these values are nearly identical to (—) pregnant control blood lead levels values. Blood lead concentrations were not evaluated in 30 mg Pb/liter treated pregnant rats. N = 8 for all points except 200 mg Pb/liter pregnant rats, where N = 24. The cross bars represent SEM. From McCauley *et al.* (1982).

1.5. Mechanisms of Resistance to Lead Encephalopahy in the Mature Brain

As described in Section 1.3, effects on cellular energy metabolism appear to be important in the pathogenesis of lead encephalopathy. Recent studies suggest that the brain becomes resistant to lead toxicity by mechanism(s) that sequester the metal in extramitochondrial (nontoxic) sites (Holtzman *et al.*, 1984; C. DeVries et al., in preparation). These studies were based on the following observations: (1) the *in vitro* effects of lead are the same in immature and mature cerebellar or cerebral mitochondria (Holtzman *et al.*, 1978b); (2) the cerebellar lead concentrations are the same in immature (encephalopathic) and mature (encephalopathy resistant) lead-fed animals (Holtzman *et al.*, 1982); and, (3) cerebellar mitochondria from animals fed lead from 14 days of age contain more lead than cerebral mitochondria from these animals and than cerebellar mitochondria from lead-fed adults (Holtzman *et al.*, 1980a).

The differences in sensitivity to lead toxicity in different brain regions or with maturation are not related to differences in the regional concentrations of lead. Cerebellar lead concentrations in pups fed encephalopathic lead doses for two weeks beginning at 14–18 days of age are between 20–30 µg Pb/g protein (Holtzman *et al.*, 1982). The concentrations of cerebral lead in these animals also are in this range. The greater differences between cerebellar and cerebral lead concentrations in the younger animals may be related to the presence of red cell-bound lead in the more extensive cerebellar hemorrhages found in the pups fed from 14 days compared with pups fed lead from 18 days.

In animals fed high daily lead doses from 20 days of age, in which the only light microscopic changes are patchy edema, and in animals fed lead from 24 days of age, in which there are no histopathologic indications of lead toxicity by light microscopy, cerebellar lead concentrations are the same as those of younger pups. Correcting for the higher blood lead levels in the older animals, the cerebellar parenchymal lead concentration may be, at most, 25% lower in the animals fed 3200 µg lead/g body weight from 24 days of age compared with animals fed 1600 µg lead/g body weight from 18 days of age. Again, the red cell-bound lead in the hemorrhagic immature cerebellum makes up at least part of this difference (Kochen and Greener, 1977).

The cellular and subcellular distributions of lead are very different in the cerebellum, compared with the cerebral cortex, of animals fed an encephalopathic lead dose for two weeks beginning at 18 days of age (Table VIII). The cerebellums of these animals show hemorrhage, extracellular proteinaceous edema, and necrosis of granular neurons by electron microscopy (C. DeVries, *et al.*, in preparation). Astrocytes, capillary endothelium, and Purkinje neurons generally appear normal. Lead is found, by electron microprobe elemental analysis, diffusely in small quantities in the cytoplasm, nuclei, mitochondria, and lysosomes of granular neurons. Astrocytes show a similar, but less consistent, distribution of lead. The smaller number of capillary endothelial cells, which were studied, also contain diffuse low levels of lead, while no lead could be measured in Purkinje cells.

In contrast to the cerebellums of rat pups fed lead from 18 days of age, the cerebral cortical sections from these animals do not show evidence of cell necrosis by electron microscopic examination (C. De Vries *et al.*, in preparation). Specifically, small neurons appear normal and show only low levels of lead in an occasional mitochondrion (Table VIII). Astrocytes also appear normal except for the frequent dense inclusions in lysosomes and nuclei and occasionally in the cytoplasm (Holtzman *et al.*, 1984). These inclusions contain large quantities of lead by microprobe analysis. Outside these inclusions, lead is not demonstrable in the cytoplasm or nucleus, but it can be present in low concentrations in an occasional astrocyte mitochondrion. Cytoplasmic and nuclear inclusions are occasionally found in pericytes, microglia, and oligodendroglia. Capillary endothelial cells, in the cerebral cortex as in the cerebellum, show diffuse low levels of lead. Similar inclusions and patterns of lead distribution are found in the cerebral cortex of rat pups fed lead beginning at 24 days of age.

These observations on the cellular distribution of lead in the encephalopathy-resistant rat pup brain are similar to those reported in other cell types. Cytoplasmic and intranuclear inclusions have been observed in the epithelial cells of the renal proximal tubules of lead-fed rats (Goyer *et al.*, 1970; Choi and Richter, 1972; Goyer and Moore, 1974). These authors proposed that localization of lead in the nucleus protects these cells by preventing diffusion of lead to cytoplasmic sites, such as mitochondria, which are susceptible to the toxic effects of the metal (Goyer and Moore, 1974). In an important recent study, lead-binding proteins were found in kidney and brain, but not liver or lung (Oskarsson *et al.*, 1982). These proteins may be involved in the transport and binding of lead in intracellular sites, since they are present in those tissues in which lead-containing inclusions are found. Their presence specifically in astrocytes could be the mechanism for binding and localizing intracellular lead. By this mechanism, the astrocyte may protect the more vulnerable neurons by taking up most of the parenchymal lead. Toews *et al.* (1978) found, by atomic absorption spectroscopy, similar concentrations of lead in swollen and vacuolated cerebellar capillaries from 5-day-old lead-fed rat pups and in normal-

TABLE VIII. **Subcellular Distribution of Lead in the Cerebellums and Cerebrums of Rat Pups Fed Lead Nitrate for Two Weeks Beginning at 18 Days of Age**[a,b]

	Cerebellum	Cerebrum
Neurons (small)		
Cytoplasm	+	−
Mitochondria	+	±
Lysosomes	+	−
Nuclei	+	−
Astrocytes		
Cytoplasm	−	+ ±
Mitochondria	±	±
Lysosomes	±	+ + or +
Nuclei	+	+ ±

[a] Data taken from Holtzman *et al.* (1983).
[b] −, no lead by EDX
+, diffuse, above background
+ +, lead-containing densities, well above background

407

HEAVY METAL
TOXICITY AND
ENERGY
METABOLISM IN
THE DEVELOPING
BRAIN

appearing capillaries isolated from 20-day-old lead-fed pups. Consistent with this observation, Holtzman *et al.* (1983) found a similar pattern of lead distribution in the "sensitive" and "resistant" endothelium suggesting that these cells develop a resistance to the presence of intracellular lead. In the encephalopathy-resistant brain, the adjacent astrocytes may add protection by preventing excessive intracellular lead accumulation.

OTHER METALS WITH AGE-DEPENDENT SUSCEPTIBILITY AND CNS TOXIC MECHANISMS POTENTIALLY SIMILAR TO THOSE OF LEAD

2. Introduction

2.1. Mercury

The industrial use of mercury compounds as catalysts for organic synthetic processes led to the contamination of Minamata Bay. The outbreak of methyl mercury poisoning in Japan in the early 1950s emphasized that mercury and particularly methyl mercury are extremely hazardous neurotoxicants (Chang, 1977). Alkyl mercury compounds are readily absorbed through the pulmonary and gastrointestinal systems and cause damage to the CNS in human subjects and experimental animals. Neuropathological alterations in the fine structure of adult animals after mercury intoxication has received extensive coverage, but much less information is available concerning neuropathology in the neonate (Change, 1977). As pointed out by these authors, this is somewhat ironic since among the 111 cases of Minamata disease originally detected, 19 were newborns.

Methyl mercury rapidly traverses the placenta and a single, low teratogenic dose of methyl mercury results in a fourfold greater concentration of mercury in the fetal brain than in the maternal brain (Matsumoto *et al.*, 1967). Degenerative changes in the neonatal rat nervous system have been described by Chang *et al.* (1977) following prenatal exposure to a nonteratogenic dose of methyl mercury. A prominent finding is focal disruption on the nuclear envelope. The blood–brain barrier system in the neonate may be more vulnerable than in adults to methyl mercury making the former highly vulnerable. Swollen capillary endothelial cells, myelin figures, and large areas of focal degradation also are seen.

These pathologic alterations are consistent with observations previously reported by Chang and collaborators for neonatal liver and kidney, indicating that developing organs are more vulnerable to mercury and its derivatives than are corresponding adult tissues. Although the subcellular distribution may vary somewhat in different organs, in general, subcellular mercury is largely distributed between the mitochondrial fraction, microsomal fraction, and supernatant, whereas little is present in the nuclear fraction (Chang, 1977). In brain, the cellular distribution after methyl mercury intoxication is: neurons of calcarine cortex > Purkinje cells of cerebellum > anterior horn motor neurons >

409

HEAVY METAL
TOXICITY AND
ENERGY
METABOLISM IN
THE DEVELOPING
BRAIN

granule cells of cerebellum. After mercuric chloride administration, the distribution had the following rank order: dorsal root ganglion neurons > Purkinje cells of cerebellum > anterior horn motor neurons > calcarine cortical neurons > granule cells of cerebellum. The amounts of mercury found in glia and Schwann cells are about the same in all regions.

In spite of widespread effects of mercury on the central and peripheral nervous system in animals and man, the molecular mechanisms(s) are not well established. Electrophysiological studies of skeletal muscle receptor indicate that methyl mercury acts both presynaptically and postsynaptically (von Burg and Landry, 1976; Juang, 1976). Eldefrawi et al. (1975) demonstrated that methyl mercury prevents acetylcholine interaction with nicotinic receptors. In subsequent studies, Shamoo and collaborators (Fig. 5) (Abd-Elfattah et al., 1981) showed both mercuric ions and methyl mercury strongly inhibit the binding of ^{3}H-QNB to muscarinic ACh receptors in rat brain synaptosomes. Mercuric ions were 350 times as potent as methyl mercury in this respect. Evidence was presented for involvement of sulfhydral (SH) groups based on the reversibility of the inhibition by D-penicillamine (Fig. 6). In contrast with nicotinic receptors that contain essential disulfide linkages, muscarinic receptors have free SH groups available. To explain their observations, the authors propose the model shown in Fig. 7.

At the neuromuscular junction, mercuric chloride and the organic mercury compound, mersalyl, first increase and later decrease both spontaneous and stimulated release of acetylcholine (Binah et al., 1978). Lower concentrations of $HgCl_2$ cause only an inhibitory effect. It has been suggested that the increase of transmitter release is associated with an interaction with SH groups in the presynaptic membrane (Manalis and Cooper, 1975; Juang, 1976). Binah et al. (1978) (Table IX) offer an alternative explanation, suggesting that mercurials affect regulatory mechanisms of intracellular Ca^{2+} ions, i.e., mitochondrial

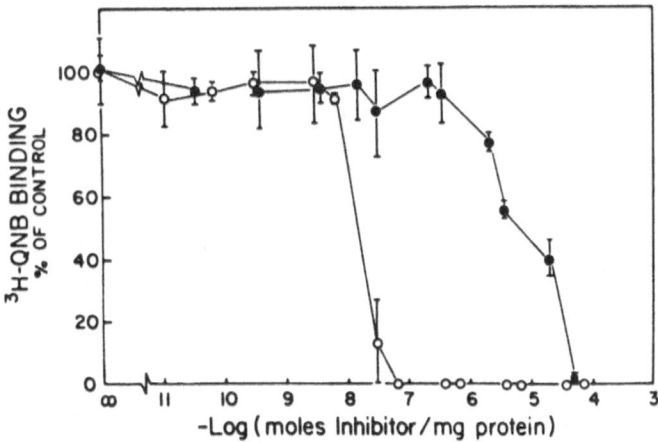

FIGURE 5. Inhibition of [^{3}H]-QNB binding to rat brain muscarinic receptors by methyl mercury (●) and mercuric chloride (○). Results were based on milligrams of protein to eliminate variations between separate experiments. Each point represents the mean of three to five different experiments ± standard error of the mean. From Abd-Elfattah et al. (1981).

calcium transport and calcium transport of synaptosomal vesicles. This interpretation is based in part on the observed inhibition of mitochondrial transport by 0.5 μM HgCl₂ or 5 μM mersalyl. Mercurials are thought to act directly on the calcium carrier since, at those concentrations, mitochondrial oxidative phosphorylation is not impaired. Higher concentrations appear to affect most mitochondrial function (Southard et al., 1974). The reduction in transmitter release following the observed initial increase could result from a build up of mercuric compounds inside nerve terminals.

Studies relating organic mercury encephalopathy to energy metabolism have not been described in the central nervous system of the developing animal. In adult rat, Verity et al. (1975) developed a model of subacute methyl mercury (MeHg) intoxication by daily intragastric administration of 10 mg of MeHg/kg. Synaptosomes isolated from brain during a latent phase (6–10 days) show no significant change in respiratory control ratio (RCR). During the neurotoxic phase, a significant decline in RCR is observed in cerebral synaptosomes. Even greater effects on respiration were noted in synaptosomes from the cerebellum when stimulated by 2,4-dinitrophenol employing glutamate as substrate. In vitro, at 5–15 μM MeHg, there is an increase in state 4 respiration, inhibition of state 3, but no change in 2,4-dinitrophenol uncoupled respiration with pyruvate malate. Above 25 μM, considerable inhibition of electron transfer was observed and cytochrome oxidase was inhibited by 50%. Verity et al. (1975)

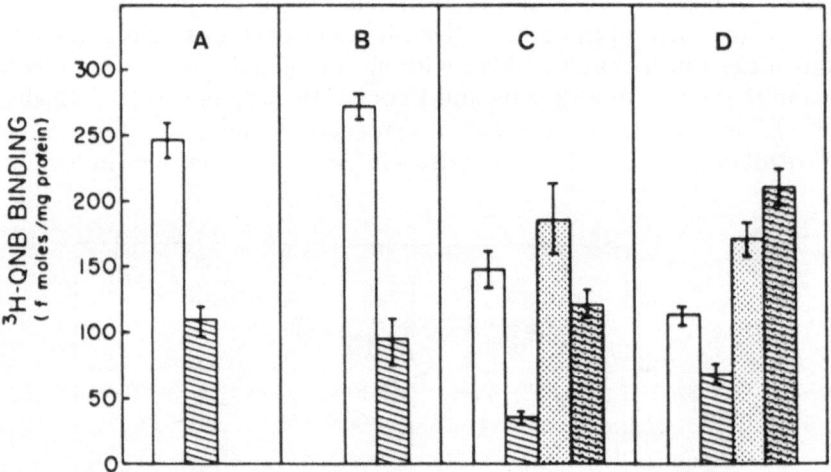

FIGURE 6. Regeneration of rat brain muscarinic receptor by D-penicillamine after inhibition with mercuric chloride. Rat brain lysed synaptosomes (5 mg of protein) samples went through different treatments from panels A to D. Aliquots were taken from each sample in each treatment for the [³H]-QNB binding assay and protein determination. Panel A represents [³H]-QNB binding to rat brain synaptosomes treated with 5×10^{-5} M mercuric chloride for 1 hr (▨) compared with control (□). Panel B shows the same effect as in A after washing the free and loosely bound mercuric cations. Panel C shows the effect of D-penicillamine (10^{-3} M) treatment (for 1 hr) on [³H]-QNB binding of mercuric-untreated (▦) and mercuric-treated (▨) washed synaptosomes in comparison with [³H]-QNB binding to control (□) and mercuric-treated (▨) washed synaptosomes. Panel D represents the same procedure as in C except that all samples were washed to remove the free D-penicillamine and the soluble mercuric-D-penicillamine complexes from the incubation medium. From Abd-Elfattah et al. (1981).

previously had shown that MeHg inhibits protein synthesis in synaptosomes in both a cycloheximide-sensitive and a smaller, chloramphenicol-sensitive compartment. From their study on inhibitors of mitochondrial function, they suggested a close relationship between mitochondrial respiration and extramitochondrial protein synthesis. In a more recent paper (Cheung and Verity, 1981), they showed that addition of MeHg to a cell-free protein-synthesizing system, which contains brain microsomes from neonatal rats, inhibits [³H]leucine incorporation. In the presence of MeHg, there was a dose-dependent decline in ATP. This decline in ATP is thought to account for inhibition of protein synthesis by mercurials.

Direct *in vitro* effects of HgCl₂ on renal cortical mitochondrial function have been described by Weinberg *et al.* (1982). These studies were performed prior to studies concerning the effects of HgCl₂ *in vivo*. At a threshold level of 2 mM Hg²⁺, or 1 nmole Hg²⁺/mg mitochondrial protein, there was marked stimulation of state 4 respiration and a mild inhibition of state 3, but a significant increase in ADP translocation (atractyloside-insensitive form). In addition, there was a stimulation of oligomycin-sensitive mitochondrial ATPase. These effects could be prevented or reversed by either dithiothreitol (DTT) or

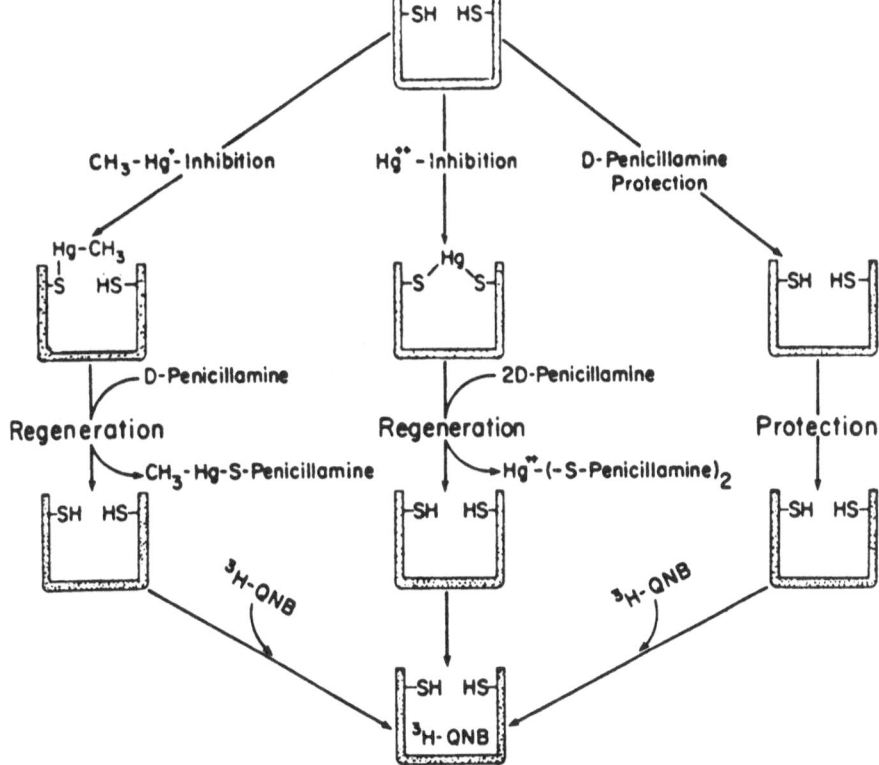

FIGURE 7. A hypothetical model of the interaction of methyl mercury and mercuric chloride with muscarinic receptors. The possible role of essential −SH groups in [³H]-QNB binding sites and their relative sensitivities to mercurials and regeneration of muscarinic receptors by D-penicillamine are shown in the model. From Abd-Elfattah *et al.* (1981).

the presence of serum albumin, although the effects could not be prevented by the presence of other divalent metal ions or easily washed off by nonprotein-containing solutions. Following exposure of adult Sprague-Dawley rats to 5 mg/kg of $HgCl_2$ subcutaneously, kidney cortical mitochondria were isolated 1 to 3 hr later. The renal cortex and renal cortical mitochondria contained 1.87 and 0.72 nmole of Hg^{2+}/mg protein, respectively. Unlike the *in vitro* effects, *in vivo* exposure resulted in decreased ADP intake by isolated mitochondria. At 2 hr, there was inhibition of state 3 respiration and 2,4-dinitrophenol-stimulated respiration as well. These effects were not reversed by DTT or by albumin. These results implicate compromised mitochondrial bioenergetics, but also suggest cellular processes being involved in addition to those observed with *in vitro* exposure of mitochondria to Hg^{2+} ions.

2.2. Cadmium

In addition to kidney effects on systemic blood pressure through an angiotensin II mechanism, cadmium is especially harmful to the neonatal CNS (Gabbiani *et al.*, 1967). Acute exposure studies to $CdCl_2$ in a QS-outbred strain of mice 1, 8, 15, and 22 days of age were reported by Webster and Valois (1981). Histological studies 24 hr after exposure of 1-day-old mice to 2 mg Cd^{2+}/kg reveal petechial hemorrhages, edema, and cellular pyknosis throughout the brain. At day 8 (4 mg/kg) and day 15 (6 mg/kg), similar damage is observed, but is confined to the cerebellum. By postnatal day 22 (8 mg/kg Cd^{2+}), the brain appears to be unaffected. However, animals allowed to survive 6–8 weeks after cadmium exposure at the older ages do show behavioral abnormalities (Webster and Valois, 1981). Electron microscopic studies on 1-day-old animals indicate petechial hemorrhages occur 2 hr after dosing and damage to capillaries increases over the next 6 hr. Degenerative changes in nerve cells appear at this later time. Less severe endothelial vacuolization has been recognized in fetal brains from dams exposed to cadmium late in gestation (Rohrer *et al.*, 1978).

Chronic cadmium toxicity in rats leads to increased motor activity and as with manganese toxicity, to an increase in tyrosine hydroxylase (striatum) and

TABLE IX. The Effects of $HgCl_2$ on the Calcium Uptake by Rat Brain Mitochondria[a,b]

t (min)	Control (no $HgCl_2$)	0.01 μM $HgCl_2$	0.1 μM $HgCl_2$	1 μM $HgCl_2$	2 μM $HgCl_2$	5 μM $HgCl_2$
3	0.46	0.48	0.48	0.26	0.02	0.2
9	0.92	1.02	1.27	0.47	0.14	0.3
15	1.0	1.10	1.61	0.58	0.25	0.27

[a] From Binah *et al.* (1978).
[b] Results expressed as nmole/Ca taken up/mg mitochondrial protein per t as mentioned. The incubation mixture contained: 80 mM KCl, 4 mM $MgCl_2$, 10 mM Tris-HCl pH 7.4, 10 mM K-succinate, 2 mM ADP(K), 1 mM K_2PO_4, 2.5 μM $^{45}CaCl_2$, about 1 mg mitochondrial protein/ml of reaction mixture and $HgCl_2$ as stated. The results are the average of at least 4 different mitochondrial preparations for each concentration of $HgCl_2$ tested (S.D. of each value not above ±0.05 nmoles Ca/mg protein per t).

413

*HEAVY METAL
TOXICITY AND
ENERGY
METABOLISM IN
THE DEVELOPING
BRAIN*

tryptophane hydroxylase (midbrain) (Rastogi *et al.*, 1977). Muscarinic receptor binding also is impaired (Hedlund *et al.*, 1979). *In vitro* cadmium ions inhibit (Na^+,K^+)ATPase at an $IC_{50} = 5$ µM. Only at relatively high concentrations (200 µM) is there significant cadmium inhibition of choline uptake *in vitro*. Lai *et al.* (1980b) reported similar inhibition of choline uptake by aluminum (III) and manganese (II) at high concentration although neither ion was as potent as Cd^{2+} ion as in inhibitor of (Na^+,K^+)ATPase. Aluminum ions are somewhat more potent as an inhibitor of choline uptake $(Al^{3+}\ IC_{50} = 123$ µM vs. Cd IC_{50} $= 363$ µM), but much less potent versus (Na^+,K^+)ATPase (Al $IC_{50} = 8.3$ mM vs. Cd $IC_{50} = 5$ µM). Cadmium at this concentration has no effect on choline uptake. All three cations $(Cd^{2+}, Mn^{2+}, Al^{2+})$ show selective inhibitory effects on the reuptake of the catecholamines norepinephrine and dopamine and on 5-hydroxytryptamine (Lai *et al.*, 1980a,b, 1983); only cadmium selectively inhibits synaptosomal phospholipid methylation (Wong and Lim, 1981).

2.3. Tin

Neurotoxic effects of organotin compounds are known to occur in humans and animals as well. There have been no human fatalities reported in the use of inorganic tin compounds. The use of a preparation containing diethyl tin iodide (stalinon) for dermatologic problems in France caused more than 100 deaths and a large number of nonfatal exposures following its ingestion (Barnes and Stoner, 1959). According to Nordberg (1980), the preparation contained about 10% triethyltin (TET) thought to be the causative agent. There is considerable uptake of TET into the brain with resulting white matter edema. Reiter *et al.* (1981) have shown that a single dose (306 mg/kg) of TET given at a critical period (5 days postnatal) produces permanent alterations in both brain weight and behavior, effects that appear to be age specific. Trimethyltin also is neurotoxic, acting primarily in the limbic system and in particular on hippocampal CA3-CA4 pyramidal cells. It produces a variety of somatosensory and behav-

TABLE X. ATP Hydrolysis in the Presence of Various Inhibitors of Oxidative Phosphorylation[a,b]

Addition	ATP hydrolysis (nmole of P_i/min per mg of protein)
None	6.9 ± 1.0 (18)
Rotenone (1 nmole/mg of protein)	42.5 ± 1.3 (29)
Amytal (0.8 mM)	46.6 (1)
Cyanide (0.5–1.0 mM)	46.2 ± 2.8 (6)
Nitrogen	44.5 (2)
Trimethyltin (20 µM)	86.5 ± 2.1 (3)
Trimethyltin (20 µM) + rotenone (1 nmole/mg of protein)	86.4 ± 3.2 (7)
Triethyltin (1.0 µM)	73.3 ± 2.0 (6)
Triethyltin (1.0 µM) + rotenone (1 nmole/mg of protein)	86.9 ± 3.9 (7)

[a] From Aldridge and Street (1971).
[b] The reaction was started by the addition of mitochondria. For experiments with cyanide or under N_2 stoppered flasks were used. The results are expressed as means ± S.E.M. with numbers of observations in parentheses.

ioral effects according to Howell *et al.* (1982). The neurotoxicity following exposure to TET varies as a function of age. Rats exposed to TET as neonates are hyperactive as adults, whereas rats exposed as adults are hypoactive when tested under the same paradigm. Neonatal exposure produce irreversible effects, whereas older rats recover within a month after exposure ceased. Finally, neonatal exposure results in reduced brain weight, but TET produces an increase in brain size in adults.

It is known that trialkyl tin compounds interfere with oxidative phosphorylation and energy production (Aldridge and Street, 1971). These authors showed that both trimethyltin (TMT) and TET combine with a site 1 binding site in rat liver mitochondria which could account for the observed inhibition of mitochondrial ATP synthesis with either NAD-linked substrate-supported respiration (pyruvate) or succinate as substrate. In addition, oxygen uptake with either substrate, which normally is stimulated by uncoupling agents, is inhibited. ATP synthesis linked to the oxidation of reduced "Cytochrome–C," ATP hydrolysis, and oxygen uptake are reduced to zero at low TET and TMT concentrations as binding of these compounds to site 1 approaches 100%. ATP hydrolysis was increased by either TMT or TET (Table X).

The effects of organotin compounds on mitochondrial respiration show some similarities to those of oligomycin. High oligomycin concentrations do not inhibit oxygen uptak stimulated by 2,4-dinitrophenol as do the tin compounds (Fig. 8). The oligomycin effects are slow in onset and progressive, whereas tin compounds rapidly reach steady state according to Aldridge and Street (1971). Finally oligomycin inhibits to the same degree sites 1, 2, and 3, whereas organotin compounds require higher concentrations to inhibit ascorbate-linked phosphorylation compared with the other sites (Figs. 9,10).

In brain, Cremer (1970) has described the selective inhibition of glucose

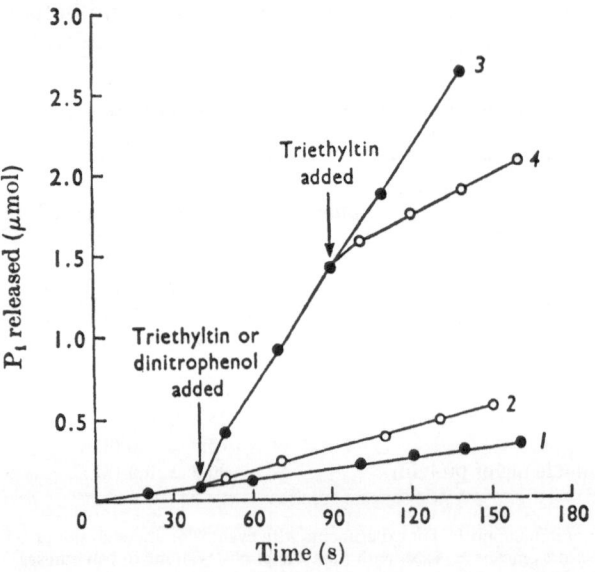

FIGURE 8. Effect of 2,4-dinitrophenol and triethyltin on ATP hydrolysis by mitonchondria. The reaction was started by the addition of mitochondria (2.86 mg of protein). Rotenone (1.2 nmole/mg of protein) was present in all flasks, and the concentration of 2,4-dinitrophenol was 30 μM and that of triethyltin 0.25 μM. Curve 1, control; curve 2, triethyltin added; curve 3, 2,4-dinitrophenol added; curve 4, 2, 4-dinitrophenol + triethyltin added. From Aldridge and Street (1971).

415

*HEAVY METAL
TOXICITY AND
ENERGY
METABOLISM IN
THE DEVELOPING
BRAIN*

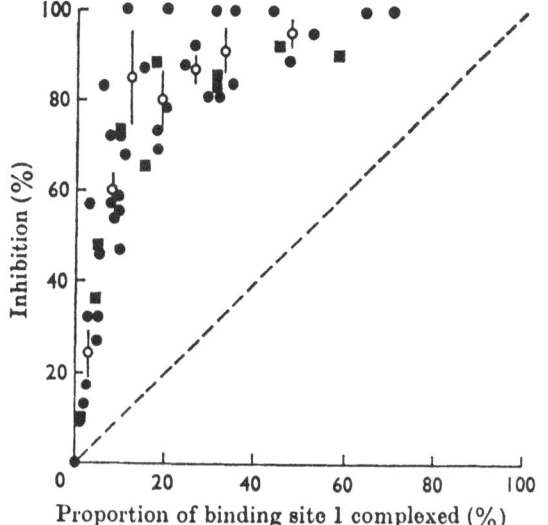

FIGURE 9. Inhibition of oxidative phosphorylation and the complexing by triethyltin and trimethyltin of binding site 1. The results are taken from manometric experiments previously published (Aldridge, 1958; Aldridge and Street, 1964). Individual results for (●) triethyltin and (■) trimethyltin are shown and (○) the grouped results by the means with the vertical lines indicating the SEM. The broken line indicates a 1:1 relationship. From Aldridge and Street (1971).

oxidation by TET *in vivo*. Young rats (7 to 8 weeks of age) were given 10 mg of TET sulfate/kg by intraperitoneal injection. Two hours later injected animals and controls received either [2-^{14}C]glucose or [1-^{14}C]acetate followed by decapitation 2–30 minutes later. The incorporation of [^{14}C]glucose into glutamate, glutamine, GABA, and aspartate were greatly decreased (Table XI,XII). The incorporation of [^{14}C]acetate was unaffected. The results are interpreted as showing an action by TET on pyruate oxidation since there was no inhibition of brain glycolysis (Table XIII).

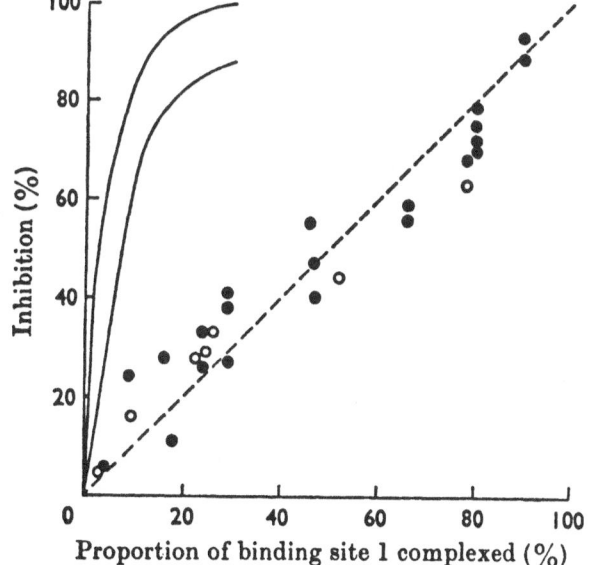

FIGURE 10. Inhibition of phosphorylation linked to oxidation of ascorbate and complexing by triethyltin of binding site 1. The results are from nine experiments, each with a different preparation of mitochondria in a medium containing ascorbate (40mM) and NNN′ N′ - tetramethyl-p-phenylenediamine (1 mM) at 25 °C (●) and at 37 °C (○). The broken line indicates a 1:1 relationship and the continuous lines are taken from Fig. 9. From Aldridge and Street (1971).

TABLE XI. Labeling of Amino Acids from [2-^{14}C]Glucose (Frozen Brain)[a,b]

	Control			Triethyltin		
	μmole/g of tissue	Specific radioactivity (dpm/μmole)	Relative specific radioactivity	μmole/g of tissue	Specific radioactivity (dpm/μmole)	Relative specific radioactivity
Glutamate	10.07 ± 1.76 (4)	4070 ± 470	1.0	8.0 (2)	1800	1.0
Aspartate	2.25 ± 0.21 (3)	2940 ± 150	0.73	2.3 (2)	—	—
γ-Aminobutyrate	1.67 ± 0.25 (4)	2480 ± 420	0.61	1.39 (2)	730	0.41
Glutamate	4.30 ± 1.06 (3)	1410 ± 310	0.35	5.30 (2)	530	0.30

[a] Adapted from Cremer (1970).

[b] The conditions were as described in Cremer (1970) except that the guillotined heads fell into liquid N$_2$ and the frozen brains were removed for amino acid analyses. All animals were killed at 3 min after receiving [2-^{14}C]glucose. The relative specific radioactivity is the specific radioactivity of the amino acid/the specific radioactivity of glutamate.

TABLE XII. Specific Radioactivity of Amino Acids in the Brains of Rats after the Intravenous Injection of [1-^{14}C]Acetate[a,b]

Time (min)	Animals	Total radioactivity (dpm/gram of brain)	Specific radioactivity (dpm/μmole)			
			Glutamate	Aspartate	γ-Aminobutyrate	Glutamine
3	Control	28340 ± 4020 (4)	830 ± 79 (4)	525 ± 99 (4)	408 ± 114 (3)	2920 ± 383 (4)
3	Triethyltin	35390 ± 1950 *(3)	949 ± 146 (3)	573 ± 158 (3)	321 ± 59 (3)	3540 ± 354 (3)
10	Control	28920 ± 1800 (4)	1149 ± 118 (4)	814 ± 98 (4)	663 ± 75 (4)	2430 ± 663 (4)
10	Triethyltin	34110 ± 3920 (4)	1048 ± 108 (4)	758 ± 101 (4)	638 ± 29 (4)	3210 ± 582 (4)

[a] From Cremer (1970).

[b] Rats (200 g body wt.) received 15 μCi of [1-^{14}C]acetate by intravenous injection. Where indicated, rats were given triethyltin sulphate (10 mg/body wt.) by intraperitoneal injection 2 hr before the radioactive acetate. Animals were guillotined and the head fell into liquid N$_2$. Analyses of amino acids in the frozen brains were determined. Means values ± SD are given with the number of determinations in parentheses. The probabilities that values for the animals given triethyltin were significantly different from the control means were tested and found to be not significant except where marked * $P < 0.05$.

417

HEAVY METAL
TOXICITY AND
ENERGY
METABOLISM IN
THE DEVELOPING
BRAIN

TABLE XIII. [2-^{14}C]Glucose Metabolism in the Blood and Brain of Fed Rats With and Without Triethyltin Sulphate[a,b]

Animals	Time (sec)	Blood (dpm/ml)	Brain (dpm/g)	Blood glucose (μmole/ml)	Blood glucose (dpm/μmole)	Brain glucose (μmole/g)	Brain glucose (dpm/μmole)	Brain lactate (μmole/g)	Brain lactate (dpm/μmole)	Brain glutamate (μmole/g)	Brain glutamate (dpm/μmole)
Control	30	363700 ± 43440 (8)	68020 ± 17950 (8)	6.10 ± 0.39 (8)	59700 ± 7320 (8)	0.242 ± 0.048 (8)	52160 ± 10500 (8)	3.67 ± 0.63 (8)	10400 ± 3490 (8)	7.79 ± 0.39 (7)	410 ± 140 (7)
Triethyltin	30	431600 ± 60555 (5)	56710 ± 6395 (5)	8.79 ± 1.07 (5)	49150 ± 4840 (5)	0.746 ± 0.204 (5)	33370 ± 4970 (5)	3.37 ± 0.71 (5)	7430 ± 1495 (5)	5.41 ± 0.36 (4)	203 ± 50 (4)
Test of significance		NS	NS	$P < 0.001$	$P < 0.01$	$P < 0.001$	$P < 0.005$	NS	NS	$P < 0.001$	$P < 0.005$
Control	90	321475 ± 47720 (8)	98240 ± 17250 (8)	6.21 ± 0.59 (8)	51880 ± 6650 (8)	0.323 ± 0.056 (7)	33740 ± 10770 (5)	4.03 ± 0.26 (5)	9350 ± 1790 (7)	8.63 ± 1.17 (8)	1870 ± 330 (7)
Triethyltin	90	341260 ± 17520 (6)	87610 ± 15300 (6)	7.41 ± 1.32 (6)	47410 ± 10630 (6)	0.714 ± 0.206 (5)	22870 ± 5300 (5)	3.67 ± 0.71 (6)	9210 ± 1750 (6)	7.26 ± 1.65 (6)	1070 ± 198 (6)
Test of significance		NS	NS	NS	NS	$P < 0.005$	NS	NS	NS	NS	$P < 0.001$
Control	180	277640 ± 49700 (5)	142530 ± 6995 (9)	5.65 ± 0.54 (9)	47100 ± 7730 (5)	0.319 ± 0.109 (16)	33820 ± 2600 (9)	3.88 ± 0.80 (16)	12480 ± 7540 (9)	8.76 ± 1.18 (16)	4660 ± 850 (9)
Triethyltin	180	297500 ± 34850 (5)	107910 ± 9870 (9)	7.63 ± 1.43 (9)	44795 ± 11760 (5)	0.949 ± 0.306 (13)	29290 ± 5535 (9)	3.59 ± 0.52 (13)	9810 ± 2955 (7)	6.66 ± 0.69 (12)	2400 ± 725 (7)
Test of significance		NS	$P < 0.001$	$P < 0.005$	NS	$P < 0.001$	NS	NS	NS	$P < 0.001$	$P < 0.001$

[a] Adapted from Cremer (1970).

[b] Rats (200 g body wt.) received 10 μCi of [2-^{14}C]glucose by intravenous injection. Where indicated, rats were given triethyltin sulphate (10 mg/kg body wt.) by intraperitoneal injection 2 hr before the radioactive glucose. Animals were guillotined and the heads fell into liquid N_2. Analyses of the metabolites in the frozen brains were determined. Mean values ± SD are given with the number of determinations in parentheses. The probabilities that values for the animals given triethyltin were significantly different from the control means at a particular time are given as a test of significance: NS is not significant with $P > 0.05$.

2.4. Manganese

Manganese poisoning results in degenerative changes in basal ganglia. Numerous reports of manganese poisoning especially in the mining of manganese ore have been reviewed (Chandra and Seth, 1980). Symptoms of poisoning superficially resemble parkinsonism with insidious onset. Early symptoms include apathy, anorexia, and asthenia, often involving behavioral excitement. As the process evolves, there is an intermediate phase of clumsiness, monotonous speech, poor articulation, and fixed facial expression. In a few months, the symptoms are exaggerated and there is muscle rigidity, tremor, and autokinesis. The similarity in manganese poisoning to parkinsonism and the involvement of the neurotransmitter dopamine was recognized by the late George Cotzias (1974) and formed the basis of therapeutic approaches that he pioneered. Ironically, dopa does not ameliorate manganese symptoms.

To compare the effects of manganese on growing versus adult animals, Chandra and Seth (1980) exposed suckling rats by intubating nursing dams from the second day of parturition for a period of 30 days with 15 mg/kg per day of $MnCl_2 4H_2O$. The activity of succinic dehydrogenase, ATPase, and monamine oxidase in brains of pups and dams were monitored and related to brain manganese contents.

The brain contents of manganese increased by 64% in pup brains by 15 days. There were no significant differences in the manganese concentrations in control and manganese-exposed dam brains. After 30 days, there was increase of threefold in pup brains and a 56% rise in the exposed dam brains.

Although monoamine oxidase activity increased in both instances, there was a greater increase in dam brains. The activity of succinic dehydrogenase and ATPase decreased to a much greater degree in manganese-exposed pups compared with dams. The marked accumulation of manganese in pups may be due to a greater degree of absorption and a poorly developed blood–brain barrier in the neonate. Similar differences in brain $^{54}Mn^{2+}$ were noted in neonatal mice compared with adults by Cotzias *et al.* (1974) after intraperitoneal injection. These authors also reported a complete absence of metal elimination. An increased synthesis of brain dopamine and norepinephrine was observed in early stages of Mn^{2+} toxicity which could explain early behavioral changes observed in animal and in man. Such correlations of change in amine level and behavior have been reported (Chandra *et al.*, 1979).

TABLE XIV. A Summarizing Table

	Enhanced toxicity to immature brain	Inhibition of cellular energetics	Neurotransmitter systems affected	
			ACh	Dopamine
Lead	+	+	+	+
Mercury	+	+	+	?
Cadmium	+	?	+	+
Tin	+	+	?	?
Manganese	?	+	?	+

419

HEAVY METAL
TOXICITY AND
ENERGY
METABOLISM IN
THE DEVELOPING
BRAIN

3. Summary

In this review, we have discussed a group of heavy metals that share the following characteristics: greater toxicity in the young compared with the adult brain, cellular aerobic energy metabolism as a probable important site of action in the pathogenesis of the CNS toxicity, and correlative effects of the metals on neurotransmitter metabolism and behavioral toxicity. These effects are summarized in Table XIV. These characteristics of metal-induced toxicity are most extensively studied in the animal model for lead encephalopathy. Although the analogies of the age-dependent toxic effects of these metals on the brain and the role(s) of these proposed sites of action may break down in detail, such parallels could be important in guiding future research. In particular, careful studies on the critical maturational stages at which metal exposure and resultant cellular or molecular effects result in encephalopathic changes are necessary to understand the pathogenesis of these metal toxicities. Such studies may also be important in the further understanding of the role of these systems (energy metabolism and neurotransmitters) in the development of the complex functioning of the CNS. Finally, an understanding of the critical exposure time and mechanisms of pathogenesis in these animal models of metal-induced encephalopathies may be important in guiding future human toxicological research and public health policies.

References

Aghajanian, G. R., and Bloom, F. E., 1967, The formation of synaptic junctions in developing rat brain: Quantitative microscopy study, Brain Res. 6:716–727.

Aldridge, W. N., 1958, The biochemistry of organotin compounds, Biochem. J. 69:367–376.

Aldridge, W. N., and Street, B. U., 1964, Oxidative phosphorylation. Biochemical effects and properties of trialkyltins, Biochem. J. 91:287–297.

Aldridge, W. N., and Street, B. W., 1971, Oxidative phosphorylation. The relation between the specific binding of trimethyltin and triethyltin to mitochondria and their effects on various mitochondrial functions, Biochem. J. 124:221–234.

Abd-Elfattah, Anwar S., and Shamoo, A. E., 1981, Regeneration of a functionally active rat-brain muscarinic receptor by D-penicillamine after inhibition with methylmercury and mercuric chloride, Mol. Pharmacol. 20:492–497.

Aronstam, R. S., Abood, L. S., and Hoss, W., 1978, Influence of sulfhydryl reagents and heavy metals on the functional state of the muscarinic acetylcholine receptor in rat brain, Mol. Pharmacol. 14:575–586.

Ash, G. R., and Bygrave, F. L., 1977, Ruthenium red as a probe in assessing the potentials of mitochondria to control intracellular calcium in liver, FEBS Lett. 78:166–168.

Aub, J. C., Fairhall, L. T., Minot, A. S., and Resnikoff, P, 1925, Lead poisoning, Medicine 4:1–250.

Barnes, J. M., and Stoner, H. B., 1959, The toxicology of tin compounds, Pharmacol. Rev. 11:211–231.

Beattie, A. D., Moore, M. R., Goldberg, A., Finlayson, M. J. W., Graham, J. F., Mackie, E. M., Main, J. C., McLaren, D. A., Murdock, R. M., and Stewart, G. T., 1975, Role of chronic low-level lead exposure in the aetiology of mental retardation, Lancet 1:589–592.

Binah, O., Meire, U., and Rahamimoff, H., 1978, The effects of Hg Cl$_2$ and mersalyl on mechanisms regulating intracellular calcium and transmitter release, Eur. J. Pharmacol. 51:453–457.

Bornschein, R. L., Fox, D. A., and Michaelson, I. A., 1977, Estimation of daily exposure in neonatal rats receiving lead via dam's milk, Toxicol. Appl. Pharmacol. 40:577–587.

Buell, S., Fosmire, G., Ollerich, D., and Sandstead, H., 1977, Effects of postnatàl zinc deficiency on cerebellar and hippocampal development in the rat, Exp. Neurol. **55**:199–210.

Bull, R. J., Stanaszek, P. M., O'Neill, J. J., and Lutkenhoff, S. D., 1975, Specificity of the effects of lead on brain energy metabolism for substrates donating a cytoplasmic reducing equivalent, Environ. Hlth. Perspect. **12**:89–95.

Bull, R. J., Lutkenhoff, S. D., McCarty, G. E., and Miller, R. G., 1979, Delays in the postnatal increase of cerebral cytochromes in lead exposed rats, Neuropharmacology **18**:83–92.

Byers, R. K., and Lord, E. E., 1943, Late effects of lead poisoning on mental development, Am. J. Dis. Child. **66**:471–483.

Campbell, J. B., Woolley, D. E., Virjarjan, V. K., and Overmann, S. R., 1982, Morphometric effects of postnatal lead exposure on hippocampal development of the 15-day-old rat, Dev. Brain. Res. **3**:595–612.

Carroll, P. T., Sibergeld, E. K., and Goldberg, A. M., 1977, Alterations of central cholinergic function by chronic lead acetate exposure, Biochem. Pharmacol. **26**:397–402.

Chandra, S. V., and Seth, P. K., 1980, Manganese encephalopathy: Effect of manganese exposure on growing versus adult rodents, in: Advances in Neurotoxicology (L. Manzo, ed.), Pergamon Press, Oxford, United Kingdom, pp. 49–55.

Chandra, S. V., Girja, S., and Saxena, D. K., 1979, Manganese-induced behavioral dysfuncton and its neurochemical mechanism in growing mice, J. Neurochem. **33**:1217–1221.

Chang, L. W., 1977, Neurotoxic effects of mercury—A review, Environ. Res. **14**:329–373.

Chang, L. W., Reuhl, K. R., and Lee, G. W., 1977, Degenerative changes in the developing nervous system as a result of in utero exposure to methyl mercury, Environ. Res. **14**:414–423.

Cheung, M., and Verity, M. A., 1981, Methyl mercury inhibition of synaptosome protein synthesis: Role of mitochondrial dysfunction, Environ. Res. **24**:286–298.

Chipelinsky, A. B., and de Lores Arnaiz, G. R., 1970, Levels of cytochromes in rat brain mitochondria during postnatal development, Biochem. Biophys. Acta **197**:321–323.

Choi, D. D., and Richter, G. W., 1972, Lead poisoning: Rapid formation of intranuclear inclusions, Science **177**:1194–1195.

Clasen, R. A., Hartmann, J. F., Storr, A. J., Coogan, P. S., Pandolfi, S., Laing, I., Becker, R., and Hass, G. M., 1974, Electron microscopic and chemical studies of the vascular changes in edema of lead encephalopathy. A comparative study of the human and experimental disease, Am. J. Pathol. **74**:215–239.

Cotzias, G. C., Papavasiliou, P. S., Mena, I., Tang, L. C., and Miller, S. T., 1974, Manganese and catecholamines, in: Advances in Neurology (F. H. McDowell and A. Barbeau, eds.), Raven Press, New York, pp. 235–244.

Cremer, J. E., 1970, Selective inhibition of glucose oxidation by triethyltin in rat brain in vivo, Biochem. J. **119**:95–102.

Davison, A. N., and Dobbing, J., 1968, The developing brain, in: Applied Neurochemistry, F. A. Davis Company, Philadelphia, pp. 253–286.

De la Burde, B., and Choate, M. S., 1975, Early asymptomatic lead exposure and development at school age, Behav. Pediatr. **87**:638–642.

Dienel, G., Ryder, E., and Greengard, O., 1977, Distribution of mitochondrial enzymes between the perikaryal and synaptic fractions of immature and adult rat brain, Biochem. Biophys. Acta **496**:484.

Dobbing, J., and Sands, J., 1970a, Growth and development of the brain and spinal cord of the guinea pig, Brain Res. **17**:115–123.

Dobbing, J., and Sands, J., 1970b, Timing of neuroblast multiplication in developing human brain, Nature (London) **226**:639–640.

Eldefrawi, M., Eldefrawi, A., and Shamoo, A., 1975, Molecular and functional properties of the acetylcholine receptor, Ann. N.Y. Acad. Sci. **264**:183–202.

Eldefrawi, M., Mansour, N., and Eldefrawi, A., 1977, Interactions of acetylcholine receptor with organic mercury compounds, Adv. Exp. Biol. **84**:449–459.

Gabbiani, G., Baric, D., and Deziel, C., 1967, Toxicity of cadmium for the central nervous system, Exp. Neurol. **18**:154–160.

Gmerek, D. E., McCafferty. M. R., Oneill, K. J., Melamed, B. R., and O'Neill, J. J., 1981, Effect of inorganic lead on rat brain mitochondrial respiration and energy production, J. Neurochem. **36**:1109–1113.

421

HEAVY METAL
TOXICITY AND
ENERGY
METABOLISM IN
THE DEVELOPING
BRAIN

Goldstein, G. W., 1977, Lead encephalopathy: The significance of lead inhibition of calcium uptake by brain mitochondria, *Brain Res.* **136**:185–188.

Goldstein, G. W., Ashbury, A. K., and Diamond, I., 1974, Pathogenesis of lead encephalopathy. Uptake of lead and reaction of brain capillaries, *Arch. Neurol.* **31**:382–389.

Goldstein, G. W., Wolensky, J. S., and Csejtey, J., 1977, Isolated brain capillaries: A model for the study of lead encephalopathy, *Ann. Neurol.* **1**:235–239.

Golter, M., and Michaelson, I. A., 1975, Growth behavior and brain catecholamines in lead-exposed neonatal rats: A reappraisal, *Science* **187**:359–361.

Govoni, S., Memo, M., Spano, P. F., and Trabucchi, M., 1979, Chronic lead treatment differentially affects dopamine synthesis in various rat brain areas, *Toxicology* **12**:343–349.

Goyer, R. A., and Moore, J. F., 1974, Cellular effects of lead, in: *Advances in Experimental Medicine and Biology*, Volume 48 (M. Friedman, ed.), Plenum Press, New York, pp. 447–462.

Goyer, R. A., May, P., Cates, M., and Krigman, M. R., 1970, Lead and protein content of isolated intranuclear inclusion bodies from kidneys of lead-poisoned rats, *Lab. Invest.* **22**:245–251.

Greengard, P., and McIlwain, H., 1955, Metabolic response to electrical pulses in mammalian cerebral tissues during development, in: *Biochemistry of the Developing Nervous System* (H. Waelsch, ed.), Academic Press, New York, pp. 251.

Gregson, N. A., and Williams, P. L., 1969, A comparative study of brain and liver mitochondria from newborn and adult rats, *J. Neurochem.* **16**:617–626.

Hedlund, B., Gammara, M., and Barfai, T., 1979, Inhibition of striatal muscarinic receptors *in vivo* by cadmium, *Brain Res.* **168**:216–218.

Himwich, H. E., 1951, *Brain Metabolism and Cerebral Disorders*, Williams and Wilkins, Baltimore, Maryland.

Holtzman, D., and Hsu, J. S., 1976, Early effects of inorganic lead on immature rat brain mitochondrial respiration, *Pediatr. Res.* **10**:70–75.

Holtzman, D., Hsu, J. S., and Mortell, P., 1978a, The pathogenesis of lead encephalopathy in the rat pups: Effects of maternal $PbCO_3$ feedings from birth, *Pediatr. Res.* **12**:1077–1082.

Holtzman, D., Hsu, J. S., and Mortell, P., 1978b, *In vitro* effects of inorganic lead on isolated rat brain mitochondrial respiration, *Neurochem. Res.* **3**:195–206.

Holtzman, D., Herman, M. M., Desautel, M., and Lewiston, N., 1979, Effects of altered osmolality on respiration and morphology of mitochondria from the developing brain, *J. Neurochem.* **33**:453–460.

Holtzman, D., Herman, M. M., Shen Hsu, J., and Mortell, P., 1980a, The pathogenesis of lead encephalopathy. Effects of lead carbonate feedings on morphology, lead content, and mitochondrial respiration in brains of immature and adult rats, *Virchows Arch. A. Pathol. Anat. Histol.* **387**:147–164.

Holtzman, D., Obana, K., and Olson, J., 1980b, Ruthenium red inhibition of *in vitro* lead effects on brain mitochondrial respiration, *J. Neurochem.* **34**:1776–1778.

Holtzman, D., Hsu, J. S., and Desautel, M., 1981, Absence of effects of lead feedings and malnutrition on mitochondrial and microsomal cytochromes in the developing brain, *Toxicol. Appl. Pharmacol.* **58**:48–56.

Holtzman, D., De Vries, C., Nguyen, H., Jameson, N., Olson, J., Carrithers, M., and Bensch, K., 1982, Development of resistance to lead encephalopathy during maturation in the rat pups, *J. Neuropathol. Exp. Neurol.* **41**:652–663.

Holtzman, D., De Vries, C., Nguyen, H., Olson, J., and Bensch, K., 1984, Maturation of resistance to lead encephalopathy: Cellular and subcellular mechanisms, *Neurotoxicology* (in press).

Howell, W. E., Walsh, T. J., and Dyer, R. S., 1982, Somatosensory dysfunction following acute trimethyl tin exposure, *Neurobehav. Toxicol. Teratol.* **4**:197–201.

Hrdina, P. D., Peters, D. A. V., and Singhal, R. L., 1976, Effects of chronic exposure to cadmium, lead, and mercury on brain biogenic amines in the rat, *Res. Commun. Chem. Pathol. Pharmacol.* **15**:483–493.

Juang, M., 1976, Electrophysiological study of the action of methylmercuric chloride and mercuric chloride on the sciatic nerve sartorius muscle preparation of the frog, *Toxicol. Appl. Pharmacol.* **37**:339–348.

Kiraly, E., and Jones, D. G., 1982, Dendritic spine changes in rat hippocampal pyramidal cells after postnatal lead treatment: A Golgi study, *Exp. Neurol.* **77**:236–239.

Kochen, J. A., and Greener, Y., 1977, Brain lead levels in hemorrhagic lead encephalopathy, *Pediatr. Res.* **11**:563.

Kostial, K., and Vouk, V. B., 1957, Lead ions and synaptic transmission in the superior cervical ganglion of the cat, *Br. J. Pharmacol. Chemother.* **12**:219–222.

Krigman, M. R., and Hogan, E. L., 1974, Effects of lead intoxication on the postnatal growth of the rat nervous system, *Environ. Health Perspect.* **7**:187–199.

Krigman, M. R., Mushak, P., and Bouldin, T. W., 1977, An appraisal of rodent models of lead encephalopathy, in: *Neurotoxicology*, Volume 1 (L. Roizin, H. Sheraki, and N. Grcervic, eds.), Raven Press, New York, pp. 299–302.

Lai, J. C. K., Lim, L., and Davison, A. N., 1980a, Inhibition of rat brain synaptosomal dopamine uptake by Cd^{2+}: A kinetic study, *Biochem. Soc. Trans.* **8**:68.

Lai, J. C. K., Guest, J. F., Leung, T. K. C., Lim, L., and Davison, A. N., 1980b, The effects of cadmium, manganese and aluminum on sodium-potassium-activated and magnesium-activated adenosine triphosphatase activity and choline uptake in rat brain synaptosomes, *Biochem. Pharmacol.* **29**:141–146.

Lai, J. C. K., Wong, P. C. L., and Lim, L., 1983, Structure and function of synaptosomal and mitochondrial membranes: Elucidation using neurotoxic metals and neuromodulatory agents, in: *Neural Membranes* (G. Y. Sun, N. Bazan, J-Y Wu, G. Porcellati, and A. Y. Sun, eds.), The Humana Press, New Jersey, pp. 355–374.

Landrigan, P. J., Whitworth, R. H., Balok, R. W., Stachling, N. W., Barthel, W. F., and Rosenblum, B. F., 1975, Neuropsychological dysfunction in children with chronic low-lead absorption, *Lancet* **1**:708–712.

Lin-Fu, J. S., 1973, Vulnerability of children to lead exposure and toxicity, *N. Engl. J. Med.* **289**:1229–1233.

Lin-Fu, J. S., 1980, Lead poisoning and undue lead exposure in children: History and current status, in: *Low Level Lead Exposure: The Clinical Implications of Current Research* (H. L. Needleman, ed.), Raven Press, New York, pp. 5–16.

Lowe, D. A., Richardson, B. P., Taylor, P., and Donatsch, P., 1976, Increasing intacellular sodium triggers calcium release from bound pools, *Nature (London)* **260**:337–338.

Lucchi, L., Memo, M., Airaghi, M. L., Spano, P. F., and Trabucci, M., 1981, Chronic lead treatment induces in rat a specific and differential effect to dopamine receptors in different brain areas, *Brain Res.* **213**:397–404.

Maker, H. S., Lehrer, G. M., and Silides, D. J., 1975, The effect of lead on mouse brain development, *Environ. Res.* **10**:76–91.

Manalis, R. A., and Cooper, G. P., 1975, Evoked transmitter release increased by inorganic mercury at frog neuromuscular junction, *Nature (London)* **257**:690–691.

Mandel, P., Rein, H., Harth-Edel, S., and Mardell, R., 1965, Distribution and metabolism of ribonucleic acid in the vertebrate central nervous system, in: *Comparative Neurochemistry* (D. Richter, ed.), Pergamon Press, New York, pp. 149–163.

Matsumoto, H., Suzuki, A., Monta, C., Nakamura, K., and Solki, S., 1967, Preventive effect of penicillamine on the brain effect of fetal rat poisoned transplacentally with methyl mercury, *Life Sci.* **6**:2321–2330.

McCauley, P. T., Bull, R. J., and Lutkenhoff, S. D., 1979, Association alterations in energy metabolism with lead-induced delays in rat cerebral cortical development, *Neuropharmacology* **18**:93–101.

McCauley, P. T., Bull, R. J., Tonti, A. P., Lutkenhoff, S. D., Meister, M. V., Doerger, J. U., and Stober, J. A., 1982, The effect of prenatal and postnatal lead exposure on neonatal synaptogenesis in rat cerebral cortex, *J. Toxicol. Environ. Health* **10**:639–651.

Michaelson, L. A., 1973, Effects of inorganic lead on levels of RNA, DNA and protein on developing neonatal rat brain, *Toxicol. Appl. Pharmacol.* **26**:539–548.

Modak, A., Weinstein, S., and Stavinoha, W. B., 1975, Effect of chronic ingestion of lead on the central cholinergic system in rat brain regions, *Toxicol. Appl. Pharmacol.* **34**:340–347.

Moore, C. L., 1971, Specific inhibition of mitochondrial Ca^{2+} transport by ruthenium red, *Biochem. Biophys. Res. Commun.* **42**:298–305.

Needleman, H. L., Bunnoe, C., Leviton, A., Reed, R., Peresie, H., Mher, C., and Barrett, P., 1979, Deficits in psychologic and classroom performance of children with elevated dentine lead levels, *N. Engl. J. Med.* **300**:689–695.

423

HEAVY METAL
TOXICITY AND
ENERGY
METABOLISM IN
THE DEVELOPING
BRAIN

Neidle, A., van den Berg, C. J., and Grynbaum, A., 1969, The heterogeneity of rat brain mitochondria isolated on continuous sucrose gradient, *J. Neurochem.* **16**:225–234.

Nordberg, G. F., 1980, Neurotoxic effect of metals and their compounds, in: *Advances in Neurotoxicology* (L. Manzo, ed.), Pergamon Press, Oxford, and New York, pp. 3–13.

Oskarsson, A., Squibb, K. S., and Fowler, B. A., 1982, Intracellular binding of lead in the kidney: The partial isolation and characterization of postmitochondrial lead binding components, *Biochem. Biophys. Res. Commun.* **104**:290–298.

Parr, D. R., and Harris, E. J., 1976, The effect of lead on the calcium-handling capacity of rat heart mitochondria, *Biochem. J.* **159**:289–294.

Patel, A. J., Michaelson, I. A., Cremer, J. E., and Balazs, R., 1974, The metabolism of [^{14}C]glucose by the brain of suckling rats intoxicated with inorganic lead, *J. Neurochem.* **22**:581–590.

Pentschew, A., and Garro, F., 1966, Lead encephalo-myelopathy of the suckling rat and its implications on the porphyrinopathic nervous disease, *Acta Neuropathol.* **6**:266–278.

Pihl, R. O., and Parkes, M., 1977, Hair element content in learning disabled children, *Science* **198**:204–206.

Pounds, J. G., Wright, R., Morrison, D., and Casciano, D. A., 1982a, Effect of lead on calcium homeostasis in the isolated rat hepatocyte, *Toxicol. Appl. Pharmacol.* **63**:389–401.

Pounds, J. G., Morrison, D., Wright, R., Casciano, D. A., and Shadelock, J. G., 1982b, Effect of lead on calcium-mediated cell function in the isolated rat hepatocyte, *Toxicol. Appl. Pharmacol.* **63**:402–408.

Press, M. F., 1977a, Lead encephalopathy in children, *Am. J. Pathol.* **84**:485–488.

Press, M. F., 1977b, Lead encephalopathy in neonatal Long-Evans rats: Morphological studies, *J. Neuropathol. Exp. Neurol.* **34**:169–193.

Rahamimoff, R., Erulkar, S. D., Alnaes, R., Rotshenker, S., and Rahamimoff, H., 1976, Modulation of transmitter release by calcium ions and nerve impulses, *Cold Spring Harbor Symp. Quant. Biol.* **40**:107–116.

Rastogi, R. B., Merali, Z., and Singhal, R. J., 1977, Cadmium alters behaviour and the biosynthetic capacity for catecholamines and serotonin in neonatal rat brain, *J. Neurochem.* **28**:789–794.

Reijnierse, G. L., Veldstra, H., and van den Berg, C. J., 1975, Subcellular localization of γ-amino-butyrate transaminase and glutamate dehydrogenase in adult rat brain, *Biochem. J.* **152**:469–475.

Reiter, L. W., 1982, Age related effects of chemicals on the central nervous system, in: *Banbury Report, #11, Environmental Factors in Human Growth and Development* (V. R. Hunt, M. Kate Smith, and D. Worth, eds.), Cold Spring Harbor Laboratory, Cold Spring Harbor, New York, pp. 245–265.

Reiter, L. W., Heavner, G. B., Dean, K. F., and Ruppert, P. H., 1981, Developmental and behavioral effects of early postnatal exposure to triethyltin in rats, *Neurobehav. Toxicol. Teratol.* **3**:285–293.

Rodier, P. M., Reynolds, S. S., and Roberts, W. N., 1979, Behavioral consequences of interferences with CNS development in the early fetal period, *Teratology* **19**:327–336.

Rohrer, S. R., Shaw, S. M., and Lamar, C. H., 1978, Cadmium induced endothelial cell alterations in the fetal brain from prenatal exposure, *Acta Neuropathol.* **44**:147–149.

Ryder, E., 1980, Enzymatic profile of mitochondria isolated from selected brain regions of young adult and one year old rats, *J. Neurochem.* **34**:1550–1552.

Salganicoff, L., and Koeppe, R. E., 1968, Subcellular distribution of pyruvate carboxylase, diphosphopyridine nucleotide, and triphosphopyridine nucleotide isocitrate dehydrogenase and malate enzyme in rat brain, *J. Biol. Chem.* **243**:3416–3420.

Samson, F. E., Balfour, W. M., and Jacobs, R. J., 1960, Mitochondrial changes in developing rat brain, *Am. J. Physiol.* **199**:693–696.

Sauerhoff, M. W., and Michaelson, I. A., 1973, Hyperactivity and brain catecholamines in lead exposed developing rats, *Science* **182**:1022–1024.

Shih, T. M., and Hanin, I., 1978, Effects of chronic lead exposure on levels of acetylcholine and choline and an acetylcholine turnover rate in rat brain areas *in vivo*, *Psychopharmacology* **58**:263–269.

Silbergeld, E. K., 1973, Effects of lead on neuromuscular function: in vivo evidence for site of action at presynaptic level, *Fed. Proc.* **32**:275.

Silbergeld, E. K., 1977, Interactions of lead and calcium on the synaptosomal uptake of dopamine and choline, *Life Sci.* **20**:309–318.

Silbergeld, E. K., and Adler, H. S., 1978, Subcellular mechanisms of lead neurotoxicity, *Brain Res.* **148**:451–467.

Silbergeld, E. K., and Goldberg, A., 1974, Lead-induced behavioral dysfunction: An animal model of hyperactivity, *Exp. Neurol.* **46**:146–157.

Silbergeld, E. K., and Goldberg, A. M., 1975, Pharmacological and neurochemical investigations of lead induced hyperactivity, *Neuropharmacology* **14**:431–444.

Silbergeld, E. K., Wolinsky, J. S., and Goldstein, G. W., 1980, Electron probe microanalysis of isolated brain capillaries poisoned with lead, *Brain Res.* **189**:369–376.

Southard, J., Nitiseqojo, P., and Green, D. E., 1974, Mercurial toxicity and the perturbation of the mitochondrial control system, *Fed. Proc.* **33**(10):2147–2153.

Sterling, G. H., O'Neill, K. J., McCafferty, M. R., and O'Neill, J. J., 1982, Effect of chronic lead ingestion by rats on glucose metabolism and acetylcholine synthesis in cerebral cortex slices, *J. Neurochem.* **39**:592–596.

Suzuki, K., 1980, Special vulnerabilities of the developing nervous system to toxic substances, in: *Experimental and Clinical Neurotoxicology* (P. S. Spencer and H. H. Schaumburg, eds.), Williams and Wilkins, Baltimore, Maryland, pp. 48–61.

Toews, A. D., Kolber, A., Hayward, J., Krigman, M. R., and Morrell, P., 1978, Experimental lead encephalopathy in the suckling rat: Concentration of lead in cellular fractions enriched in brain capillaries, *Brain Res.* **147**:131–138.

Tyler, D. B., and Van Harreveld, A., 1942, The respiration of the developing brain, *Am. J. Physiol.* **136**:600–603.

Verity, Y. A., Brown, W. J., and Cheung, M., 1975, Organic mercurial encephalopathy: *In vivo* and *in vitro* effects of methyl mercury on synaptosomal respiration, *J. Neurochem.* **25**:759–766.

Van den Berg, C. J., Krzalic, L. T., Mela, P., and Waelsch, H., 1969, Compartmentation of glutamate metabolism in brain. Evidence for the existence of 2 different tricarboxylic acid cycles in brain, *Biochem. J.* **113**:281–290.

Van den Berg, C. J., Matheson, D. F., Ronda, G., Reijnierse, G. L. A., Blokhuis, G. G. D., Kroon, M. C., Clarke, D. D., and Garfinkel, D., 1975, in: *Metabolic Compartmentation and Neurotransmission*, (S. Berl, D. D. Clarke, and D. Schneider, eds.), Plenum Press, New York, pp. 515–543.

Von Burg, R., and Landry, T., 1976, Methylmercury and skeletal muscle receptor, *J. Pharm. Pharmacol.* **28**:458–551.

Von Rossum, G. D. W., and Kapoor, S. C., 1985, Effects of inorganic lead *in vitro* in ion exchanges and respiratory metabolism of rat kidney cortex (in preparation).

Webster, W. S., and Valois, A. A., 1981, The toxic effects of cadmium on the neonatal mouse CNS, *J. Neuropathol. Exper. Neurol.* **40**:247–257.

Weinberg, J. J., Harding, P. G., and Humes, H. D., 1982, Mitochondrial biogenesis during the initiating of mercuric chloride-induced renal injury, *J. Biol. Chem.* **257**:60–67.

Winick, M., and Noble, A., 1965, Quantitative changes in DNA, RNA, and protein during prenatal and postnatal growth in the rat, *Develop. Biol.* **12**:451.

18

General Anesthesia

RICHARD C. WIGGINS

1. Introduction

Neural function is coupled to energy metabolism. Understanding the anatomical and chemical basis of signal conduction in nerve fibers is an area of great scientific accomplishment and intensive ongoing research. Relationships between energy metabolism and different behavioral states is enormously more complicated because of the complex neuroanatomical basis of mental function. Anesthesia is a clinically important mental state or condition, which is analyzed from the perspective of energy metabolism in this chapter.

Because of the revolutionary benefits of surgery without the sensation of pain, physicians have seized on a number of chemical compounds, which when administered to their patients produce a useful state of unconsciousness. Chemical substances that produced general anesthesia vary greatly in their chemical structure and properties and include solids, gases, and volatile liquids, as well as the inert noble gases under several atmospheres of pressure. In addition to general anesthetics, certain compounds, which when administered selectively to a tissue, will block nerve conduction to produce a region of local anesthesia. Local anesthetics act by reversibly blocking nerve membrane sodium channels, and thus propagation of the action potential fails (Taylor, 1959). The actions of local anesthetics are well described and reviewed in numerous sources including chapters in Katz (1975) and Fink (1980). The mechanisms of general anesthetics are not known, which is not surprising since the basis of consciousness and unconsciousness is unknown.

Although no compelling unified theory of general anesthesia has emerged, the behavioral state has been studied from a variety of physiological, biochemical, and biophysical approaches. Comprehensive reviews are numerous (Eger, 1974; Gordon, 1975; Katz, 1975; Fink, 1980). Possibly sleep, hypnosis, anesthesia, and coma have important and instructive neurophysiological and neu-

RICHARD C. WIGGINS • Department of Neurobiology and Anatomy, University of Texas Medical School at Houston, Houston, Texas 77025.

roanatomical similarities and differences that warrant comparative study. Unfortunately, our knowledge in these areas is fragmentary and largely incomplete. The following brief considerations of these behaviors may serve as a useful introduction to the more detailed description of anesthesia.

1.1. Sleep

Humans typically spend about one third of their lives in a complex and dynamically inconstant mental condition inadequately called sleep. Surprisingly little is known of its mechanisms and function, and different states have been described (Jouvet, 1967). The sleeping brain is psychologically active, although without the sensory alertness and conscious control experienced in the waking state. External stimuli lead to arousal. There are periodic, local changes in blood flow that may reflect important changes in energy status.

In man and the other primates, there may be a distinct early stage of sleep characterized by a particular pattern of electrical activity. Other stages, which are more characteristic of all mammals, include slow-wave sleep and activated (or paradoxical) sleep.

There are no absolute behavioral criteria of slow-wave sleep since the typical EEG of this stage may also appear in the awake animal. Characteristically, one finds synchronized cortical activity with spindles and/or high-voltage slow waves. Although the localized character of the EEG is varied, activity in the frontal and associated areas may be synchronous with activity in the mesencephalic reticular formation and pyramidal tract. Depressed cortical activity is apparently associated with localized, decreased cerebral blood flow and temperature.

Activated sleep (paradoxical) is associated with dreams (which may be subject to memory), rapid eye movements, localizations, and so forth, atonia of the antigravity postural muscles, decreased blood pressure possibly with intermittent hypertensive phases, and respiratory irregularity. The distinctive behavioral features of activated sleep are considered definitive in determining its duration. The paradoxical nature of activated sleep derives from the early assumption that this was a light stage of sleep (because of the marked behavioral activity and restlessness associated with it). By physiological criteria, this paradoxical sleep may represent a deeper stage of sleep than slow-wave periods. Interestingly, cerebral blood flow increases during activated sleep (Reivich et al., 1968; Townsend et al., 1973) either from increased metabolic activity of the brain tissue or from vasodilation, and there may be an increase in cerebral temperature at the onset of this stage. Assuming that this increased blood flow reflects increased energy metabolism, then paradoxical sleep may represent a mental condition in which metabolism in increased above the normal level (Siesjo, 1978).

The relationship of these stages of sleep may be fundamentally important in identification of the mechanism of sleep. Are these stages different manifestations, stages, or levels of one process, or are they a result of qualitatively different mechanisms operating in the brain? In the event sleep represents a modulation of different mechanisms, waxing and waning in their respective influences, then their regulation and neuroanatomical basis will likely take a great deal of effort to reveal. As with many areas of neuroscience, the period of useful whole brain studies is probably over.

1.2. Hypnosis

This artificial mental state is possibly behaviorally induced and because of the increased responsiveness to suggestion, may be clinically useful in combating pain. The brain is psychologically active, but altered from normal. Although nothing of substance is known of its neurophysiological and neuroanatomical basis, it is probably not associated with any substantial differences in brain metabolism. The condition is distinct from sleep, although the patient is passive and shows a normal awake EEG. How this state relates to the equally nebulous concepts of the conscious and unconscious mind is entirely speculation, although waking, sleep, and hypnoid possibly constitute three distinct states (Barolin, 1982). Most likely the condition is behaviorally rather than physiologically based; however, speculations are varied. The subject has not received any detailed, contemporary analysis in the basic neuroscience literature.

1.3. Coma

This abnormal condition is characterized by severely depressed psychological functions, resulting in the absence of awareness. External stimulation is not sufficient to gain arousal of conscious activity. Coma is an active area of research with numerous reviews (Duffy and Plum, 1981; Plum and Posner, 1980). Reversible coma may be associated with normal or increased levels of ATP and phosphocreatine, whereas a decline in these energy metabolites signals severe brain damage. Possibly the onset of coma is somewhat analogous to a "switching down" of synaptic activity or placing the brain in a "standby" configuration. This hypothesis would account for the normal levels of energy metabolites, or even a build-up (as production gains on utilization). However, left untreated, the normal energy balance deteriorates to produce lasting damage. Unresponsive, sleeplike coma is of limited duration (about one month). After this time, the brain-damaged patient's eyes may open, giving a wakeful appearance, although without recognizable psychological activity or alertness.

Extensive brain damage may lead to brain death, in which case the organ is incapable of maintaining body homeostasis. The condition is distinct from the vegetative state of chronic coma since, in the latter condition, the brain may maintain homeostasis. Coma in its various forms may be progressive stages of brain damage or different neuroanatomical conditions. Brain death clearly represents a diffuse, severe form of damage including the cerebral hemispheres as well as the brain stem. Tissue damage in coma is regionally more restricted, and the extent and degree of neuropathology probably governs the manifestation of coma as well as the potential for recovery.

2. Energy Consumption during Mental Activity

The normal brain consumes energy at a maximum, or near maximum, rate. Although physiologically abnormal conditions may alter the overall rate of energy consumption, different states of mental activity apparently do not. However, this does not preclude significant changes in local energy metabolism within specific brain regions. In fact, local blood flow in defined regions is apparently subject to dynamic regulation and changes during various mental activities such as hypnosis, reading, and so forth (Scheinberg, 1975). Although such restricted changes may reflect metabolic fluctuation in specific brain regions, these typically have no significant effect on the total brain rate. The rates of cerebral blood flow, glucose consumption, and oxygen consumption are generally thought to be coupled, each accurately reflecting the tissue metabolic rate.

Physiological variation during normal conditions, such as sleep, or during abnormality, such as coma or anesthesia, may produce significant overall changes in the brain metabolic rate. Paradoxical sleep (see Section 1.1) may represent a unique condition of increased metabolic rate in the normal brain. Interestingly, certain types of schizophrenia (with catatonic excitation and stupor) may also show an increased cerebral metabolic rate (Hoyer and Oesterreich, 1975). Severe and prolonged coma presumably results from profound brain dysfunction and represents an extensive case of depressed energy metabolism. Anesthesia may be usefully thought of as a transient, reversible form of coma. Although it is often thought that anesthesia always depresses cerebral oxygen consumption, this is not correct as evidenced by the characteristics of dissociative anesthesia (see Section 3.3). It is more generally known that certain anesthetics (see Section 3.2) uncouple blood flow and oxygen consumption so that the former may not serve as a useful measure of metabolism in the anesthetized preparation.

General anesthetics act on the central nervous system and traditionally these are thought to abolish the sensation of pain by depressing neural activity and the brain metabolic rate. More recently agents such as ketamine, which produce mental "dissociation," have been successfully employed as anesthetics (Corssen et al., 1968; Pender, 1971; Garfield et al., 1972) without mental or metabolic depression. Unlike the case with barbiturate anesthesia during which the brain becomes unresponsive to various stimuli including pain, in the case of ketamine anesthesia the brain is at least partly responsive to pain signals (Miyasaka and Domino, 1968; Tamasy et al., 1975; Sparks et al., 1975); however, it does produce a profound analgesia within a mentally awake but "disconnected state." Clearly, without a vastly more complete picture of the neuroanatomical basis of the relevant behavioral or mental states, we can not begin to determine the different mechanisms of depressant- versus catatonic-acting anesthetics. As an example, the relatively obscure chloral derivatives may produce a useful anesthesia by selective depression of cortical pathways to produce sedation (Balis and Monroe, 1964). However, from a metabolic point of view, it is clear that effective anesthesia can be produced without central nervous depression. The traditional view that anesthesia is a function of depressed neural activity is an artifact of our limited knowledge of compounds acting as depressants. With the emergence of new compounds of the phencyclidine group, our views of anesthetic actions may well turn from metabolic depression toward the selective neuroanatomical pathways involved in pain, response to pain, and memory of it. Anesthesia may result from depressing the sensation of pain or from mental dissociation from the sensation, and possibly either of the mechanisms may be brought about by depressant or excitatory agents acting selectively on various excitatory or inhibitory pathways.

Since it is apparent that metabolic generalizations about anesthetics presently have limited value, the following sections describe the actions of specific compounds. The present review is structured to be largely comparative rather than comprehensive. Additional detail may be found in Siesjo (1978).

3. Differential Effects of Anesthetic Compounds on the Brain Metabolic Rate

Siesjo (1978) provides a comprehensive review of the actions of anesthetic compounds, including critical analysis of the various techniques employed. The present review compares major effects of representative anesthetics to demonstrate that different anesthetic agents have profoundly different effects; the anesthetized brain is not a single condition. For each compound, the cerebral metabolic rate for oxygen ($CMRO_2$) and cerebral blood flow (CBF) are discussed and contrasted (glucose consumption is presumably linked in all cases to the $CMRO_2$).

3.1. Barbiturates

The $CMRO_2$ is unaffected by low, nonanesthetic dosages of barbiturate. A progressive increase in thiopental (Pentothal®), for example, to produce unconsciousness also produces a progressive drop in the $CMRO_2$ in humans (Ketty et al., 1947–48; Himwich et al., 1946–47; Pierce et al., 1962; Wechsler et al., 1951) and in laboratory animals (Homberger et al., 1946). Similarly, McCall and Taylor (1952) show that a drop in $CMRO_2$ using phenobarbital is associated with the transition from drowsiness to actual unconsciousness. Provided Pa_{CO_2} is maintained, the reduction in $CMRO_2$ is accompanied by a comparable drop in CBF (Scheinberg, 1975).

It is generally believed that the consumption of oxygen and glucose are coupled. Therefore, barbiturates probably depress the cerebral metabolic rate for glucose equally with the $CMRO_2$ (Hawkins et al., 1974). The effect on glucose consumption is not uniform in different brain regions (Sokoloff, 1975) and gray matter is more severely affected than white.

3.2. Halothane

This inhalation anesthetic produces a depressed anesthetic state similar to that observed in barbiturate-induced anesthesia. In anesthetic concentration (1%), halothane inhalation is associated with a drop in the brain $CMRO_2$. Although at least part of the metabolic rate reduction is linked to the associated drop in body temperature (Cohen *et al.*, 1964; Wollman *et al.*, 1964), the major part of the $CMRO_2$ reduction is directly resultant from the effects of halothane (Siesjo, 1978).

Halothane uncouples CBF from brain metabolic activity, presumably causing vasodilation by acting directly on the blood vessels (Scheinberg, 1975). Although depressed brain metabolic activity is generally linked with a reduction in CBF, halothane and other inhalants such as chloroform, cyclopropane, ether, and trichloroethylene prove that the rule is not universal in the anesthetic state.

3.3. Ketamine

Clearly the state of general anesthesia is not dependent on a loss of consciousness as demonstrated by ketamine. This anesthetic induces a state described as catatonic detachment (Winters, 1972), produced either directly, or indirectly, by general excitation of the central nervous system (Wong and Jenkins, 1974). Presumably the brain sensations of stimuli, including those normally conveying pain, are not blocked and reach cortical receiving areas; however, these inputs are dissociated from perception by unknown mechanisms.

The brain $CMRO_2$ during ketamine anesthesia is most likely unchanged (Kreuscher and Grote, 1967) or possibly slightly increased (Dawson *et al.*, 1971). Cerebral blood flow is apparently increased.

4. Conclusions and Speculations

The example of ketamine and related compounds is proof that general anesthesia does not require CNS depression. Anesthetics that do produce metabolic depression (i.e., barbiturates, halothane, etc.) seem to do so by lowering energy demand, not by rendering energy reserves insufficient. For example, energy substrates such as ATP, phosphocreatine, and so forth are present in either normal or slightly elevated concentrations in halothane- and pentobarbital-anesthetized animals (review by Siesjo, 1978). Either these anesthetics (1) block phosphorylation and oxidation, thereby depriving neurons of energy, or (2) they inhibit neural function, thereby blocking Krebs cycle, phosphorylation, and oxidation. An increase in ATP, phosphocreatine, glucose, and glycogen can be observed briefly before the loss of consciousness (McCandless and Wiggins, 1981), indicating reduced utilization before loss of consciousness. Energy reserves are present in a similar "standby" condition in reversible coma (see Section 1.3) indicating a "switching down" of synaptic activity.

However, the example of ketamine shows that overall brain metabolism and the anesthetic state are not necessarily coupled. Quite possibly there is a neuroanatomical specificity to anesthesia, and that both general depressants as well as specifically acting compounds may induce anesthesia. Sokoloff (1977) demonstrates that functional activity in discrete brain regions is coupled to local rates of glucose utilization. Hawkins and Biebuyck (1980) show that glucose utilization during halothane anesthesia is generally depressed, but not in all brain regions and not uniformly in the affected areas. Most likely certain brain regions begin to respond to halothane at lower concentrations than other regions. Thus, not only is the nerve synapse possibly the most sensitive site of anesthetic action, but unequal effects in different brain regions most likely produce an altered consciousness and/or anesthesia. Without doubt the utility of whole brain energy studies is at an end with regard to understanding anesthesia. Further work must have meaningful neuroanatomical specificity.

References

Balis, G. V., and Monroe, R. R., 1964, The pharmacology of chlordose, *Psychopharmacologia* **6**:1–30.

Barolin, G. S., 1982, Experimental basis for a neurophysiological understanding of hypnoid states, *Eur. Neurol.* **21**:59–64.

Cohen, P. J., Wollman, S. C., Alexander, P. E., Chase, P. E., and Behar, M. G., 1964, Cerebral carbohydrate metabolism in man during halothane anesthesia, *Anesthesiology* **25**:165–191.

Corssen, G., Miyasaka, M., and Domino, E. F., 1968, Changing concepts of pain control during surgery: Dissociative anesthesia with CI-581, *Anesth. Analg.* **47**:746–759.

Dawson, B., Michenfelder, J. D., and Theye, R. A., 1971, Effects of ketamine on canine cerebral blood flow and metabolism: Modification of pain administration of thiopental, *Anesth. Analg. Ann. Res.* **50**:443–447.

Duffy, T. E., and Plum, F., 1981, Seizures, coma, and major metabolic encephalopathies, in: *Brain Neurochemistry* (G. J. Siegel, R. W. Albers, B. W. Agranoff, and R. Katzman, eds), Little Brown and Co., Boston, pp. 693–718.

Eger II, E. I., 1974, *Anesthetic Uptake and Action*, 1st ed., Williams and Wilkins Co., Baltimore.

Fink, B. R. (ed.), 1980, *Molecular Mechanisms of Anesthesia, Progress in Anesthesiology*, Volume 2, Raven Press, New York.

Garfield, J. M., Garfield, F. B., Stone, J. G., and Hopkins, D., 1972, A comparison of psychologic responses to ketamine and thiopental-nitrous oxide-halothane anesthesia, *Anesthesiology* **36**:329–338.

Gordon, E., (ed.), 1975, *A Basis and Practice of Neuroanesthesia, Monographs in Anesthesiology*, Volume 2, (A. R. Hunter, ed.), Excerpta Medica, Amsterdam.

Hawkins, R. A., and Biebuyck, J. J., 1980, Regional brain function during graded halothane anesthesia, in: *Progress in Anesthesiology*, Volume 2 (E. R. Fink, ed.), Raven Press, New York, pp. 145–156.

Hawkins, R. A., Miller, A. L., Cremer, J. E., and Veech, R. L., 1974, Measurement of the rate of glucose utilization by rat brain *in vivo*, *J. Neurochem.* **23**:917–923.

Himwich, W. A., Homberger, E., Maresca, R., and Himwich, H. E., 1946–47, brain metabolism in main: Anesthetized and in "Pentothal" narcosis, *Am. J. Psychiatry* **103**:689–696.

Homberger, E., Himwich, W. A., Etsten, B., York, G., Maresca, R., and Himwich, H. E., 1946, Effect of pentothal anesthesia on canine cerebral cortex, *Am. J. Physiol.* **147**:343–345.

Hoyer, S., and Oesterreich, K., 1975, Blood flow and oxidative metabolism of the brain in patients with schizophrenia, *Psychiatr. Clin.* **8**:304–313.

Jouvet, M., 1967, Neurophysiology of state of sleep, in: *The Neurosciences* (G. C. Quanton, T. Melnechuls, and F. O. Schmitt, eds.), The Rockefeller University Press, New York, pp. 529–544.

Katz, R. L. (ed.), 1975, *Molecular Mechanisms of Anesthesia, Progress in Anesthesiology*, Volume 1, Raven Press, New York.

Kety, S. S., Woodford, R. B., Harmel, M. H., Freyham, F. A., Appel, K.E., and Schmitt, C. F., 1947–48, Cerebral blood flow and metabolism in schizophrenia: The effect of barbiturate semi-narcosis, insulin coma and electroshock, *Am. J. Psychiatry.* **104**:765–770.

Kreuscher, H., and Grote, J., 1967, Die wirkung des phencyclidin-derivatives ketamine (Cl 581) auf die durchblutung und saverstoffaunahme des gehirns beim hund, *Der Anaesthetist* (Berl.) **16**:304–308.

McCall, M. C., and Taylor, H. W., 1952, Effect of barbiturate sedation on the brain in toxemia of pregnancy, *J. Am. Med. Assoc.* **149**:51–54.

McCandless, D. W., and Wiggins, R. C., 1981, Cerebral energy metabolism during the onset and recovery from halothane anesthesia, *Neurochem. Res.* **6:**1319–1326.

Miyasaka, M., and Domino, E. F., 1968, Neuronal mechanisms of ketamine-induced anesthesia, *Int. J. Neuropharmacol.* **7:**557–573.

Pender, J. W., 1971, Dissociative anesthesia, *J. Am. Med. Assoc.* **215:**1126–1130.

Pierce, E. C., Jr., Lambertsen, C. L., Deutsch, S., Chase, P. A., Linde, H. W., Dripps. R. P., and Price, H. L., 1962, Cerebral circulation and metabolism during thiopental anesthesia and hyperventilation in man, *J. Clin. Invest.* **41:**1664–1671.

Plum, F., and Posner, J. B., 1980, *The Diagnosis of Stupor and Coma*, 3rd ed., F. A. Davis Co., Philadelphia.

Reivich, M. E., Isaacs, G., Evants, E., and Kety, S. S., 1968, The effect of slow wave sleep on regional cerebral blood flow in cats, *J. Neurochem.* **15:**301–306.

Scheinberg, P., 1975, Cerebral blood flow, in: *The Nervous System*, Volume 2, *The Clinical Neurosciences* (D. B. Tower, ed.), Raven Press, New York, pp. 147–156.

Siesjo, B. K., 1978, *Brain Energy Metabolism*, John Wiley and Sons, Chichester.

Sokoloff, L., 1977, Relation between physiological function and energy metabolism in the central nervous system, *J. Neurochem.* **29:**13–26.

Sokoloff, P., 1975, Influence of functional activity on local cerebral glucose utilization, in: *Brain Work*, Alfred Benzon Symposium (D. H. Ingrar and N. A. Lassen, eds.), Munksgaard, Copenhagen, pp. 385–388.

Sparks, D. L., Corssen, G., Aizenman, B., and Black, J., 1975, Further studies of the neural mechanisms of ketamine-induced anesthesia in the Rhesus monkey, *Anesth. Analg.* **54:**189–195.

Tamasy, V., Koranyi, L., and Tekeres, M., 1975, EEG and multiple unit activity during ketamine and barbiturate anesthesia, *Br. J. Anaesth.* **47:**1247–1251.

Taylor, R. E., 1959, Effect of procaine on electrical properties of squid axon membrane, *Am. J. Physiol.* **196:**1071–1078.

Townsend, R. E., Prinz, P. N., and Obrist, W. D., 1973, Human cerebral blood flow during sleep and waking, *J. Appl. Physiol.* **35:**620–625.

Wechsler, R. L., Dripps, R. D., and Kety, S. S., 1951, Blood flow and oxygen consumption of the human brain during anesthesia produced by thiopenthal, *Anesthesiology* **12:**309–314.

Winters, W. D., 1972, Epilepsy or anesthesia with ketamine, *Anesthesiology* **36:**309–312.

Wollman, H., Alexander, S. C., Cohen, P. J., Chace, P. E., Melman, E., and Behr, M. E., 1964, Cerebral circulation of man during halothane anesthesia, *Anesthesiology* **25:**180–184.

Wong, D. H. W., and Jenkins, L. C., 1974, An experimental study of the mechanisms of action of ketamine on the central nervous system, *Can. Anaesth. Soc.* **21:**57–67.

19

Neurochemical Effects of Viral Infections in the Central Nervous System

K. W. RAMMOHAN, A. A. FAROOQUI, and L. A. HORROCKS

1. Introduction

In recent years it has become apparent that dysfunction of virus-infected cells occurs by a variety of mechanisms. Cells can harbor viruses and be unaffected, with regard to vital functions of cell division or cell differentiation. Similarly, virus persistence has been observed *in vivo* in the absence of any clinical illness, and these observations collectively raise the question as to what leads to cellular dysfunction in the course of a viral infection. Correlations of changes in specific molecular events to alterations in cellular function have recently been studied in a number of viral infections. Although a complete review of all viral systems studied in this regard is beyond the scope of this review, basic mechanisms highlighted by specific viral infections will be discussed such that many mechanisms by which viruses cause central nervous system (CNS) dysfunction will become evident.

2. General Considerations

Viral infections lead to cellular dysfunction or death by any of the following mechanisms.

2.1. Direct Cytolytic Effect

This can be secondary to impairment of vital cell functions and usually occurs late in infection after sufficient viral replication has been established.

K. W. RAMMOHAN • Department of Neurology, Ohio State University, Columbus, Ohio 43210. *A. A. FAROOQUI and L. A. HORROCKS* • Department of Physiological Chemistry, Ohio State University, Columbus, Ohio 43210.

Such impairment can involve cellular machinery for protein synthesis or respiration or can alter cell membrane permeabilities such that death of the cell with release of infectious virus occurs.

Impairment of host-cell protein synthesis occurs with a variety of viral infections. Often, the impairment is selective in that while protein synthesis of viral proteins occurs, there is almost complete cessation of host-cell protein synthesis. This is exemplified by infections of susceptible cells by vesicular stomatitis virus (Huang and Wagner, 1965). Almost complete inhibition of host-cell protein synthesis occurs within minutes and parallels the onset of synthesis of viral polypeptides. The specific mechanisms of impairment of protein synthesis are unknown. It has been suggested that incubation of susceptible cells with viruses results in early alterations of membrane permeability to protein toxins. In studies using HeLa cells infected with encephalomyocarditis virus, it was shown that permeability to alpha-sarcin (a protein of 16,800 mol. wt. that inhibits protein synthesis) occurred within minutes, whereas uninfected cells were impervious to this protein (Fernandez-Puentes, 1983). Also, attachment of virus to specific cellular receptors was shown to be necessary for this virus to induce cell membrane alterations (Fernandez-Puentes, 1983).

2.2. Immune-Mediated Cellular Dysfunction

It is well established that the immune response to a viral infection can result in immune-mediated destruction of not only infected cells, but also uninfected cells that have become sensitized during the initial virus infection. Such "autoimmune" phenomena are known to mediate delayed demyelination in the postinfectious encephalomyelitis syndromes (Weiner et al., 1973).

A variety of immune mechanisms, some virus specific and others nonspecific, cause immune-mediated injury. Thus, although interferon has been traditionally associated with protection from virus infections, sometimes the ability to mount a greater interferon level seems to be associated with greater morbidity and mortality in experimental animals infected with lymphocytic choriomeningitis virus (Riviere et al., 1980). The specific role of interferon in causation of disease was demonstrated by the ability of anti-interferon globulin to abrogate virus-induced disease and death (Riviere et al., 1980). Complement-mediated injury, antibody-dependent cytotoxicity, and T-cell-mediated cytotoxicity are all various immunological mechanisms by which virus-infected cells undergo destruction. Immune-mediated mechanisms for destruction of virus-infected cells have been the subject of other reviews (Moller, 1981) and therefore will not be discussed here.

2.3. Cellular Virus Persistence and Impairment of Cell Differentiation

It is known that viruses can infect cells and remain latent. Although viral gene products are seldom expressed by these cells, hybridization studies using nucleic acid probes can readily detect the presence of viral genome. The effects of the latent virus on cells are not known, but may be important for oncogenic viruses where transformation of the cell is associated with the presence of viral genome in the transformed cells.

435

NEUROCHEMICAL
EFFECTS OF
VIRAL
INFECTIONS IN
THE CENTRAL
NERVOUS
SYSTEM

Viral infection, resulting in a persistently infected state with expression of some or all of the viral gene products, occurs *in vitro* and *in vivo*. Such viruses may be defective in their ability to effectively replicate and produce infectious progeny, but often do not appear to impair survival of the cell. In the persistently infected state, cells appear normal and can be maintained *in vitro* indefinitely. Although effects of virus infection are not readily apparent, it is now known that specific differentiation functions of a cell can be affected. Examples pertaining to the nervous system are discussed below.

3. Neurochemistry of Virus-Induced Demyelination

Demyelination induced by viruses is classified under myelinoclastic disorders, wherein the myelin, which is normal prior to virus infection, breaks down during the course of the virus encephalopathy. Significant alterations occur in the chemical composition of myelin immediately preceding and during demyelination. Primary demyelination from infection of oligodendroglia or secondary demyelination due to loss of neurons can occur, depending on the specific cell types infected in the brain.

Most of the information regarding chemical changes in the brain in virus-induced demyelination is available from studies of subacute sclerosing panencephalitis (SSPE) caused by measles virus or chronic infection of the nervous system of dogs with canine distemper virus. In SSPE, although the myelin itself appears normal ultrastructurally, abnormal chemical changes have been noted (Norton *et al.*, 1966). The lipid to protein ratio was normal, but more cholesterol and less cerebrosides and total phospholipids were found (Norton *et al.*, 1966). A marked decrease of ethanolamine glycerophospholipids including plasmalogens was noted. Cholesterol esters were plentiful in the white matter. Cholesterol esters are normally absent in mature white matter and their presence is usually indicative of active demyelination (Morell *et al.*, 1972).

In BGM cells (African green monkey kidney) persistently infected with measles virus, there was a marked increase in the incorporation of radiolabeled fatty acids into the neutral lipid fraction (Anderton *et al.*, 1983). Compared with uninfected cells, the increase was up to two-fold for palmitic, stearic, and oleic acids and 8- to 14-fold for arachidonic acid. The lipid metabolism of BGM cells acutely infected with measles virus was unmodified. The radiolabeled fatty acids incorporated into the neutral lipids in persistently infected cells were principally associated with the triacylglycerol fraction (Anderton *et al.*, 1983). Although there was no significant difference in the incorporation of radioactivity into the total phospholipid of uninfected and persistently infected BGM cells, there was a large decrease in radiolabeled arachidonic acid incorporated into phosphatidylethanolamine and to a lesser extent phosphatidylcholine and phosphatidylinositol in persistently infected cells (Anderton *et al.*, 1983). Stearic acid incorporation was also reduced in phosphatidylcholine and phosphatidylethanolamine fractions of persistently infected cells.

A variety of abnormalities of the hydrolytic enzymes have been demonstrated in virus-induced demyelination and in multiple sclerosis, a disorder suspected to be of viral origin (Allen, 1983). In canine distemper demyelinating encephalomyelitis, the severity of demyelination is related to increased activities of plasmalogenase and acid proteases and smaller increases in neutral proteases and β-glucuronidases (McMartin *et al.*, 1972). Elevated plasmalogenase activity precedes cellular invasion and the activation of lysosomal enzymes (Horrocks *et al.*, 1978). Increased activities of cathepsins A and B, β-galactosidase and β-glucuronidase occur in measles virus encephalitis, Sindbis virus encephalitis, and Semliki Forest virus encephalitis (Bowen *et al.*, 1974). Comparison of the neuropathological changes and activities of hydrolytic enzymes indicates that β-galactosidase and cathepsin A might be useful indices of neuronal degeneration and cellular infiltration into white matter, respectively (Bowen *et al.*, 1974). Furthermore, increased β-glucuronidase activity of infected tissue was a sensitive indicator of cellular reaction to injury, probably reflecting proliferation of glial cells. An increase of lysosomal enzymes early in infection prior to morphological evidence of cell damage has been reported in Semliki Forest virus infections (Millson and Bountiff, 1973) and mice infected with scrapie (Millson and Bountiff, 1973), and probably reflects damage mediated through the release of these enzymes.

4. Neurochemical Changes in Subacute Spongiform Encephalopathy

Alterations in lysosomal enzymes have also been implicated in what are now known as slow virus diseases of the nervous system. These disorders, all of which are caused by transmissible agents, are characterized by long incubation periods (often years) and result in a slow but relentless deterioration of the nervous system that is uniformly fatal. Conventional viruses such as measles RNA genome and papovaviruses DNA genome are among a variety of agents that cause such disorders, but in a number of disorders that affect man and other mammals, the agent remains poorly characterized. Kuru and Creutzfeldt-Jakob disease (CJD) in man, scrapie in sheep, and transmissible mink encephalopathy (TME) in the mink all belong to the group of disorders termed subacute spongiform encephalopathies. The pathological picture of spongiform changes observed by light microscopy is common to all these disorders and forms the basis for their common classification. Characteristically, there is a lack of inflammation (hence the term "encephalopathy") in the affected brain, and singularly there is a lack of immune response to this agent, at least by the conventional techniques of assessment of immune response to a pathogen.

Neurochemical changes in the brain have been studied in CJD in man and brains from experimental animals that were inoculated with brain tissue from patients with CJD and kuru. Suzuki and Chen (1966) analyzed brains from three patients with various cases of CJD and compared them with controls. As the disease progressed, the lipid to protein ratio was increased in gray and white matter. Quantitative changes in the lipids were also observed. The amount of cholesterol remained constant, but the total glycolipid content increased. Phospholipids and gangliosides markedly decreased in the brains of CJD patients. The amount of RNA and proteolipid was also decreased. No changes were

437

*NEUROCHEMICAL
EFFECTS OF
VIRAL
INFECTIONS IN
THE CENTRAL
NERVOUS
SYSTEM*

reported in glycosaminoglycans. Recently, studies by Tamai *et al.* (1978) showed an increase in the lipid content in CJD brains mainly due to an increase in total cholesterol unlike the previous findings by Suzuki and Chen (1966), who attributed this increase to the glycolipids.

Alterations in the ganglioside composition have also been reported in brains of patients with CJD (Yu *et al.*, 1974; Yu and Manuelidis, 1976). Chimpanzees inoculated with brain tissue from patients with CJD and kuru had an abnormal content and pattern of gangliosides. An abnormal distribution of long-chain bases in gangliosides obtained from brains of patients with CJD was the most remarkable observation by Tamai *et al.* (1978). The C-20 sphingosine in gangliosides was markedly reduced. The ratio of C-20 to C-18 sphingosine was 0.03 and 0.23 in diseased gray matter from two cases compared with the normal ratio of 0.84. Severe neuronal depletion and accompanying astrocytosis has been implicated as a possible mechanism for the ganglioside changes found in experimental CJD (Yu and Manuelidis, 1976). However, recent studies of isolated neurons, astrocytes, and oligodendroglia have shown ganglioside patterns similar to those of the whole brain, and therefore the explanation for the ganglioside changes is not readily apparent (Hamberger and Svennerholm, 1971; Norton *et al.*, 1975). It is now known that cultured cells transformed *in vitro* by conventional oncogenic viruses show an altered ganglioside metabolism (Hakamori, 1973; Richardson *et al.*, 1975; Fishman and Brady, 1976). It is possible that similar changes occur in CJD brains *in vivo*.

A variety of alterations in brain enzymes have been reported to occur in brains from patients and animals with subacute spongiform encephalopathies. In mice with experimental scrapie, a selective increase in the levels of a number of glycosidases including β-glucuronidase was observed prior to the onset of clinical or morphological signs of cellular damage (Millson and Bountiff, 1973). Recent studies on brain glycosidases in CJD also indicate increased activities of β-glucuronidase and β-galactosidases in gray matter (Annuziata and Federio, 1981). The role of these hydrolytic enzymes in the pathogenesis of these disorders remains unknown, but their alterations prior to visible evidence of cell damage could implicate them in the pathogenesis of these conditions, rather than being the result of injury or cell death.

Besides these changes in hydrolytic enzymes, alterations in the activities of NAD diaphorase, acetylcholinesterase, and glycosyl transferases have also been reported to occur in CJD- and scrapie-infected mouse brains (Friede and DeJong, 1964). Thus Freide and DeJong (1964) reported significant decreases in NAD diaphorase levels in various regions of CJD brains. Using microhistochemical techniques, diminished activities of acetylcholinesterases have been demonstrated in brains obtained at postmortem from CJD patients (Pope *et al.*, 1964). Increased activities of glycosyl transferases occur in brains from scrapie-infected animals (Suckling and Hunter, 1974). Altered activities of glycosyl transferases have also been reported in virally transformed cells and brain explants from scrapie-infected mice *in vitro* (Haig and Clarke, 1971). Whether the alterations in glycosyl transferases observed in scrapie are a result of a virus-induced transformation of brain cells remains to be determined.

Alterations in the plasma membrane and synaptosomal membranes occur in patients and animals with subacute spongiform encephalopathies (Lampert *et al.*, 1971). Ruptured plasma membrane and curled fragmented membranes contained in membrane-bound vacuoles have been identified ultrastructurally in these brains, and have been suspected to harbor the infectious agent (Lampert *et al.*, 1971). Electron spin studies in scrapie suggested early alterations in protein distribution within the synaptosomal membranes and rigid membranes in the late stages of this disorder (Viret *et al.*, 1981). By contrast, brain explants from hamsters infected with scrapie showed a significantly greater percentage of cells capable of capping with concanavalin A and suggested greater fluidity of cell membranes (Rutter *et al.*, 1981). These differences may well be related to differences in the cell populations studied in different stages of these disorders. Their role in the pathogenesis of the disease remains speculative.

Correlation of the various chemical alterations with structural and functional changes in CJD has been done by microchemical studies of various layers of the cortex from brains of patients with CJD by Bass *et al.* (1974). Neuronal loss preceding astrogliosis could be shown by analysis of the DNA and RNA content of various layers of the cortex. Although the DNA and RNA contents were normal in the unaffected superficial cortical layers, RNA content per cell was significantly lower in layers 3, 5, and 6 where patchy neuronal loss with minimal astrogliosis was observed. The diminished RNA content per cell was attributed to the marked decrease in the ribonucleoprotein particles and cisternae of the endoplasmic reticulum in neurons that had normal Golgi apparatus, mitochondria, and neurotubules. The decreased cerebroside content was probably due to secondary demyelination due to neuronal death. The mechanism of functional alterations that occurs from these structural changes remains speculative. Gangliosides have an important role in ion transport and transmitter release functions. Changes in ganglioside metabolism could interfere with neurotransmitter release and result in neuronal dysfunction (Clowes *et al.*, 1972). Hamsters with scrapie show hypersensitivity to central serotonergic drugs and have significantly decreased brain serotonin content unlike control animals (Rohwer *et al.*, 1981; Goudsmit *et al.*, 1981). Whether these changes are related to ganglioside alterations is not known.

5. Virus-Induced Alterations of Homeostasis

A novel way by which viruses cause disease has recently been described in experimental infection of mice with lymphocytic choriomeningitis virus (Oldstone *et al.*, 1982). Following neonatal inoculation of a relatively noncytopathic strain of this virus (Armstrong strain 1371), stunted growth and early death of C3H/St mice occurred (Doyle *et al.*, 1980). The cause of death was uncertain, but believed to be metabolic since histopathological examinations of all organs were normal. Viral antigen was readily demonstrated in the growth-hormone-producing cells of the pituitary and impairment of production of growth hormone was observed *in vivo* (Oldstone *et al.*, 1982). Hypoglycemia concomitant with the growth hormone deficiency was identified. The specific mechanism whereby cell differentiation and secretion of growth hormones was impaired remains unidentified.

Experimental virus-induced endocrinopathies are known to occur with Coxsackie virus (diabetes mellitus), encephalomyocarditis virus (diabetes mellitus), and reovirus type-1 (polyendocrinopathy). Direct cytolytic effects or autoimmune-mediated damage of selected cells form the basis of these virus-induced dysfunctions (Craighead, 1975; Onodera et al., 1981; Haspel et al., 1983). Although Coxsackie virus has been implicated in human juvenile diabetes mellitus (Yoon et al., 1979), viral etiologies for other human endocrinopathies have not been studied.

439

NEUROCHEMICAL
EFFECTS OF
VIRAL
INFECTIONS IN
THE CENTRAL
NERVOUS
SYSTEM

6. Virus-Induced Impairment of Synaptic Transmission

Impairment of synaptic transmission has been postulated to occur in some neurotropic viral infections. Following the intracerebral inoculation of hamster neurotropic measles virus into weanling BALB/c mice, the majority of animals exhibit signs of an acute encephalopathy that is usually fatal. The infected brains have abundant viral antigen, but lack necrosis or inflammation. The mechanism of CNS dysfunction remains unknown, but localization of viral antigen to the postsynaptic terminals could be readily demonstrated (Pottelsberghe et al., 1979). A similar postsynaptic localization of viral antigen or viral particles has also been demonstrated for the JHM strain of coronavirus and rhabdovirus (Knobler et al., 1981; Greenstein et al., 1981). Although viral antigen is seldom observed presynaptically, a transsynaptic spread of virus has been postulated. Cultured neurons infected with vescicular stomatitis virus (VSV) and maintained in vitro in the presence of antiviral antibodies demonstrate virus budding from postsynaptic sites directly onto coated pits and vesicles on presynaptic membrane (Dubois-Dalcq et al., 1980). It is well known that coated pits and vesicles are involved with reuptake of synaptic vesicle membrane and trophic factors, and therefore it is possible that competition for uptake mechanisms by viruses could result in dysfunction of synaptic transmission.

Alterations in the metabolism of neurotransmitters occur in the wake of viral infections. An increased content of homovanillic acid and 5-hydroxyindole acetic acid has been reported in mouse brain during infections with herpes simplex, vaccinia, pseudorabies, rabies, and neuroinfluenza viruses (Lycke and Roos, 1968, 1972; Lycke et al., 1970). Increased rates of turnover of dopamine and serotonin have been observed. Similar observations have also been reported in Venezuelan equine encephalitis in which a significant increase in the concentration of brain dopamine and homovanillic acid was observed (Bonailla et al., 1975). The mechanism of virus-induced turnover of monoamines is not understood, but the rise in monoamine concentrations could be accounted for by the increase in activity of tyrosine hydroxylase (Levitt et al., 1965), the rate-limiting step in catecholamine synthesis.

7. Virus-Induced Alterations of Neuronal Plasma Membranes and Cell Surface Receptors

A variety of neurotropic viruses induce cell fusion in vitro and in vivo. Examples include visna virus, myxovirus, and paramyxovirus. Viruses inactivated with ultraviolet irradiation are capable of inducing cell fusion, a phenomenon known as "fusion from without." By contrast, "fusion from within" results from alterations in cell membranes induced by changes that occur in an infected cell during virus replication.

Several mechanisms have been described for fusion of biological membranes. Fusion of protein-free phospholipid vesicles has been induced by alteration of divalent cation concentrations (Ca^{2+} or Mg^{2+}) (Papahadjopoulos *et al.*, 1977). Lysolecithin has been shown to induce fusion of cell membranes of chick erythrocytes (Martin and MacDonald, 1976). Changes in lysophospholipid concentration and phospholipase activity have been studied in cell lines infected with measles virus (Suzuki and Matsumoto, 1982). There was a close correlation of the magnitude of cell fusion and the release of lysosomal phospholipases A_1 and A_2 into the cytosol. Hydrolysis of endogenous phospholipids in the plasma membrane to the lysophospholipids was followed by incorporation of lysophosphatidylcholine into the cells and conversion to triacylglycerol, phosphatidylcholine, and phosphatidylethanolamine during cell fusion. An extensive accumulation of triacylglycerol was also observed. The precise role of the viral glycoprotein (fusion protein) in mediating fusion of the lipid bilayer that has undergone perturbation following the changes in the phospholipids remains to be defined. Nevertheless, it is clear that viruses can modulate the lysosomal phospholipase activities and marked changes in lysophospholipids occur concomitant with the process of cell fusion.

Impairment of transmembrane signal transfer by hormones or neurotransmitters occurs following infection of neurons or glia *in vitro* with a variety of viruses such as measles, canine distemper, rabies, and lymphocytic choriomeningitis (Halbach and Koschel, 1979; Munzel and Koschel, 1982). Generation of 3'5'-cyclic adenosinemonophosphate (cAMP) prior to and following establishment of persistent infection with measles virus was studied in C6 rat glioma lines following stimulation with β-adrenergic agonists. Although a 1000-fold increase of 3'5'-cAMP occurred in uninfected cell lines, less than half that increase was generated in infected cell lines. The defect was shown to be secondary to loss of functional adenylate cyclase units rather than to alterations of activity of single enzyme units (Munzel and Koschel, 1982). The effects of persistent virus infection on the β-adrenergic receptor remain to be studied. Similarly, alterations in the cholinergic system have been shown in neuroblastoma cell lines persistently infected with lymphocytic choriomeningitis virus (Oldstone *et al.*, 1977).

Cellular glycoproteins on cell surfaces that mediate various normal biological functions can sometimes serve as receptors for viruses. Acetylcholine receptors on cultured myotubules in the mature muscle were shown to mediate attachment and entry of rabies virus (Lentz *et al.*, 1982). The binding and subsequent infection by rabies virus could be blocked by pretreatment of muscle with bungarotoxin that binds irreversibly with acetylcholine receptors. This provides a mechanism by which the virus replicates in the muscle and spreads through the neuromuscular junctions into peripheral nerves and possibly extend transsynaptically into the central nervous system.

Infection of susceptible cells by viruses can sometimes lead to induction on cell surfaces of new receptors not previously present. Herpes virus infection can result in the expression of receptors that bind the Fc fragment of immunoglobulins. The biological significance of such receptors is unknown.

8. Reye's Syndrome

441

NEUROCHEMICAL
EFFECTS OF
VIRAL
INFECTIONS IN
THE CENTRAL
NERVOUS
SYSTEM

Reye's syndrome is a metabolic encephalopathy seen primarily in children (Reye et al., 1963). The etiology of Reye's syndrome is unknown, but viruses have been implicated because the disease most frequently follows upper respiratory infections with influenza or varicella (DeVivo and Keating, 1976). In addition to virus, toxins (exogenous or endogenous), genetic predisposition, and combinations of these putative factors have all been considered as etiologic components of Reye's syndrome (DeVivo and Keating, 1976). The neuropathologic observations in Reye's syndrome are rather constant and relatively nonspecific (DeVivo and Keating, 1976; Trauner, 1982). Virtually all of the metabolic abnormalities in Reye's syndrome can be explained on the basis of liver dysfunction (Trauner, 1982). Elevated serum transaminases, hyperammonemia, and prolongation of prothrombin time are the most characteristic features of this encephalopathy (DeVivo and Keating, 1976; Trauner, 1982). The other biochemical abnormalities include lactic acidemia, elevations in serum concentration of creatine phosphokinase, hyperaminoacidemia, and organic acidemia (Trauner, 1982). Brain concentrations of the putative neurotransmitter octopamine are markedly increased, whereas dopamine and norepinephrine concentrations are decreased (Trauner, 1982). Reye's syndrome patients may have increased serum concentrations of short-chain fatty acids (Trauner et al., 1975) that are capable of producing coma (Zieve et al., 1974). In fact, intravenous infusion of the short-chain fatty acid octanoate into rabbits results in the major clinical, biochemical, and pathological changes found in humans with Reye's syndrome (Trauner, 1982; Trauner and Adams, 1981). Thus, fatty acids are potential endogenous toxins in this encephalopathy, either alone or in combination with ammonia (Trauner and Adams, 1981; Brown and Madge, 1972).

Several of the hepatic and cerebral histologic and functional abnormalities in Reye's syndrome are consistent with the view that a damaging free radical or peroxidative mechanism is involved (Jurkowitz et al., 1974; Partin et al., 1979; Lotscher et al., 1979, 1980). Further, many of the metabolic observations made in Reye's syndrome suggest that pathways normally involved in either peroxidative or free radical generation are stimulated. For example, the high levels of serum urate in cases of this disorder (Aprille et al., 1980) suggest activation of xanthine oxidase pathway since the lactic acidosis in Reye's syndrome patients is not high enough to cause decreased renal clearance of the urates. This pathway is known to produce lipid peroxidation in biological membranes under appropriate conditions (Lai and Piette, 1977, 1978) and has been shown to be increased in other pathological conditions (Tubaro et al., 1980). Another nearly constant feature of Reye's syndrome is a rise in free fatty acids and an increase in hepatic triacylglycerols (Pollack et al., 1975; Manunes et al., 1974; Ogburn et al., 1982). This, together with the observed increase in liver peroxisomes (Bradel and Reiner, 1975) may give rise to hepatic peroxide or free radicals since peroxisomal beta-oxidation of long-chain fatty acids utilizes an NADPH oxidase that reduces oxygen directly to hydrogen peroxide (deDuve, 1978). Parenthetically, it is also known that hepatic peroxisomes metabolize long-chain fatty acids, but cannot break them down further than butyryl-CoA or perhaps hexanoyl-CoA (Osmundsen et al., 1979). Normally, liver mitochondria have no difficulty in oxidizing short-chain fatty acids, but in the event of mitochondrial damage and increased peroxisomal activity due to endogenously produced deleterious agents or toxins, fatty acids, possibly of short- and medium-chain length, could accumulate. These fatty acids, as mentioned earlier, along with ammonia can be extremely toxic to osmotically sensitive cell compartments. In fact, it is well known that fatty acids (e.g., arachidonate) can cause brain cortical swelling under proper conditions (Chan *et al.*, 1980; Kontos *et al.*, 1980). This swelling may be due to the free fatty acid itself, lipoxygenase or cyclooxygenase products, or to nonenzymic free radical peroxidation. The latter possibilities are most attractive since it is known that nonsteroidal anti-inflammatory agents can stimulate lipid peroxidation in the liver (Pennington and Smith, 1978).

References

Allen, I. V. 1983, Hydrolytic enzymes in multiple sclerosis, in: *Progress in Neuropathology*, Volume 5, (H. M. Zimmerman, ed.), Raven Press, New York, pp. 1–17.

Anderton, P., Wild, T. F., and Zwingelstein, G., 1983, Measles-virus-persistent infection in BGM cell, *Biochem. J.* **214**:665–670.

Annuziata, P., and Federio, A., 1981, Brain glycosidases in Creutzfeldt-Jakob disease, *J. Neurol. Sci.* **49**:325–328.

Aprille, J. R., Austin, J., Costello, C. E., and Royal, N., 1980, Identification of the Reye's syndrome "serum factor", *Biochem. Biophys. Res. Commun.* **94**:381–389.

Bass, N. H., Hess, H. H., and Pope, A., 1974, Altered cell membranes in Creutzfeldt-Jakob disease, *Arch. Neurol.* **31**:174–182.

Bonailla, E., Ryder, S., and Hernandez, H., 1975, Venezuelan equine encephalomyelitis virus infection: Effect on monoamine metabolism of mouse brain, *J. Neurochem.* **25**:529–530.

Bowen, D. M., Flack, R. H. A., Martin, R. O., Smith, C. B., White, P., and Davison, A. N., 1974, Biochemical studies on degenerative neurological disorders. 1-Acute experimental encephalitis, *J. Neurochem.* **22**:1099–1107.

Bradel, E. J., and Reiner, C. B., 1975, The fine structure of hepatocytes in Reye's syndrome, in: *Reye's syndrome* (J. D. Pollack, ed.), Grune and Stratton, New York, pp. 147–158.

Brown, R. E., and Madge, G. E., 1972, Fatty acids and mitochondrial injury in Reye's syndrome, *N. Engl. J. Med.* **286**:787–788.

Chan, P. H., Fishman, R. A., Lei, J. L., and Quan, S. C., 1980, Arachidonic acid induced swelling in uncirbated rat brain slices, *Neurochem. Res.* **5**:629–639.

Clowes, A. W., Cherry, R. J., and Chapman, D., 1972, Physical effects of tetanus toxin on model membranes containing ganglioside, *J. Mol. Biol.* **67**:49–57.

Craighead, J. E., 1975, The role of viruses in the pathogenesis of pancreatic disease and diabetes mellitis, *Prog. Med. Virol.* **19**:162–207.

deDuve, C., 1978, A reexamination of the physiological role of peroxisomes, in: *Tocopherol, Oxygen and Biomembrane* (C. deDuve and O. Hayaishi, eds.), Elsevier, Amsterdam, pp. 351–361.

DeVivo, D. C., and Keating, J. P., 1976, Reye's syndrome, *Adv. Pediatr.* **22**:175–230.

Doyle, L. B., Doyle, M. V., and Oldstone, M. B. A., 1980, Susceptibility of newborn mice with H-2^k background to lymphocytic choriomeningitis virus infection, *Immunology* **40**:589–596.

Dubois-Dalcq, M., Hooghe-Peters, E. L., and Lazzarim, R. A., 1980, Antibody-induced modulation of rhabdovius infection of neurons *in vitro*, *J. Neuropathol. Exp. Neurol.* **39**:507–522.

Fernandez-Puentes, G., 1983, Permeability of alpha sarcin virus-infected cells, *Mol. Cell. Biochem.* **50**:185–191.

Fishman, P. H., and Brady, R. O., 1976, Biosynthesis and function of ganglioside, *Science* **194**:906–915.

Friede, R. L., and DeJong, R. N., 1964, Neuronal enzymatic failure in Creutzfeldt-Jakob disease, *Arch. Neurol.* **10**:181–195.

Goudsmit, J., Rohwer, R. G., Silbergeld, E. K., and Gadjusek, D. C., 1981, Hypersensitivity to central serotonin receptor activation in scrapie-infected hamsters and the effect of serotonergic drugs on scrapie symptoms, *Brain Res.* **220**:372–377.

Greenstein, J. I., Baron-Van Evercooren, A., Lazzarini, R. A., and McFarland, H. F., 1981, Infection of the central nervous system produced by the R1 vesicular stomatitis virus, *Lab. Invest.* **44**:487–495.

Haig, D. A., and Clarke, M. C., 1971, Multiplication of the scrapie agent, *Nature (London)* **234**:106–107.

Hakamori, S., 1973, Glycolipids of tumor cell surface, *Adv. Cancer Res.* **18**:265–315.

Halbach, M., and Koschel, K., 1979, Impairment of hormone dependent signal transfer by chronic SSPE virus infection, *J. Gen. Virol.* **42**:615–619.

Hamberger, A., and Svennerholm, L., 1971, Composition of gangliosides and phospholipids of neuronal and glial cell enriched fractions, *J. Neurochem.* **18**:1821–1829.

Haspel, M. V., Onodera, T., Prabhakar, B. X., Horita, M., Suzuki, H., and Notkins, A. L., 1983, Virus-induced autoimmunity: Monoclonal antibodies that react with endocrine tissue, *Science* **220**:304–306.

Horrocks, L. A., Spanner, S., Mozzi, R., Fu, S. C., D'Amato, R. A., and Krakowka, S., 1978, Plasmologenase is elevated in early demyelinating lesions, *Adv. Exp. Biol. Med.* **100**:423–438.

Huang, A. S., and Wagner, R. R., 1965, Inhibition of cellular RNA synthesis by nonreplicating vesicular stomatitis virus, *Proc. Natl. Acad. Sci. U.S.A.* **54**:1579–1582.

Jurkowitz, M., Scott, K. M., Altshuld, R. A., Merola, A. J., and Brierley, G. P., 1974, Retention and loss of energy coupling in aged heart mitochondria, *Arch. Biochem. Biophys.* **165**:98–113.

Knobler, R. L., Dubois-Dalcq, M., Haspel, M. V., Clay Smith, A. P., Lampert, P. W., and Oldstone, M. B. A., 1981, Selective localization of wild type and mutant mouse hepatitis virus (JHM Strain) antigens in CNS tissue by fluorescence, light and electron microscopy, *J. Neuroimmunol.* **1**:81–92.

Kontos, H. A., Wei, E. P., Povlishock, J. T., Dietrick, W. D., Magiera, C. J., and Ellis, E. F., 1980, Cerebral arteriolar damage by arachidonic acid and prostaglandin G-2, *Science* **209**:1242–1244.

Lai, C. S., and Piette, L. M., 1977, Hydroxyl radical production involved in lipid peroxidation of rat liver microsomes, *Biochem. Biophys. Res. Commun.* **78**:51–59.

Lai, C. S., and Piette, L. M., 1978, Spin-trapping studies of hydroxyl radical production involved in lipid peroxidation, *Arch. Biochem. Biophys.* **190**:27–38.

Lampert, P., Hooks, J., Gibbs, J. R., C. J., and Gajdusek, D. C., 1971, Altered plasma membranes in experimental scrapie, *Acta Neuropathol.* **19**:81–93.

Lentz, T. L., Burrage, T. A., Smith, A. L., Crick, J., and Tignor, G. H., 1982, Is the acetylcholine receptor a rabies virus receptor, *Science* **215**:182–184.

Levitt, M., Spector, S., Sjoerdsma, A., and Udenfriend, S., 1965, Elucidation of the rate-limiting step in norepinephrine biosynthesis in the perfused guinea-pig heart, *J. Pharmacol. Exp. Ther.* **148**:1–8.

Lotscher, H., Winterhalter, K. H., Carafoli, E., and Richter, C., 1979, Hydroperoxide can modulate the redox state of pyridine nucleotides and calcium balance in rat liver mitochondria, *Proc. Natl. Acad. Sci. U.S.A.* **76**:4340–4344.

Lotscher, H., Winterhalter, K. H., Carafoli, E., and Richter, C., 1980, Hydroperoxide induced loss of pyridine nucleotides and release of calcium from rat liver mitochondria, *J. Biol. Chem.* **255**:9325–9330.

Lycke, E., and Roos, B. E., 1968, Effect on the monoamine-metabolism of the mouse brain by experimental Herpes simplex infection, *Experientia* **24**:687–689.

Lycke, E., and Roos, B. E., 1972, The monoamine metabolism in viral encephalitides of the mouse II. Turnover of monoamines in mice infected with Herpes simplex virus, *Brain Res.* **44**:603–613.

Lycke, E., Modigh, K., and Roos, B. E., 1970, The monoamine metabolism in viral encephalitides of the mouse 1. Virological and biochemical results, *Brain Res.* **23**:235–246.

Manunes, P., DeVries, G., Miller, H., and David, R. B., 1974, Fatty acids in Reye's syndrome, *Pediatr. Res.* **8**:436.

Martin, F. J., and MacDonald, R. C., 1976, Phospholipid exchange between bilayer membrane vesicles, *Biochemistry* **15**:321–327.

McMartin, D. N., Koestner, A., and Long, J. F., 1972, Enzyme activities associated with the demyelinating phase of canine distemper, *Acta Neuropathol.* **22**:275–287.

Millson, G. C., and Bountiff, L., 1973, Glycosidases in normal and scrapie mouse brain, *J. Neurochem.* **20**:541–546.

Moller, G., 1981, MHC restriction of antiviral immunity, *Immunol. Rev.* **58**:5–157.

Morell, P., Bornstein, M. R., and Norton, W. T., 1972, Diseases of myelin, in: *Basic Neurochemistry* (W. R. Albers, G. J. Siegel, R. Katzman, and B. W. Agranoff, eds.) Little Brown and Co., Boston, pp. 497–516.

Munzel, P., and Koschel, K., 1982, Alteration in phospholipid methylation and impairment of signal transmission in persistently paramyrovirus-infected rat glioma cells, *Proc. Natl. Acad. Sci. U.S.A.* **79**:3692–3696.

Norton, W. T., Poduslo, S. E., and Suzuki, K., 1966, Subacute sclerosing leukoencephalitis. II. Chemical studies including abnormal myelin and abnormal ganglioside pattern, *J. Neuropathol. Exp. Neurol.* **25**:582–590.

Norton, W. T., Abe, T., Poduslo, S. E., and DeVries, G. H., 1975, The lipid composition of isolated brain cells and axons, *J. Neurosci. Res.* **1**:57–75.

Ogburn, P. L., Sharp, H., Lloyd-Still, J. D., Johnston, S. B., and Holman, R. T., 1982, Abnormal polyunsaturated fatty acid patterns of serum lipids in Reye's syndrome, *Proc. Natl. Acad. Sci. U.S.A.* **79**:908–911.

Oldstone, M. B. A., Holmstoen, J., and Walsh, R. M., 1977, Alterations of acetylcholine enzymes in neuroblastoma cells persistently infected with lymphocytic choriomeningitis virus, *J. Cell. Physiol.* **91**:459–472.

Oldstone, M. B. A., Sinha, Y. N., Blount, P., Tishon, A., Rodriquez, M., Wedel, R. V., and Lampert, P. W., 1982, Virus-induced alterations in homeostasis: Alterations in differentiated functions of infected cells *in vivo*, *Science* **218**:1125–1127.

Onodera, T., Toniolo, A., Ray, U. R., Jenson, A. B., Knazek, R. A., and Notkins, A. L., 1981, Virus-induced diabetes mellitus: Polyendocrinopathy and autoimmunity, *J. Exp. Med.* **153**:1457–1473.

Osmundsen, H., Neat, C. E., and Norum, K. R., 1979, Peroxisomal oxidation of long chain fatty acids, *FEBS Lett.* **909**:292–296.

Papahadjopoulos, D., Vail, W. J., Newton, C., Nir, S., Jacobson, K., Poste, G., and Lazo, R., 1977, Studies on menbrane fusion III. The role of calcium-induced phase changes, *Biochim. Biophys. Acta* **465**:579–598.

Partin, J. C., Bove, K., Partin, J. S., and Schubert, K., 1979, Liver and muscle ultrastructure in Reye's syndrome, in: *Reye's Syndrome* (J. F. S. Crocker, ed.), Grune and Stratton, New York, pp. 217–232.

Pennington, S. N., and Smith, C. T., 1978, Indomethacin stimulation of lipid peroxidation and chemiluminescence in rat liver microsomes, *Lipids* **13**:636–643.

Pollack, J. D., Cramblett, H. G., Flynn, D., and Clark, D., 1975, Serum and tissue lipids in Reye's syndrome, in: *Reye's Syndrome* (J. D. Pollack, ed.), Grune and Stratton, New York, pp. 227–243.

Pope, A., Hess, H. H., and Lewin, E., 1964, Microchemical pathology of the cerebral cortex in presenile dementias, *Trans. Am. Neurol. Assoc.* **89**:15.

Pottelsberghe, C. V., Rammohan, K. W., McFarland, H. F., and Dubois-Dalcq, M., 1979, Selective neuronal, dendritic, and postsynaptic localization of viral antigen in measles-infected mice, *Lab. Invest.* **40**:99–108.

Reye, R. D. C., Morgan, G., and Baral, J., 1963, Encephalopathy and fatty degeneration of the viscera, *Lancet* **2**:749–752.

Richardson, C. L., Baker, S. R., Morre, D. J., and Keena, T. W., 1975, Glycosphingolipid synthesis and tumorigenesis, *Biochim. Biophys. Acta* **417**:175–186.

Riviere, Y., Gressor, I., Guillon, J. C., Bandu, M. T., Ronco, P., Morel-Maroger, L., and Verroust, P., 1980, Severity of lymphocytic choriomeningitis virus disease in different strains of suckling mice correlates with increasing amounts of endogenous interferon, *J. Exp. Med.* **152**:633–640.

Rohwer, R. G., Neckers, L. M., Trepel, J. B., Gajdusek, C. D., and Wyatt, R. J., 1981, Serotonin concentrations in brain and blood of scrapie infected and normal hamsters and mice, *Brain Res.* **220**:367–371.

Rutter, G., Asher, D. M., Rohwer, R. G., Gibbs, Jr., C. J., and Gadjusek, D. C., 1981, Increased concanavalin A capping in cells from brains of scrapie-infected hamsters, *Arch. Virol.* **68**:129–133.

445

*NEUROCHEMICAL
EFFECTS OF
VIRAL
INFECTIONS IN
THE CENTRAL
NERVOUS
SYSTEM*

Suckling A. J., and Hunter, G. D., 1974 Glycosyl transferase activity in normal and scrapie-affected mouse brain, *J. Neurochem.* **22**:1005–1012.

Suzuki, K., and Chen, G., 1966, Chemical studies on Jakob-Creutzfeldt disease, *J. Neuropathol. Exp. Neurol.* **15**:396–408.

Suzuki, Y., and Matsumoto, M., 1982, Release of lysosomal phospholipase A_1 and A_2 into cytosol and rapid turnover of newly-formed lysophosphatidylcholine in FL cells during fusion from within induced by measles virus, *J. Biochem. (Tokyo)* **92**:1683–1692.

Tamai, Y., Kojima, H., Ikuta, F., and Kumanishi, T., 1978, Alteration in the composition of brain lipids in patients with Creutzfeldt-Jacob disease, *J. Neurol. Sci.* **35**:59–76.

Trauner, D. A., 1982, Reye's syndrome, *Trends Neurosci.* **5**:131–133.

Trauner, D. A., and Adams, H., 1981, Intracranial pressure elevations during octanoate infusion in rabbits: An experimental model of Reye's syndrome, *Pediatr. Res.* **15**:1097–1099.

Trauner, D. A., Nyhan, W. L., and Sweetman, L., 1975, Short chain organic acidemia and Reye's syndrome, *Neurology* **25**:290–293.

Tubaro, E., Lotti, B., Cavallo, G., Croce, C., and Borelli, G., 1980, Liver xanthine oxidase increase in mice in three pathological models, *Biochem. Pharamacol.* **29**:1939–1943.

Viret, J., Dormont, D., Molle, D., Court, L., Leterrier, F., Cathala, F., Gibbs, Jr., C. J., and Gadjusek, D. C., 1981, Structural modifications of nerve membranes during experimental scrapie evolution in mouse, *Biochem. Biophys. Res. Commun.* **101**:830–836.

Weiner, L. P., Johnson, R. T., and Hernden, R. N., 1973, Viral infection and demyelinating disease, *N. Engl. J. Med.* **288**:1102–1105.

Yoon, J. W., Austin, M., Onodera, T., and Notkins, A. L., 1979, Virus-induced diabetes mellitus: Isolation of a virus from the pancreas of a child with diabetic ketoacidosis, *N. Engl. J. Med.* **300**:1173–1179.

Yu, R. K., and Manuelidis, E. E., 1976, Ganglioside alterations in guinea pig brains at end stages of experimental Creutzfeldt-Jacob disease, *J. Neurol. Sci.* **35**:15–23.

Yu, R. K., Ledeen, R. W., Gadjusek, D. C., and Gibbs, C. J, 1974, Ganglioside changes in slow virus disease: Analysis of chimpanzee brains affected with Kuru and Creutzfeldt-Jakob agents, *Brain Res.* **70**:103–112.

Zieve, R. J., Zieve, L. Y., Doizaki, W. M., and Gilsdorf, R. B., 1974, Synergism between ammonia and fatty acids in the production of coma: Implication for hepatic coma, *J. Pharmacol. Exp. Ther.* **191**:19–26.

Index

447